LESSONS UNLEARNED

LESSONS UNLEARNED

THE U.S. ARMY'S ROLE IN CREATING THE FOREVER WARS IN AFGHANISTAN AND IRAQ

PAT PROCTOR

UNIVERSITY OF MISSOURI PRESS

COLUMBIA

Library of Congress Cataloging-in-Publication Data

Names: Proctor, Pat, 1971- author.
Title: Lessons Unlearned : the U.S. Army's role in creating the forever
    wars in Afghanistan and Iraq / Pat Proctor.
Other titles: U.S. Army's role in creating the forever wars in Afghanistan
    and Iraq
Description: Columbia : University of Missouri Press, 2020. | Series:
    American military experience | Includes bibliographical references and
    index.
Identifiers: LCCN 2019034808 (print) | LCCN 2019034809 (ebook) | ISBN
    9780826221940 (hardcover) | ISBN 9780826274373 (ebook)
Subjects: LCSH: Low-intensity conflicts (Military science)--United States.
    | United States. Army--Organization--Evaluation. | United States.
    Army--Operational readiness--Evaluation. | United States--History,
    Military--Case studies. | Intervention (International law)--Case
    studies. | Low-intensity conflicts (Military science)--Case studies. |
    Military doctrine--United States. | Strategy--United States.
Classification: LCC U240 .P765 2020 (print) | LCC U240 (ebook) | DDC
    355.02/180973--dc23
LC record available at https://lccn.loc.gov/2019034808
LC ebook record available at https://lccn.loc.gov/2019034809

THE AMERICAN MILITARY EXPERIENCE SERIES
JOHN C. MCMANUS, SERIES EDITOR

The books in this series portray and analyze the experience of Americans in military service during war and peacetime from the onset of the twentieth century to the present. The series emphasizes the profound impact wars have had on nearly every aspect of recent American history and considers the significant effects of modern conflict on combatants and noncombatants alike. Titles in the series include accounts of battles, campaigns, and wars; unit histories; biographical and autobiographical narratives; investigations of technology and warfare; studies of the social and economic consequences of war; and in general, the best recent scholarship on Americans in the modern armed forces. The books in the series are written and designed for a diverse audience that encompasses nonspecialists as well as expert readers.

Selected titles from this series:

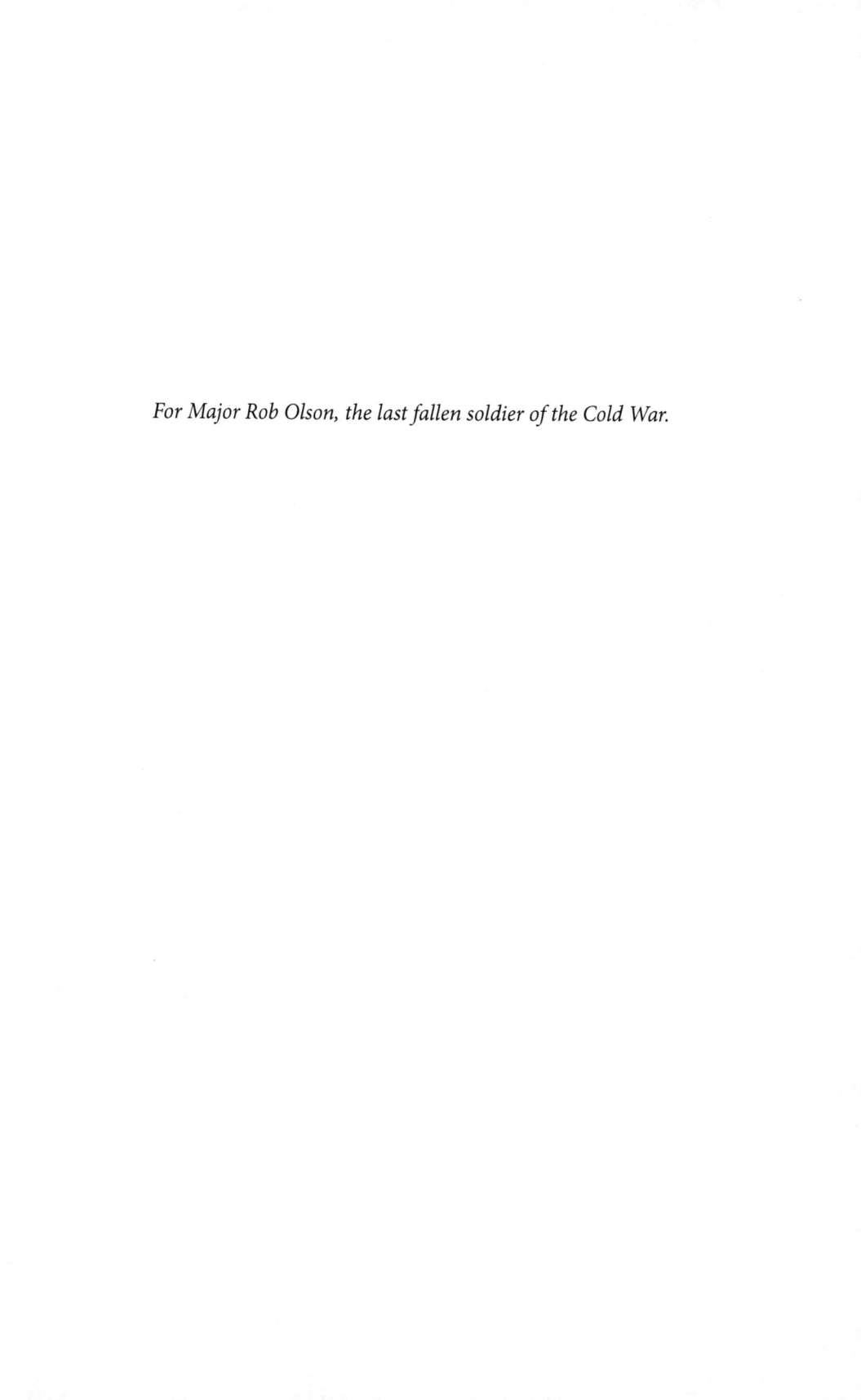

For Major Rob Olson, the last fallen soldier of the Cold War.

CONTENTS

LIST OF FIGURES

xi

ACKNOWLEDGMENTS

A PROJECT OF this scale does not happen without the help of a great many people. I have done my best to mention all who contributed to this work in meaningful ways, but I am certain to have overlooked some of those who helped me. To those persons I apologize in advance.

I will begin by thanking the Department of History at Kansas State University for graciously honoring me with a James C. Carey Fellowship; the resources the fellowship provided were invaluable in my completing this work. I would especially like to thank Don Mrozek, whose wise counsel and advice—not only in the early stages of the book but throughout my academic career—profoundly influenced the shape this book ultimately took.

I would also like to thank a number of people at the U.S. Army War College for their help. Jim Kievit and Chris Mason offered keen insights; my many discussions with them as well as with Larry Goodson and Brett Weigle, both of whom helped me conceptualize this project, helped me immensely. Likewise, my frequent discussions with Conrad Crane; I can't thank him enough for giving so generously of his time and in sharing his views with me—not to mention hunting down so many helpful documents and resources for me. I found the advice of and reviews by Tony Echevarria helpful as I began this work. I also benefited from my many discussions with the members of Seminar 16—especially Col. Santosh Dhakal, Col. Chris Hackler, and Col. Maurice Sipos.

The staffs of the libraries and archives connected with the U.S. Army War College made valuable contributions. I would especially like to thank Gail Silkett at the U.S. Army War College Library and Rich Baker, Tom Buffenbarger, Steve Bye, Rodey Foytik, and Shannon Schwaller at the archives of the U.S. Army Heritage and Education Center. The service all of you provide to our Army is indispensable, if underappreciated.

The team at the U.S. Army Peacekeeping and Stability Operations Institute provided much-needed assistance. My special thanks here go to Col. Jay

Liddick, whose feedback was very important as I wrote the first draft, and the assistant director, William J. Flavin, for his assistance.

I also wish to thank a number of people at the U.S. Army Command and General Staff College at Fort Leavenworth, Kansas—in particular, the Department of Military History for pointing me toward great resources, listening to my ideas, and providing invaluable feedback. Special thanks go to Geoff Babb, Tom Hanson, Sean Kalic, and Jim Willbanks for their time and assistance.

Several others at Fort Leavenworth made important contributions to this work. Michael Browne and Elizabeth Dubuisson at the Ike Skelton Combined Arms Research Library came to the rescue and were incredibly helpful filling in some important last-minute gaps in my research. I would also like to thank Brad Cox and Kelvin Crow for the important work they do at the U.S. Army Combined Arms Center History Office in preserving and archiving the history of the U.S. Army.

Numerous other people have my deep gratitude for providing me with materials that proved indispensable. Thank you to Paul Herbert and Andrew Wood at the First Division Museum at Cantigny Park, not only for the materials but for the excellent advice on this book and the service you provide to our Army and our nation. Thanks go as well to Tammy S. Schultz of the Marine Corps War College for her help and her scholarship and to Lynn Gamma and Tammy Horton at the U.S. Air Force Historical Research Agency at Maxwell Air Force Base, Alabama, for helping me find key documents at a critical time.

I would also like to thank a number of former senior Army leaders who gave selflessly of their time as I researched this book: Gen. (Ret.) Eric Shinseki; Gen. (Ret.) Gordon Sullivan; Lt. Gen. (Ret.) James Dubik; Lt. Gen. (Ret.) Jon Riggs; Lt. Gen. (Ret.) Roger Schultz; Brig. Gen. (Ret.) John S. Brown; and Col. (Ret.) John Gingrich. While I am critical in this book of some of your decisions and actions, I know that you always had the best interests of the Army and the nation at heart and I honor your incredible service; your tireless dedication to our nation and its defense is beyond reproach. My thanks to you and to all of the men and women who have served our nation in uniform for your service.

Two people have been instrumental in helping this book to see the light of day. Thank you to Andrew Davidson, editor in chief of the University of Missouri Press, for seeing the value in this work. And a special thanks goes to my literary agent, Grace Freedson. Your tireless effort and faith in me and my work are always a source of inspiration.

And, of course, thank you to my beautiful wife, Aree, and my patient and understanding children, Amy and Jonathan, for your silent courage in enduring as an Army family. Soldiers are blessed to frequently be thanked for their service to the nation, but Americans sometimes forget that soldiers don't serve alone. So, on behalf of a grateful nation, thank you for *your* service!

And, finally, thank you to God, through Whom all things are made possible.

The views expressed in this book are those of the author and do not reflect those of the University of Missouri Press or the official policy or position of the Department of the Army, the Department of Defense, or the U.S. government.

LESSONS UNLEARNED

INTRODUCTION

I LOVE THE U.S. ARMY. After all, I spent nearly my entire adult life—three decades—serving the nation in its ranks. But sometimes love means speaking hard truths.

And this is the painful truth. The Army claims that its purpose is to fight and win the nation's wars.[1] Yet the past three decades—since the end of the Cold War—have proven that the Army is only capable of fighting and winning the nation's battles. Fighting and winning battles requires killing people and breaking things to impose one's will on an opponent's military. In these activities, since the end of the Korean War, the Army has proven itself without peer. But winning wars requires more than winning battles; it also requires using violence or the threat thereof to impose one's will on an opponent's government and people. And in these activities—again, since the end of the Korean War—the U.S. Army has proven itself largely incompetent.

By way of evidence one need look no further than America's present wars. The United States has been engaged in the war in Afghanistan—the longest war in the nation's history—since October 2001, nearly eighteen years as of this writing. One might be forgiven for thinking that the United States has only been engaged in Iraq since mid-2014, when the Islamic State in Iraq and Syria (ISIS) began its conquest of much of Sunni Arab Iraq. But a more clear-eyed examination reveals that America has been engaged in a singular, continuous war in Iraq since the coalition invaded the country in early 2003, with only a brief, three-year respite during which nearly all U.S. forces temporarily withdrew.[2] The war has now expanded into Syria and shows no sign of ending soon because, while the Army is without peer in high-intensity conflict, it lacks the low-intensity conflict competencies required to bring such wars to a successful conclusion.

What's worse is that this represents a deliberately engineered incompetence. After the fall of the Soviet Union, the senior leadership of the Army refused to acknowledge that the world had changed—that the danger of a great power war against the Soviet Union in western Europe had given way to a future of low-intensity conflicts—and reshape itself in response to this new strategic reality. Despite overwhelming evidence—repeated deployments to low-intensity conflicts throughout the 1990s and a growing chorus of critics warning that these types of operations represented the future of warfare—the senior leaders of the Army continued to stubbornly march their organization toward ever-greater capacity to fight a great power war that never came. Thus, the Army found itself tragically unprepared when the wars in Afghanistan and Iraq began.

Historically, Americans have been reluctant to blame their Army for its incompetence. Many laid the blame for the deteriorating situation in Afghanistan on the George W. Bush administration, claiming the president and his advisers had taken their "eye off the ball" there, distracted by a war of choice in Iraq. President Bush is also frequently blamed for making the decision to invade Iraq based on spurious intelligence about weapons of mass destruction. Others blame Secretary of Defense Donald Rumsfeld for his decision to drastically cut the number of troops deployed in the initial invasion, or they point to presidential envoy Paul Bremer for disbanding the Iraqi Army, which the Army expected would provide security after the fall of Saddam Hussein's regime. Yet others blame President Barack Obama for withdrawing all U.S. forces from Iraq in 2011 rather than leaving an adequate residual force to continue to train and support the Iraqi Army.[3]

Certainly each of these decisions by America's civilian leadership contributed to the ongoing military disasters in Afghanistan and Iraq, but none of them change the fact that the U.S. Army was deliberately unprepared for the prolonged low-intensity conflicts that it faced in each country. Moreover, this unpreparedness significantly compounded whatever strategic challenges the United States already faced in prosecuting these conflicts. For this reason, the senior leaders of the Army deserve a large share of the blame for the disastrous forever wars in Afghanistan and Iraq from which the United States still struggles to extricate itself.

From the end of the Cold War to the beginning of the Iraq War, the Army marched relentlessly toward ever-greater capacity to fight a peer competitor in a high-intensity conflict despite the growing body of evidence that it would almost certainly not have to fight this kind of war again. The United States was dealt a humiliating defeat by militias in Somalia, yet the Army refused to change. A U.S. military intervention in Haiti failed to produce political change on the ground, but rather than reflecting on this failure, the

Army dumped the conflict on the United Nations (UN) and went home. The Army's inability to forge a political settlement in Bosnia-Herzegovina and Kosovo trapped the United States in a decades-long quagmire, but Army leaders refused to institutionalize the lessons of these conflicts. A growing chorus of observers, inside and outside the Army, warned that these low-intensity conflicts were the new face of warfare in the twenty-first century, but the Army's senior leaders—steeped in a culture that emphasized preparation to fight high-intensity conflicts over all other activities—continued to develop expensive, high-tech weapons to fight a third world war.

Thus, when the United States was attacked on September 11, 2001, and the World Trade Center towers fell, the stage was set for a slow-motion military disaster. The apparent "cheap win" in the first days of the war in Afghanistan through the use of special operations forces (SOF) and air-power further validated transformers—those advocating a high-intensity-conflict, information-age transformation of the Army—in their conviction that technology could supplant numbers. The Army that invaded Iraq in March 2003 was ill-prepared for the character of warfare that it ultimately faced. While the depleted Iraqi Army rapidly melted before the advance of the vastly superior U.S. Army, it did not disappear. Instead it hid among the population, evading America's high-tech surveillance and precision strike capabilities. Once Saddam's regime was toppled, the Iraqi Army reemerged, not as a conventional military threat but as an insurgency that severely challenged the halting U.S. efforts to establish a new Iraqi government. Other adversaries also emerged, including Shiite militias, Sunni Iraqi Islamists, and foreign terrorist groups. Back in Afghanistan the war that had seemed to be all but won in 2002 likewise transformed into a grueling battle against al-Qaeda and Taliban insurgents.[4]

An eleventh-hour gamble by President Bush—a troop "surge"—seemed to put the war in Iraq on the path toward conclusion, but President Obama's later attempt to replicate the Iraq surge in Afghanistan failed to stem the tide of violence in that country.[5] And, as it turned out, the war in Iraq was not over, either; the Army withdrew from Iraq at the end of 2011, only to return in 2014 to combat the reemergence of a Sunni insurgency in the form of ISIS. No president would again dare to withdraw forces from either theater before a political solution to the conflict was absolutely secure. But no political settlement has yet emerged and, as of this writing, America has spent nearly eighteen years paying the price for its army's intentional unpreparedness to fight the war on terror.

I use the word *intentional* because I contend that the senior leaders of the Army stubbornly insulated their organization from the lessons that might otherwise have been learned from the low-intensity conflicts of the 1990s. Of course, some units and individuals clearly did draw on their experience from these interventions. For instance, the 1st Infantry Division, which deployed from Germany to Iraq in the first rotation of units into the war, had just returned from Kosovo in 2002 and many of its leaders had participated in earlier deployments to Bosnia-Herzegovina. As division commander Maj. Gen. John Batiste would later comment, "We understood well that combat was important, but so was stability and support operations. And the notion that building relationships and changing attitudes and giving people alternatives to the insurgency was also terribly important."[6]

But where units were able to draw on the experiences from the Army's 1990s interventions it was incidental, because the Army had failed to institutionalize the lessons from these conflicts and integrate them into its training, education, and organizations. Thus, while the 1st Infantry Division could draw on its experiences in the Balkan States and the 101st Airborne Division could benefit from the academic background of its commander, Maj. Gen. David Petraeus,[7] the overwhelming majority of units had little or no training or experience in low-intensity conflict and were unprepared to combat the insurgencies that emerged in Afghanistan and Iraq.

Moreover, for every Batiste or Petraeus there were dozens of Army leaders who made the situation worse. The Army's stubborn focus on high-intensity conflict before September 11 produced leaders like Lt. Col. Nate Sassaman, who tacitly encouraged violence against Iraqi civilians and then covered it up when it resulted in deaths in 2004, and Col. Michael Steele, of *Black Hawk Down* fame, who launched massive, counterproductive air assault raids throughout Saddam's home province in 2006 (which resulted in numerous civilian casualties) instead of seeking a political settlement to the Sunni insurgency. It also produced senior leaders like Lieutenant Generals Dan McNeill and David Barno in Afghanistan, who remained laser-focused on hunting down the remnants of al-Qaeda throughout 2003 and 2004 while Taliban insurgents reconquered much of southern Afghanistan virtually unopposed.[8]

Of course the Army did, eventually, institutionalize low-intensity conflict proficiency. The tale of how General Petraeus and a team of "insurgents" wrote a new counterinsurgency doctrine and implemented this vision in Iraq and Afghanistan has become legend. But by February 2007, when Petraeus took command of all forces in Iraq, more than three thousand U.S.

troops had already died in that war. And nearly 1,200 U.S. troops would die in Afghanistan before Gen. Stanley McChrystal arrived in June 2009 to implement the new counterinsurgency doctrine there.[9] At least equally important, this supposed turnaround in Army thinking on low-intensity conflict has not resulted in "wins"—enduring political solutions—in either Afghanistan or Iraq. As of this writing, American troops are still fighting and dying in both countries.

One could reasonably argue that the Army failed to fully implement the doctrine enshrined in Petraeus's Field Manual (FM) 3-24, *Counterinsurgency*. For instance, despite the manual's repeated insistence that in a counterinsurgency it is the Army's job to identify and solve the country's political problems, the Army abdicated its political role in both countries to provincial reconstruction teams led by U.S. Department of State diplomats and containing a hodgepodge of Department of Defense, Department of Justice, U.S. Agency for International Development, and other U.S. government agency officials.[10] One could also argue that this doctrine came too late; because the Army was unprepared to fight low-intensity conflicts at the beginning of each war, the situation had so deteriorated by the time counterinsurgency doctrine did arrive that the situation in both countries was beyond recovery.

Doing It Again

This book is not intended merely as a simple history or as an exercise in laying blame for past sins, however. It is instead an intervention, because the U.S. Army is in the process of making the same mistake again.

By 2006 the Army's failure in Afghanistan and Iraq had silenced the "transformers" who had dominated the debate over the direction of the Army at least since the end of the Cold War, creating space for General Petraeus and his "insurgents" to institutionalize counterinsurgency in Army doctrine and training. But now that the crisis has abated—yet, notably, before either war has actually been brought to a successful conclusion—Army transformers have reemerged and undone most of these gains.

Just as it did after Vietnam, the Army is intentionally forgetting its hard-won lessons from Afghanistan and Iraq. This effort began as early as 2010, when Bernard I. Finel, former professor of strategy at the National War College, made the ridiculous suggestion that the Army could have avoided the counterinsurgency in Iraq if it had simply unilaterally withdrawn after Saddam was captured in 2003. Likewise, Col. Craig Collier wrote that counterinsurgency advocates were "reluctant to admit that killing the enemy actually worked" and insisted that "killing or capturing an insurgent consistently and quantifiably had a more positive impact than anything else we did."[11]

The effort to forget low-intensity conflict truly took off at the end of 2011, after the Army withdrew from Iraq (though that withdrawal proved to be only temporary). At both live training at the Army's National Training Center (NTC) at Fort Irwin, California, and in virtual "warfighter exercises" run by the Army's Mission Command Training Program from Fort Leavenworth, Kansas, the Army began to abandon training in low-intensity conflict and to instead return to training in high-intensity conflict.[12] At first the Army seemed almost apologetic about these moves toward restoring proficiency in high-intensity conflict. A 2013 article in *Army* magazine, the official publication of the Association of the United States Army, assured its readers that the Army would not abandon the lessons of Afghanistan and Iraq; the "reintroduction of conventional force-on-force training engagements" in training rotations at the NTC would be combined with "wide area security operations that include COIN [counterinsurgency] elements and a few extra wrinkles."[13]

But by late 2013 the effort to expunge low-intensity conflict from the collective Army consciousness had become explicit. Leading the ideological charge was Col. (Ret.) Gian Gentile. Embittered by the belief that "the myth of the counterinsurgency narrative"—the rise of General Petraeus and his "insurgents"—denigrated the sacrifices made by him and his soldiers in Baghdad in 2006, Gentile penned *Wrong Turn*, a 208-page assault on counterinsurgency doctrine in general and FM 3-24 and its authors in particular. He dismissed FM 3-24's "Paradoxes of Counterinsurgency Operations" as "a jumble of dreamy statements that bordered on some mixture of philosophy, theory, and military operational history." Seizing on the common refrain of counterinsurgency advocates that "an army 'can't kill its way to victory,'" Gentile painted counterinsurgency doctrine as rejecting the use of violence and echoed Collier's claim that it was killing insurgents that actually won the war—a war that is still on going, as of this writing, six years later.[14]

Incredibly, in 2017—despite the fact that the United States was once again fully engaged in two low-intensity conflicts in Afghanistan and Iraq—the Army completely abandoned low-intensity conflict proficiency and returned to full-time preparation for a third world war. The newest edition of the Army's capstone doctrine, FM 3-0, *Operations*, pays implicit lip service to low-intensity conflict, acknowledging that "the U.S. Army must be manned, equipped, and trained to operate across the range of military operations." Yet in the same sentence, the manual insists that "large-scale ground combat against a peer threat represents the most significant readiness requirement." While it fails to name either Afghanistan or Iraq in its rejection of the importance of proficiency in low-intensity conflict, FM 3-0 makes clear its authors' view that the Army was trapped in counterinsurgencies in the two countries because it failed to "consolidate gains" during the brief high-intensity conflict phase of each war. The Army could avoid such low-intensity conflicts in the future, the manual explains, by "exploitation" to destroy "every part of an enemy's ability to resist," ensuring "that enemies cannot transition a conventional military defeat into a protracted conflict that negates initial successes."[15] In other words, this manual contends that the Army could have avoided having to fight a low-intensity conflict in Afghanistan or Iraq in the first place if, during the high-intensity conflict phase of each conflict, it had simply killed every person—combatant or noncombatant—who might potentially resist later.

The manual insists that the United States is in a race to recapture its dominance in high-intensity conflict capacity before the next war with "Russia, China, Iran, [or] North Korea." In the foreword, the commander of the U.S. Army Combined Arms Center, Lt. Gen. Michael Lundy, warns that these countries "already have overmatch or parity, a challenge the joint force has not faced in twenty-five years." For Lundy, the Army's focus on counterinsurgency had been a costly distraction: "As the Army and the joint force focused on counter-insurgency and counter-terrorism at the expense of other capabilities, our adversaries watched, learned, adapted, modernized and devised strategies that put us at a position of relative disadvantage in places where we may be required to fight."[16]

The senior leaders of the Army seem to actually believe that a great power war is imminent. Lundy warns, "The proliferation of advanced technologies; adversary emphasis on force training, modernization, and professionalization; the rise of revisionist, revanchist, and extremist ideologies; and the ever-increasing speed of human interaction makes large-scale ground combat more lethal, *and more likely*, than it has been in a generation." While the authors of FM 3-0 see Iran and North Korea as threats, they insist that the future holds—and the Army must prepare for—"large-scale combat operations against a peer threat": China or Russia.[17]

At first blush, the U.S. Army's construction—ongoing at the time of this writing—of Security Force Assistance Brigades (SFABs) would appear to indicate that it is still taking seriously the requirement to engage in low-intensity conflicts. After all, the outgoing chief of staff of the Army, Gen.

Mark Milley, claimed that the SFABs are an acknowledgment that the Army is "likely to be involved in train, advise, and assist operations for many years to come." Yet these brigades are not designed as much to make the Army better at low-intensity conflict as to allow the rest of the Army to look the other way and continue to get better at high-intensity conflict. According to Brig. Gen. Brian Mennes, director of Force Management for the Army G-3/5/7, SFABs are intended to help the Army "to reduce . . . the demand for combat advising from conventional brigade combat teams,"[18] presumably so that they can continue to prepare for high-intensity conflicts.

SFABs are also a backdoor way of providing additional forces for a great power war. As Mennes has noted, "In a time of national emergency, SFABs provide options for the Army to grow BCTs [brigade combat teams] rapidly."[19] This second purpose, creating a basis for the rapid mobilization of combat forces for a high-intensity conflict, is repeated throughout the Army's rhetoric on SFABs. As Lt. Col. Jonathan Thomas, also from the Force Management Directorate, said of this expansibility, "The SFAB will provide a cadre of officers and [noncommissioned officers] who will facilitate the regeneration of an SFAB into a full-blown brigade combat team." The Army's deputy chief of staff for operations and training, Lt. Gen. Joseph Anderson, called the SFAB a "standing chain of command for rapidly expanding the Army."[20]

Incredibly, despite the end of the Cold War and the collapse of the Soviet Union in 1991, the numerous low-intensity conflicts the U.S. Army faced in the 1990s, and America's disastrous and ongoing interventions in Afghanistan and Iraq, the Army remains obsessed with the delusion that a war against a peer competitor is imminent.

There Will Never Be Another Great Power War

There is not going to be another great power war—at least as long as nuclear weapons remain the dominant feature of the strategic landscape.

In his *History of the Peloponnesian War*, writing of the great power war between Athens and Sparta, ancient historian Thucydides warned that war, "far from staying within the limit to which a combatant may wish to confine it, will run the course that its chances prescribe."[21] The idea that the United States could fight a war with Russia only in the Baltic States or a war with China only in the South China Sea flies in the face of millennia of world history; as soon as either party began to lose, they would open hostilities in another theater, which would be met by reciprocal escalation until the world was plunged into another world war. It is as true today as it was in the fifth century BC; there is no such thing as a limited war between great powers, and this is especially true in the nuclear age.

It is revealing that the Army has embraced a different reading of Thucydides, one expressed in the troubled exhortations of Harvard University professor Graham Allison, who has spoken at the invitation of the U.S. Military Academy at West Point and the U.S. Army War College. Army general Joseph Votel put Allison's book *Destined for War: Can America and China Escape Thucydides's Trap?* on U.S. Central Command's 2018 reading list. In the book, Allison warns that in twelve of the sixteen cases that he examined, the emergence of a new great power precipitated a great power war.[22] Yet he completely ignores the fact that in two of the remaining four cases—namely, the rise of the Soviet Union and the reemergence of a unified Germany—the rise of a great power did not result in war. Both of these instances occurred in the nuclear age. In fact, while Allison fails to acknowledge it, none of the cases that did result in war occurred after the advent of nuclear weapons.

Nuclear weapons, the cornerstone of America's deterrence against China and Russia—a capability upon which the United States spends $25 billion or more annually—precludes the possibility of a direct war between great powers for the foreseeable future. This is hardly a novel insight. Deterrence theorists have been, to various degrees, reaching the same conclusion for more than sixty years. Military professionals concur; General Sir Rupert Smith of the British Army wrote in his book *The Utility of Force* that, because of the advent of nuclear weapons, "war as battle in a field between men and machinery, war as a massive deciding event in a dispute in international affairs: such war no longer exists."[23]

Yet despite the overwhelming improbability of a great power war, and having experienced nearly three decades of unrelenting low-intensity conflicts (including those in Somalia, Haiti, Bosnia-Herzegovina, Kosovo, Afghanistan, and Iraq) punctuated by only a handful of brief high-intensity conflicts against second-rate powers like Iraq and Serbia (decidedly *not* peer competitors), the Army is once again preparing for a world war.

Meanwhile, there are real and dramatic costs to the Army's chronic inability to fight low-intensity conflicts successfully. The Costs of War project, led by Brown University professor Neta Crawford, estimated in late 2016 that the failure to bring the conflicts in Afghanistan and Iraq to a close cost the United States $4.79 trillion.[24] But this figure pales in comparison to the intangible strategic costs to the nation: the loss of influence with our allies in Europe, the Middle East, and the Pacific and the surrender of international political power to China and Russia. America's inability to effectively intervene in low-intensity conflicts has encouraged regional competitors and great powers to engage in subversion, proxy wars, and outright aggression in places like Lebanon, Georgia, Ukraine, Syria, and Yemen. If the United States does not develop the capacity to fight and win in low-intensity conflicts, its military power will continue to become increasingly irrelevant to international politics and American foreign policy will hold ever-diminishing weight in the world.

By no means do I recommend in this book that the Army completely abandon its considerable capacity to engage in high-intensity conflict. Instead, I contend five things. First, the high-intensity conflicts that the Army will face in the future will be infrequent, short, and fought against national militaries less capable than that of the United States. Second, the Army will much more frequently face low-intensity conflicts that, even if fought well, will be long, difficult, and manpower-intensive. Third, the demands of low-intensity conflict are vastly different from those of high-intensity conflict. Fourth, no one Army unit can be trained, manned, and equipped to be good at both low- and high-intensity conflict. Fifth, and finally, to successfully meet the challenges of these first four realities, the Army must bifurcate, with a few, small, highly lethal units trained, manned, and equipped to fight the brief, infrequent high-intensity conflicts the United States will face while the larger remainder of the Army is trained, manned, and equipped to effectively fight the low-intensity conflicts that will constitute the majority of operations in which it will be engaged for the foreseeable future.

Unlike the work of some low-intensity conflict observers at the time and since, this study does not find fault with the decision made by Army leaders to return to competency in high-intensity conflict after the Vietnam War. One could legitimately make the case that the Army did so with too much zeal, throwing out the proverbial baby with the bathwater; in their drive to recapture the Army's competency in high-intensity conflict, senior leaders oversaw a purge of the lessons of the Vietnam War from Army doctrine and training. Yet with more than 150 Soviet divisions and 600,000 Soviet soldiers facing them on the other side of the Iron Curtain,[25] the senior Army leaders of the 1970s and 1980s can be forgiven for believing that another great power war was at least possible, if not probable.

Likewise, this book does not blame the Army for its unpreparedness to wage the low-intensity conflicts of the 1990s. The Army had very reasonably focused on high-intensity conflict for the previous two decades. Furthermore, relatively tiny U.S. military interventions in Central America in the 1980s created the perception that the Army could safely leave low-intensity conflict to SOF. And, finally, at the dawn of the 1990s, just as the Cold War was ending, U.S. military interventions in Panama and the Persian Gulf created the misperception that the future might predominantly hold high-intensity conflicts. For all these reasons, it is not surprising that the Army was unprepared for the realities of the 1990s, a decade in which it found itself engaging in a string of low-intensity conflicts in heretofore unlikely places like Somalia, Haiti, Bosnia-Herzegovina, and Kosovo.

The Army's undeniable incompetence in fighting low-intensity conflicts became increasingly hard to excuse, however, as the 1990s wore on and the Army refused to learn and institutionalize the lessons of each conflict, one after the other. Moreover, a growing mountain of evidence and a growing chorus of low-intensity conflict observers made it clear that the Army could expect to find itself fighting predominantly low-intensity conflicts for the foreseeable future.

The decision, then, by senior leaders in the 1990s to keep the Army on its path toward ever-greater preparedness for a great power war in the face of this growing body of evidence that it was the wrong course lies at the heart of their culpability in the disastrous forever wars in Afghanistan and Iraq.

Yet, again, this book is not just a history but an intervention. It is an attempt to try to understand why Army transformers ignored the arguments that low-intensity conflict observers were making, how transformers discredited the compelling arguments that these observers were making, and how they kept the Army from building its capacity to engage in such conflicts. By understanding why and how the Army of the 1990s refused to prepare to fight low-intensity conflicts on the eve of the war on terror, it might be possible to prevent the Army of today from drifting toward a similar disaster.

Methodology, Definitions, and Frameworks

As such, this book takes the form of an intellectual history. It is the history of both low- and high-intensity conflict thought in the U.S. Army in the period between the end of the Gulf War in 1991 and the beginning of the Iraq War in 2003. During this time, low-intensity conflict observers both inside and outside the Army argued that the military interventions of the 1990s ably demonstrated that the Army was ill-prepared for such conflicts and warned that the future held many more such conflicts.[26] Steeped in an Army culture that emphasized preparation to fight high-intensity conflicts over all other activities, Army transformers—the senior leadership of the Army chief among them—ignored these warnings and continued the Army's transformation toward an ever-more deployable, high-tech, networked force built to fight high-intensity conflicts against conventional adversaries.[27]

Throughout this book, the term *low-intensity conflict* is used to refer to operations ranging from humanitarian assistance and disaster relief outside the United States or its territories to counterguerrilla and counterinsurgency operations. What these operations all have in common—what characterizes them as low-intensity conflicts—is that they involve an organized military force (in the case of this book, the U.S. Army) using violence or the threat of violence to compel, coerce, or influence an irregular force (a militia, insurgent group, or other armed band) and/or a civilian population to accept a political order.

The use of the term *low-intensity conflict* is an imperfect choice, but forty years of confused Army and joint doctrine have so muddled the categorization of the "range of military operations" or the "spectrum of conflict" that this book consciously avoids using any of the various terms—such as *operations other than war*, *military operations other than war*, or *stability and support operations*—that were in vogue throughout the 1990s. Other terms, such as *small wars* or *small-scale contingencies*, obscure the fact that such operations might be quite large (as was the case at the height of the wars in Afghanistan and Iraq) yet still have a character quite distinct from that of high-intensity conflicts. Of course, *low-intensity conflict* comes with its own historical baggage, being the term of choice for such operations in the late 1980s and early 1990s, but this history is more distant and less likely to evoke preconceptions in present-day readers.

Through the book, the term *high-intensity conflict* is used to describe conventional combat between two organized military forces of industrial-age or greater technological capability. This book consciously uses this term rather than *war, conflict, major combat operations*, or the currently fashionable *large-scale combat operations* for the same reason as explained above.

It is important to note that both low- and high-intensity conflict *are* war; as nineteenth-century theorist Carl von Clausewitz wrote, "war is . . . an act of force to compel our enemy to do our will."[28] One must concede that the use of the threat of violence to compel, coerce, or influence and the inclusion of a civilian population as a target of violence or the threat of violence in my definition of low-intensity conflict does stretch Clausewitz's definition of war. But to the degree that intimidation with weapons in close proximity to the target of that intimidation is a form of violence, the definition—however imperfect—does qualify low-intensity conflict as a form of war.

It is also important to acknowledge that there are a few activities in which the Army engages that are outside the scope of low- or high-intensity conflict and, generally, outside the scope of war as defined by Clausewitz. The first is what is, in today's military parlance, referred to as "defense support of civil authorities."[29] These activities include but are not limited to humanitarian aid and disaster relief inside the United States or its territories.[30]

The other activity in which the Army participates that is neither low- nor high-intensity conflict is *deterrence*, the use of the threat of force by one organized military force (in this case, the Army) to dissuade the government of another country from using its own organized military force to engage in violence. Deterrence can be intended to dissuade a country from attacking the United States, a third country, or—as in the case of Kosovo before the beginning of the North Atlantic Treaty Organization air war—its own people.[31] Acknowledging the above murkiness created by the inclusion of the threat of violence in the definition of low-intensity conflict, deterrence is not war in the Clausewitzian sense unless it fails and a high- or low-intensity conflict ensues.

Since the end of the Vietnam War, the U.S. Army has repeatedly engaged in a much more elaborate taxonomical categorization of low-intensity conflict than is presented above. In fact, by the 1990s the practice of ever-more surgically dissecting low-intensity conflict into subcategories had become something of an academic discipline, consuming scores of journal articles and whole chapters of Army doctrine. In the interest of clarifying rather than obfuscating the nature of low-intensity conflict, this book will not engage in such hairsplitting.

Instead, it will explore the various taxonomies that the Army applied to low-intensity conflict between 1990 and 2003. Such exercises were, at their core, attempts to separate "war" from "operations other than war." And these distinctions were far from academic; they lie at the heart of the Army's failure to learn and institutionalize the lessons of its 1990s low-intensity conflicts before the beginning of the war on terror. Because the Army saw its ultimate purpose as fighting and winning the nation's wars, and it classified all activities except high-intensity conflict as "operations other than war,"[32] it could safely dismiss these lessons. In other words, the senior leaders of the Army consciously decided not to institutionalize the lessons of its 1990s low-intensity conflicts because they believed they did not constitute war and, thus, were someone else's job.

The U.S. Army institutionalizes operational lessons in six ways, each constituting a deeper level of commitment than the one before it: doctrine, education, training, essential tasks, organization, and incentives.

Doctrine is the distillation of lessons from history and military theory into a written body of knowledge.[33] While one could (and this book will) reasonably criticize the quality of the doctrine produced, the Army did respond to the post–Cold War increase in low-intensity conflicts by producing new doctrine.

Likewise, early in the 1990s the Army added a few hours of education on low-intensity conflict to the curriculum for its majors at the U.S. Army Command and General Staff College (CGSC)—which educated more than eleven thousand officers between 1991 and 2001.[34] Later in the decade, however, these hours of instruction were almost completely subsumed into other topics.

Beyond doctrine and education, the Army made no significant moves to institutionalize low-intensity conflict. It did not direct that any of its units conduct routine training on low-intensity conflict or add low-intensity conflict elements to the training environments at its training centers—the NTC at Fort Irwin and the Battle Command Training Program (which would later become the Mission Command Training Program) at Fort Leavenworth. The Joint Multinational Readiness Center in Germany and the Joint Readiness Training Center at Fort Polk, Louisiana, did eventually begin training on low-intensity conflict, but this training was only given to those units about to deploy to such conflicts in Bosnia-Herzegovina or Kosovo.

Likewise, the Army did not add low-intensity conflict tasks to the mission-essential task lists for any of its existing units. It did create two organizations—the Army–Air Force Center for Low Intensity Conflict and the Peacekeeping Institute—to capture the lessons of low-intensity conflicts, but it never created any units specialized to prosecute these conflicts. At the point of their creation in the late 1980s, the Army's light infantry divisions had been envisioned as low-intensity conflict forces. Yet by the time they were fully organized and equipped and began training, their original purpose had been subverted and they were focused squarely on high-intensity conflict tasks.

And, finally, the Army never offered incentives to its service members for demonstrated proficiency, knowledge, or experience in low-intensity conflict in terms of additional pay or preferential consideration for promotion or assignments.

Since Army units were not given low-intensity conflict essential tasks, they almost never trained on them, no such specialized units were created, and the Army provided no incentives to encourage its soldiers to learn about or participate in low-intensity conflicts, soldiers were only exposed to such conflicts if they were among the few units to deploy to one of them or if they chose to read the minimal—and problematic—doctrine produced on the topic. Admittedly, those midgrade officers who did attend the CGSC in the early 1990s received a few hours of education on low-intensity conflict during their year at the school. For many of them, this was their only exposure to the subject throughout their entire career before the beginning of the war on terror. Moreover, for most, this education was not reinforced by any practical experience.

Throughout the 1990s the Army's thinking on its low-intensity conflict experiences was roughly split into two camps. A group of low-intensity conflict observers believed that the Army should increase its capacity to engage in such conflicts by institutionalizing the lessons of these experiences. On the other hand, mainstream Army thought was represented by a group of conventionally minded transformers who believed the Army should harness the promise of the so-called revolution in military affairs— the name given to the dramatic impact that emerging telecommunications technology was having on warfare[35]—in order to continuously improve the Army's capacity to engage in high-intensity conflicts.

Thought in these two camps diverged widely, but there were three points of convergence. First, almost immediately following the end of the Cold War, both camps agreed that, as a consequence of the collapse of the Soviet Union, the Army could expect to face many more low-intensity conflicts well into the foreseeable future. Low-intensity conflict observers believed that this fact demanded that the Army increase its capability to fight these types of conflicts. Transformers, on the other hand, saw this fact as a nuisance, an obstacle to be overcome or a tax to be paid; the Army had to continue its march toward ever greater capacity to fight what they believed was the inevitable high-intensity conflict to come because, while least likely, it was the most dangerous contingency.

Later in the 1990s another point of convergence between low-intensity conflict observers and Army transformers began to emerge: asymmetry. Both camps agreed that asymmetry entailed a potential future adversary avoiding the Army's strengths and instead developing asymmetric capabilities to exploit the Army's weaknesses. Low-intensity conflict observers foresaw enemies avoiding the U.S. military's precision strike and other overwhelming conventional capabilities using low-tech, low-intensity conflict means. They advocated building the Army's capacity to fight low-intensity conflicts in order to defeat this form of asymmetry. Army transformers, on the other hand, expected adversaries to employ early-entry denial measures, cyberwarfare, precision strikes, and other technological means in asymmetric ways to defeat the Army's conventional advantages.

At the end of the 1990s both the transformers and the low-intensity conflict observers began to focus on urban operations. The transformers were the first to become concerned with operations in cities because their experiments and wargames had begun to indicate that these environments presented an asymmetric challenge in that conducting operations in urban environments necessarily nullified the Army's advantage in intelligence collection and precision targeting. Yet because transformers saw urban environments merely as complex terrain in which it was difficult for forces to move, see, or communicate, they considered cities as places to be bypassed or, if that was not possible, besieged. They contemplated technological means to eventually negate the advantages conferred to adversaries by urban terrain. To the extent that Army transformers considered the populace of cities at all, it was as obstacles to detecting adversaries or employing fires against them.

Low-intensity conflict observers, on the other hand, understood that the true challenge urban environments posed was that they were densely populated settings, places full of civilians who could not simply be bypassed or ignored—and whose presence made it more difficult for U.S. soldiers to find and attack adversaries hiding (and operating) among them. These observers also realized the inescapable truth that because the civilians in cities represent the political objective of any military operation, the Army would have to be prepared to operate among them. Such observers insisted that, in light of the problems of urban operations, the Army should increase its low-intensity conflict capacity to meet these challenges.

A historian should always be suspicious of "models" or "universal laws" that purport to describe, predict, or direct human events. But as one examines the Army's 1990s interventions in Somalia, Haiti, Bosnia-Herzegovina, and Kosovo, it is difficult to avoid seeing a repeating pattern. In each case Army units faced the unfamiliar challenges of low-intensity conflict, learned the best they could on the job, and gradually improved their performance in response to these lessons. Yet because the Army refused to institutionalize these hard-won lessons, they were lost at the end of each conflict. At the beginning of the next intervention, Army units had to start all over again, beginning at the first rung as they once more scaled the ladder toward greater understanding of how to fight and win low-intensity conflicts.

The first rungs on this ladder toward greater understanding and capacity to engage in low-intensity conflicts were the deployability, sustainability, and rotation of Army units; combined, these concepts represented the basic problem of how to get units to a low-intensity conflict and sustain them with troops, weapons, food, and fuel. As Army units learned about deployability and sustainability, they also discovered the degree to which they had to rely on contractors to provide these services.

The next rung on the ladder was the realization that the Army was not manned, equipped, or trained to properly execute low-intensity conflict and therefore had to reorganize, reequip, or retrain to meet the challenge at hand. And it is here that Army transformers—and occasionally misguided but well-meaning low-intensity observers—first became obstacles to learning. Some refused to acknowledge the Army's lack of competence in fighting low-intensity conflicts, instead contending that the tasks required to execute such conflicts were merely lesser included tasks of high-intensity conflict proficiency and therefore required no specialized personnel, equipment, or training: a unit good at high-intensity conflict would necessarily be good at low-intensity conflict. Others argued that the Army shouldn't be participating in such conflicts in the first place because it was not the Army's job to do so. As an extension of this argument, some maintained that training and operating in low-intensity conflicts detracted from a unit's ability to execute its primary mission: high-intensity conflict. Some argued that such conflicts should be the purview of SOF rather than the conventional Army.

All of these cognitive obstacles to understanding low-intensity conflict fall apart under closer examination. The dismal performance of Army units—trained solely for high-intensity conflict—in the first years of the wars in Afghanistan and Iraq ably demonstrated that low-intensity conflict tasks are *not* lesser included tasks of high-intensity conflict proficiency. And the idea that such conflicts are an illegitimate use of the Army is completely ahistorical; as a great many historians have pointed out, the majority of the Army's history has been dominated by low-intensity conflicts;[36] great power wars like the Spanish-American War, World War I, and World War II are the historical anomalies.

The relegation of low-intensity conflict to SOF deserves special attention, since this misconception has come back in to vogue in the Army's recent return to a focus on high-intensity conflict. The notion that such conflict is the domain of SOF is arguably an artifact of the era of President Ronald Reagan and the tiny U.S. military interventions in the low-intensity conflicts in Central America in the 1980s. The seemingly positive results of these interventions encouraged Congress to pass the Nunn-Cohen Amendment to the 1987 National Defense Authorization Act. This amendment gave SOF proponency for many of the tasks commonly present in low-intensity conflict.[37] Senior Army leaders eagerly used the amendment as license to finally and completely abandon any remaining attention to such conflict.

The amendment proved to be little more than wishful thinking, however. The Army's low-intensity conflict interventions since the passage of Nunn-Cohen—those in Somalia, Haiti, Bosnia-Herzegovina, Kosovo, Afghanistan, and Iraq chief among them—clearly reveal two things. First, SOF are simply too small to do the tasks supposedly assigned to them by Nunn-Cohen on a scale that is useful for the vast majority of low-intensity conflicts. Second, because SOF are too small, the conventional Army is inevitably forced to execute these tasks itself. The Army's interpretation of the Nunn-Cohen Amendment has proven itself impossible to implement in the real world. In practice, both SOF and the Army have been forced by the facts of war to ignore this statute (or, rather, the statute as it was interpreted by the Army).

Any assertion that the Army should not participate in low-intensity conflicts because it is forbidden from doing so by the Nunn-Cohen Amendment is patently ridiculous. First, the actual language of the amendment says no such thing; it simply gives SOF proponency for these tasks. Second, and more important, even if it were the case that the amendment forbade the Army from executing low-intensity conflict tasks, the Army and SOF have been breaking this law routinely for three decades.

In the 1990s, as units and interested observers overcame these cognitive obstacles and proceeded up the ladder of understanding of low-intensity conflict to the next rung, they realized that the challenges of such conflict were very different from those of high-intensity conflict, and that the former required mastery of new competencies. Some of these tasks included training indigenous or host nation military and police forces, integrating with SOF, and interoperating and coordinating with multinational forces and intergovernmental organizations such as the UN.

At the next rung, units began to understand that the essential character of low-intensity conflicts—whether humanitarian aid and disaster relief or counterinsurgency—was that they occurred in and among civilian populations. At this level of learning, units began to wrestle with the importance of rules of engagement, the nuanced application of force, and the need for non-lethal capabilities. They also began to understand the importance of public affairs and psychological or information operations in influencing the population and the need for so-called civil-military operations—coordinating with local or host nation governments, civil relief organizations, other U.S. government agencies, and nongovernmental organizations. Army units also occasionally experienced complications arising from their reliance on contractors operating in and among a populace.

At this level the primary cognitive obstacle among well-meaning low-intensity conflict observers was the desire to apply empirical measurements to a units' actions—measures of effectiveness or performance. This obsession with chasing statistics and measuring progress toward "winning" distracted units from pursuing the true aim of a low-intensity conflict: achieving a political settlement.

In fact, the next, very important—and very contentious—rung in this ladder of learning was the realization that the primary objective of low-intensity conflicts was to reach a sustainable political solution. As a corollary to this, observers sometimes also reached the realization that *any* conflict, whether high- or low-intensity at its outset, would inevitably devolve into a low-intensity one, requiring Army units to engage the political dimension of that conflict to bring it to a successful conclusion.

The first cognitive obstacle to fully acknowledging the political dimension of low-intensity conflict was presented by well-meaning but misguided observers: the desire to protect U.S. soldiers from harm. Units and commanders often sacrificed the ability to interact with the populace and effectively engage the political dimension of the conflict because of their excessive concern for force protection. Other observers insisted that the politics of low-intensity conflict was not the Army's job; it should be outsourced to the host nation, intergovernmental organizations such as the UN, allies, or other U.S. government agencies such as the Department of State.

Each of these cognitive obstacles, like the others before them, collapsed under scrutiny. Units overemphasized force protection because they didn't understand that the political dimension was their responsibility; failing to risk exposure in order to engage this dimension extended the duration of the war and put U.S. forces at even greater risk—not to mention putting the mission itself in jeopardy of failure. If the host nation could have engaged the political dimension without the U.S. Army, the Army wouldn't have been deployed to the low-intensity conflict in the first place. And the UN or U.S. allies couldn't affect a political solution on behalf of the Army because they seldom shared all of America's interests and goals.

The suggestion that the political dimension of low-intensity conflicts should be relegated to the State Department—represented abroad by U.S. ambassadors and their tiny country teams—deserves special scrutiny. This idea has persisted through the wars in Afghanistan and Iraq and continues to hold sway among many military observers today. It proceeds from a misconception that some U.S. statute requires the military to subordinate itself to the State Department. Moreover, this idea has been perpetuated by the Army's consistent practice since the end of the Cold War of dumping the political dimension of low-intensity conflicts on the State Department rather than engaging in this messy business itself.

First, there is no statutory requirement for the Army to subordinate itself to the State Department once a joint task force is established in a conflict zone. As Title 22 of the U.S. Code, Section 3927(a), states, "Under the direction of the President, the chief of mission [ambassador] to a foreign country . . . shall have the full responsibility for the direction, coordination, and supervision of all government executive branch employees in that country (except for Voice of America correspondents on official assignment and employees under the command of a United States area military commander.)"[38] U.S. military doctrine clearly communicates the consequences of this statute to Army leaders in Joint Publication (JP) 3-24, *Counterinsurgency*, which states, "The COM [chief of mission] directs, coordinates, and supervises all government executive branch employees in that country *except* for Service members and employees under the command of the JFC [joint force commander]."[39] In other words, once a joint task force is established for a military operation (as was the case in each low-intensity conflict examined in this book), the commander of that task force (the senior military commander in the conflict) and all of the forces and civilian officials under his command are no longer under the authority of the ambassador, the embassy, or the State Department.

Second, the Army's low-intensity conflict experiences since the end of the Cold War—chiefly in Somalia, Haiti, Bosnia-Herzegovina, Kosovo, Afghanistan, and Iraq—clearly show that the Department of State is incapable of engaging the political dimension of such conflicts on the Army's behalf. The country team is tiny in comparison to the deployed military force in even a modest low-intensity conflict and lacks the mobility, weaponry, and armor protection required to operate in and among the populace without military support. Much more important, however, the State Department is simply incapable of successfully engaging the political dimension of such a conflict. Low-intensity conflict environments differ from nonviolent, peaceful environments in that politics and force or the threat thereof are inextricably intertwined; in a low-intensity conflict, the ability to wield force is required to impose a political settlement on a party to the conflict. Civilian government agencies such as the State Department can't do this tough political work for the Army because they lack the ability to wield force. Unless Army leaders are prepared to take orders from State Department officials on whom to kill, they must be resigned to engaging the political dimension of low-intensity conflicts themselves.

Some observers posited another cognitive obstacle to the Army's responsibility to engage the political dimension of low-intensity conflict. They suggested that the U.S. government should form some new type of "other" force to address the politics of such conflict so that the Army didn't have to do so. Another manifestation of this cognitive obstacle was the idea, posed by both low-intensity conflict observers and Army transformers, that the Army should prevent *mission creep*—a term that came into vogue after the U.S. military intervention in Somalia[40]—by artificially limiting the tasks and missions it accepts in a low-intensity conflict.

These were certainly recipes for the Army to avoid engaging in the politics of low-intensity conflicts. Yet they also produced forever wars that did not conclude with a political settlement. These cognitive obstacles stemmed from the same root: a refusal to acknowledge that these conflicts were, in fact, war, which Clausewitz described as "an act of force to compel our enemy to do our will." The idea that the United States would build some "other" force to engage in the politics of low-intensity conflict was simply another way of expressing the fallacy that politics wasn't the Army's job. And units that believed that they were victims of mission creep had begun the

23

low-intensity conflicts refusing to accept that the political dimension was their responsibility; in other words, it wasn't that they were victims of mission creep but that they were simply annoyed because, to their minds, the political dimension of the conflict had been illegitimately thrust upon them.

The final rung on the ladder toward complete understanding and competency in low-intensity conflict was the realization that in order to achieve a political settlement the Army had to take sides in the conflict and back a winner. In virtually every instance in which the Army is committed to a low-intensity conflict, U.S. strategy, statements by the president, or the simple facts on the ground make it abundantly clear which faction the U.S. civilian leadership would prefer prevail. By engaging the political dimension of the conflict—through discussions with the various military and civilian leaders in the conflict and the nuanced application of force or the threat of force to coerce compliance—the Army is indeed capable of tilting the outcome of the conflict in favor of that faction. After all, this is why the federal government dispatches the Army to intervene in conflicts around the world in the first place.

While a very few observers inside and outside the Army did acknowledge the political dimensions of low-intensity conflict and the necessity that the Army engage in it, the cognitive obstacles to this final, complete understanding of such conflict were simply too great; while smaller constituent units in each conflict in the 1990s occasionally achieved this level of understanding, the Army as a whole never did so.

The first and most profound obstacle to reaching this level of understanding was the American predilection for self-determination, with its roots in the U.S. revolutionary past and the Wilsonian ideals first expressed at the end of World War I. It was hard for the Army to abandon these principles— even when using force on another country "to compel our enemy to do our will." Thus the Army repeatedly—most prominently in Haiti and Iraq—used the threat of violence or actual violence to depose a political group and then, incredibly, began to act impartially toward all parties, including its former enemies, and tried to include every faction in a postwar political framework. And where Army units refused to engage in the political dimension of conflicts, the State Department pursued this course in its stead.

A corollary to the confused bent toward supposed enlightened neutrality was the desire to preserve the "legitimacy" of the military intervention— primarily with the people of the country in question but also with the international community and the American people.[41] In reality, this was just another symptom of the Army's refusal to acknowledge that low-intensity conflict is in fact war, the purpose of which is to impose a political solution on a country's people and political leadership against their will. International audiences judge the legitimacy of a U.S. military intervention according to their respective interests; criticisms of the Army's tactics or whether it has chosen the "right" side in imposing a peace will always be driven by these interests.

The American people judge the legitimacy of the Army's actions in a particular conflict in accordance with the means used—whether the violence applied is proportionate or excessive—and, at least equally important, whether those means are successful. Once its military forces are committed to a conflict, the American people are very much less concerned with whether all of the factions in that conflict are treated equally and impartially than they are with whether the United States wins the war.

And it has been a painfully long time since the U.S. Army conclusively won a war.

NOTES

1. Headquarters, U.S. Department of the Army, *Operations*, FM 100-5 (1993), 13-0–13-8.

2. One could even make a compelling case that the war in Iraq is, in fact, America's longest war, having begun with the liberation of Kuwait in 1991 and continued as a low-intensity conflict for the succeeding two and a half decades, with a brief, monthlong return to high-intensity conflict in 2003 and a brief, three-year respite beginning at the end of 2011 and lasting until mid-2014.

3. 153 Cong. Rec., H5081 (daily ed., March 1, 2007) (statement of Rep. Woolsey); Fallows, "The Right and Wrong Questions"; Nicolaus Mills, "Punished for Telling Truth about Iraq War," CNN, March 20, 2013, http://www.cnn.com/2013/03/20/opinion/mills-truth -teller-iraq/index.html; Kaplan, "Who Disbanded the Iraqi Army?; Alice Fordham, "Fact Check: Did Obama Withdraw from Iraq Too Soon, Allowing ISIS to Grow?," *Morning Edition*, NPR, December 19, 2015, https://www.npr.org/2015/12/19/459850716 /fact-check-did-obama-withdraw-from-iraq-too-soon-allowing-isis-to-grow.

4. Millett and Maslowski, *For the Common Defense*, Kindle, locations 12781–84; Lawrence, *America's Modern Wars*, Kindle, locations 5594–95. See also Gordon and Trainor, *The Endgame*; Gordon and Trainor, *Cobra II*; Ricks, *Fiasco*; Filkins, *The Forever War*; Woodward, *State of Denial*; and Baker and Hamilton, *Iraq Study Group Report*.

5. Barack Obama, "Remarks by the President in Address to the Nation on the Way Forward in Afghanistan and Pakistan," December 1, 2009, White House Office of the Press Secretary, https://obamawhitehouse.archives.gov/the-press-office/remarks-president-address-nation -way-forward-afghanistan-and-pakistan; Hastings, "The Runaway General."

6. John Batiste, interview with Steve Bowman, December 7, 2005, part 1, video, Bowman Materials, Colonel Robert R. McCormick Research Center, First Division Museum at Cantigny.

7. Batiste interview; Ricks, *Fiasco*, 227–90; Khatchadourian, "The Kill Company."

8. Ricks, *Fiasco*, 284–88; Rashid, *Descent into Chaos*, Kindle, 240–62.

9. Sara Wood, "Petraeus Takes Command of Multinational Force Iraq," U.S. Department of Defense, February 12, 2007, http://archive.defense.gov/news/newsarticle.aspx?id=3032; "U.S. Casualties in Iraq," Global Security, n.d., accessed December 4, 2017, https://www.globalsecurity.org/military/ops/iraq_casualties.htm; "Operation Enduring Freedom: Fatalities by Year and Month," iCasualties.org, n.d., accessed December 4, 2017, http://icasualties.org/OEF/ByMonth.aspx; Associated Press, "New U.S. Commander Aims to Turn Afghan Tide," NBC News, June 15, 2009, http://www.nbcnews.com/id/31363630/ns/world_news-south_and_central_asia/t/new-us-commander-aims-turn-afghan-tide/#.WiVkQkqnGUk. See also Crane, *Cassandra in Oz*; Ricks, *The Gamble*; Kaplan, *The Insurgents*; Mansoor, *Surge*; Robinson, *Tell Me How This Ends*; and Broadwell and Loeb, *All In*.

10. Headquarters, U.S. Department of the Army, *Counterinsurgency*, FM 3-24, 1-10, 1-23, 2-2, 5-1–5-2; Christoff, *Provincial Reconstruction Teams*.

11. Fitzgerald, *Learning to Forget*, Kindle, 50; Finel, "An Alternative to COIN"; Craig Collier, quoted in Fitzgerald, *Learning to Forget*, Kindle, 201.

12. Nowowiejski, "Regaining the Edge," 26–28.

13. Steele, "The National Training Center," 26–37.

14. Gentile, *Wrong Turn*, Kindle, locations 197, 206, 7–8, 175, 240.

15. Headquarters, U.S. Department of the Army, *Operations*, FM 3-0 (2017), ix, 1-39.

16. Michael D. Lundy, "Foreword," in Headquarters, U.S. Department of the Army, FM-3-0 (2017), n.p.

17. Lundy, "Foreword," n.p., emphasis in the original; Headquarters, U.S. Department of the Army, FM 3-0 (2017), 1-3.

18. Gary Sheftick, "First Security Force Assistance Brigade Training for Deployment," Army News Service, October 16, 2017, https://www.army.mil/article/195178/first_security_force_assistance_brigade_training_for_deployment.

19. Brian Mennes, cited in Sheftick, "First Security Force."

20. David Vergun, "Soldier Lethality, Mobile Networks Key for Army Future Readiness, Chief of Staff Says," Army News Service, January 22, 2018, https://www.army.mil/article/199181/soldier_lethality_mobile_networks_key_for_army_future_readiness_chief_of_staff_says; C. Todd Lopez, "Security Force Assistance Brigades to Free Brigade Combat Teams from Advise, Assist Mission," U.S. Army, May 18, 2017, https://www.army.mil/article/188004/security_force_assistance_brigades_to_free_brigade_combat_teams_from_advise_assist_mission, brackets in the original; "New Academy Will Train NCOs for Security Force Assistance Brigades," brackets in the original.

21. Thucydides, *The History of the Peloponnesian War*, Kindle, 129.

22. Lionel Beehner, "Dr. Graham Allison on the Likelihood of War with China," Modern War Institute, November 8, 2017, https://mwi.usma.edu/dr-graham-allison-likelihood-war-china/; "28th Annual Strategy Conference, 'The Changing Character of War,'" Strategic Studies Institute, U.S. Army War College, n.d., accessed August 26, 2019, https://ssi.armywarcollege.edu/events/details.cfm?q=184; MWI Staff, "War Books, Special Edition: Centcom Commander's 2018 Reading List," Modern War Institute, December 4, 2017, https://mwi.usma.edu/war-books-special-edition-centcom-commanders-2018-reading-list/; Allison, *Destined for War*, Kindle, locations 68–71.

23. Schelling, *Arms and Influence*, 1–34; Morgenthau, "International Relations," in *The Restoration of American Politics*, 174; Morgenthau, "Atomic Force and Foreign Policy," in *The Restoration of American Politics*, 156; Smith, *The Utility of Force*, 3. See also Mearsheimer, "Nuclear Weapons and Deterrence"; and Congressional Budget Office, *Projected Costs of U.S. Nuclear Forces*.

24. Soergel, "War on Terror."

25. European Security Study, *Strengthening Conventional Deterrence in Europe*, 86.

26. Because the views held by this group never enjoyed the official sanction of the U.S. Army, the camp described here as low-intensity conflict observers was amorphous and leaderless. But among its most significant members from 1991 to 2003 were Steven Metz, Robert Baumann, Roger Spiller, Lester Grau, Jacob Kipp, Conrad Crane, and Andrew Terrill.

27. The views of the Army transformers were very much the mainstream of Army thought between the end of the Cold War and the beginning of the war on terror. As such, this group's chief advocates were the leaders of the Army—the sequential chiefs of staff, from Gen. Carl Vuono, to Gen. Gordon Sullivan and Gen. Dennis Reimer, to Gen. Eric Shinseki.

28. Clausewitz, *On War*, Kindle, 75.

29. Headquarters, U.S. Department of the Army, FM 3-0 (2017), 1-16.

30. Admittedly, this definition becomes murkier should a declared insurrection cause the U.S. Army to engage in violence or the threat of violence against armed Americans or civilians.

31. Nuclear deterrence and the enforcement of no-fly zones are very particular variants of deterrence, but varieties in which the Army does not directly participate (because it lacks nuclear arms and fixed-wing attack aircraft).

32. Sullivan and Dubik, "Land Warfare in the 21st Century," iii.

33. John Spencer, "What Is Army Doctrine?" Modern War Institute, March 21, 2016, https://mwi.usma.edu/what-is-army-doctrine/.

34. Combined Arms Center History Office, *1991 Annual Command History*, 249; Combined Arms Center History Office, *2001 Annual Command History*, 13.

35. Shimko, *The Iraq Wars*, Kindle, locations 1062-71.

36. Schadlow, *War and the Art of Governance*, 1-13; Sturgill, *Low-Intensity Conflict in American History*, 5; Boot, *The Savage Wars of Peace*, xvii–xxiv; Echevarria, *Reconsidering the American Way of War*, 1-2.

37. National Defense Authorization Act for Fiscal Year 1987, Pub. L. No. 99–661, November 14, 1986, Title XIII, Part B.

38. 22 U.S.C. 3927(a), quoted in Joint Warfighting Center, *Joint Task Force Commander's Handbook for Peace Operations*, IV-6.

39. Joint Staff, *Counterinsurgency*, JP 3-24, V-1, emphasis added.

40. Stewart, *The United States Army in Somalia*, 12–14.

41. For the notion of the "legitimacy" of military intervention, see Headquarters, U.S. Department of the Army, *Operations in a Low-Intensity Conflict*, FM 7-98, chap. 1, sec. 2, 1-7.

1 PHOENIX OR ICARUS?

THE U.S. ARMY's journey from its defeat in the Vietnam War in 1973 to its large role in the stunning victory of the Gulf War in 1991—and the defeat of Saddam Hussein's army in Kuwait and Iraq was indeed stunning—has taken on the air of a creation myth within the Army. In the Gulf War, the Iraqi Army, in heavily fortified defensive positions, was soundly defeated by an attacking force of roughly equal numbers. The Iraqis suffered thirty thousand deaths, to the coalition's 292. The Iraqi Army had been decimated by a monthlong, high-tech, precision bombing campaign followed by a GPS- and computer-enabled, lightning, one-hundred-hour ground war.

But as stunning as this victory was, President George H. W. Bush's unfortunate declaration that "the specter of Vietnam has been buried forever in the desert sands of the Arabian Peninsula" was tragically premature.[1] A little over a decade later, the Army would be helplessly entangled in grueling insurgencies that it was unprepared to fight; the most capable military force in the history of the world was humbled by ill-trained and ill-equipped guerillas in the mountains of Afghanistan and in the dusty streets of Iraq.

Phoenix: Vietnam to the Gulf War

The legend of the Army's long road from defeat in the Vietnam War to triumph in the Gulf War was fixed in the Army psyche by autobiographies like Gen. Norman Schwarzkopf's *It Doesn't Take a Hero* and Gen. Colin Powell's *My American Journey*.[2] Both books recounted the tale of the Army's emergence from the "dark years" of post-Vietnam chaos to dominance of high-intensity conflict in the first war against Iraq.

With the benefit of hindsight—and in the shadow of the seventeen years of unrelenting low-intensity conflict since the beginning of the war on terror— it seems most apt to describe the period between the end of the Vietnam War and the end of the Gulf War as one in which the baby was thrown out with the bathwater. While we now know that the North Atlantic Treaty

Organization (NATO) would never have to defend against an invasion of Western Europe by the Warsaw Pact states, it is hard to blame the senior leaders of the Army for rebalancing their organization and better training it to face the Soviet Union in high-intensity conflict. After all, at the end of the Vietnam War, the Soviets had more than 150 total divisions and 600,000 or more men stationed across Eastern Europe.[3]

But Gen. William DePuy, Army chief of staff Gen. Creighton Abrams, and their successors did much more than rebalance the Army. Operating under the misguided assumption that the Army could simply choose not to engage in future low-intensity conflicts as it saw fit, its senior leaders actively expunged the lessons of Vietnam, thereby preventing the Army from preserving any of the costly and hard-won capacity it had gained to effectively fight such conflicts.

This culture of focus on high-intensity conflict to the exclusion of other types of operations pervaded the post–Vietnam Cold War Army, one that found itself engaged in a number of low-intensity conflicts—Nicaragua and El Salvador chief among them—in the nearly two decades between the end of U.S. involvement in Vietnam and the fall of the Soviet Union. But steeped in a culture that considered low-intensity conflict outside its purview, the Army in each case ignored lessons it might otherwise have learned.

The word *counterinsurgency* arguably has its roots in the early 1960s, as President John F. Kennedy and like-minded observers considered the supposed new communist strategy of using "wars of national liberation" rather than high-intensity conflict to subvert and communize the Third World.[4] Counterinsurgency thinking prompted the formation of the U.S. Army Special Forces—the Green Berets. It also seduced the Kennedy administration, and subsequently that of President Lyndon B. Johnson, into growing U.S. involvement in the Vietnam War. But while the Army did make significant tactical innovations throughout the course of the war, it never truly institutionalized the precepts of counterinsurgency. In fact, the U.S. Marine Corps was much more aggressive than the Army—through its combat action platoons—in implementing counterinsurgency tactics at the local level. Army general William Westmoreland, until 1968 the commander of Military Assistance Command–Vietnam (MAC-V), the senior military command in Vietnam, actually tried to discourage the Marines from focusing so heavily on counterinsurgency.[5]

On March 22, 1968, President Johnson announced that General Westmoreland would be replaced as the commander of MAC-V by his deputy, Gen. Creighton Abrams. When Abrams succeeded Westmoreland, he dramatically changed the Army's strategy in Vietnam.[6]

Historians studying the evolution of counterinsurgency doctrine and tactics in Vietnam frequently characterize Abrams's strategy as one of "Vietnamization"—building the capacity of the Army of the Republic of Vietnam (ARVN)—and pacification, classic counterinsurgency tactics at the local, provincial, and national levels to build the capacity of the South Vietnamese government. Likewise, this strategy is frequently contrasted with the firepower-intensive search-and-destroy tactics that had prevailed under General Westmoreland's tenure.[7]

As is often the case, the truth is more complicated. The strategy General Abrams could feasibly pursue in Vietnam in 1968 and beyond was severely circumscribed by the American public's growing impatience with a war that had featured direct U.S. military intervention since 1964, and resulted in the death of a great many Americans, seemingly without significant progress. As a result, MAC-V was under intense pressure to keep casualties low and bring U.S. troops home.[8] Abrams's response was to move away from a strategy of attrition and toward one focused less on attempting to win the war outright and more on creating space and time for the ARVN and government of South Vietnam to grow to the point that they could take over the conduct of the war and free the U.S. military to leave. Abrams would focus the Army's efforts on interdicting the flow of supplies from North Vietnam into South Vietnam and, to the extent he could, insulating the rural populace from the influence of the insurgency, the National Liberation Front (NLF, popularly known as the Viet Cong) and the North Vietnamese Army (NVA). At the same time, the United States would engage in its policy of Vietnamization to build the capacity of the ARVN and draw down U.S. forces.[9]

While this strategy did ultimately succeed in extricating the United States from Vietnam, it failed to achieve the stated policy objective of three presidents—Kennedy, Johnson, and Richard M. Nixon—over the previous decade: a viable and independent South Vietnam able to defend itself against subversion from within or aggression from North Vietnam. The Nixon administration finally achieved peace by acquiescing to the NVA remaining in South Vietnam after the U.S. withdrawal.[10] As a result, two years after the withdrawal, amid a roar of accusations and counteraccusations in Washington, DC, Saigon fell to the North Vietnamese on April 30, 1975.[11] To the shock of the nation and the world, America had lost what to that time had been its longest war.

In the course of U.S. involvement in Vietnam the war had been, at various times and places, a low-intensity conflict, a high-intensity conflict, or both simultaneously. Before the direct intervention of U.S. military forces against the NLF and North Vietnam, which began in 1964, the war was largely a low-intensity conflict. As the escalation of U.S. ground troops was matched by an escalation from North Vietnam, the Army began to face NVA regiments and divisions in high-intensity conflict. At the same time, it continued to face a low-intensity conflict against the NLF for control of the rural population in large swaths of South Vietnam. And, occasionally—most notably during the siege of Khe Sanh,[12] the Cambodian incursion, the Laotian incursion, and the spring 1972 NVA offensive, the Vietnam War was decidedly high-intensity in character. Despite this ambiguity, the Army took away many lessons on low-intensity conflict during and immediately following the war.

Due to the length of the war, the Army did wrestle with the problem of rotation of personnel into and out of theater. While troops were initially deployed as whole units to Vietnam, as the war dragged on, soldiers began yearly rotations into and out of country on an individual basis. This practice had its precedence in the Korean War as early as 1951.[13] The policy had been implemented to reduce the incidence of "combat fatigue" that had afflicted veterans of World War II and Korea, but it came with significant drawbacks. For one, units were forced to contend with a constant influx of inexperienced soldiers. Worse, rotations hurt unit cohesion, as a nearly continuous turnover of personnel repeatedly filled small units that had built close bonds through the trials of combat with strangers. To make matters worse, while soldiers rotated every year, battalion commanders rotated into and out of their units every six months. The logic behind this policy was that it would produce the maximum number of lieutenant colonels and colonels with combat experience. In practice, though, it was terribly disruptive, crippling the Army's effectiveness in Vietnam as each unit was continually forced to overcome the mistakes of inexperienced leaders. As James Kitfield quotes an anonymous general, "It was . . . almost as if the services were using Vietnam to train officers for the next war, as opposed to fighting the one very much at hand."[14]

By 1968 the Army was keenly aware that it was not trained, manned, or equipped to fight the low-intensity conflict components of the Vietnam War. General Abrams himself would tell Gen. Cao Van Vien, chairman of the South Vietnamese General Staff, "You know, we've been to [the Command and General Staff College at Fort] Leavenworth or something, and had all those lessons and books. And I don't remember anybody talking about the stuff you and I are talking about today. They didn't have any lectures on that—anything! And they don't have F.M.s [field manuals] about that. And here we are, we're all mixed up in it, supposed to be helping."[15] This disconnect between what was being taught in training and education in the United States and what Army units were actually doing in Vietnam did not end with the Army's leadership; soldiers were not being prepared, either.[16]

One low-intensity conflict capability that did not grow as much—and in fact regressed—over the course of the Vietnam War was the skills involved in training host nation forces to fight. Before the direct intervention of U.S. forces in the war in 1964, the American role in Vietnam was primarily an advisory one.[17] One effort that did flourish under General Westmoreland was to raise Regional Force (RF) companies and Popular Force (PF) platoons in rural areas. But this effort was spearheaded by the ARVN and approached almost as busywork while U.S. forces did the heavy lifting of fighting the NLF and NVA. By the time General Abrams took over for Westmoreland in 1968, Vietnamization was little more than a euphemism for U.S. withdrawal and a handover of the war back to the ARVN.[18]

Many observers also came to realize the importance of the populace to the low-intensity conflict in Vietnam. A number of outside observers, including Secretary of Defense Robert McNamara and Marine lieutenant general Victor Krulak, were critical of Westmoreland's attrition strategy, advocating a more population-centric approach. When Abrams took command of all U.S. forces in Vietnam in 1968, he made population security a centerpiece of his strategy. This "one war" approach included protecting the population from the NLF to allow civil-military efforts to proceed. Abrams's new strategy also explicitly addressed a more nuanced application of force, imposing limits on the amount of firepower that should be employed in populated areas.[19]

For those observers who acknowledged the need to put the population at the center of operations in a low-intensity conflict, the civil-military effort, commonly referred to as pacification, was a central concern. Westmoreland largely ignored pacification in favor of large, set-piece operations to find, fix, and destroy the NLF and NVA. A study in 1966, the Program for the Pacification and Long-Term Development of South Vietnam, not surprisingly concluded that pacification was being neglected, insisting that the war could not be won without winning the civil-military effort at the "village, district and provincial levels."[20] By 1967 Westmoreland did have a more robust pacification effort in place in South Vietnam. Notably, this effort—the Civilian Operations and Revolutionary Development Support (CORDS) program—incorporated other government agencies such as the U.S. Agency for International Development and was led by Ambassador Robert Komer, who reported directly to the general.[21] These efforts accelerated under General Abrams.

Yet population-centric operations also spawned an unhealthy focus on metrics and empirical measures of success. One such measure was the Hamlet Evaluation System, which graded villages on a broad array of measures to arrive at an A through E score of their fealty to the national government, with the ultimate goal of determining U.S. progress in winning the war.[22] Similar measures were implemented for numbers of bombs dropped and patrols conducted. These measures were not uncontroversial at the time. Intelligence officer William Colby, director of the pacification effort in Vietnam, called the pacification measures "fairly soft," and Brian Jenkins of the RAND Corporation warned that a measure of U.S. military activity did not necessarily indicate the United States was "doing better" in the war itself.[23]

Of course, the most maligned measure of progress during the war was the tally of enemy killed—the infamous body count. A 1971 U.S. Army War College study that surveyed around 450 junior and midgrade officers who had served in Vietnam argued that this obsession with measuring attrition was corrosive to the integrity of the Army and its leaders.[24]

Throughout the war, observers inside and outside the Army arrived at the conclusion that the low-intensity conflict was peculiarly political in nature. In fact, this realization underpinned much of the post-1968 pacification effort; the theory was that by addressing the underlying grievances of the rural populace the U.S. military and the government of South Vietnam could turn the people away from supporting the insurgency. The effort resulted in an impressive number of wells dug and schools and roads built, but did little to quell the insurgency.[25]

After the war, observers concluded that the Army had failed to win because it had not sufficiently addressed the political dimension of the conflict. In 1975, U.S. Army War College commandant Maj. Gen. DeWitt Smith lamented the irony that the Army could "win so frequently, and so well, in a war-fighting sense, and yet lose a war in a strategic or political sense."[26] In a 1979 study by the BDM Corporation commissioned by the War College, the authors wrote that the Army took an approach to the war that was "heavily biased towards the materiel and technological end of the scale and slights the psychological and political element."[27]

But while the BDM authors did understand the necessity to engage in the political dimensions of low-intensity conflict, they fell into the cognitive trap of looking for someone else to take that responsibility away from the Army. The study was skeptical of the utility of the Army in "banana wars" and concluded that the host nation government—not the Army—bore the responsibility to engage in the political dimensions of the conflict.[28]

In fact, throughout the Vietnam War, there were many observers—most of the senior leaders inside the Army chief among them—who disputed the Army's responsibility to engage in the politics of the low-intensity conflict. As the commander of the 1st Infantry Division, Maj. Gen. William DePuy was notorious for refusing to engage in the political dimension of the low-intensity conflict being waged in the towns and villages within his area of operations, instead relying on an unrelenting application of overwhelming firepower in a high-intensity conflict against the NVA and NLF. He was also notorious for firing battalion commanders and midgrade officers who disagreed with this approach.[29] Maj. Gen. Charles P. Brown, when he arrived as commander of I Field Force Vietnam, was horrified to find the 173rd Airborne Brigade supporting RF/PF militias—"sitting on their ass holding hands with RF'ers and the PF'ers"—and engaging in the local politics of the villages in their sector instead of waging a high-intensity campaign to destroy the NVA and NLF.[30]

Notably, while the Army was far from coming to a consensus on the question of whether it should engage in the political dimensions of low-intensity conflict, there was almost a unanimous—in fact an innate—understanding that the Army had to take sides in the conflict. Every instrument of U.S. national power was engaged in support of the government in Saigon—whomever that happened to be at any given time. After the murder of President Ngo Dinh Diem in a U.S.-sanctioned coup in November 1963, a parade of governments came and went through the presidential palace in Saigon, but each was consistently backed by the U.S. Department of State and the U.S. military.[31] And at the provincial and local levels, both the South Vietnamese government and the ARVN could count on military support—and, later in the war, civil-military support—from the United States. This was probably a symptom of the age in which the Vietnam War took place; it was, after all, the Cold War, and the United States was engaged in a crusade to "defend" Saigon against "aggression," a perceived attempt by Hanoi to spread communism, ostensibly directed by Beijing and Moscow.[32]

While the U.S. Army learned many lessons during and immediately after the Vietnam War, its senior leaders ultimately chose not to institutionalize them. In fact, they deliberately turned away from these experiences, choosing instead to return to preparation for a high-intensity conflict in Europe against the Soviet Union.

In justifying its wholesale rejection of the lessons of Vietnam, the Army presented a number of arguments for why the lessons lacked value. Many of these arguments echoed the popular American sentiment that it had been a mistake to intervene in Vietnam in the first place.[33] Another corollary to this idea, frequently argued by advocates of abandoning the lessons of Vietnam, was the idea that the war was unwinnable regardless of potential actions that could have been taken by the Army, either because of the strategic impossibility of fighting a limited war against an opponent willing to pay any price for victory or because of the poor strategic decisions made by civilian leaders in Washington. Others argued that the U.S. military in general and the Army in particular should not be committed to wars such as the one in Vietnam. This idea first found official endorsement from President Nixon in 1969 with the so-called Nixon Doctrine and its exhortation that countries should be responsible for their own "internal security."[34] Underlying this doctrine was the premise that fighting for one's own freedom from subversion was somehow transformative or necessary for the gains of counterinsurgency to be enduring.

The Army's relationship with the American people whom it is charged to protect has always been a complicated one, and this relationship was seldom more complicated than it was in the immediate aftermath of Vietnam. While many veterans of the war may have privately felt betrayed by the American people, the dominant narrative among midgrade and senior leaders in the Army was that the American people had lost confidence in the war because America's civilian leaders had failed to enlist or sustain their support. Gen. Charles Brown insisted that if the Army were to be committed to war, the American people needed "to be convinced that there is a reason and it is necessary." This sentiment—that the president had committed the United States to war without enlisting the support of the American people—was the driving motivation behind Army chief of staff Gen. Creighton Abrams's effort after the war to reorganize the Army according to the Total Force initiative. This restructuring placed many of the Army's critical capabilities in the Army Reserves and National Guard. As a consequence, the Army as a whole could not be deployed to war without mobilizing its reserve components—a move President Johnson had refused to make before the Vietnam War because it would have required the consent of Congress.[35]

The U.S. Army Training and Doctrine Command (TRADOC), established under the command of General DePuy in mid-1973, immediately following the end of the U.S. involvement in Vietnam, stood at the center of the Army's rejection of the lessons of Vietnam. DePuy was a veteran of World War II who had seen tough fighting in Normandy.[36]

Twenty years after the Second World War, as director of operations for General Westmoreland at MAC-V, DePuy was the architect of the air-mobile attrition tactics that would mark the first three years of the Vietnam War, tactics he would put into practice himself as the commander of the 1st Infantry Division.[37] In his position as vice chief of staff of the Army, DePuy ensured his influence would extend well beyond his time in the Army, nurturing the careers of like-minded officers—frequently called DePuy's "fair-haired boys"—who would lead the U.S. military over the subsequent two decades. These were men like future generals Colin Powell and Max Thurman.[38]

As TRADOC commander, DePuy oversaw more than a rejection of the lessons of Vietnam; he would lead the Army through a redefinition of what forms of warfare were inside and outside the purview of the Army. From its establishment onward, TRADOC was dedicated to putting the experience of Vietnam behind the Army and getting it back to what it saw as its proper role: high-intensity conflict.[39]

This urgency to rebuild the Army's high-intensity conflict prowess was born of a perception that the United States had fallen behind the Soviets in its degree of preparedness for war. Gen. Don Starry, initially deputy commander and later commander of TRADOC, worried that the United States had been distracted by Vietnam. Starry would later say that the Soviets, waiting east of the Iron Curtain, "understood what was happening in U.S. Army Europe and elected to take advantage of the situation." He would add, in the 1980s, "We no longer enjoy any advantages, conventional or integrated."[40]

DePuy began to reshape Army doctrine almost immediately after establishing TRADOC, and the model he used in this reformation was the 1973 Yom Kippur War. On October 6, 1973, only a few months after the creation of TRADOC, Egypt and Syria, supported by Arab allies, including Iraq and Jordan, launched a surprise attack against Israel. Protected by highly effective Soviet air defense systems, Egypt crossed the Suez Canal and made impressive gains, moving deep into the Sinai Peninsula. The Syrians coordinated their own attack from the north with the Egyptian advance and, likewise using surface-to-air missile batteries to cover their advance, spilled into the occupied Golan Heights. While they were initially taken by surprise and for a time understandably stunned by the speed and boldness of the enemies' actions, once the Israelis were able to fully mobilize, the tide of the war turned. In the south the Egyptians were pushed back across the Suez.

In the north the Israelis drove deep inside Syria to within shelling range of Damascus. By October 22, a little less than two weeks after the start of the war, a ceasefire was in place and the war was over. In the north, the Syrians and their Iraqi and Jordanian allies had lost nearly 1,300 tanks to the Israeli's 250 tanks, and nearly 3,500 Syrians were dead to the Israelis' 800. In the south, the Egyptians lost around 1,000 tanks and as many as 12,000 dead to the Israelis 400 tanks and nearly 3,000 dead.[41]

The Yom Kippur War had a dramatic impact on General Starry when he surveyed the battlefield only a few months later in January 1974.[42] In less than two weeks, both sides in the conflict had lost more tanks than the United States lost in all of Europe during World War II. There were technical lessons to be realized: relatively cheap antitank weapons had overcome the traditional dominance of tanks and armor. But more dramatic were the strategic lessons of the Yom Kippur War: the lethality of modern, high-intensity conflict, the speed with which the new weaponry and tactics translated into decision on the battlefield, and the distances and velocities at which modern war occurred had all seemed to have changed the character of warfare.[43]

Both DePuy and Starry saw the shadows of NATO and the Warsaw Pact in the Yom Kippur War. They found lessons in the Israelis' initial unpreparedness for the conflict, noting that it was not until Israel could fully mobilize its reserves that it was able to turn the tide. But Israelis could hop in cars and drive to the front; the United States had to traverse the Atlantic Ocean to reinforce Europe. DePuy and Starry also found doctrine and training lessons. Seeing a Cold War allegory in this Arab-Israeli war, they believed that it was the Israelis' Western-style freedom of action and independent thinking that had allowed their commanders to react rapidly and outmaneuver the Arab commanders, who were cobbled by their forced adherence to the rigid Soviet doctrine. DePuy and Starry also concluded that on the new, more lethal modern battlefield, the Army could not afford to "learn on the job," as they believed it had in the early days of World War I, World War II, and Korea. The Army had to train and prepare to win the first battle of future wars. DePuy saw a new direction for the Army in this conflict: an outnumbered but tactically and technologically superior defender was able to repel and counterattack a numerically superior but tactically and technologically inferior opponent—and with devastating results.[44]

Within the Department of Defense, the strategy to replicate this dynamic in the confrontation between East and West would be called the "offset strategy." The West would offset the East's superior numbers with superior technology and tactics.[45] DePuy envisioned a new doctrine for the Army that leveraged the supposedly better training and more innovative thinking of the American soldier, and superior American technology, to outmaneuver and keep the Soviets at bay, while American firepower attacked the enemy throughout the depth of its formations. This new doctrine would be called active defense.[46]

Active defense was codified in the Army's capstone doctrine, Field Manual (FM) 100-5, *Operations*, in July 1976. The manual explicitly and unapologetically cast aside the lessons of Vietnam in favor of a renewed focus on high-intensity conflict. The Army had to "*win the first battle of the next war*," and that war would be unequivocally high-intensity in character. The manual focused on "Battle in Central Europe against the Warsaw Pact," but "the principles" it espoused would be equally valid "to military operations anywhere in the world" against any type of force from "highly modern mechanized forces" to "light, irregular units in a remote part of the less developed world."[47]

DePuy was explicit about FM 100-5's rejection of the lessons of Vietnam. In a letter to the Army chief of staff, Gen. Fred Weyand, DePuy wrote, "This manual takes the Army out of the rice paddies of Vietnam and places it on the Western European battlefield against the Warsaw Pact." He would later say that the manual's objective was to "get the Army reoriented from a light infantry war to its main mission in Europe which was an entirely different kind of a war."[48] DePuy drove home his point by stating that, in his opinion, Vietnam had been a dangerous distraction from the Army's "main mission."

If the 1976 edition of FM 100-5 was a codification of the rejection of the lessons of Vietnam, then Col. Harry Summers's book *On Strategy: A Critical Analysis of the Vietnam War* (1982) was a rewriting of those lessons. Summers had been a researcher at the Strategic Studies Institute of the U.S. Army War College and was tasked by the commandant with taking the findings of the earlier BDM Corporation study—that the United States had failed in Vietnam because it had failed to engage the political dimensions of the low-intensity conflict—and turning them into a book-length Army publication. Summers's final work was diametrically opposed to the findings of the BDM study,[49] immediately endearing him and his work to the senior leadership of the Army.

On Strategy started promisingly enough, with the now famous—and perhaps apocryphal—anecdote of a conversation between Summers and a North Vietnamese colonel in which Summers said "You never defeated us on the battlefield" and the North Vietnamese colonel replied "That may be so . . . but it is also irrelevant."[50] The anecdote implied that Summers might focus his analysis on the Army's misguided obsession with attrition warfare rather than solving the political problem in South Vietnam.

Instead the book was a 240-page assault on the idea that the U.S. Army had any responsibility to think about the politics of war at all, let alone engage in the political dimensions of the low-intensity conflict in Vietnam. Summers quoted nineteenth-century Prussian military theorist Carl von Clausewitz dozens of times and then proceeded to completely misunderstand his dictum that war is "an act of violence meant to force the enemy to do our will."[51]

According to Summers, the Army had misunderstood the "true nature of the Vietnam war,"[52] foolishly engaging in a low-intensity conflict when it should have been more aggressively killing its way to victory. He explicitly rejected the notion of counterinsurgency, saying that the defeat in Vietnam sprung not from "the degree to which we pursued counterinsurgency doctrine, but out of the doctrine itself." He pilloried those Army leaders who had embraced counterinsurgency while praising North Vietnamese general Van Tien Dung, whom he quoted as saying, "The basic law of war was to destroy the enemy's armed forces." For Summers, counterinsurgency was a "fad," a "dogma" that had distracted the Army from its true task in Vietnam, using the full force of its might to engage in a high-intensity conflict to destroy the North Vietnamese. Counterinsurgency doctrine had "stultified military thinking." In Summers's view, this was why the United States had lost the war.[53]

The senior leaders of the Army loved *On Strategy*. U.S. Army War College commandant—and later director of the Joint Staff—Maj. Gen. Jack Merritt hailed the book as "firmly on the mark." The War College received numerous letters of praise for the book from other senior leaders across the Army. *On Strategy* remained on the Army chief of staff's reading list until 2004 and was required reading at the Command and General Staff College (CGSC), the Naval War College, and the National Defense University. Naturally, Summers used his book as the text for his own class on the Vietnam War at the U.S. Army War College, infecting a whole generation of senior leaders of the Army with this dismissive view of the value of proficiency in low-intensity conflict.[54]

While DePuy and Starry had created a new doctrine to restore the Army's dominance in high-intensity conflict, it was not immediately clear that they would have an Army capable of implementing the strategy; in the aftermath of the Vietnam War, the Army faced deep-seated problems. In July 1973, over the emphatic objections of senior military leaders, President Nixon had ended the draft. Now overseeing a new all-volunteer force, the senior leaders of the Army wrestled with how to bring talent into a profession whose reputation had been badly damaged by the war.[55] By 1979 the percentage of Army recruits with high school diplomas dipped to its lowest level since immediately after the end of the Vietnam War. At the same time, leaders within the Army were still contending with chronic drug and alcohol abuse within the ranks, much of which had begun during the war.[56]

When President Jimmy Carter entered office in 1977, he began a unilateral de-escalation of military spending that alarmed the nation's senior military leaders. While the 3.6-million-man Soviet Army had increased its number of tanks by 35 percent and its number of artillery pieces by 40 percent, the active component of the Army had shrunk from 1.57 million men in 1968 to 785,000 by 1974 and suffered from a relative numerical disadvantage in every class of weapon system.[57] Moreover, while the Army had sixteen active divisions of about sixteen thousand men on paper, many of their critical functions had been moved to the reserve components under General Abrams's Total Force initiative. Carter's secretary of defense, Harold Brown, would warn the president that "basically what we have is a hollow Army."[58]

All was not bleak, however, for senior Army leaders focused on restoring America's high-intensity conflict prowess. While they are generally associated with the military buildup under the administration of President Ronald Reagan, a number of new weapons programs, including those of the M1 tank, the Bradley Infantry Fighting Vehicle, and an array of precision-guided munitions, saw significant development during Carter's presidency. And in 1979 the budget cuts subsided as the Carter administration finally submitted a budget that represented a modest increase in defense spending, which grew to $157.5 billion, an increase of $16 billion.[59]

Two big foreign policy shocks at the end of 1979 would force further increases in defense spending. First the U.S. client regime in Iran was overthrown by the Iranian Revolution, the U.S. embassy in Tehran was overrun by protesters, and fifty-two American diplomats and citizens were taken hostage, beginning the infamous Iran hostage crisis, which, to the dismay of the Carter administration and the American people, dragged on for more than a year.[60] Then, in late December 1979, the Soviet Army accepted a supposed "invitation" from the government in Kabul to occupy Afghanistan, assassinated President Hafizullah Amin, and installed a puppet regime there. Most Americans blamed Carter for the perceived loss of America's standing in the world, and his approval ratings plummeted. But as his popularity waned, the Department of Defense reaped the windfall; the 1981 defense budget swelled by nearly $26 billion, to $173.9 billion, with plans for further increases over the next four years. In 1981 and 1982 alone, military pay and allowances grew on average by as much as one-third.[61]

The spending increases did little to rescue the president's standing in the polls. Matters got even worse for Carter in 1980 when an attempt to rescue the hostages in Iran ended in tragedy. Dubbed Operation Eagle Claw and commanded by Army major general James Vaught, the multiservice operation would be launched from the supercarrier USS *Nimitz* with an intermediate staging base in the Iranian desert called Desert One. The operation suffered from a muddled chain of command and interservice rivalry. The U.S. military had hastily cobbled together a special operations capability—especially in aviation assets—that had not existed before the conception of the rescue operation. The lack of training and rehearsals resulted in catastrophe: eight service members were killed when two helicopters collided at Desert One on the dusty night of the operation. An investigation of the debacle was largely subsumed by politics within the Joint Staff,[62] but the episode would eventually fuel a debate over reform of the U.S. military's special operations forces.

Amid a stagnant economy and foreign policy woes, President Reagan succeeded Carter in a sweeping electoral victory. Immediately upon entering office in January 1981, Reagan embarked on the largest peacetime defense buildup in U.S. history. To lead this military revitalization, Reagan tapped Secretary of Defense Caspar Weinberger, a veteran of Douglas MacArthur's staff during World War II and a deft Washington insider. Weinberger would later be joined by a DePuy protégé, Lt. Gen. Colin Powell, serving as national security adviser.[63]

Weinberger pushed through a 25 percent increase in defense spending in 1981 and a 20 percent increase in 1982. All told, the defense budget would see $330 billion in real growth over Reagan's first term, representing an increase of 56 percent over Carter's 1980 budget. This massive increase in defense spending was squarely focused on the high-intensity conflict threat presented by the Soviet Union. Reagan and Weinberger envisioned an American military that was, as Weinberger would say, "capable of defending all theaters simultaneously." While Europe remained the central front, 1981 National Security Decision Documents 32 and 238 expressed the administration's desire to meet Soviet aggression anywhere it might appear: in Central America, Europe, the Middle East, or South Asia.[64]

With bigger defense budgets came a turnaround in the manning of the Army. Increased pay and benefits and legislative action to increase the quality of recruits gave a boost to a renewed recruiting effort under the leadership of DePuy protégé Maj. Gen. Max Thurman.[65] By the mid-1980s, under the new slogan, "Be All That You Can Be," recruiting quality began to improve dramatically.[66]

This manning turnaround was accompanied by an impressive reequipping of the Army. Since the 1970s the Army had been working to field its so-called Big Five weapon systems: the M1 Abrams main battle tank, the Bradley Infantry Fighting Vehicle. the Apache attack helicopter, the Black Hawk utility helicopter, and the Patriot air defense system. By the early 1980s many of these systems were beginning to enter the Army's inventory; between 1981 and 1984, the Army funded 2,929 new M1 Abrams tanks, 2,200 Bradley Fighting Vehicles, and 171 Apache helicopters.[67]

Along with this massive buildup in military capability came increased quality in training. The so-called training revolution is generally associated with General DePuy, but his main contribution was really the refocus of Army doctrine on high-intensity conflict. The primary architect for the training revolution was DePuy's deputy chief of staff for training, Maj. Gen. Paul Gorman.[68] Gorman's first innovation was to implement a system called tactical engagement simulation, which more accurately counted which vehicles and individuals were "casualties" in force-on-force wargame–style exercises. The innovation met with fierce resistance from senior Army leaders, General DePuy chief among them, but when the system was implemented in training in Germany, it revealed startling shortfalls in individual and small unit training readiness.[69]

Gorman's solution borrowed heavily from sister services. In reaction to appalling losses in the early days of the Vietnam War, the U.S. Navy had implemented a new dogfighting training center colloquially known as Top Gun. During this training, a cadre of expert instructors replicated enemy fighter pilots in realistic dogfighting exercises. Every exercise was followed by an intensive after-action review. The first aviators to graduate from the course began arriving in Vietnam in 1969, and the casualty rates among aviators dropped dramatically. The U.S. Air Force would follow suit after the war, when it finally dispassionately examined its own performance in Vietnam. In 1975 it established the Red Flag training exercises, complete with "aggressor squadrons." The Air Force expanded on the Top Gun concept by including ground attack, command and control, and electronic warfare aircraft to its training center's curriculum.[70]

Gorman envisioned a similar ground warfare training center for the Army. That vision would find form—again, over the objections of many senior leaders in the Army—in the National Training Center (NTC) at Fort Irwin, California, a 650,000-acre simulation of World War III in the Mojave Desert. Beginning in 1982,[71] mechanized Army battalions and later whole brigades "deployed" to the training center to fight the "Krasnovians"—an opposing force that used Warsaw Pact doctrine and tactics—in simulated combat adjudicated by the Multiple Integrated Laser Engagement System, a kind of laser tag that tallied the lethality and effects of modern weapon systems. Every battle was watched by observer-controllers and followed by an after-action review in which units critiqued and improved their ability to fight outnumbered and win.[72]

Since the Vietnam War, the U.S. Army had imagined a future battlefield where perfect intelligence and highly accurate, precision strike munitions would finally make irrelevant the tyranny of terrain—that absolute equalizer of ground forces since the brave three hundred Spartans at Thermopylae and before. In 1968, having seen the promise of laser-guided bombs in Vietnam, the new chief of staff of the Army, General Westmoreland, believed that the U.S. military had undergone a "quiet revolution in ground warfare" that would one day promise the ability to "destroy anything we locate through instant communications and the almost instantaneous application of highly lethal firepower."[73] Westmoreland told the Association of the United States Army in an October 1969 speech that he could "see battlefields that are under 24-hour real or near real time surveillance of all types" in which units would be linked "through the use of data links, computer assisted intelligence evaluation, and automated fire control." While these capabilities would not be realized in the ten years that Westmoreland predicted in 1969,[74] by the early 1980s rapid advances in computer and telecommunications technology were beginning to make his vision seem within reach.

The Soviet Union had been watching Western technological develop ments—including the U.S. use of laser-guided munitions in the latter days of the Vietnam War—with growing concern. Moreover, while Soviets theorists had drawn many of the same conclusions as the U.S. Army had from the Yom Kippur War, they were particularly alarmed by the impact that radar detection systems and precision-guided weapons had had across the battlefield. Worried that these technological advances might negate their numerical advantage in Europe—the exact aim of the U.S. military's "offset strategy"—these theorists began to talk about a "military-technical revolution" occurring in the West. American military theorists studying their adversaries' perceptions of emerging technology began to call these same technological advances a revolution in military affairs (RMA).[75]

As the Reagan military buildup began to restore the U.S. military's confidence and reverse the West's perceived disadvantage in the European balance of power, the senior leaders of the Army began to think beyond active defense to a more offensive doctrine. AirLand Battle, unveiled in 1982,[76] envisioned the Army rapidly responding to Soviet aggression in Europe with a counteroffensive that attacked the Soviets throughout the depth of their formation. This new concept of "deep battle" or an "extended battlefield" included the use of emerging technologies to find and destroy every echelon of the Soviet Army in a coordinated and near-simultaneous fashion.[77]

This new doctrine was enshrined in the 1982 edition of FM 100-5, which was written by a team led by Lt. Col. Huba Wass de Czege of the CGSC. The manual noted that "wide-ranging surveillance, target-acquisition sensors, and communications that provide intelligence almost immediately" had changed the character of warfare on the modern battlefield. AirLand Battle played to Americans' perceptions of themselves, particularly their belief that they held an advantage over the Soviets in allowing subordinates to "act independently within the context of an overall plan." As the new FM 100-5 insisted, "Improvisation, initiative, and aggressiveness are the traits that have historically distinguished the American soldier." Placing advanced technology in the hands of empowered subordinates would allow the Army to "move fast, strike hard, and finish rapidly."[78]

While the Army spent the 1970s and early 1980s in a headlong rush toward ever-greater capability to fight and win in a high-intensity conflict against the Soviet Union, the conversation over low-intensity conflict doctrine, training, and practice was largely neglected.

Because low-intensity conflicts did not enjoy official emphasis from the senior leadership of the Army, doctrine in this area atrophied. That doctrine that did exist on military intervention in such conflicts reflected the Nixon Doctrine's preference for support to host nation activities over direct intervention. The 1967 edition of FM 31-16, *Counterguerrilla Operations*, remained the only dedicated Army doctrine on counterinsurgency until it was updated in 1981. The 1972 edition of FM 31-23, *Stability Operations*, insisted that the "the US role in any counterinsurgency mission must be primarily advisory." In 1974 the Army replaced "stability operations" with a new term that better emphasized the Army's indirect role in internal conflicts: FM 100-20 bore the title *Internal Defense and Development*.[79]

Low-intensity conflict was also subverted as a topic of conversation in the professional journals of the Army's educational institutions. With the exception of Lt. Col. Donald Vought's "Preparing for the Wrong War?" in a 1976 installment of the CGSC's professional journal, the *Military Review*, there was barely any discussion of low-intensity conflict in the publication. In fact, after the end of the Vietnam War, the *Military Review* didn't really wrestle with the subject of the war itself again until it dedicated an issue to the topic in 1989.[80]

While the Army may have been disinterested in low-intensity conflict, President Reagan's anticommunist foreign policy forced it to engage in these types of conflicts around the world. Reagan engaged in military interventions in Nicaragua, El Salvador, Lebanon, and Grenada. He also committed the U.S. military to support peacekeeping missions in Honduras, Chad, and the Philippines and counternarcotic operations and support to counterinsurgency efforts in Bolivia and Colombia. In addition, Reagan committed money, observers, and logistical support to a number of United Nations (UN) peacekeeping missions worldwide.[81]

At first blush, this activist foreign policy seemed to reveal a fundamental difference between the Army and the president on what, specifically, the lessons of Vietnam were. While the Army had taken from Vietnam an aversion to military intervention in low-intensity conflicts, Reagan would say the true lesson of Vietnam was that "if we are forced to fight, we must have the means and the determination to prevail or we will not have what it takes to secure the peace." But despite this talk of resolve and determination, in practice Reagan's Cold War interventions were indirect, limited, and cheap in terms of manpower, casualties, and treasure.[82] This reluctance to intervene directly on behalf of allies and against adversaries reflected the Nixon Doctrine, but it also echoed intellectual trends inside the Army on low-intensity conflict doctrine toward an indirect approach that relied on the host nation to do the messy job of engaging in a conflict's political dimensions.

After Reagan took office, he saw Nicaragua as a perfect test case for his foreign policy—an aggressive reassertion of military containment of communist expansion and a rollback of communist gains in Latin America. Nicaragua had become a Soviet client state and was supporting a communist insurgency in El Salvador by smuggling arms and ammunition through neighboring Honduras and by sea. Reagan began providing support to an anti-Sandinista insurgency—the *contrarrevolución*, whose proponents were known as the Contras—and began subverting the Sandinista government.[83]

Honduras was the base of operations for supporting both the insurgency in Nicaragua and the counterinsurgency in El Salvador. In 1983 the Army established the Regional Military Training Center, run by U.S. Army Special Forces, ostensibly to train the Honduran and Salvadoran Armies to protect themselves from Nicaragua. U.S. Southern Command (SOUTHCOM) deployed 4,500 service members to support these efforts. U.S. Army units conducted training exercises along the Honduran border with Nicaragua, supposedly to deter Nicaraguan aggression against Honduras but in practice to interdict supplies moving from Nicaragua to the insurgency in El Salvador. The U.S. government also aided Honduras in providing sanctuary and training to the Contras.[84]

El Salvador was also key to the Reagan foreign policy objective of stopping communist "aggression" in Central America. As such, the Reagan administration called for the deployment of trainers to Central America to help the Salvadoran Army combat the growing communist insurgency. But Congress suffered from what was, at the time, commonly referred to as the Vietnam Syndrome—an aversion to U.S. military intervention in low-intensity conflicts. Consequently, Congress was resistant to military intervention in Central America; it did authorize intervention, but severely restricted the number of U.S. troops that could participate and the activities in which they could engage.[85] American forces could not participate directly as combatants in the Salvadoran Civil War; instead, they supported the Salvadoran military with advice and training. Moreover, the U.S. force committed to the effort never exceeded fifty-five service members, with U.S. Army Special Forces (the famed Green Berets) as its core.[86] Thus, the Reagan-era model for U.S. military intervention in counterinsurgency—small teams of Special Forces supporting a host nation's counterinsurgency efforts—was not so much a doctrine as a constraint placed on the Reagan administration by Congress.

In many ways, the U.S. military intervention in El Salvador did more harm than good for the Army's understanding of its ability to engage in low-intensity conflicts. The effort was judged a success,[87] the ultimate obstacle to learning. And those lessons that the Army did learn from the Salvadoran Civil War were the wrong ones: that it could outsource low-intensity conflict to special operations forces (SOF) and the political dimensions of such conflict to the host nation government.

The biggest clash of President Reagan's anticommunism campaign in the Western Hemisphere would not be a low-intensity conflict in Central

America, but a decidedly high-intensity conflict on the Caribbean island of Grenada. With the encouragement of their Warsaw Pact and Cuban advisers, elements within the government of Grenada had murdered the country's leftist strongman and his associates. Seeing a possible communist expansion in the making, Reagan sought and received the acquiescence of Grenada's neighbors and the British governor general of Grenada to intervene. Ostensibly to rescue several hundred American medical students on the island, Reagan launched Operation Urgent Fury on October 25, 1983. A force of six thousand Americans, with U.S. Army Rangers, Army paratroopers, and the Marines at its core, defeated the Grenadian military and its communist advisers and deposed the government in a six-day campaign. Journalists representing the international media groused at their rough treatment during the operation—they were not allowed onto the island until the operation was nearly over, and then only in press pools—but most observers generally judged Operation Urgent Fury a success.[88]

Grenada yielded few lessons germane to low-intensity conflict. The entire population of the tiny island was fewer than 100,000, meaning there were only fifteen Grenadians for every American service member participating in the operation;[89] the people would have been easy to control and secure even if they had not been jubilant to have their autocratic and ineffective government deposed.[90]

The primary lessons taken from this operation related to the communications and cooperation between services during the operation.[91] One of the most famous, if apocryphal, stories of communications issues during the operation—immortalized in the 1986 movie *Heartbreak Ridge*—involved a service member having to call back to the United States using a pay phone and credit card to get an air strike in support of his unit.[92] In the most serious communication snafu during the operation, marines accidentally called in an air strike on Army paratroopers, mortally wounding one soldier, because they had no communications with the Army forces on the island. There were many other glaring issues with the cooperation between services for the operation.[93]

These issues with interoperability between the services would eventually lead to congressional legislation that would have a far-reaching impact on the Army—and its ability to engage in low-intensity conflicts.

A 1949 amendment to the National Security Act established the Department of Defense as overseer of the individual services and the secretary of defense and chairman of the Joint Chiefs of Staff as advisers to the president.[94] By the early 1980s three "commanders in chief" (CINCs)—the CINCPAC, a Navy admiral in the Pacific region; the CINCEUR, an Army general in Europe; and the CINCSOUTH, an Army general in Central and South America—oversaw U.S. military forces in their respective regions. In 1983, as the Middle East became an increasingly critical part of the world for U.S. national security, the Rapid Deployment Joint Task Force there was transformed into U.S. Central Command (CENTCOM) with its own CINCCENT.[95]

Yet, as was evidenced in Operation Eagle Claw in Iran and Operation Urgent Fury in Grenada, the lines of command and control from the CINCs to each service's forces within their region were by no means clear. The service chiefs continued to wield a great deal of power over forces from their respective services operating in each of these regions around the world and continued to provide advice on national security issues directly to Congress and the president.[96]

The problems with interservice coordination during Operation Urgent Fury gave ammunition to advocates of reform. The reform effort did not, however, truly gain momentum until senior senator and retired Air Force major general Barry Goldwater took over as chairman of the Senate Armed Services Committee. Eager to leave a lasting mark on national security before his retirement, Goldwater ushered a reform bill through Congress that proposed sweeping changes to the balance of power between the Department of Defense and the individual services. The fight was fierce, with senator and former secretary of the Navy John Warner leading a Navy revolt against reform. But ultimately the 1986 Goldwater-Nichols Defense Reorganization Act became law.[97]

Goldwater-Nichols was a dramatic reassertion of Congress's power over the U.S. military. It recognized the chairman of the Joint Chiefs of Staff as the principal uniformed adviser to the president on matters of national security. It demanded that officers of each service receive "joint" (interservice) education and serve in joint assignments as a prerequisite for promotion to higher ranks in their respective services. It increased the size of the Joint Staff at the expense of the service staffs and gave regionally aligned commanders more autonomy from their services by making them answerable to the secretary of defense rather than the Joint Chiefs of Staff. The act gave CINCs "authoritative direction over all aspects of military operations" within their regions, breaking the power that the service chiefs had previously held over their component commanders in each region.[98]

Ever since the disastrous Operation Eagle Claw, the chief of staff of the Army, Gen. Edward C. "Shy" Meyer, had become the leading voice in a chorus of observers inside and outside the military calling for the consolidation of U.S. SOF capabilities under a unified command with the power—at the time only held by the services—to train, man, and equip SOF units.[99] The services strongly objected, seeing such a move as robbing them of troops and resources. Many reformers also advocated the creation of a SOF CINC with power analogous to the regional commands to direct operations, though in this case on a worldwide scale. Flush with its rediscovery of its power over the U.S. military, Congress followed the Goldwater-Nichols Defense Reorganization Act with a legislative effort led by senators William Cohen and Sam Nunn. In 1987—again, over the services' objections—Congress created U.S. Special Operations Command (SOCOM). U.S. Army Special Forces, Navy SEALs, and a number of other, smaller SOF organizations would be consolidated under this new command.[100]

Low-intensity conflict figured prominently in the Nunn-Cohen Amendment to the 1987 Defense Authorization Act. The amendment would have an especially dramatic impact on the future of the Army's low-intensity conflict proficiency. First, the bill created the position of assistant secretary of defense for special operations and low-intensity conflict and the Low-Intensity Conflict Board inside the National Security Council.[101]

But this legislation also shifted primacy for training on and executing low-intensity conflicts from the Army and Marine Corps to SOCOM. Nunn-Cohen defined "special operations activities" as including, among other capabilities, foreign internal defense, civil affairs, psychological operations, counterterrorism, and humanitarian assistance. While the definition of low-intensity conflict in Field Circular (FC) 100-20, *Low Intensity Conflict*, also included "peacekeeping operations" and a wide assortment of other activities called "peacetime contingency operations," the "capabilities" specified in the Nunn-Cohen Amendment as "special operations activities" were listed as "operations" or capabilities in the Army's interim low-intensity conflict doctrine. Most of the senior leadership, eager to relinquish responsibility for low-intensity conflict, took this as license to cede this responsibility to SOF. The Army considered itself relieved by Congress of any responsibility to prepare for or resource itself to fight low-intensity conflicts.[102] This was, of course, exactly what the Army had wanted; ever since the end of the Vietnam War it had been actively ignoring low-intensity conflicts in favor of ever-greater capacity to execute high-intensity conflicts.

Before Nunn-Cohen, in response to the Reagan-era increase in the number of low-intensity conflicts in which the U.S. military was engaged, the Army had made its first tentative steps toward more low-intensity conflict proficiency since the end of the Vietnam War. The primary effort toward this end was the development of the light infantry division. While such a division was originally envisioned as a high-intensity conflict force, in 1984 the new chief of staff of the Army, Gen. John Wickham, insisted that they should be focused on low-intensity conflict. As a result, the mission statement for such a division would include, "Rapidly deploys to defeat enemy forces in Low Intensity Conflict and, when properly augmented, reinforces US forces committed to a mid-high intensity conflict." Six of the sixteen U.S. Army divisions were to be light infantry divisions. In 1986, Joint Publication (JP) 1-02, *Unified Action Armed Forces*, charged the services to man, train, and equip for "the effective prosecution of war and military operations short of war." The Army answered this call with TRADOC Pamphlet (PAM) 525-44, *US Army Operational Concept for Low-Intensity Conflict*, which attempted to define the characteristics of this form of warfare and the requirements to fight it. The 1986 edition of FM 100-5 also, for the first time since the Vietnam War, addressed low-intensity conflict. While it continued to insist that the "overriding mission of US forces is to deter war," the manual acknowledged that the Army would also have to fight "on the unique battlefields of Low Intensity Conflict. . . against irregular or unconventional forces, enemy special operations forces, and terrorists."[103]

The Department for Joint and Combined Operations at the CGSC had already started working on an update to FM 100-20, now retitled *Low-Intensity Conflict*, when JP 1-02 was published.[104] In response to the joint publication, the Army published its progress on the manual as the interim FC 100-20 in 1986. This circular was by no means perfect; it reflected the Reagan-era approach to counterinsurgency, with a heavy reliance on the host nation and an indirect role for Army forces. But it was in many ways ahead of its time; it acknowledged that the host nation would only call on support from U.S. ground troops if its internal war had reached such "serious proportions" that the direct intervention of Army forces in the conflict might be required. And in March 1986 the Joint Chiefs of Staff established the Army–Air Force Center for Low Intensity Conflict (A-AFCLIC) to study this form of warfare and produce further doctrine on the subject.[105]

Yet after the 1987 Nunn-Cohen Amendment relieved the Army of responsibility for low-intensity conflict, the Army reversed its first tentative steps toward a focus on such conflict. The 1988 and 1993 updates to PAM 525-44 were never published.[106] The chief of staff of the Army, Gen. Carl Vuono, directed the commander of the Combined Arms Center, Lt. Gen. Gerald Bartlett, to outsource production of FM 100-20 to the newly formed

A-AFCLIC. Thus, while Fort Leavenworth, Kansas, retained a proponent office for low-intensity conflict, A-AFCLIC became the epicenter of Army thought on the subject. While this agency was able to produce revised doctrine—FM 100-20 / Air Force Publication 3-20, *Military Operations in Low Intensity Conflict*—rather quickly, post-Nunn-Cohen resistance from across the Army to including low-intensity conflict in the Army's doctrine at all held up publication until 1990. The chief of staff of the Army himself directed A-AFCLIC to more forcefully emphasize the "indirect employment of US forces in LIC [low-intensity conflict]."[107] Gen. Max Thurman held up publication because he wanted low-intensity conflict better tied to the tenets of AirLand Battle—a decidedly high-intensity conflict doctrine. The final FM 100-20, *Military Operations in Low Intensity Conflict*, published in 1990,[108] differed significantly from the earlier version, reflecting the Army's new self-imposed "supporting" role to SOF in low-intensity conflict.

Nunn-Cohen also reversed the Army's course on the development of the light infantry division as a low-intensity conflict force. The old guard of the Army had rebelled against the idea of committing light infantry divisions to low-intensity conflict as their primary purpose ever since they were first conceived. General DePuy, by now retired, wrote a letter to General Wickham in which he complained that leaders of the 7th Infantry Division—the testing ground for the new concept—"do not know exactly what they are to do in LIW [low-intensity warfare]" and that their training for low-intensity conflict was dulling their edge as a weapon in high-intensity conflict. Doctrine writers were also subverting the purpose of the light infantry division as a low-intensity conflict force; the topic was barely addressed in FC 71-101, *Light Infantry Division Operations*.[109]

Nunn-Cohen gave official sanction to these efforts to subvert the purpose of the light infantry division. In the early 1980s the Joint Readiness Training Center (JRTC)—intended as a light infantry counterpart to the heavy forces' NTC—was conceived with an explicit purpose of "training conventional units in nonconventional warfare techniques which primarily derive from conventional unit participation in low intensity conflict." In a briefing on the progress of building the JRTC before Nunn-Cohen, Combined Arms Center commander Vuono told planners that they should "not dwell too much on use of" the light infantry division "in mid/high intensity" but rather focus on low-intensity conflict. But after Nunn-Cohen, the purpose of the JRTC began to drift toward high-intensity conflict; in 1987, JRTC planners claimed they would still incorporate civilian populaces in the training but wanted the training to cover the entire spectrum of conflict from low to high intensity. By the time the JRTC was established at Fort Chaffee, Arkansas, that focus had shifted completely toward high-intensity conflict. The scenarios in which light infantry units engaged at the JRTC were against a Soviet-style, motorized enemy. Light infantry units at the JRTC only faced guerilla-style forces in rear areas as part of their base defenses. Observers from A-AFCLIC who traveled to the JRTC in 1988 and 1989 returned frustrated by the fact that "the JRTC focus remains on force-on-force combat training exercises" and "no effort is being made to train leaders and units in the political restrictions of low intensity environment or how to operate when there are constraints on the use of force."[110]

This neglect was directly related to the Nunn-Cohen legislation, which the Army believed had relieved it of responsibility for low-intensity conflict. One of the A-AFCLIC observers added that, at the JRTC, "counterinsurgency tasks are being looked upon only as tasks for SOF and not for conventional forces."[111]

After Nunn-Cohen, a renewed disinterest in low-intensity conflict pervaded the Army. Since the end of the Vietnam War, the amount of time dedicated to low-intensity conflict instruction at the CGSC had steadily declined until it constituted only eight of nearly five hundred total hours of instruction in the 1981–82 academic year. This trend would be reversed in the early and mid-1980s, as the Reagan administration pursued a more muscular foreign policy in Central America, until low-intensity conflict constituted nearly 10 percent of the curriculum by academic year 1988–89.[112] But this total of thirty-three course hours—while a dramatic improvement from the beginning of the decade—was still a tiny fraction of the total curriculum. Moreover, this course featured many of the obstacles to clearer understanding of low-intensity conflict that would hinder units in the next decade. "Emphasis," the course materials explained, "is given to the requirements for host nation leadership and multi-agency participation." Meanwhile, the U.S. Army War College cut the number of low-intensity conflict electives available to its students from sixteen to six over this same period.[113] Nunn-Cohen and the relegation of low-intensity conflict doctrine to A-AFCLIC had considerably muted the discussion of such conflict inside the Army in the late 1980s.

This is not to say that discussions of low-intensity conflict were absent from Army discourse. The School of Advanced Military Studies was established at the CGSC in the mid-1980s. With Col. Huba Wass de Czege, the author of the 1982 edition of FM 100-5, as its first director, its goal was to teach a select group of Army majors—chosen from within the student population of the CGSC—the art of operational-level planning. This school had initially dedicated itself solely to high-intensity conflict, but in 1987 it, too, began to include seventeen hours (about two days) of instruction on low-intensity conflict in its ten-month curriculum.[114]

In the summer of 1987 General Vuono assumed the office of chief of staff of a completely revitalized Army. The active force stood at nearly 800,000 soldiers.[115] As a result of the wildly successful "Be All That You Can Be" campaign overseen by Thurman and generous benefits authorized by Congress, Vuono stood atop the most educated and best trained Army the United States had ever fielded. Retention incentives and a completely revamped noncommissioned officer (NCO) education system had made the Army NCO corps the envy of the world. The Army was also the best equipped in the world; it was wrapping up fielding of the Big Five weapon systems, the most advanced of their kind in the world. An equally impressive array of new munitions, like the precision Copperhead artillery round and helicopter-launched Hellfire missile and the devastating Multiple Launch Rocket System, were also being fielded and integrated into Army doctrine and training.[116]

This new Army was squarely focused on the Soviet Union and the threat to America's allies in Western Europe; as General Vuono would say in 1989, "The Soviets continue to present the most dangerous threat to our interests and allies throughout the world." Nearly 200,000 of the active Army's 770,000 soldiers were stationed in Europe. Two corps, four armored divisions, five separate maneuver brigades or regiments, and an array of support units stood at the frontier of freedom, ready to hold the Soviets at bay while the rest of the U.S. Army's twelve active and ten reserve component divisions mobilized, deployed to Europe, drew from their pre-positioned stocks, and rolled into the fight. To ensure its readiness, the Army practiced this envisioned mobilization every year in massive return of forces to Germany (REFORGER) exercises conducted with its NATO allies.[117]

But this would represent the zenith of U.S. post-Vietnam military readiness. The congressional and public consensus for a more muscular foreign policy evaporated almost as quickly as it had formed. Amid accusations of defense contractors' fraud, waste, and abuse—toilet seats, coffeemakers, and hammers bought at many hundreds of dollars apiece—defense budgets began to fall. The national debt had doubled since Reagan had taken office, and debt service payments had tripled. In 1985 a new, more fiscally conservative Congress passed the Balanced Budget and Emergency Deficit Control Act, and defense spending began to fall at the rate of 3 percent every year after 1985; the defense budget would drop 10 percent between 1986 and 1989.[118]

And while the focus of the Army in the late 1980s remained squarely on a high-intensity conflict in Western Europe, there were dramatic events occurring elsewhere in the world that might have caused the Army to question this stance.

After the untimely death of his patron, Panamanian president General Omar Torrijos, in 1981, General Manuel Noriega, chief of the intelligence services of the Panamanian Defense Force, rose to the presidency of his country through a campaign of terror and subversion of the democratic process. For nearly a decade, Noriega successfully balanced the competing interests of Cuba and the United States, Latin American drug cartels and nationalists, and the broader electorate within his own country. But his regime came crashing down when he ignored the results of a May 1989 election and fanned the flames of nationalist violence against Americans, culminating in the murder of a U.S. marine by the Panamanian police. Fed up, new president George H. W. Bush launched Operation Just Cause in December 1989.[119]

In many ways, the invasion of Panama was a "redo" of the invasion of Grenada. U.S. Army Rangers and paratroopers from the 82nd Airborne Division descended on the country, while Navy SEALs seized sensitive sites across the isthmus. But this was a much larger operation, also involving the 5th and 7th Infantry Divisions and the 193rd Infantry Brigade. The war also featured several new capabilities of the rebuilt, post–Reagan era military, including the AH-64 Apache attack helicopter and a proliferation of night vision devices that allowed the operation to be prosecuted primarily at night.[120]

The war was also the first major post-Goldwater-Nichols U.S. military operation. The Army and Air Force worked relatively seamlessly together to simultaneously attack twenty-seven different targets. Air Force aircraft also provided close air support to Army units in contact with the Panamanian Defense Force throughout the operation. The synergy was devastating: in an eight-day campaign (December 20–28, 1989), the Panamanian military lost 322 troops to the U.S. military's twenty-three. But this does not measure the true disintegrative effect that the coordinated attacks had on the Panamanian Defense Force; the U.S. military successfully destroyed the Panamanians' ability to fight as an army. And, of course, on January 3, 1990, Noriega was taken into custody.[121] In this way the war presaged the dramatic victory of the Gulf War that would come soon after.

This war also presaged the three decades of low-intensity conflict that awaited the Army. The postconflict phase of Operation Just Cause presented a number of factors that should have worked to U.S. advantage. Panama was only a short flight from the United States. The language of the country was Spanish, a familiar language—if not a first one—for many U.S. soldiers. Moreover, the majority of the Panamanian population wanted Noriega gone. And, finally, there was already a duly elected government of Panama likewise glad to see Noriega gone and eager to take the reins of power.[122]

Yet, for a number of reasons, the U.S. Army seemed surprised by the rapidity with which the conflict transitioned from a high- to a low-intensity one. First, while the Army had adroitly analyzed the Panamanian Defense Force and the Noriega regime, they had failed to consider the effect that the war would have on Panamanian society. There was extensive damage to infrastructure during the high-intensity phase of the war. Whether they were released by Noriega or simply escaped from Panama's dilapidated prisons, convicts spilled onto the streets after the invasion. Moreover, the vacuum created by the disintegration of Panamanian security forces allowed criminal gangs known as dignity battalions to prey on the people of Colon and Panama City; as many as three hundred civilians were either killed in the crossfire between military forces or murdered by criminals. While SOUTHCOM had directed the senior Army headquarters—the XVIII Airborne Corps—to build and execute a plan to stabilize postconflict Panama, this Army staff failed to dedicate any combat forces to the task, instead looking to specialty troops such as military police and civil affairs troops to control a population of 2.3 million people spread across a country the size of South Carolina. Looting and vandalism spread as the United States struggled to come to grips with the turmoil. Damage and losses during the crisis topped $1 billion.[123]

A number of lessons emerged from the chaos of the postconflict phase in Panama. The first lesson related to the Army's ability to deploy forces for such conflicts. While the Army had no problem generating combat forces for the brief high-intensity conflict phase, the civil affairs and engineer troops needed to deal with postconflict governance and reconstruction tasks were primarily resident in the Army's reserve components. The Army's civil-military efforts languished while they waited for these troops to be mobilized and arrive.[124]

The Army also discovered that it was not properly trained or manned for the challenges of low-intensity conflict. Military police carried a disproportionate proportion of the responsibility for the postconflict phase of this operation, being asked to do everything from training the Panamanian police, to guarding prisoners of war (POWs) and safeguarding refugees, to imposing law and order on the chaotic streets of Panama City. They found that they were both untrained and chronically undermanned for the character and scale of the low-intensity war they found themselves waging in Panama. When reserve component military police finally arrived in the conflict and began to embed themselves in Panamanian National Police precincts, their numbers were still insufficient; this effort had to be augmented by U.S. Army Special Forces to fill the gap in numbers.[125]

The requirement to establish the Panamanian National Police proved especially challenging for the U.S. Army. There were almost no facilities or equipment available to stand up this police force, and those resources that were available were in chronic disrepair. Vehicles, weapons, and radios had either been damaged or looted during the war. The delay in deployment of engineers to the conflict meant damaged facilities could not be repaired, as those few engineers on the ground in Panama at the beginning of the war were dedicated to repairing the electricity infrastructure. The U.S. Army ended up transferring weapons captured from the Panamanian Defense Force to the National Police, solving one problem but creating another. The Army spent in excess of $1 million shipping Dodge sport utility vehicles to Panama to augment the one hundred Chevrolet Corsicas purchased by the Panamanian government to outfit its police. It is not at all clear how much the United States paid for fuel and spare parts to get the Panamanian police on the street to establish law and order. These logistical problems only compounded the endemic incompetence and corruption rampant in the police force long before the U.S. invasion.[126]

Observers also noted that the postconflict phase of the war in Panama required the U.S. Army to operate in and among the people. And the civil-military effort was similarly riddled with difficulties. There was a shortage of civil affairs personnel. This only contributed to the already chronic problems with the Army's effort to coordinate with other U.S. government agencies; all of these agencies had been excluded from the initial planning, and the U.S. embassy staff in Panama City was critically undermanned.[127]

Army Reserve Lt. Col. Rudolph C. Barnes Jr. elevated the discussion of lessons from Panama by highlighting that terminating the conflict required that the Army engage in its political dimension. He observed that the objectives of the invasion—deposing strongman Noriega and assisting the fledgling government of Guillermo Endara in assuming control of the country—required more than just security; it required that the government be able to "provide essential services" to the Panamanian people.

This in turn, however, required that U.S. military activities be linked to the "political, economic and covert military activities" taking place in the country. Barnes also highlighted the "dichotomy" whereby, before and after the war, these activities were "controlled by the US ambassador," but during the war itself, "the military chain of command preempted the ambassador's control of military activities." In perhaps the first invocation of an idea that would during the Iraq War two decades later come to be known as military diplomacy, Barnes wrote that Army leaders needed to act as "warrior diplomats" to achieve "close coordination between military and diplomatic activities."[128]

Barnes also arrived at the precise reason why the Army could not out-source its responsibilities to handle the political dimension of a low-intensity conflict to the State Department. He observed that while the Army did need all of the skills it was lacking in the postconflict phases of Operation Just Cause, "even in unconventional LIC environments, organized violence is essential to achieve politico-military objectives." Nevertheless, Barnes still managed to fall into another cognitive trap; he suggested that the military's primary purpose should be to "help indigenous forces achieve politico-military objectives,"[129] reinforcing the Nixon Doctrine's misconception that the Army could outsource the political dimensions of low-intensity conflict to the host nation government.

Unfortunately, lessons that might otherwise have been learned from the postconflict phase of the war in Panama were not retained because of two dramatic events that took place at the dawn of the 1990s: the fall of the Soviet Union and the Gulf War. The former stole attention away from the challenges the U.S. Army was facing in Panama at a critical time when learning might otherwise have occurred. The latter reassured the Army that high-intensity conflict competency should still have primacy in the post–Cold War world, short-circuiting a debate over the doctrine, training, manning, and equipping of the Army for the future.

The speed with which the Eastern Bloc and the Soviet Union collapsed stunned the American national security establishment. The first signs came when the communist regimes in Czechoslovakia and Romania collapsed before popular dissent and, unlike earlier popular uprisings in the communist world, were not crushed by the Red Army. Instead the Soviet Union began to reduce its presence across Eastern Europe and downsize the Warsaw Pact armies. In fact, the Soviets even showed renewed interest in arms reduction talks.[130]

Across the Eastern Bloc, Soviet premier Mikhail Gorbachev's new policy of glasnost promised a gentle transformation of communist society, but even these efforts collapsed in ruin in late 1989 when an emigration crisis in East Germany spiraled out of the Kremlin's control and ended with the disintegration of the Iron Curtain as, one by one, the countries of the Eastern Bloc broke free of Soviet control. Germany reunified in September 1990, and in November 1990 the Warsaw Pact—the focus of U.S. national security strategy for four decades—dissolved.[131]

Gorbachev's efforts to salvage the Soviet Union as a political entity collapsed soon after. A coup in August 1991 succeeded in arresting Gorbachev but collapsed because the Russian Army would not fire on civilians who rallied to support Boris Yeltsin, the leader of a new semiautonomous Russian republic. Soon after, the Soviet Union was replaced by a commonwealth of independent states and the Soviet empire passed into history.[132]

The U.S. national security establishment was mixed on its opinion of how the West should react to this dramatic turn of events in the East. In response to Warsaw Pact de-escalation, President Bush accepted a postponement of NATO nuclear and conventional modernization and a 4.7 percent decrease each year in allied defense spending. He also proposed modest, phased reductions in U.S. active forces, especially those supporting Europe. As the full scale of the changes in Eastern Europe became clear, the chairman of the Joint Chiefs of Staff, Gen. Colin Powell, proposed a reduction of up to one-third in force structure, a downsizing to 1.8 million service members. The services immediately opposed what they saw as draconian cuts.[133]

There were those inside and outside the Army who believed that Russia still presented a mortal threat to the United States and needed to be contained, and the Bush administration was chief among them. The former director of the J-5 (plans and strategy) of the Joint Staff argued in the pages of the journal *Parameters* that the Russians still had a considerable nuclear arsenal and "the shadow of residual Soviet Power" remained a danger if "ages-old enmities" were to be unleashed by "the receding tides of the Cold War." Perhaps most important is that only months before the failed Soviet coup, General Vuono, the outgoing chief of staff of the Army, acknowledged, "The evolving environment call[s] for a revised military strategy for the United States," though he also warned, "The future of the Soviet Union itself is by no means certain, and Europeans view with great uneasiness their giant Eurasian neighbor as it lurches about in search of its destiny."[134]

Not everyone agreed. Army lieutenant colonel James Dubik, a School of Advanced Military Studies student in 1990–91, would write that even in light of the large, capable, conventional military that the Soviets retained as they approached collapse, "the Soviet threat to the United States is insignificant." He cited reform pressures and internal instability as reasons why the Soviets would not use their considerable military might against the West in the foreseeable future. He also contended—presciently, given that this exact event would occur two months later—that "even if a conservative coup seized control of the central government, one could question—for internal economic and political reasons—whether she would pose a significant threat to the United States." Dubik concluded, "The aggressive, ideological and militarily menacing Soviet Union need not be contained any more [*sic*]." Lieutenant Colonel Barnes was even more blunt. He wrote that events in Eastern Europe had "emasculated the Soviet Union as a threat to the United States," adding, "Traditional Cold War strategies based on conventional war between NATO and Warsaw Pact nations in Europe are now obsolete."[135]

Congress clearly believed that the threat posed by the Soviets had been significantly reduced. In 1990 it had already demanded a cut in defense spending of 13 percent over the subsequent five years. But as events in Eastern Europe unfolded, Congress was eager to reap a "peace dividend" from the end of the Cold War. Over the course of the Bush administration, the defense budget would shed $50 billion, with much deeper cuts looming after the 1992 elections.[136]

The other dramatic event that short-circuited a deeper analysis of the lessons of the low-intensity conflict in Panama—and a more extensive reevaluation of the Army's focus after the fall of the Soviet Union—was the Gulf War. Iraqi dictator Saddam Hussein had amassed a massive army and an even more massive debt in the eight-year Iran-Iraq War, which had ended—with both sides claiming victory—in 1988. Looking for a way to sustain his force and pay off his considerable war debt, Saddam eyed tiny, oil-rich Kuwait to his south. He shocked the world on August 2, 1990, when he concluded a minor border and oil-drilling dispute by invading and annexing his tiny neighbor. Within six days Iraq had swept through Kuwait and was poised on the border with Saudi Arabia—a vital American ally in the Middle East. Worried about the threat to the Saudis, and reminded by British prime minister Margaret Thatcher of the supposed "lessons of Munich"—that the appeasing of a dictator (Adolf Hitler) after his illegal invasion and annexation of a neighboring country (Czechoslovakia) had led to World War II—Bush declared on August 5 that he was prepared to use force to restore the status quo ante.[137]

The challenge seemed daunting. The Iraqi Army was the fourth largest in the world (the U.S. Army was third)—with nearly a million men, 5,700 tanks, and 3,700 artillery pieces. The Iraqis would have months to dig in and fortify their position in Kuwait before the coalition of allies could launch their military response. Simulations of the conflict—which the Army had been running in command post exercises for years—indicated that the United States might suffer tens of thousands of casualties. And there was also the specter of the use by Saddam of chemical weapons, which he had already done to devastating effect during his war with Iran.[138]

King Fahd of Saudi Arabia acceded to the U.S. request to send troops and amass for the operation on Arabian soil, and over the course of the next two months the XVIII Airborne Corps—consisting of two light divisions and one heavy division, a Marine expeditionary force, an armored cavalry regiment, and an array of supporting units—arrived for Operation Desert Shield, the defense of Saudi Arabia. One hundred twenty thousand American troops with 700 tanks, 1,400 armored fighting vehicles, and 600 pieces of artillery were joined by 32,000 troops and 400 tanks from across the Arab world.[139]

The operation to liberate Kuwait—dubbed Operation Desert Storm— would require more forces. After an announcement by President Bush on November 8, 1990, the United States deployed the VII Corps, with an additional three heavy divisions and an additional armored cavalry regiment from Europe. The Marines added two divisions to this force while the British added one of their own armored divisions and the French added a light armored division. Saudi Arabia's neighbors—including the exiled Kuwaitis—also added forces. Ultimately, twenty-three nations from across the globe would join the coalition to liberate Kuwait, contributing everything from ground combat units to hospital and chemical decontamination units, naval forces, airpower, and money. Turkey massed its army on Iraq's northern border to open a northern front to the war and also allowed its airfields to be used by the U.S. Air Force. In the end it would take the United States six months to move more than 500,000 troops into theater to face what the U.S. military believed were as many as 550,000 Iraqi troops in and around Kuwait.[140]

CENTCOM commander Gen. Norman Schwarzkopf's plan for the Gulf War was to dismantle the Iraqi Army and national command infrastructure with airpower before the invasion. There was no small amount of disagreement within the Air Force as to whether to focus on the Iraqis' strategic command and control network, as RMA theorists led by Air Force Col. Jack Warden advocated, or whether to concentrate on Iraqi ground forces, as the U.S. Army desired. In the end, the air war would be a hybrid of the two. Once airpower had fragmented and weakened the Iraqis by as much as 50 percent,[141] Western ground forces would destroy the Iraqi ground force with a sweeping flanking movement through Iraq to cut off the Iraqi forces in Kuwait from Baghdad. Marine forces would attack along the coast and threaten an amphibious landing to keep Iraqi forces fixed in Kuwait. The Arab coalition forces—spearheaded by the exiled Kuwaiti forces—would drive into Kuwait City to liberate the capital.[142]

The war itself was a stunning display of the technological revolution that had reshaped the U.S. military. The thirty-eight-day air campaign—dubbed Instant Thunder in an obvious dig at the gradual, limited Vietnam air campaign, Rolling Thunder—began on January 17, 1991, establishing uncontested control of the skies and paralyzing the Iraqi air defense, infrastructure, and command network. The heretofore untested AH-64 Apache attack helicopters joined a dizzying array of high-tech Air Force weapons. While the Air Force deconstructed the Iraqi Army from the air, the U.S. Army used aerial observation and counterbattery radars in conjunction with precise, computer-aimed artillery fires to destroy the Iraqis' forward positions and artillery.[143]

On February 24, the four-day ground war began with the XVIII Airborne Corps launching a helicopter-borne air assault attack 176 kilometers deep into Iraq. Under a swarm of Army, Marine, and coalition helicopters, the corps ground forces rushed forward to seal off the Iraqis' right flank while the fifty-thousand-vehicle coalition force of the VII Corps—preceded by Air Force A-10 tank killers, Marine AV-8 Harriers, and Army Apaches—launched its armored main assault toward the eight divisions of the Iraqis' elite Republican Guard. American planners had expected a much more grueling fight, but the Iraqi Army, having just been punished by a month of the most intense and precise aerial dismemberment in the history of warfare, disintegrated before the American attack in only one hundred hours. As many as 100,000 Iraqis deserted before the ground war even began.[144]

Despite the coalition being outnumbered, the results of the war were stunningly lopsided in its favor. The Air Force lost twenty in battle, the Army and Marines lost 122 (thirty-five of which were killed in friendly fire incidents), and the Navy lost six. The U.S. allies lost ninety-two. The Iraqis lost between ten and twenty thousand, with one to two thousand of those being civilians. An additional eighty-six thousand Iraqis were taken prisoner during the war.[145] Individual battles in the war were even more stunning; in a sharp, four-hour battle at the Al-Rumaylah oil fields, the U.S. Army's 24th

Infantry Division (Mechanized) destroyed seven hundred Iraqi vehicles and took three thousand prisoners while losing only one M1 Abrams tank and having only one soldier wounded.[146]

Unfortunately, the surviving Iraqi Army that escaped Kuwait was still sufficient to keep Saddam in power by crushing Shiite Arab and Kurdish rebellions later in 1991.[147]

The nature of the war militated against the U.S. Army learning lessons about low-intensity conflict, though it did learn many lessons about deployability. First, it discovered that the United States lacked the strategic mobility assets required to deploy anywhere except Western Europe in response to an invasion by the Soviet Union. Senior leaders knew that in the volatile post–Cold War world they could not predict to which theater they might have to deploy next. Between August 1990 and February 1991, aircraft moved 500,000 troops and 600,000 tons of supplies to the Persian Gulf, while ships moved an additional three thousand troops, 3.4 million tons of equipment and supplies, and 6.1 million tons of fuel. Many observers also noted the alarming gap in time between the first unit to deploy—the 82nd Airborne Division—and the arrival of heavier, follow-on units.[148] They feared the next adversary might not wait quietly for six months while the United States built up forces to fight a war.

The senior leadership of the Army began to envision pre-positioned fleets of vehicles and stockpiles of supplies and equipment around the world, both afloat and on the ground, prepared for the next conflict wherever it might occur.[149] In the aftermath of the Gulf War, General Carl Vuono described a post–Cold War strategy with "some forward-deployed land and air forces [and] pre-positioned equipment afloat and ashore" but with a primary reliance on "the projection of power from within the continental United States to trouble spots around the world." He concluded, "The United States must build the capacity to project this power throughout the world. [Deployability] thus becomes a sine qua non for all Army forces." But this future war was still envisioned as a high-intensity conflict. Vuono set the Army's goal as deploying "a multi-division US corps, to include a capability for forcible entry, substantial armored forces, and sufficient sustainment anywhere in the world in one month."[150]

The Iraq War was too short for the Army to learn anything about rotating forces in and out of theater, but it did learn lessons about sustaining itself during the long buildup before the ground war. Specifically, it learned about the critical need for increased transportation capability, both from the United States to theater and for the distribution of supplies and equipment within the theater. It also discovered that it would be highly dependent on civilian contractors to provide everything from laundry, supply, and transportation support to maintenance of the Army's new high-tech computer systems.[151]

The Army also learned important lessons about the Total Force initiative that Gen. Abrams had implemented after the Vietnam War. The U.S. military ultimately activated nearly 230,000 reservists to round out its units and provide the vital support functions resident only in the reserve components, the largest mobilization of the reserves since the Korean War. Yet only 46 percent of mobilized reservists would actually make it to the theater before the war had ended. About one-fifth of mobilized reservists did not deploy because of training, physical maladies, civilian employment, or having dependents. Three U.S. Army National Guard armored brigades that were mobilized for the conflict could not meet basic training standards and thus were not deployed to the conflict. In response, the Army's senior leaders envisioned a new regime of readiness and mobilization training to prepare the Army for short-notice deployment for the next war.[152]

Unfortunately, the Army learned a number of negative lessons from the Gulf War that would become obstacles to learning about low-intensity conflict over the course of the subsequent decade. Chief among these was the idea that future conflicts would continue to be of a high-intensity nature, meaning that the Army did not need to significantly reshape itself for the post–Cold War world. General Vuono wrote after the Gulf War that even though the Eastern Bloc was collapsing, there were still "rising challenges elsewhere in the world." These challenges, he predicted, would be high-intensity ones. The "ominous set of threats emerging in the international environment," he noted, was characterized by "the proliferation of advanced weapons and the rise of major military powers throughout the world." These threats justified continued preparation to fight high-intensity—as opposed to low-intensity—conflicts.[153]

As a corollary to this misconception, many observers took away from Desert Storm that the Army could avoid the messy, postconflict, low-intensity conflict phases of a war by using the so-called Powell Doctrine to limit the scope of the Army's mission. General Powell himself would say in a *Foreign Affairs* article two years after the war that to avoid low-intensity conflicts, "clear and unambiguous objectives must be given to the armed forces" by policy-makers. Retired general Don Starry would say after the war that, because "the political goal of the operation was made clear at the outset," the president obtained "public support among US and allied 'people at home'" and, most important, "when the agreed-upon political aim had been accomplished, forces were redeployed" and the Army was spared having to "relearn the hard lessons of those earlier limited wars," Vietnam chief among them.[154]

In answer to those who argued that these limited objectives had left the war unfinished or the Shiite Arab Iraqis at Saddam Hussein's harsh mercy, the Army had a curt—if simplistic—response. After the end of Gulf War, a Center for Army Lessons Learned briefing for the new chief of staff of the Army, Gen. Gordon Sullivan, began with the heading "WE WON!!!" and the subheading "DON'T SNATCH DEFEAT FROM THE JAWS OF VICTORY."[155]

The dramatic lopsidedness of the victory also convinced many observers that the RMA had so fundamentally changed warfare that many of the rules that had previously governed it no longer applied. The hopes of RMA advocates shaped the Gulf War before it even began, having a significant impact on the planning and prosecution of the air war. But the Army's performance in the Gulf War also engendered talk of a revolution in warfare. With the exception of the Patriot air defense system—which did not perform as well as expected against Iraqi Scud missiles—the Big Five systems that had been fielded throughout the 1980s had proven their worth beyond the wildest dreams of the Army's senior leadership. Despite prewar derision of the AH-64 Apache, it turned out to be a particular standout performer during Desert Storm. And the training revolution and dramatic improvement in the quality of American soldiers recruited into the force after the advent of the all-volunteer force had put these weapons in the hands of the best trained and educated Army in the history of the United States. Less publicized but equally important to the Army's success was its employment of precision-guided munitions and GPS receivers.[156]

While the Army possessed new, perhaps even revolutionary capabilities, it still lagged in several key areas. Its stockpile of precision-guided munitions was small and was nearly exhausted by the Gulf War. Moreover, the Army still lacked sufficient ground and aerial sensors. It was eager to get unmanned aerial vehicles fielded down to the division level. The Army also lacked an automated means to pass intelligence from the corps level down to the individual battalions, companies, and platoons on the ground that needed that intelligence in order to act or fully exploit the U.S. military's new precision capabilities to target and destroy enemy forces. The ultimate goal was real-time intelligence in the hands of commanders on the battlefield at every level.[157]

The Army also faced considerable challenges determining where, precisely, its own forces were on the battlefield and communicating that information across its forces. The Army and Marines both also lacked sufficient computer and communication capability to pass instructions or position information to their units.[158] The Air Force's new Joint Surveillance Target Attack Radar System (JSTARS) allowed the Army to find moving vehicles throughout the depth of the battlefield, as AirLand Battle had envisioned, but the Army's inability to see itself continued to make attack without visual contact hazardous; thirty-five of the 146 U.S. service members killed in the war were killed by friendly fire.[159]

The dramatic scale of the victory in the Gulf War yielded as much hubris as it did knowledge gained. President Bush would tell numerous audiences, "The specter of Vietnam has been buried forever in the desert sands of the Arabian Peninsula."[160] That specter would, of course, return to haunt the Army a little more than a decade later.

Icarus: The Army's Forever Wars

The same U.S. Army that had emerged transformed and triumphant from the ashes of the Vietnam War—having outlasted the Soviet Army and prevailed in the desert sands of the Arabian Peninsula in a stunning victory against the massive, Soviet-style Iraqi Army—would be humiliated not much more than a decade later by ragtag bands of insurgent fighters in Afghanistan and Iraq. Both wars seemed to begin with the same promise as the Gulf War—that the RMA would deliver to the United States a swift victory, with few casualties and no messy, postwar, low-intensity conflict with which to contend. Yet after brilliant initial victories in each theater, the war on terror devolved into disastrous forever wars that have lasted nearly two decades and continue, as of this writing, with no end in sight.

The Army would finally learn important lessons from the long slog of low-intensity conflicts that each of these wars turned into and remain. Tragically, these lessons came too late for the 4,200 or more Americans who died in Afghanistan and Iraq before the Army developed and implemented an effective low-intensity conflict doctrine.[161] Moreover, by the time this doctrine finally did arrive, the situation had so deteriorated in both countries as to make victory difficult, if not impossible, to achieve. Perhaps most important, the Army never truly, fully embraced all of the consequences of this new doctrine; it ultimately outsourced the political dimensions of the low-intensity conflict in each country to the State Department and refused to back the designated winner, condemning the United States to seemingly interminable wars in both theaters.

Al-Qaeda, the terrorist organization founded by Saudi-born jihadist Osama bin Laden, declared war on the West in the late 1990s. But it was not until one late summer day in 2001 that bin Laden finally got the undivided attention of the United States. A nineteen-man al-Qaeda terrorist cell flew two commercial airliners into the World Trade Center in New York City, flew another into the side of the Pentagon, and hijacked a fourth that crashed into a field in rural Pennsylvania after the passengers confronted the hijackers on their plane. As the dust settled over the smoldering ruins of Ground Zero in Lower Manhattan, nearly three thousand people were pulverized or inciner-ated in the worst terrorist attack ever to take place on American soil.[162]

On September 11, 2001, George W. Bush faced the first real crisis of his eight months as president. At the end of the day, after he returned to the Oval Office, he told Americans in an evening address that he shared their "quiet, unyielding anger." He made clear against whom America would direct its anger: "We will make no distinction between the terrorists who committed these acts and those who harbor them." He also explained how the nation would respond: "America and our friends and allies join with all those who want peace and security in the world and we stand together to win the war against terrorism."[163]

The country's immediate reaction to the shock of 9/11 was to use military force. This sentiment was reflected as well in Congress, which reconvened on September 13 to express its outrage and demand retaliation.[164] When Bush spoke to the American people again on September 21, he insisted that America would use every instrument of national power, making no bones about the fact that, although it might appear to be an unconventional war it was entering, America was indeed going to war. He warned that this new war would be "a lengthy campaign unlike any other we have ever seen. It may include dramatic strikes visible on TV and covert operations secret even in success."[165]

Not surprisingly, the Taliban regime in Afghanistan was defiant in the face of American threats. And al-Qaeda was downright belligerent.[166]

Immediately following 9/11, the entire weight of the U.S. national security establishment focused on planning military action to eliminate the al-Qaeda threat in Afghanistan, and the president almost immediately concluded that this plan must involve ground troops. America's diplomatic effort in the immediate aftermath of 9/11 focused exclusively on enlisting allies in this military effort. The first air strikes of the war in Afghanistan began on October 7, 2001, after the Taliban failed to yield to U.S. demands to turn over bin Laden.[167]

Once the war in Afghanistan began, the American military quickly swept the Taliban from power. U.S. Army Special Forces marched with the armies of the Northern Alliance, a coalition of non-Pashtun ethnic minorities that had fought the Taliban regime for years. SOF coordinated the militia forces and provided them with close air support from U.S. bombers. Meanwhile, the Army Rangers seized key facilities to allow regular Army forces to enter the country. Throughout 2001 there were never more than 5,200 American troops in Afghanistan.[168] Yet by November 12, 2001, the combination of U.S. firepower and Northern Alliance pressure had driven the Taliban out of the Afghan capital of Kabul. Enlisting the reluctant support of the Southern Alliance, a confederation of non-Taliban Pashtuns in southern Afghanistan, the coalition routed the Taliban from their last stronghold and spiritual capital in Kandahar on December 7, 2001, only two months after the war began.[169]

While SOF and Afghan Allies pursued Taliban fighters, the United States began to build up its conventional presence in Afghanistan. Engineers, communications specialists, and headquarters units flowed into Bagram Airfield and two other air bases. They were joined by a British Marine battalion and a Canadian infantry battalion. The Third Army, under the command of Lt. Gen. Paul Mikolashek, deployed forward to Kuwait to manage the flow of material and troops into Afghanistan. Maj. Gen. F. L. "Buster" Hagenbeck and his 10th Mountain Division headquarters deployed to the theater to form Combined Force Land Component Command (CFLCC) Forward in Afghanistan.[170]

Noticeably, only a handful of conventional U.S. combat formations had been deployed to Afghanistan. Steeped in an Army culture laser focused on high-intensity conflict, CENTCOM commander Gen. Tommy Franks was largely ignorant of low-intensity conflict. In fact, unlike some of his subordinates, Franks had never participated in a peacekeeping operation or any other low-intensity conflict.[171] It apparently never occurred to him that the Army might eventually face a postinvasion low-intensity conflict in theater, and CENTCOM had no real plan for postinvasion Afghanistan. Instead Franks focused solely on the initial objective of routing the Taliban from power. Believing that American airpower and Northern and Southern Alliance forces were sufficient to do the job, Franks directed that only a few U.S. conventional forces be deployed to—yet none employed directly in combat in—Afghanistan in the first days of the war. As a result, no U.S. forces were available to block mountain passes or stiffen the resolve of America's erstwhile Pashtun Southern Alliance allies. Osama bin Laden, Ayman al-Zawahiri, and most of the al-Qaeda senior leadership—as well as Mullah Omar and the Afghan Taliban leadership—used local relationships and bribery of Southern Alliance leaders to escape through mountainous Tora Bora into Pakistan.[172]

In keeping with the Bush administration's aversion to nation building—and Franks's unawareness of the possibility of a low-intensity conflict—the U.S. military's strategy in Afghanistan was a "light footprint." A tiny U.S. conventional force of no more than fourteen thousand troops finally arrived in Afghanistan, but when it did, it completely ignored the necessity to build a government in Kabul in favor of incessant combat operations to hunt down and kill Taliban and al-Qaeda remnants in the hinterlands.[173] In March 2002 Major General Hagenbeck assembled a force consisting of two U.S. light infantry brigades, various coalition infantry units, 1,200 Afghan fighters with U.S. Army Special Forces advisers, about eight hundred service members from a hodgepodge of SOF elements, and a task force of attack and utility helicopters from the 101st Airborne Division and attacked an al-Qaeda stronghold in the Shah-e-Kot Valley near Gardez. Without artillery support and with a muddled chain of command and control between U.S. Army, U.S. Air Force, SOF, and coalition and Afghan forces, Operation Anaconda was beset with difficulties even before it made contact with the considerable al-Qaeda and Taliban force it was sent to challenge. When the smoke cleared, just less than two weeks later, fifteen coalition troops were dead, including eight Americans, and most of the Taliban and al-Qaeda leaders had slipped away into Pakistan.[174]

The method by which the United States had fought the war had sown the seeds of future troubles for the coalition in Afghanistan. The Northern Alliance, with which the United States had allied to depose the Taliban, was ethnically distinct from and terribly unpopular with Afghan Pashtuns, who constituted the largest ethnic group in Afghanistan and nearly half the total population. The Northern Alliance was also terribly unpopular with the Pakistani government, whose help the United States would desperately need to sustain its effort in Afghanistan. The other U.S. ally, the Southern Alliance—almost exclusively non-Taliban Afghan Pashtuns—was tainted by its cooperation with the United States and the Northern Alliance and rejected by most Afghan Pashtuns. Moreover, the Southern Alliance had already been rejected by the Pakistanis in favor of the Taliban during the Afghan

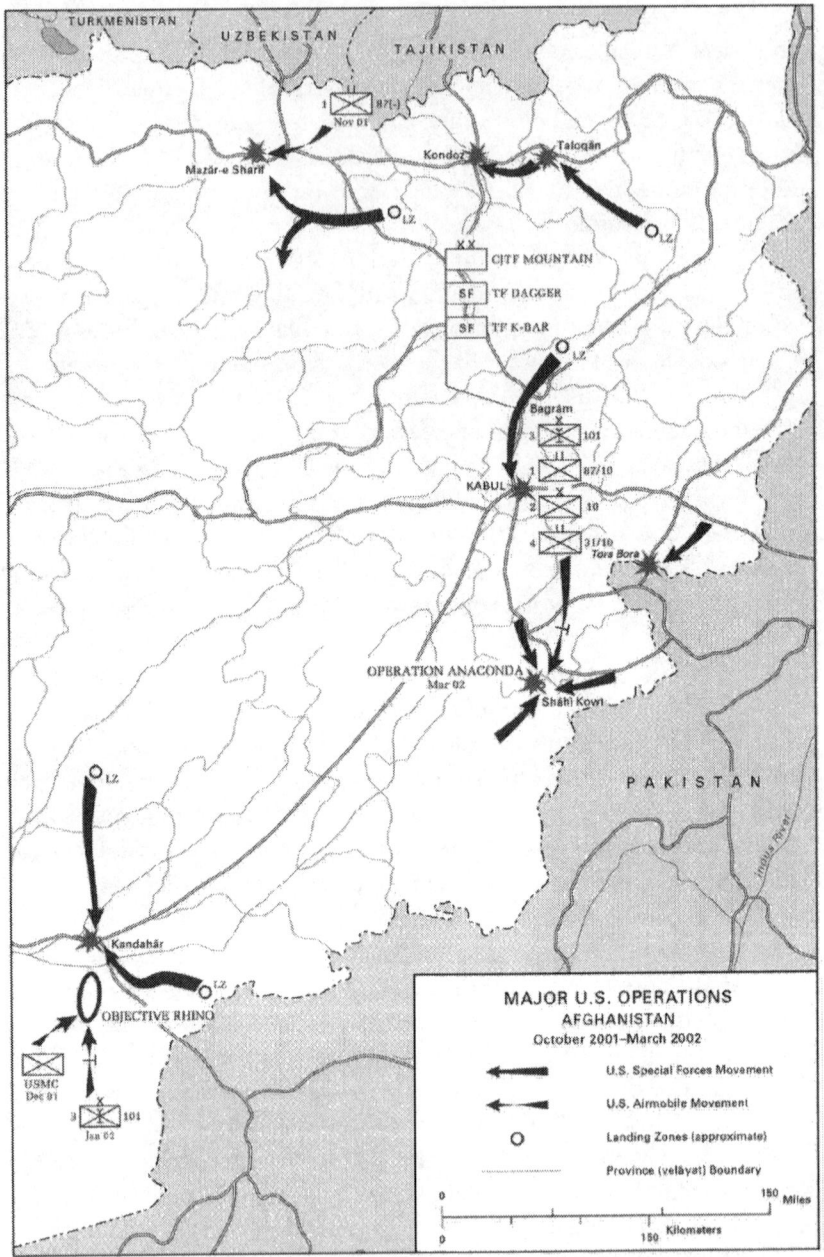

Figure 1.1. Operation Enduring Freedom. *Source:* R. Cody Phillips, *Bosnia-Herzegovina: The U.S. Army's Role in Peace Enforcement Operations, 1995–2004.* Reprinted in Allan R. Millett and Peter Maslowski, *For the Common Defense,* Kindle, locations 12791–92.

Civil War, before U.S. forces ever arrived.[175] Meanwhile, the U.S. Army, busy chasing al-Qaeda and Taliban remnants across southern Afghanistan, was doing nothing to help its Afghan allies establish political control of the country.

Despite these bad omens, America, after briefly savoring the images of liberated Afghans celebrating in the streets of Kabul, had already moved on to its next target in the war on terror.

When President Bush gave his first State of the Union Address in January 2002, he enunciated a new strategy for the post-9/11 world. He named the countries about which he was most concerned and described the danger they represented. By pursuing weapons of mass destruction and supporting terrorists, Iran, Iraq, and North Korea constituted an "axis of evil" that must be confronted.[176]

In the wake of this ambitious speech, the war in Afghanistan seemed to be all but over. In June 2002 a *loya jirga*, a traditional Afghan gathering of senior leaders and selected delegates, chose an interim government. In the West the event was considered an unqualified success. The selection of delegates by local shuras was considered sufficiently democratic, and the nomination of the eventual head of the interim government, Pashtun Southern Alliance leader Hamid Karzai, by the former Afghan ruler, King Mohammed Zahir Shah, seemed to lend legitimacy to the process.[177]

The UN was much less enthusiastic, however, about the administration's plans for Iraq, the first target in Bush's axis of evil. In August 2002 the administration was able to convince the UN Security Council to hold Iraq in "'material breach' of disarmament obligations" from the Gulf War. But the Security Council's resolution on the conflict fell short of authorizing force.[178]

There was little support in the United States or abroad for military action without a UN resolution authorizing force. Still, Congress enthusiastically authorized military action, even if the UN failed to endorse an invasion.[179]

Leaders of the Muslim world, despite the fear of Saddam they had harbored since his invasion of Kuwait, were still urging Bush to reconsider an invasion of Iraq. They feared that a prolonged campaign in Iraq would ignite the streets in their own countries. Turkey, a NATO ally, refused to allow the United States to use its soil to launch a ground attack into Iraq, though it did finally acquiesce to the use of its airspace for overflights.[180]

In the end the Bush administration chose to go ahead with the invasion of Iraq without a Security Council resolution authorizing force. After the invasion began, the United States did get the Security Council to vote unanimously to provide humanitarian aid to the Iraqi people, but even then it refused to commit to the postwar reconstruction of the country. Once the invasion began, support for the war in the United States surged to 75 percent, but fully half the American people believed that the UN should run postwar reconstruction of Iraq, a role the Security Council had already flatly rejected.[181]

Saddam Hussein's plan for the defense of Iraq reflected his very real fear of a coup. Iraq's regular Army divisions were positioned in an outermost ring around the periphery of the country, while the more elite Republican Guard divisions defended the inner ring around Baghdad. And the Special Republican Guard division, under the leadership of Saddam's most loyal Ba'ath Party generals, guarded the sensitive palaces and offices in downtown Baghdad.[182]

CENTCOM commander Tommy Franks's plan for the invasion of Iraq—dubbed COBRA II, after the Allied plan for the breakthrough of German defenses in Normandy—initially looked very much like the plan for the Gulf War. But under pressure from Secretary of Defense Donald Rumsfeld, troop numbers were reduced sharply, reflecting Rumsfeld's conviction—seemingly validated by Franks's plan for the war in Afghanistan—that the technology of the RMA had supplanted the need for massive numbers of U.S. troops. The 4th Infantry Division, commanded by Maj. Gen. Raymond Odierno, would attack Iraq from the north, through Turkey. The U.S. V Corps, with the 3rd Infantry Division and 101st Airborne Division (a helicopter-borne air assault division commanded by Maj. Gen. David Petraeus) would attack from Kuwait toward Baghdad, west of the Euphrates River. The I Marine Expeditionary Force under Maj. Gen. James Mattis would attack from Kuwait toward Baghdad from east of the Euphrates. And the British would seize Basra in southern Iraq. U.S. Army Special Forces would embed with the Kurdish Peshmerga Army in the north and find targets for coalition air strikes before the invasion and then search for weapons of mass destruction in Iraq's western deserts after the invasion began.[183]

The attack on Iraq began in earnest on March 19, 2003, with a failed "decapitation" air strike intended to kill Saddam. Despite this setback, the operation was a miracle of tactical precision. Defying expectations of weeks of preparatory air strikes similar to those that had preceded the Gulf War's ground invasion, the U.S. ground offensive in the Iraq War began only two days after the start of the formal air campaign. Coalition air forces would continue to attack Iraqi targets throughout the invasion, flying as many as forty-one thousand sorties. Fully two-thirds of the U.S. bombs dropped were precision-guided munitions.[184]

All did not go totally according to plan. Turkey would not allow the 4th Infantry Division to attack from its soil, so the 173rd Airborne Brigade jumped into the Kurdish north to fix the Iraqis in a second front. Once inside Iraq, coalition forces faced a threat they had not anticipated: the Fedayeen Saddam, an irregular force wearing civilian clothes and employing guerilla tactics. These fanatical Saddam loyalists—reinforced by perhaps thirty-five thousand convicts released from Iraqi prisons—harassed convoys, ambushed isolated units, and turned bridges and mosques into strongpoints.[185] A sandstorm a few days into the invasion seemed to threaten the operation's success as well, but the technology of the RMA allowed the Air Force JSTARS to continue to find and attack Iraqi Army targets. New infrared scopes and computerized self-locating systems allowed M1 Abrams Tanks and M2 Bradley Fighting Vehicles to spot and destroy Iraqi armor through the poor visibility without fear of casualties from friendly fire.[186]

By April 5 the Army was executing its first "Thunder Run" through Baghdad, and by April 9 the Marines were toppling the statue of Saddam in Firdos Square in Baghdad. On May 2 President Bush delivered his infamous speech from the flight deck of the carrier USS *Lincoln*, proclaiming "Major combat operations in Iraq have ended" under a banner reading "Mission Accomplished."[187]

By high-intensity conflict measures, Operation Iraqi Freedom had been a dazzling success. An expeditionary force of slightly more than 200,000 had defeated an army more than twice its size and toppled Saddam's regime. COBRA II had depended in large measure on the mass surrender or desertion of the Iraqi Army. In this regard, all went according to plan.[188] But as turned out, these armed, military-age men did not march quietly into prisoner of war camps. Instead they melted into the populace, only to reemerge as an even greater threat after the end of "major combat operations."

The planning for Phase IV of COBRA II, the postconflict reconstruction of Iraq, had been an afterthought during the buildup to the war. Caught between Secretary of Defense Rumsfeld and the chief of staff of the Army, Gen. Eric Shinseki, the number of troops available to secure Iraq—a country of thirty-eight million people spread across 400,000 square miles—dwindled from 300,000 to 100,000 as the invasion approached.[189] Third Army/CFLCC planners were forced to assume that the Iraqi Army would be reconstituted after the war and fill this gap, an assumption endorsed by General Franks

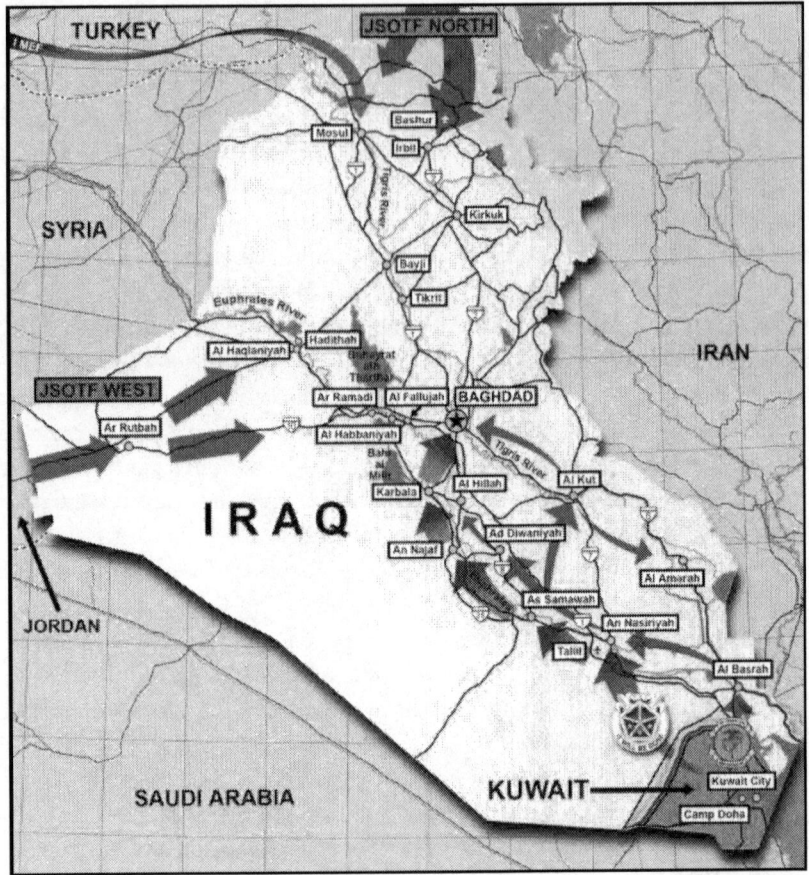

Figure 1.2. Operation Iraqi Freedom. *Source:* Gregory Fontenot, E. J. Degen, and David Tohn, *On Point: U.S. Army in Operation Iraqi Freedom.*

before the invasion. But rehearsals of Phase IV never occurred, and the operation was eventually deemed so complex as to require its own plan, named ECLIPSE II—in homage to the plan for the occupation of postwar Germany. The writing of this plan did not begin until three days before the attack on Iraq began.[190]

Not surprisingly, given the absence of a fully developed and rehearsed plan, things rapidly spun out of control in Iraq. Within hours of the toppling of Saddam's statue, rampant looting and violence began to spread throughout Baghdad. Shiite militia groups—backed by Iran—started forming in the south and east of Iraq and waging a campaign of reprisals against Ba'athist holdouts.[191]

Paul Bremer, head of the Coalition Provisional Authority, began his tenure in May 2003 by disbanding the Iraqi Army, the force U.S. military planners had been counting on to impose stability. With that move, more than 400,000 combat-trained, military-age men were suddenly aggrieved and unemployed. Sunni Arab Iraqis, a minority in Iraq who had been the privileged class in Saddam's regime,[192] began a spirited resistance to the American occupation. To make matters worse, many high-level members of the regime had escaped into neighboring Syria with much of Saddam's wealth, money they were using to fund the Sunni insurgency in Baghdad and the Sunni Arab region in northwest Iraq.[193]

And Syria was the source of another problem. During the invasion of Iraq, the U.S. military captured about two hundred non-Iraqi "foreign fighters," mostly from Saudi Arabia, who had come to join the jihad against the infidel invaders. By May 2003, according to Rumsfeld, foreign fighters from across the Arab world were pouring into Iraq through Syria in "busloads." And these fighters were not just operating as lone agitators; according to the CFLCC deputy commander, Maj. Gen. William Webster, foreign fighters were offering organized resistance to the U.S. occupation.[194]

Undeterred by these ominous signs, General Franks pulled his CFLCC headquarters out of Iraq, put V Corps in charge of the reconstruction, and retired from the Army on July 7, 2003. Lt. Gen. William Wallace, who had led V Corps during the invasion, relinquished command to Lt. Gen. Ricardo Sanchez, the most junior three-star general in the Army. Sanchez had been a brigade commander in peacekeeping operations in Kosovo and had worked in SOUTHCOM dealing with other government agencies and low-intensity conflicts.[195] Yet with no plan for the occupation and no Iraqi forces to impose order, these experiences did not translate into success in Iraq. In a single day in October 2003, while the staff of Sanchez's new Combined Joint Task Force 7 (CJTF-7) argued about whether to call the new enemy in Iraq an insurgency, five suicide car bomb attacks destroyed targets across Baghdad, including the International Committee of the Red Cross and the UN headquarters for Iraq. This was America's bloody introduction to Abu Musab al-Zarqawi and al-Qaeda in Iraq.[196]

With no plan in place for the occupation, and no doctrine or experience on which to rely, each military commander on the ground had to do what he thought best in the moment. Major General Petraeus, commander of the 101st Airborne Division in Mosul, had earned his doctorate in international relations with his 1987 dissertation on the "lessons of Vietnam." Armed with this academic background, he led his division to considerable success in Mosul.[197] But other commanders, raised in an army trained to fight the Soviet hordes on the plains of northern Europe, struggled to understand the conflict in which they found themselves engaged and created many more insurgents than they killed or captured with their heavy-handed tactics. In the 3rd Infantry Division in Baghdad, soldiers used tanks to crush the cars of looters. Odierno's 4th Infantry Division destroyed buildings from which they received fire from insurgents, rounded up the families of suspected insurgents or even whole villages, and implemented tactics reminiscent of the worst practices of the Vietnam War, such as unobserved harassment and interdiction fires. One of the division's battalion commanders, Lt. Col. Nate Sassaman, turned a blind eye as his soldiers tossed captured military-age men into the Tigris River, allegedly killing at least one.[198]

By the end of 2003 nearly twice as many Americans had fallen to the insurgency in Iraq as had died during the initial invasion. The only bright spots in an otherwise dark year were the death of Saddam Hussein's sons in a firefight in July 2003 and the capture of Saddam himself in Ad-Dawr, Iraq, across the Tigris River from his hometown of Tikrit in December 2003.[199]

Yet Saddam's capture did nothing to quell the insurgency. On March 31, 2004, a Blackwater Security convoy was ambushed in the city of Fallujah, a hotbed of insurgent activity in the Sunni Arab province of Al Anbar. The American public awoke that morning to scenes of the charred bodies of four American contractors, having been beaten and burned, hanging from a bridge on the edge of the city. The 1st Marine Regiment moved into the city and a five-day battle ensued, during which the Marines made significant progress in killing insurgents and dismantling insurgent infrastructure.[200] Yet through means that are still not entirely clear, insurgents were able to communicate with pan-Arab news agencies like Al Jazeera and put their own spin on the military operation. Political pressure mounted and, on April 9, 2004, the office of Paul Bremer announced a "unilateral suspension of offensive operations" in Fallujah. Three weeks later, an Iraqi-brokered handover of the city was initiated and the Marines withdrew. Forty Americans and hundreds of Iraqis were dead, and the city was soon firmly back in insurgent hands. The Marines would be forced to return in November 2004 to clear the city once again, this time at the cost of fifty-four more Americans killed and nearly ten times that number wounded.[201]

On April 28, 2004, things went from bad to worse for the U.S. military in Iraq. CBS's *60 Minutes II* aired a story about abuses at the U.S.-run Abu Ghraib prison. The story included graphic photos of U.S. soldiers, including a woman, humiliating naked Iraqi men.[202]

In reaction to the incidents at Abu Ghraib, the Arab streets exploded in outrage and violence. Al Jazeera and other pan-Arab satellite television stations continued to feature around-the-clock coverage of the story. Al-Zarqawi released a video in which he beheaded captured aid worker Nick Berg to avenge the Abu Ghraib abuses.[203] The steady stream of foreign fighters that had begun in 2003 became a flood. Many foreign fighters captured after 2004 cited video of alleged atrocities shown on networks like Al Jazeera as their motivation to join the war in Iraq.[204]

Iran was also playing a disruptive role in Iraq by encouraging firebrand cleric Muqtada al-Sadr and arming his militia, Jaysh al-Mahdi (JAM). In March 2004, after Sunni insurgents killed two hundred pilgrims celebrating the Shiite festival of Ashura in Karbala, al-Sadr began an uprising across southern Iraq with its epicenter in Al-Najaf. The result was a weeks-long battle between more than three thousand coalition troops and JAM across Diwaniya, Karbala, Al-Najaf, and several other southern cities.[205] After the battle, JAM remained a force in the Shiite south and the Sadr City neighborhood of Baghdad, but otherwise appeared to recede from Iraqi politics.

In June 2004 Gen. George Casey replaced Lt. Gen. Ricardo Sanchez, and CJTF-7 gave way to Multinational Force–Iraq in Baghdad. Lt. Gen. David Petraeus took over the nascent effort to stand up a new Iraqi Armed Forces under a new headquarters, the Multi-National Security Transition Command–Iraq (MNSTC-I, pronounced "min-sticky"). Over Petraeus's private objections, Casey acceded to CENTCOM commander Gen. John Abizaid's desire to get American troops off of the streets of Iraq and into large, well-defended forward operating bases (FOBs).[206]

Abizaid's reasoning betrayed a fundamental misunderstanding of the nature of low-intensity conflict. He saw U.S. forces as an "antibody in Iraqi society," believing that their visible presence made victory harder to achieve. This was a manifestation of the Army's predilection to put protection of its soldiers before engaging the populace.[207] But when coupled with the impulse to get Iraqi forces on the street in place of Americans, it betrayed a deeper motive, the desire to relieve American troops of the responsibility for engaging in the political dimensions of low-intensity conflict.

Rather than engaging in the political dimensions of the Iraq War to end the conflict, the U.S.-led coalition rushed to create a legitimate host nation government to which it hoped it could hand over the whole mess. On July 13, 2003, Paul Bremer had established the Iraqi Governing Council—filled with exiles like Ahmed Chalabi—that Iraqis roundly rejected. Rather than identify the leaders that did have legitimacy and engage them to forge a political solution, the Coalition Provisional Authority pinned its hopes on elections.[208]

The elections reflected the deep sectarian and ethnic divides in Iraq. In January 2005, Sunni Arab Iraqis largely boycotted the election of a transitional national assembly. They did participate in the late-2005 referendum on the Iraqi constitution, but opposed it by 96.9 percent. It was al-Qaeda in Iraq, however, that finally lit the fuse on the sectarian powder keg. In March 2006 the group destroyed the golden dome of the Al-Askari Mosque—easily among the most sacred sites in Shia Islam—in Samarra, the heart of the so-called Sunni Triangle.[209]

Within hours of the destruction of the mosque, al-Sadr's long-dormant JAM—a fifty-thousand-man militia—was on the street, attacking Sunni mosques across Baghdad. A daytime curfew finally silenced huge riots across Iraq, but violence continued.[210] By March, Iraqi prime minister Ayad Allawi was warning that Iraq was in the throes of a civil war. By April, his voice had been joined by those of the Egyptian president and the Saudi foreign minister. By June, more than fourteen thousand Iraqis had been killed just in sectarian violence since the Samarra mosque bombing. By July, one hundred Iraqis were dying every day. Members of the Badr Brigade—the military wing of the Shiite Supreme Council for the Islamic Revolution in Iraq—were covertly supporting JAM from their government posts inside the Ministries of the Interior and Transportation. By October, 1.6 million Iraqis had left the country to escape the violence, while nearly as many had been forced from their homes and had become refugees within their own country. Iraq had descended into sectarian carnage. Somewhere between sixteen thousand and sixty thousand Iraqi civilians were killed in the violence in 2006 alone.[211]

Violence against coalition forces was also on the rise, the number of attacks doubling from 2004 levels to eight hundred a week.[212] By the end of 2004, nearly one thousand American service members had been killed in postinvasion Iraq, more than eight times the number killed in the initial invasion. In addition, at least eight hundred more were killed in Iraq in 2005 and again in 2006 as snipers, ambushes, truck and car bombs, and roadside bombs—improvised explosive devices—ground down the U.S. Army in Iraq. By November 2005, 160,000 U.S. troops and 23,000 coalition allies were fighting for their lives against perhaps 18,000 insurgents across Iraq.[213]

Still, the United States pursued its "light footprint" strategy, hunkering down in FOBs and commuting to the fight, anticipating a time when U.S. forces could "stand down" as Iraqi forces "stood up." In the meantime, Iraq continued to rip itself apart. Perhaps the only bright spot was the June 2006 death-by-air strike of al-Qaeda in Iraq leader Abu Musab al-Zarqawi.[214]

The war that Americans thought had ended in Afghanistan was turning against the United States as well. In summer 2002 the Taliban launched a new offensive across southeast Afghanistan. Many of the same double-dealing tribal leaders that had helped al-Qaeda escape into Pakistan were funneling U.S. weapons from the fledgling Afghan National Army into the hands of Taliban fighters, who were engaged in a guerilla campaign across Pashtun Afghanistan. Rocket and mortar attacks against U.S. bases and small arms attacks against helicopters became commonplace, as did the effective use of improvised explosive devices against ground forces.[215] In the din of the preparations for the war in Iraq, the American people hardly noticed the major reversal in Afghanistan.

The UN did, however, take notice when a second, fiercer Taliban offensive in spring 2003 established Taliban enclaves in southern Afghanistan. Under NATO auspices, the United States established the International Security Assistance Force (ISAF) of thirty thousand U.S. and coalition troops to patrol the major cities of Afghanistan.[216]

Over the course of 2004 alone, ISAF and the Afghan government spent in excess of $4 billion in U.S. aid to build the new Afghan National Army and the Afghan National Police, but this effort failed to quell the burgeoning Taliban insurgency. Despite a goal of fielding 150,000 trained Afghan troops by 2008, the Afghan National Army would not reach two-thirds of this goal by the end of 2009. And the force the coalition was building overwhelmingly consisted of Tajiks and Uzbeks from the former Northern Alliance, people who knew little more about Pashtun southern Afghanistan than did American troops.[217]

Matters continued to deteriorate. In late summer 2006, the Pakistani government reached a peace deal with the Pakistani Taliban that allowed the latter to expand its control throughout Waziristan, in the Federally Administered Tribal Areas (FATA) of northwest Pakistan. The accord created a safe haven for Afghan Taliban fighters to winter and recuperate, and the following spring, when the fighting season began, a newly regenerated Taliban flooded over southeast Afghanistan, erasing five years of progress.[218]

The U.S. war effort was right back where it had started—except that, by 2008, 295 coalition soldiers, including 155 Americans, were dead.[219]

Back in Iraq, Operation Together Forward, a joint U.S.-Iraqi operation to take back Baghdad that began in 2006, proved to be doomed from the start and managed to do nothing to stem the tide of violence in Iraq or political dissent in the United States. With the American public weary of a war in Iraq that was seemingly lost, the Democrats swept the midterm election, gaining twenty-eight House seats and six Senate seats and taking control of both houses of Congress. In a press conference days later, the president admitted that the election had been a referendum on his war policy and announced the resignation of Secretary of Defense Donald Rumsfeld, but made it clear that he was still committed to winning in Iraq.[220]

In the midst of the military disaster unfolding in Iraq, there were bright spots, and some learning did occur. The Army had eschewed the age-old model of individual replacements—with its deleterious effect on morale, unit cohesion, and the Army's experience base—in favor of unit rotations that would preserve morale and effectiveness but would also create an incredible logistical demand on the nation's strategic lift capacity and the Army's reserve components. The first rotation of soldiers into and out of Iraq was the most massive relief in place of U.S. combat forces since at least World War II; between March and June 2004 nearly 244,000 soldiers moved into and out of Iraq. The subsequent rotation, a year later, was just as massive.[221]

To sustain itself through these colossal rotations, the Army reshaped its organizations into "modular" brigades with all of the capabilities each would need for a deployment—including artillery, reconnaissance, engineering, and logistical support—incorporated into its organic organization. These brigades were then pulled through grueling, repetitive cycles of manning, equipping, training, deploying, redeploying, and recovering called Army Force Generation (ARFORGEN). To alleviate the pressure on Army communities created by all of a post's units deploying at once, the deployment of division headquarters was decoupled from the deployment of their organic brigades and these headquarters were dragged through this cycle as well. The process was toughest on activated Army Reserve and National Guard units, which by 2004–5 constituted more than 40 percent of Army personnel on the ground in Iraq.[222]

Amid growing chaos and mounting casualties, the Army quickly came to the realization that it was not manned or trained for the low-intensity conflict that had emerged in Iraq. By January 2004, artillery and armor battalions in Iraq began to be employed as convoy security, military police, or infantry. Beginning in 2004, training at the NTC became focused on the low-intensity conflict in Iraq; Arabic-speaking role-players and urban environments were added to train units on counterinsurgency tasks.[223]

The Army also realized that it needed new capabilities to succeed in the low-intensity conflict in Iraq. Chief among these was the capability to stand up Iraqi security forces. Lieutenant General Petraeus took over as commander of MNSTC-I in 2004,[224] but this effort didn't truly gain momentum until the fall of 2005, when the Army—through a more formalized process of manning and training—began to increase the quality of Military Transition Teams—Army officers and senior NCOs rounded up from across the Army and hastily organized and trained to partner with and train Iraqis. The resulting Iraqi security forces were wracked with sectarian strife—many recruits were infiltrators from Shiite militias who used their badges to victimize Sunnis—but the effort was at least effective enough to make coalition recruiting efforts a target of the insurgency.[225]

Some Army units were also learning critical lessons about the nature of low-intensity conflict—particularly the understanding that it occurs in and among the people. At the outset of the war, the 101st Airborne Division benefited from the educational background of Petraeus, its commander. In the northern city of Mosul, the division used a population-centric approach to prevent an insurgency from taking root: it lived among the population, patrolled the streets, and engaged the people. In the soldiers' barracks, posters asked, "What Have You Done to Win Iraqi Hearts and Minds Today?" Thus it was not surprising that, in July 2003, the populace of Mosul gave up the location of Saddam's sons Uday and Qusay.[226]

As these population-centric techniques spread across the Army, leaders became acutely aware that they lacked a number of capabilities needed to interact with the people of Iraq. The Army offered massive recruiting bonuses of $150,000 or more to entice Arabic speakers to join the force. Even with these massive offers on the table, by 2008 the Army had assembled less than a thousand military linguists with skills useful in Iraq or Afghanistan. Across the U.S. military, cultural awareness and languages were added to educational curricula.[227]

The Army was also building other critical specialties for operating in and among the people. Specialists in occupations such as air defense and engineering were being converted into civil affairs and psychological operations soldiers. The number of personnel in these fields would increase by one-third over the course of the war.[228]

As the Army interacted with the Iraqi populace, it slowly began to realize that it had to engage the political dimensions of the low-intensity conflict in Iraq if it were to be successful. The first area in which the Army made particular progress was in its ability to engage in civil-military operations. One innovation was better coordination with the nongovernmental organizations conducting relief activities inside Iraq. Another was the Commander's Emergency Response Program, or CERP, which allowed Army units to employ "money as a weapon system" to drive a wedge between insurgents and the populace by addressing the grievances that might make civilians sympathetic to the insurgents' cause. Rather than being relegated to SOF or a single officer or section within a conventional battalion or brigade staff, whole conventional Army units began to concern themselves with civil affairs and governance as much as with finding and killing insurgents.[229]

The Army's new appreciation of civil-military operations did lead many into the cognitive pitfall of trying to measure the immeasurable: quantifying the coalition's progress toward victory. In just one example, the 1st Cavalry Division, under the command of Maj. Gen. Pete Chiarelli, tried to quantify its efforts in Baghdad's Shiite slums of Sadr City through "SWET" (sewage, water, electricity, trash) assessments of each subdistrict.[230] This problematic tactic quickly spread across Iraq as a "best practice."

The more contentious realization the Army began to arrive at in Iraq—slowly, in discreet locations, and as early as the first years of the war—was a growing understanding that, in order to be successful, the Army must engage directly in the politics of Iraq. In the town of Tal Afar in 2005, Col. H. R. McMaster—famed for his heroics in Desert Storm and his doctoral dissertation-turned-bestseller on the Vietnam War—led his 3rd Armored Cavalry Regiment through a deliberate "clear–hold–build" counterinsurgency methodology, protecting the populace from the insurgents, isolating and destroying the insurgents, and rebuilding the governance and infrastructure of the embattled city. The key to McMaster's success was enlisting the support of civic leaders and persuading the population to reject the insurgency.[231]

Col. Sean MacFarland relieved McMaster in Tal Afar and co-opted his methodology. When MacFarland was directed to take his 1st Brigade, 1st Armored Division to Ramadi in Al Anbar Province in the Sunni Triangle, he employed McMaster's methodology with similar impressive success there. By identifying local leaders and enlisting their support, he was able to isolate and neutralize the insurgency and then establish a political order to stabilize the city.[232]

MacFarland's success in Ramadi would eventually blossom into the Sunni Awakening movement that swept Al Anbar Province. Sunni sheikhs turned against the insurgency they had once embraced and—in exchange for money and semiofficial positions within the Iraqi security forces—formed Sons of

Iraq militias to combat al-Qaeda in Iraq. It was a dangerous expedient—exchanging a present danger to the government of Iraq for a more ominous future threat—but it did ultimately stabilize Al Anbar.[233]

But for every McMaster or MacFarland, there were a dozen men like Army colonel Michael Steele. Steele had been a captain in command of U.S. Army Rangers at the fateful battle of Mogadishu in Somalia, as immortalized in the 1999 novel and 2001 movie *Black Hawk Down*. As the commander of the Rakkasans (3rd Brigade, 101st Airborne Division), he chafed at training his formation on counterinsurgency tasks before its deployment to Iraq. Once deployed to Samarra in 2006, his operations were profoundly high intensity in nature, focused on big, destructive helicopter-borne air assaults that created as many insurgents as they killed.[234]

Steele's attitudes reflected the prevailing culture of the U.S. Army in the first years of the war in Iraq. William Wallace, V Corps commander during the invasion of Iraq, argued that no one "could have or did anticipate the total collapse of this regime and the psychological impact it had on the entire nation." Wallace dismissed the Army's responsibility to predict this obvious consequence of deposing Saddam, adding, "The military did their job in three weeks." Building a new political order—the strategic objective of the invasion of Iraq in the first place—was presumably someone else's job. Even inside Petraeus's 101st Airborne Division, Wallace noted that there were "two or three infantry battalion commanders who were either not that comfortable with the nation-building aspect of things, or really weren't that enthusiastic about it."[235] One can safely assume that for the "two or three" commanders who were willing to express their discontent, there were scores more who felt the same way but were not willing to speak.

When Petraeus left Iraq, he took command of the U.S. Army Combined Arms Center, the nexus of Army doctrine, training, and education. From here he began an insurgency of his own—an effort to consolidate the lessons of Iraq into a new Army doctrine on counterinsurgency. Petraeus assembled scholars such as Lt. Col. John Nagl, another PhD who had written his dissertation on the lessons of Vietnam, and Lt. Col. (Ret.) Conrad Crane—a PhD and West Point classmate of Petraeus. Petraeus and his fellow "insurgents" believed that the Army had committed a grievous error in neglecting the counterinsurgency lessons of Vietnam. The doctrine they created in 2006 was the revolutionary manual FM 3-24, *Counterinsurgency Operations*.[236]

FM 3-24 took as its starting point population-centric operations, stating flatly, "COIN [counterinsurgency] is fought among the populace.

Counterinsurgents take upon themselves responsibility for the people's well-being in all its manifestations." It rejected the military objective of "killing every insurgent" as "impossible." Moreover, operations in pursuit of this goal were "counterproductive," as they risked "generating popular resentment, creating martyrs that motivate new recruits, and producing cycles of revenge."[237]

The manual first and foremost institutionalized the idea that counterinsurgency was primarily political in nature, attacking a phenomenon that Petraeus had observed throughout his Army career: "Though most military officers quote flawlessly Clausewitz' dictum that war is a continuation of politics by other means—many do not appear to accept fully the implications of his logic." The manual insisted, "Any successful COIN operation must address the legitimate grievances insurgents use to generate popular support."[238]

The manual was not perfect. Grounded in the Army's contemporary experience in Iraq and suffering from the legacy of the Nixon Doctrine and the Reagan-era model for counterinsurgency, it insisted that all actions must originate with and be led by a host nation—remaining silent on the circumstance that characterized the beginning of the Iraq War, in which no host nation existed. The manual also insisted that quantitative measures of effectiveness could be useful if they were "contextual to the insurgency and . . . relevant to actually measuring success," even as it noted that the Army in Vietnam had utterly failed to do this despite implementing "the single largest and most comprehensive military counterinsurgency assessment apparatus in the history of warfare."[239]

But FM 3-24 did ultimately place responsibility for engaging in the political dimensions of low-intensity conflict squarely with military commanders:

> Many insurgencies begin because groups within a society believe that they have been denied political rights. . . . Commanders should investigate whether—
> - All members of the civilian population have a guarantee of political participation.
> - Ethnic, religious, or other forms of discrimination exist.
> - Legal, social, or other policies are creating grievances that contribute to the insurgency.
>
> Commanders should also identify traditional or charismatic authority figures and what narratives mobilize political action.[240]

In this brief excerpt is both a revolution in military thought (the idea that politics in a low-intensity conflict is the domain of the U.S. Army) and the seeds of the ultimate U.S. failure to end the conflict in Iraq (the idea that the Army should build an inclusive and nondiscriminatory political order.

This latter idea constituted the primary cognitive obstacle the Army faced in reaching the most important and necessary realization to win the war in Iraq: that the Army had to back a winner in the political struggle. And this misconception was ultimately enshrined in FM 3-24. In discussing information operations, for instance, the manual told readers that "impartiality" was key. "Counterinsurgents should avoid taking sides, when possible," the manual continued. "Perceived favoritism can exacerbate civil strife and make counterinsurgents more desirable targets for sectarian violence."[241] Such exhortations betray a fundamental misunderstanding of the political dimension of low-intensity conflict in particular and war in general; war is about using violence to impose one's will on the defeated.

In the early days of 2007 President George W. Bush doubled down on his Iraq War policy. In a dramatic change in strategy that has come to be known as the "surge," Bush authorized the deployment of an additional twenty thousand troops to Baghdad and Al Anbar Province in a last-ditch effort to salvage the war. In addition to replacing Rumsfeld with Robert Gates, he also replaced Gen. George Casey with Petraeus as commander of the multinational military effort in Iraq, guaranteeing that the new doctrine laid out in FM 3-24 would be implemented there.[242] With more troops and a new security team came a new strategy: instead of watching from FOBs while Iraq ripped itself apart, U.S. forces moved out into Iraqi cities to protect the populace. "Clear–hold–build" was being taken nationwide.[243]

The surge did change the character of the war in Iraq; instead of sitting on massive FOBs and commuting to the fight, U.S. Army units moved out of their protected FOBs and into the cities to live among the populace they were charged to protect and forge political solutions to the problems in each area. In practice, this meant breaking up battalions of four hundred to nine hundred troops into companies of about one hundred troops or even platoons of thirty troops and basing them in smaller—but still hardened—joint security stations (JSSs) from which they could more quickly and easily move out to operate in and among the population.[244]

There were some important coincidental factors—conditions that were not necessarily a result of the surge—that made its success possible. The first was the Awakening movement, which spread from Al Anbar to the rest of Sunni Arab Iraq. This movement turned local insurgents—many of them Ba'ath Party loyalists or simply Sunni nationalists—away from fighting U.S. forces and toward fighting al-Qaeda in Iraq; the Awakening thus both removed a threat to U.S. forces and created an indigenous force to help U.S. forces. Equally important, by the time the surge took effect in Baghdad, much of the ethnic cleansing that had occurred in the city in 2006 and early 2007 was complete. As Lt. Col. Doug Ollivant, G-5 (chief of plans) for Multinational Division–Baghdad would put it during the surge, "the fundamental truth of the Iraqi settlement is that the sectarian civil war ended—and the Sunni lost."[245] While the change in tactics represented by the surge was necessary to quell violence in Iraq, it might not have been as successful in doing so without these factors being in place.

Still, Bush had staked his presidency on the Iraq surge, and the gamble paid off. Sectarian violence dropped dramatically, as did American casualties. More important, factions began to reach accommodation on many of the contentious issues that had once threatened to shatter Iraq along sectarian lines. With this success came a turnaround in public opinion about the war. While Americans continued to see the decision to have gone to war as a mistake, they now began to think they were seeing the light at the end of the tunnel. And by February 2009 a majority believed the United States was winning the war.[246]

But the tragedy of this success is that it came so late. By February 2007, when Petraeus returned to Iraq to institutionalize his new counterinsurgency doctrine, more than three thousand U.S. troops had already died in that war. And nearly 1,200 U.S. troops would die in Afghanistan before the strategy could be implemented there.[247] Had the Army arrived in Afghanistan and Iraq with proficiency in low-intensity conflict already institutionalized in its organization, doctrine, education, and training, many of these soldiers might still be alive.

With the course of the war seemingly reversed, President Bush could begin to chart America's withdrawal. In August 2008, in the closing days of the Bush presidency, his administration and the government of Iraqi prime minister Nouri al-Maliki reached an agreement that set a timetable for American withdrawal, one that saw forces out of Iraqi cities by the end of June 2009, combat operations ended by September 2010, and U.S. forces out of Iraq by December 2011.[248] Thus, when President Barack Obama took office in 2009, his Iraq War script had essentially been written for him by Bush; the U.S.-Iraq security agreement dictated the pace of the U.S. withdrawal from Iraq.

But there were political problems in Iraq that indicated the war might not be ended so easily. While the course of the war seemed to have been reversed, deep issues loomed for the future of the nation. First, while the U.S. Army in Iraq embraced the population-centric tactics of FM 3-24—moving into the cities to protect the population and isolate the insurgency—it never truly embraced the manual's exhortation that the military should involve itself in the politics of the conflict. Thus, even before Petraeus left Iraq in July 2010 to take command of CENTCOM, Army units began handing the politics of each district and province off to provincial reconstruction teams (PRTs), civilian teams under the direction of the State Department's Office of the Coordinator for Reconstruction and Stabilization. By 2010, PRTs were embedded in every Army brigade combat team, secured and sustained by Army forces, interacting with the civilian leaders of each province and district, and doing politics for the Army. For example, in Salah ad Din Province—the home province of Saddam, in the heart of the Sunni Triangle—Army forces were explicitly directed *not* to engage in politics and to cede all of these functions to the PRT.[249]

These PRTs were replete with problems. Built in haste as part of the surge, these teams were poorly staffed with temporary hires who often had dubious credentials to be engaging in the politics of a war zone. In just one of hundreds of examples, the State Department hired a young female nutritionist with no international relations experience—let alone reconstruction or stability experience—as the senior health adviser to the government of Salah ad Din. Even the State Department diplomats leading the PRTs were ill suited to the nuances of local politics in a low-intensity conflict environment—where violence is a political tool. As one PRT member put it, "Career Foreign Service people do not by default have the experience and leadership skills necessary to lead people in this kind of work."[250] But, more fundamentally, in a low-intensity conflict, political power derives from one's ability to wield violence; only a military force can effectively compel factions in such an environment to accept political settlements. Ultimately, PRTs failed to forge enduring political solutions because they lacked the ability to wield violence.

Yielding the political dimensions of the conflict to the State Department also undid the success of the surge at the national level. Put plainly, at the eleventh hour, the U.S. State Department refused to take sides in the war and back a winner. The 2010 parliamentary elections in Iraq failed to yield a decisive victor.[251] The Sunni-backed secular Shiite and former prime minister Ayad Allawi failed to win a clear majority but his party narrowly edged out the party of incumbent Shiite prime minister Nouri al-Maliki—winning ninety-two seats to al-Maliki's eighty-nine. An eight-month deadlock ensued. Allawi was the clear favorite of both the United States and the people of Iraq—including, critically, the Sunni Arab Iraqis whose support was essential if Iraq were to survive the withdrawal of U.S. forces. But al-Maliki was able to employ back-room deals, incumbency, and the extraconstitutional use of police forces to prevail in this power struggle. In particular, he swayed the Kurdish north with promises of settling the question of final disposition of Kirkuk, the oil-rich city on the fault line between Kurds and Sunnis, in a settlement that would almost certainly favor the former and further isolate the latter.[252]

In the midst of this political instability, the insurgency reemerged in Sunni Arab Iraq, and the remnants of al-Qaeda in Iraq launched a wave of car bomb attacks throughout Baghdad.[253] Rather than step in to assure the future of Iraq, the United States—particularly Ambassador Ryan Crocker and his successor, Chris Hill—stood by and let al-Maliki steal the election. Army generals Raymond Odierno and Lloyd Austin, who succeeded Petraeus as senior U.S. commanders in Iraq, deferred to the State Department in deciding the fate of Iraq. And in this instance, the State Department snatched defeat from the jaws of victory.[254]

Despite these ominous signs for the future of Iraq, American forces completed their withdrawal at the end of 2011. The Iraqi Council of Representatives had failed to ratify a bilateral agreement that would have protected American troops from prosecution in the Iraqi courts after 2011. In response, the United States held to the timetable established in the closing days of the Bush presidency. Making good on his pledge to end the war in Iraq, President Obama withdrew all U.S. forces.[255]

The cost of the war had been staggering. Nearly 4,500 Americans were dead and well over thirty-two thousand were wounded; nearly a trillion dollars had been spent. The cost was even more devastating for Iraq; as many as 120,000 Iraqis had been killed and millions were displaced from their homes, either inside Iraq or as refugees outside the country.[256] Americans believed that these sacrifices had bought them victory. In fact, they had only bought America a brief respite.

In 2009, in a bid to repeat the perceived success of the Iraq surge, President Obama and his new commander, Gen. Stanley McChrystal, announced the commitment of thirty thousand additional troops to reverse the course of the eight-year-old war in Afghanistan. Obama and McChrystal presented a new political-military strategy to accompany the increase in troops, a "civilian and military counterinsurgency campaign" in Afghanistan.[257]

But the policy almost immediately ran aground. Just as the surge was being implemented, General McChrystal was replaced by General Petraeus because of statements McChrystal and his staff had made to a *Rolling Stone* reporter that were critical of the administration.[258]

There were also much deeper issues that beset the Afghanistan surge. In Iraq the surge had occurred in the midst of the Sunni Awakening and at the end of a year of ethnic cleansing in Baghdad, factors that set the conditions for the success of this new approach. Three fundamental conditions in Afghanistan militated against replicating the success of the surge in Afghanistan. First, America's allies in Afghanistan were essentially the same as they had been since the beginning of the war: the Southern Alliance of non-Taliban Pashtuns and the Northern Alliance of Hazaras, Tajiks, and Uzbeks. The Southern Alliance was dominated by the Pashtun Durrani tribe, locked in a tribal conflict with the Ghilzai tribe, the other of the two largest and most powerful tribes of the Pashtun ethnic group. This conflict has been the source of warfare for the past 250 years. The Afghan monarchy was established by the chief of the Durrani tribe, Ahmad Shah Durrani, in 1747. During the Soviet occupation, the Ghilzai dominated the Afghan puppet government. After the Soviet era, the Pakistani government backed the Ghilzai-dominated Taliban, both because the Ghilzai tribe had larger numbers in FATA and because it would weaken the stronger Durrani tribe. When U.S. forces arrived in Afghanistan, fortunes reversed again: the United States helped the former Afghan king, Mohammed Zahir Shah, a Durrani, put the Durrani Hamid Karzai in power as president, supplanting the Ghilzai Taliban leader Mullah Omar.[259] Nothing about the Obama surge altered the equation in reconciling the Taliban and the Southern Alliance, which would require healing this tribal divide.

The other U.S. ally in Afghanistan, the primarily Hazara, Tajik, and Uzbek Northern Alliance, represents less than half the population, almost all of which live outside the contested areas of Pashtun Afghanistan. They are ethnically dissimilar from the Pashtuns, even speaking a different language. Moreover, this faction is politically unpalatable to the Pakistanis, who exert considerable influence over conditions in Afghanistan—not least by funding, supporting, and harboring the Taliban.[260] As with the tribal conflict within the Pashtun ethnic group, nothing about the Afghanistan surge alters these facts.

And this leads to the second factor. The Pakistanis tacitly control the four million Pashtuns who live in the ten-thousand-square-mile hinterland of rugged, mountainous terrain formerly known as FATA. (The area was merged with the Khyber Pakhtunkhwa, a province of Pakistan in mid-2018.) Through their Inter-Services Intelligence, the Pakistanis maintain close ties with the Taliban that they created in the 1990s to establish a friendly regime in Afghanistan. Through support to the Afghan Taliban in the form of money, and by providing a safe haven in which Afghan Taliban insurgents can winter and recruit for the next spring fighting season, Pakistan can, if it wishes, perpetuate the war in Afghanistan indefinitely. And Pakistan has every incentive to continue the war, as it has been an economic boon for the nation: between 2002 and 2015, the United States gave Pakistan more than $31 billion in mostly military aid.[261] And the surge does nothing to change this math.

Finally, the geography of Afghanistan is simply not conducive to the surge tactics that were so successful in Iraq. In the latter, the insurgency was primarily urban. To protect the population in Iraq, U.S. Army units moved in small elements into hardened JSSs, living in their respective areas of operation among the people they were charged to protect from the insurgents. The surge in Afghanistan likewise focused on population-centric operations. It did implement controls—especially on air strikes—that reduced civilian casualties.[262] But the challenge of Afghanistan is that the insurgency thrives in the rural areas, not in the cities, and Afghanistan is a huge country, nearly half again as large as Iraq. Just in Pashtun Afghanistan alone, there are literally thousands of tiny villages in which handfuls of people must be secured. The buildings in these villages are mud huts, not hardened structures suitable for JSSs. And the villages are connected by dirt goat trails rather than highways. Sustaining conventional Army units across such vast distances and rugged terrain would quickly overwhelm the logistics capability of the average conventional battalion.

To overcome this challenge, U.S. forces in Afghanistan instituted Village Stability Operations; U.S. Army Special Forces soldiers—specially trained to operate and survive in austere environments and train indigenous civilians to become unconventional warriors—live and work among the population in these villages to guard them against the insurgency.[263] It is the gold standard of counterinsurgency practice, but the U.S. military simply lacks the capacity to expand the practice to enough villages to make a difference; the tiny number of Special Forces A-Teams in the U.S. arsenal have been quickly overwhelmed by the number of villages and towns in need of their attention.

On May 1, 2011, in a daring early morning raid, Central Intelligence Agency and SOF elements finally found and killed al-Qaeda chief Osama bin Laden. This was the objective that President George W. Bush had used to justify the invasion of Afghanistan ten years earlier.[264] But in the decade that had passed since the beginning of the war in Afghanistan, America's mission there had expanded beyond all recognition, from establishing a stable government to eradicating the poppy harvest to expanding the rights of Afghan women. America wasn't going anywhere.

Still, President Obama did try to chart the path to a withdrawal from Afghanistan. Speaking a month after the death of bin Laden, Obama declared that U.S. forces would leave Afghanistan within three years, by August 2014. This was the timetable for withdrawal that he had set at the beginning of the Afghanistan surge.[265] The U.S. military progressively withdrew tens of thousands of troops from Afghanistan while incrementally handing over security responsibility to Afghan national security forces.

But as the deadline for U.S. withdrawal approached, Taliban fighters began to expand their control into areas once firmly held by American forces, and it became clear that the Afghan government could not hold the country alone. In May 2014 President Obama was forced to concede that a residual force of nearly ten thousand American troops would remain in the country beyond 2014. This was followed by an admission the following year that American troops would remain in Afghanistan beyond 2015, and in October 2015 Obama admitted U.S. troops would remain in Afghanistan through 2017. Then, in February 2016, Army general John Campbell, then leader of the U.S. military effort in Afghanistan, told the House Armed Services Committee that American troops would be needed in Afghanistan for at least the next five years.[266]

The truth is that the United States remains (as of this writing) trapped in this forever war in Afghanistan for many reasons. The surge strategy that the U.S. Army tried to implement there was crippled by factors that made this war distinctly different from the war in Iraq of 2007. Additionally, just as in Iraq, the Army outsourced the politics of the war to the State Department and PRTs. (PRTs actually began in Afghanistan and were exported to Iraq.)[267]

But the biggest obstacle to ending the war in Afghanistan is that—just as in Iraq—the Army refused to back a winner. Instead it abdicated its responsibility in the political dimension of the conflict in Afghanistan, handing it over to the State Department. The State Department's strategy to create a sustainable political solution for Afghanistan has been to chase "clear milestones and measures of effectiveness for Afghan reform" to build an inclusive and egalitarian society in Afghanistan that is free of corruption or cronyism and respects the rights of every Afghan.[268] The Pakistani government and the majority of Afghan Pashtuns (constituting nearly half the country's population) are never going to participate peacefully in an Afghan government controlled by America's Afghan allies: the Northern Alliance Tajiks and Uzbeks and their Southern Alliance Durrani Pashtun collaborators. Instead of trying futilely to build a Western-style liberal democracy, the United States should have—from the beginning of the war—accepted this reality, backed its allies (i.e., picked a winner), and either concentrated on building a regime in Kabul capable of occupying and subduing the Pashtun south or accepted a partition of Afghanistan into two countries—Pashtun and non-Pashtun.

Had the U.S. Army arrived in Afghanistan in 2001 with a solid, fully developed and institutionalized low-intensity conflict doctrine, it might have been able to see and implement this clear-eyed strategy for Afghanistan from the outset. But because, for the past two decades, it has been figuring out this war as it went along, the Army has pursued the unachievable goal of an inclusive and representative government,[269] a strategy that instead promises a war without end.

Reluctance to leave Afghanistan is no doubt also born of the disaster that befell Iraq after the departure of U.S. forces. Immediately following the withdrawal of U.S. troops, America's gains seemed to hold. But with the Americans no longer around to restrain him, Iraqi prime minister Nouri al-Maliki launched a campaign of repression against Sunni Arab Iraqis, looking the other way as Iraqi police shot unarmed Sunni protestors, using antiterrorism laws to round up Sunnis in mass arrests, and aligning himself with Shiite militias.[270] The former Sons of Iraq—the foot soldiers of the Awakening movement that began in Al Anbar in 2006 and spread across Sunni Arab Iraq—were a particular target of al-Maliki's persecution. Not surprisingly, Sunni Arab Iraqis resisted. As they had done in the early days of the U.S. occupation, many turned (or turned back) to al-Qaeda in Iraq. By June 2012, hundreds of Iraqis had been killed in bombings across the country and a resurgent al-Qaeda in Iraq was once more on the march.[271]

Just as America was completing its withdrawal from Iraq, an event on the other end of the Arab world that had little to do with the war on terror swept the Middle East. Unrest in Tunisia over years of oppression quickly spread to neighboring Egypt. The world watched, captivated, for eighteen days as a million Egyptians poured into Tahrir Square and other locations across their country, stubbornly resisting threats and violence until, in February 2011, they toppled the thirty-year-old regime of President Hosni Mubarak.[272] This tectonic shift—the collapse of the political order of the most populace country in the Arab world—generated a tsunami that washed over the Middle

East. Protestors poured into the streets of Algeria, Bahrain, Lebanon, Libya, Iran, Iraq, Syria, and many other Muslim countries. The international media dubbed this wave of change the Arab Spring.

America's response to the Arab Spring in Syria was shaped by its thirty-plus years of unrelenting hostility toward Iran. Washington initially saw the burgeoning civil war in Syria as an opportunity to depose the Iranian puppet regime of Bashar al-Assad. But as Assad unleashed his full military might on Syrian rebels, the United States became bogged down in an international debate over Syria's use of chemical weapons against its populace.[273] While it was focused on this threat, the United States missed a much greater threat gathering in Syria's eastern desert.

Al-Qaeda in Iraq was metastasizing. The organization took advantage of the instability caused by the Syrian Civil War to carve out a safe haven for itself from which it could attack the Iraqi government. With Syrian government forces preoccupied fighting for control of population centers in the western part of the country, al-Qaeda in Iraq, under leader Abu Bakr al-Baghdadi, seized the city of Raqqa from rival rebel groups the Free Syrian Army and al-Nusra Front. With its power secure in eastern Syria, al-Qaeda in Iraq cast off its troubled affiliation with the larger al-Qaeda organization and took on a title derived from the name of its former shadow government. It would call itself the Islamic State, but the world would come to call it the Islamic State in Iraq and Syria (ISIS).[274]

This was more than just a change of name; the Islamic State touted its victory in Raqqa as the founding of a new Islamic caliphate. As the new caliph, al-Baghdadi could claim that all Muslims were now obligated by their faith to give him their allegiance. This announcement, largely ignored in the West, drew a flood of foreign fighters from across the Muslim world. Flush with these new fighters, al-Baghdadi seized the southeastern Syrian oil fields, solidifying economic power for his new state.[275]

The new Islamic State then once more turned its attention to Iraq; it began wiping out the former leaders of the Awakening movement that had nearly destroyed al-Qaeda in Iraq during the American occupation. Once this Sunni Arab resistance was wiped out and its safe haven in Syria was secure, ISIS launched its invasion of Iraq. Mosul in northwest Iraq fell almost immediately. Ramadi and much of Al Anbar Province in western Iraq soon followed, as did Tikrit and Samarra.[276] Iraqi security forces broke and ran before this ISIS onslaught by the tens of thousands, leaving their

American-supplied equipment and armored vehicles behind. By the fall of 2015, ISIS had reached Balad and was making preparations for a siege of Baghdad. Just ahead of ISIS's attack toward the oil-rich town of Kirkuk, the Kurdish Peshmerga held a fragile line against the invaders.[277]

The United States initially refused to assist the teetering Iraqi government, but as it became clear that the survival of Iraq was threatened, the Obama administration agreed to send a meager ground force to Baghdad and Irbil on the precondition that Nouri al-Maliki be replaced by a leader more responsive to the needs of all Iraqis, not just Shiite Arabs.[278] Weeks of political wrangling ensued in Baghdad, but these conditions were eventually met and U.S. Army forces began to arrive in Iraq in mid-2014 to resume America's "advise and assist" role there—training Iraqi security forces off of the front lines while providing American airpower to assist Iraqi forces in combat with ISIS.[279]

Iran presented Iraq with no such preconditions for its help. Iranian Revolutionary Guard forces flowed into Iraq and began reforming Iraq's Shiite militias—dormant since the departure of American forces—into military units to confront ISIS. Iraqi security forces, supported by American airpower, waged a campaign of attrition that began to roll back ISIS's gains. Behind them, Iranian-backed Shiite militias waged a campaign of brutal reprisals against Sunni Arabs suspected of aiding ISIS.[280] ISIS was finally able to arrest its retreat outside the walls of Mosul and, after months of fighting, a bloody stalemate emerged.

Despite repeated attempts to extricate the United States from the wars in Afghanistan and Iraq, President Obama handed them off—along with a new war in Syria—to his successor, President Donald J. Trump, who arrived in office with an apparent mandate to dramatically expand the war on terror. Trump had campaigned on resuming waterboarding, promising to "bomb the shit" out of ISIS and kill terrorists' families.[281] Within two weeks of taking office, he made it clear that he fully intended to escalate the war against "radical Islamic terrorists," imposing a controversial ban on immigration into the United States from seven Muslim countries and authorizing a ground raid by Navy SEALs into Yemen.[282]

When President Obama left office, there were eighteen thousand American troops in Afghanistan, Iraq, and Syria. President Trump announced plans in August 2017 to raise the number of troops in Afghanistan alone to fifteen thousand. A little over a year after he took office, there were as many as 10,500 troops in Iraq and Syria—though the Pentagon has been intentionally vague about exact numbers in each country.[283] None of these wars promise to end anytime soon.

Conclusion

The U.S. Army—the most powerful, best-trained, best-educated, and best-equipped army in the history of the world, the same force that had stunned and dazzled the world with its technological and tactical prowess in the Gulf War against the world's fourth largest army—is being humiliated and defeated in the same part of the world only a little more than a decade later by poorly armed and trained guerillas. But the greatest tragedy of all is that this failure did not have to occur.

For two decades before the 1990s, America ignored the lessons of low-intensity conflict—both in Vietnam and in subsequent such conflicts that it fought or observed during this period—instead striving for ever-greater capability to fight a high-intensity conflict against the Warsaw Pact nations in northern Europe. While the Army's senior leaders were perhaps overzealous in setting this direction, it is hard to fault their decision. After all, they were facing the massive, militarily overwhelming Soviet Army, which had threatened the survival of the West for a quarter century.

But then the Soviet Union collapsed, which should have prompted a complete rethinking of this strategic direction. Instead the Gulf War convinced the Army that it should continue to prepare for high-intensity conflicts into the foreseeable future.

In the two decades after the 1990s, the Army suffered the consequences of this obsession with preparation for high-intensity conflict in two decidedly low-intensity conflicts in Afghanistan and Iraq. The Army's initial unpreparedness cost the United States dearly, both in the blood of its service members and in the trillions of dollars it has robbed from the U.S. economy. And the Army's stubborn resistance to fully embracing the lessons of low-intensity conflict, even in the face of continuing military humiliation, has trapped America in two disastrous forever wars from which it cannot extricate itself.

The cost of these wars is tragic. But the true tragedy is that the United States did not have to pay this price. The 1990s offered the Army every reason and ample opportunity to change course. Four major low-intensity conflicts—in Somalia, Haiti, Bosnia-Herzegovina, and Kosovo—should have convinced the Army that the foreseeable future held many more low-intensity conflicts, potentially of even greater scale. And each of these contingencies provided a treasure trove of lessons that the Army could have institutionalized to reshape itself. Instead, it stubbornly ignored these lessons and continued its headlong march toward ever-greater capacity to fight high-intensity conflicts.

The United States continues to pay the price for the U.S. Army's myopia to this day.

NOTES

1. George H. W. Bush, "Radio Address to United States Armed Forces Stationed in the Persian Gulf Region," transcript, American Presidency Project, March 2, 1991, http://www.presidency.ucsb.edu/ws/index.php?pid=19355.

2. Schwarzkopf and Petre, *It Doesn't Take a Hero*; Powell and Persico, *My American Journey*.

3. European Security Study, *Strengthening Conventional Deterrence in Europe*, 86.

4. Fitzgerald, *Learning to Forget*, Kindle, 37.

5. Porch, *Counterinsurgency*, Kindle, 205; Brown, *Kevlar Legions*, Kindle, 15; Krepinevich, *The Army and Vietnam*, Kindle, 2–7, 172–77.

6. Ford, *Tet 1968*, 132; Langguth, *Our Vietnam*, 493, Sorley, *Westmoreland*, 142.

7. Millett and Maslowski, *For the Common Defense*, Kindle, locations 11307–8; Krepinevich, *The Army and Vietnam*, Kindle, 190–98, 252–54.

8. Krepinevich, *The Army and Vietnam*, Kindle, 254; Kitfield, *Prodigal Soldiers*, Kindle, locations 1603–6.

9. Shaw, *The Cambodian Campaign*, 16–17; Herring, *America's Longest War*, 284–85; Proctor, *Containment and Credibility*, Kindle, 313.

10. Millett and Maslowski, *For the Common Defense*, Kindle, locations 11241–48.

11. Proctor, *Containment and Credibility*, Kindle, 394–402.

12. Langguth, *Our Vietnam*, 471–507.

13. Millett and Maslowski, *For the Common Defense*, Kindle, locations 9271–73.

14. Kitfield, *Prodigal Soldiers*, Kindle, locations 754–61, 771–87.

15. Creighton Abrams, quoted in Fitzgerald, *Learning to Forget*, Kindle, 38.

16. Kitfield, *Prodigal Soldiers*, Kindle, locations 337–38.

17. Hunt, *Pacification*, 78–130.

18. Millett and Maslowski, *For the Common Defense*, Kindle, locations 10696–703; Kitfield, *Prodigal Soldiers*, Kindle, locations 1603–6.

19. Sorley, *Westmoreland*, 86, 98–100, 169; Krepinevich, *The Army and Vietnam*, Kindle, 252–54; Millett and Maslowski, *For the Common Defense*, Kindle, locations 10919–24; Kitfield, *Prodigal Soldiers*, Kindle, locations 1409–11; Langguth, *Our Vietnam*, 575.

20. Millett and Maslowski, *For the Common Defense*, Kindle, locations 10258–60; Krepinevich, *The Army and Vietnam*, Kindle, 164–65; Fitzgerald, *Learning to Forget*, Kindle, 21–22.

21. Komer, *Bureaucracy Does Its Thing*, 106–26.

22. Komer, *Bureaucracy Does Its Thing*, 32, 116, 147; Kitfield, *Prodigal Soldiers*, Kindle, locations 1621–30.

23. Fitzgerald, *Learning to Forget*, Kindle, 25.

24. Kitfield, *Prodigal Soldiers*, Kindle, locations 1460–70.

25. Millett and Maslowski, *For the Common Defense*, Kindle, locations 10999–11007; Fitzgerald, *Learning to Forget*, Kindle, 32.

26. Downie, *Learning from Conflict*, 71.

27. Fitzgerald, *Learning to Forget*, Kindle, 51–52.

28. Fitzgerald, *Learning to Forget*, Kindle, 52–53.

29. Gole, *General William E. DePuy*, Kindle, locations 2160–69; Kitfield, *Prodigal Soldiers*, Kindle, locations 850–54, 885–88.

30. Fitzgerald, *Learning to Forget*, Kindle, 31.

31. Record, *The Wrong War*, 124–31.

32. Proctor, *Containment and Credibility*, Kindle, 95.

33. Proctor, *Containment and Credibility*, Kindle, 306.

34. Fitzgerald, *Learning to Forget*, Kindle, 38, 203; Hagopian, *The Vietnam War in American Memory*, 33; Kitfield, *Prodigal Soldiers*, Kindle, locations 1667–70; Shimko, *The Iraq Wars*, Kindle, locations 959–64; Richard Nixon, quoted in Associated Press, "Support of Nation Sought in Search for Peace," *Spokane Daily Chronicle*, November 4, 1969.

35. Fitzgerald, *Learning to Forget*, Kindle, 41, 56.

36. Fitzgerald, *Learning to Forget*, Kindle, 39–40, Kitfield, *Prodigal Soldiers*, Kindle, locations 854–62.

37. Kitfield, *Prodigal Soldiers*, Kindle, locations 850–54, 865–66, 868–72.

38. Kitfield, *Prodigal Soldiers*, Kindle, locations 1703–8, 2147–51.

39. Fitzgerald, *Learning to Forget*, Kindle, 39–40.

40. Fitzgerald, *Learning to Forget*, Kindle, 42.

41. Herzog and Gazit, *The Arab-Israeli Wars*, 225–325, esp. 306; Dunstan, *The Yom Kippur War*, 91–92.

42. Swain, "AirLand Battle," 368.

43. Kitfield, *Prodigal Soldiers*, Kindle, locations 2109–11, 2077–82, 2088–96.

44. Kitfield, *Prodigal Soldiers*, Kindle, locations 2073–77, 2107–9, 2111–13, 2120–23, 2088–96.

45. Shimko, *The Iraq Wars*, Kindle, locations 1062–79.

46. Kitfield, *Prodigal Soldiers*, Kindle, locations 2117–18; Herbert, *Deciding What Has to Be Done*, 79–85; Headquarters, U.S. Department of the Army, *Operations*, FM 100-5 (1976), 3-9, 5-7, 5-13, 14-20.

47. Headquarters, U.S. Department of the Army, FM 100-5 (1976), 1-1, 1-2, emphasis in the original.

48. Fitzgerald, *Learning to Forget*, Kindle, 43, 45.

49. Fitzgerald, *Learning to Forget*, Kindle, 53.

50. Summers, *On Strategy*, Kindle, locations 116–19.

51. Clausewitz, *On War*, Kindle, 75.

52. Summers, *On Strategy*, Kindle, location 1526.

53. Fitzgerald, *Learning to Forget*, Kindle, 53, 54–55.

54. Fitzgerald, *Learning to Forget*, Kindle, 55.

55. Kitfield, *Prodigal Soldiers*, Kindle, locations 1775–76; Shimko, *The Iraq Wars*, Kindle, locations 966–77.

56. Kitfield, *Prodigal Soldiers*, Kindle, locations 1814–17, 2780–85.

57. Kitfield, *Prodigal Soldiers*, Kindle, locations 2648–57; Brown, *Kevlar Legions*, Kindle, 16.

58. Kitfield, *Prodigal Soldiers*, Kindle, locations 2014–20, 2037–43, 2780–85.

59. Millett and Maslowski, *For the Common Defense*, Kindle, locations 11419–25, 11504–6; Kitfield, *Prodigal Soldiers*, Kindle, locations 2819–24.

60. Bowden, *Guests of the Ayatollah*, 3–7.

61. Kitfield, *Prodigal Soldiers*, Kindle, locations 2811–13, 2819–24, 3227–32.

62. Moyar, *Oppose Any Foe*, 157–92; Kitfield, *Prodigal Soldiers*, Kindle, locations 3082–85, 3090–99, 3127–31, 3211–15.

63. Busch, *Reagan's Victory*, 126; Kitfield, *Prodigal Soldiers*, Kindle, locations 3265–66; Millett and Maslowski, *For the Common Defense*, Kindle, locations 11520–27.

64. Kitfield, *Prodigal Soldiers*, Kindle, locations 3436–44; Brown, *Kevlar Legions*, Kindle, 16; White House, "U.S. National Security Strategy," National Security Decision Directive no. 32, Federation of American Scientists, May 20, 1982, https://fas.org/irp/offdocs/nsdd/nsdd-32.pdf; White House, "Basic National Security Strategy," National Security Decision Directive No. 238, Federation of American Scientists, September 2, 1986, https://fas.org/irp/offdocs/nsdd/nsdd-238.pdf.

65. Kitfield, *Prodigal Soldiers*, Kindle, locations 3227–32.

66. Shimko, *The Iraq Wars*, Kindle, locations 966–77.

67. Trybula, *"Big Five" Lessons for Today and Tomorrow*, iv; Kitfield, *Prodigal Soldiers*, Kindle, locations 3854–64.

68. Clancy and Franks, *Into the Storm*, Kindle, 152–71.

69. Kitfield, *Prodigal Soldiers*, Kindle, locations 2197–200, 2207–13.

70. Kitfield, *Prodigal Soldiers*, Kindle, locations 2240–42, 2224–28; Shimko, *The Iraq Wars*, Kindle, locations 1096–1104; Kitfield, *Prodigal Soldiers*, Kindle, locations 2233–37.

71. Kitfield, *Prodigal Soldiers*, Kindle, locations 2240–42, 2667–74, 4305–6.

72. Shimko, *The Iraq Wars*, Kindle, locations 1109–18; Brown, *Kevlar Legions*, Kindle, 16, 88.

73. Shimko, *The Iraq Wars*, Kindle, locations 946–53.

74. Fitzgerald, *Learning to Forget*, Kindle, 110.

75. Shimko, *The Iraq Wars*, Kindle, locations 247–52, 255–57.

76. Headquarters, U.S. Department of the Army, *Operations*, FM 100-5 (1982), i.

77. Shimko, *The Iraq Wars*, Kindle, locations 1265–73.

78. Kitfield, *Prodigal Soldiers*, Kindle, locations 4480–81; Headquarters, U.S. Department of the Army, *Operations*, FM 100-5 (1982), 2-9, 1-2, 2-2.

79. Fitzgerald, *Learning to Forget*, Kindle, 46–47; Birtle, *U.S. Army Counterinsurgency and Contingency Operations Doctrine*, 452.

80. Fitzgerald, *Learning to Forget*, Kindle, 55–56.

81. Millett and Maslowski, *For the Common Defense*, Kindle, locations 11602–8, 12124–34; Leonard et al., *Encyclopedia of U.S.-Latin American Relations*, 86–87, 203. See also Norton and Weiss, "Rethinking Peacekeeping."

82. Fitzgerald, *Learning to Forget*, Kindle, 61, 69–70.

83. Hines, *Joint Task Force—Bravo*, 8; Director of Central Intelligence, *El Salvador: Government and Insurgent Prospects*, 14; Travis, *Reagan's War on Terrorism in Nicaragua*, 47–53; Greg Guma, "CIA Covert Ops in Central America: Nicaragua and the Road to Contra-Gate," Global Research, June 27, 2013, http://www.globalresearch.ca/nicaragua-and-the-road-to-contra-gate/5340646.

84. Hines, *Joint Task Force—Bravo*, 8; Millett and Maslowski, *For the Common Defense*, Kindle, locations 11608–14, Guma, "CIA Covert Ops in Central America."

85. Proctor, *Containment and Credibility*, Kindle, 404–5; Peter Collins, Sam Donaldson, Charles Gibson, and Frank Reynolds, "El Salvador / United States Aid," *ABC Evening News*, Vanderbilt Television News Archive, March 10, 1983, http://tvnews.vanderbilt.edu/program.pl?ID=83041; LeoGrande, *Our Own Backyard*, 586; Sam Donaldson, Frank Reynolds, Max Robinson, and John Scali, "CIA/Casey," *ABC Evening News*, Vanderbilt Television News Archive, July 15, 1981, http://tvnews.vanderbilt.edu/program.pl?ID=72626; Bill Lynch and

Lesley Stahl, "El Salvador," *CBS Evening News*, Vanderbilt Television News Archive, February 5, 1983, http://tvnews.vanderbilt.edu/program.pl?ID=288889.

86. Ramsey, *Advising Indigenous Forces*, 83–85. See also Schwartz, *American Counterinsurgency Doctrine and El Salvador*.

87. Fitzgerald, *Learning to Forget*, Kindle, 68.

88. Millett and Maslowski, *For the Common Defense*, Kindle, locations 11649–61.

89. Stewart, *Operation Urgent Fury*, 5.

90. Millett and Maslowski, *For the Common Defense*, Kindle, locations 11649–61.

91. Kitfield, *Prodigal Soldiers*, Kindle, locations 3740–53.

92. Blake Stilwell, "That Time a Soldier Used a Payphone to Call Back to the US to Get Artillery Support in Grenada," We Are the Mighty, May 16, 2017, http://www.wearethemighty.com/articles/that-time-a-soldier-used-a-payphone-to-call-back-to-the-us-to-get-artillery-support-in-grenada; there is still considerable debate as to which services were involved or whether this incident even occurred at all.

93. Stewart, *Operation Urgent Fury*, 32; Kitfield, *Prodigal Soldiers*, Kindle, locations 3741–45, 3725–29.

94. Millett and Maslowski, *For the Common Defense*, Kindle, locations 9025–27.

95. Kitfield, *Prodigal Soldiers*, Kindle, locations 3278–85, 3343–46.

96. Kitfield, *Prodigal Soldiers*, Kindle, locations 3459–63.

97. Kitfield, *Prodigal Soldiers*, Kindle, locations 3984–90, 4130–32; Clancy and Franks, *Into the Storm*, Kindle, 628.

98. Clancy and Franks, *Into the Storm*, Kindle, 147–49; Kitfield, *Prodigal Soldiers*, Kindle, locations 5057–62.

99. Brown, *Kevlar Legions*, Kindle, 42–43.

100. Fitzgerald, *Learning to Forget*, Kindle, 79; Marquis, *Unconventional Warfare*, 112–26, 149–215; Millett and Maslowski, *For the Common Defense*, Kindle, locations 11684–87; Kitfield, *Prodigal Soldiers*, Kindle, locations 4189–94.

101. Fitzgerald, *Learning to Forget*, Kindle, 79; Kitfield, *Prodigal Soldiers*, Kindle, locations 4189–94; Hunt, "OOTW: A Concept in Flux," 3–10.

102. National Defense Authorization Act for Fiscal Year 1987, Pub. L. No. 99–661, November 14, 1986, Title XIII, Part B; Headquarters, U.S. Department of the Army, *Low Intensity Conflict*, FC 100–20, i–viii; Brown, *Kevlar Legions*, Kindle, 43.

103. Fitzgerald, *Learning to Forget*, Kindle, 72, 82; Hunt, "OOTW: A Concept in Flux," 3–10; Headquarters, U.S. Department of the Army, *Operations*, FM 100-5 (1986), 1, 4.

104. U.S. Army Combined Arms Center, *1987 Annual Historical Review*, 99–108.

105. Fitzgerald, *Learning to Forget*, Kindle, 68, 73; Armed Forces News Service, "The Center for Low Intensity Conflict Closes after 10 Years," Federation of American Scientists, June 26, 1996, https://fas.org/irp/news/1996/n19960626_960615.html.

106. Hunt, "OOTW: A Concept in Flux," 3–10.

107. Carl E. Vuono to Gerald T. Bartlett, March 13, 1987, detailing the creation of the Army–Air Force Center for Low Intensity Conflict and directing that it lead the effort to create FC 100-20, frame 326, reel 47670, microfilm collection, Records of the Army–Air Force Center for Low Intensity Conflict, U.S. Air Force Historical Research Agency; U.S. Army Combined Arms Center, *1987 Annual Historical Review*, 99–108; Army–Air Force Center for Low Intensity Conflict, "A-AF CLIC Weekly Update, 4–8 December 1989," frames 877–79, reel 47671, microfilm collection, Records of the Army–Air Force Center for Low Intensity Conflict, U.S. Air Force Historical Research Agency; Hunt, "OOTW: A Concept in Flux," 3–10; Albert M. Barnes, "FM 100-20 / AFM 2-XY Read Ahead," memorandum for commanding

general Maxwell Thurman, U.S. Army Training and Doctrine Command, on comments on FC 100-20, May 19, 1988, frames 726-27, reel 47670, microfilm collection, Records of the Army–Air Force Center for Low Intensity Conflict, U.S. Air Force Historical Research Agency.

108. Army–Air Force Center for Low Intensity Conflict, "Back Channel," memorandum drafted for commanding general Maxwell Thurman, U.S. Army Training and Doctrine Command, to send to chief of staff of the Army, Gen. Carl Vuono, about obstacles to publication of FM 100-20, n.d. [October 27, 1988], frames 1109-10, reel 47670, microfilm collection, Records of the Army–Air Force Center for Low Intensity Conflict, U.S. Air Force Historical Research Agency; William F. Furr, "FM 100-20 / AFM 2-20 Chronology of Events," October 27, 1988, frames 1185-92, reel 47670, microfilm collection, Records of the Army–Air Force Center for Low Intensity Conflict, U.S. Air Force Historical Research Agency; Headquarters, U.S. Departments of the Army and the Air Force, *Military Operations in Low Intensity Conflict*, FM 100-20, i.

109. Fitzgerald, *Learning to Forget*, Kindle, 82.

110. U.S. Army Combined Arms Center, *1987 Annual Historical Review*, 99–108, 351; Steele, "JRTC Concept Is Set," 37–41; Fitzgerald, *Learning to Forget*, Kindle, 82; Clifton Everton, "Trip Report," memorandum on trip to Joint Readiness Training Center at Fort Chafee, AR, August 14–16, 1989, August 21, 1989, frames 630-31, reel 47670, microfilm collection, Records of the Army–Air Force Center for Low Intensity Conflict, U.S. Air Force Historical Research Agency; Charles M. Ayers "Trip Report," memorandum on meeting at Fort Leavenworth, KS, on May 10, 1988, about the proposed Joint Readiness Training Center, May 11, 1988, frames 588–89, reel 47670, Records of the Army–Air Force Center for Low Intensity Conflict, U.S. Air Force Historical Research Agency.

111. Ayers, "Trip Report."

112. Fitzgerald, *Learning to Forget*, Kindle, 48; U.S. Army Combined Arms Center, *1987 Annual Historical Review*, 63.

113. U.S. Army Command and General Staff College, *United States Army Command and General Staff College Catalog, Academic Year 1988–1989*, CGSC Circular No. 351-1, May 1988, Command and General Staff College Papers, Ike Skelton Combined Arms Research Library, 60, 65, 67; Fitzgerald, *Learning to Forget*, Kindle, 48.

114. U.S. Combined Arms Center, *1987 Annual Historical Review*, 84–86.

115. Brown, *Kevlar Legions*, Kindle, 36.

116. Clancy and Franks, *Into the Storm*, Kindle, 128–49, Kitfield, *Prodigal Soldiers*, Kindle, locations 2927–30, 2959–64, 4264–69.

117. Brown, *Kevlar Legions*, Kindle, 35, 36, 93.

118. Kitfield, *Prodigal Soldiers*, Kindle, locations 3890–93; Millett and Maslowski, *For the Common Defense*, Kindle, locations 11662–74, 11759–61.

119. Millett and Maslowski, *For the Common Defense*, Kindle, locations 11846–59.

120. Crane, "Peace Dividends and Benevolent Interventions."

121. Millett and Maslowski, *For the Common Defense*, Kindle, locations 11846–59.

122. Brown, *Kevlar Legions*, Kindle, 52.

123. Yates, *The U.S. Military Intervention in Panama*, 241–71; Millett and Maslowski, *For the Common Defense*, Kindle, locations 11846–60; Cole, *Operation Just Cause*, 5; Crane, "Peace Dividends and Benevolent Interventions"; Fishel and Downie, "Taking Responsibility for Our Actions?," 66–72.

124. Crane, "Peace Dividends and Benevolent Interventions."

125. Crane, "Peace Dividends and Benevolent Interventions"; Fishel and Downie, "Taking Responsibility for Our Actions?," 66–72.

126. Fishel and Downie, "Taking Responsibility for Our Actions?," 66–72; Crane, "Peace Dividends and Benevolent Interventions."

127. Crane, "Peace Dividends and Benevolent Interventions."

128. Barnes, "The Diplomat Warrior," 55–63.

129. Barnes, "The Diplomat Warrior," 55–63.

130. Craig and Logevall, *America's Cold War*, Kindle, 322–50.

131. Craig and Logevall, *America's Cold War*, Kindle, 322–50.

132. Millett and Maslowski, *For the Common Defense*, Kindle, locations 11791–804.

133. Millett and Maslowski, *For the Common Defense*, Kindle, locations 11805–9, 11810–13.

134. Engel, *When the World Seemed New*, 440–52; Butler, "Adjusting to Post–Cold War Strategic Realities," 2–9; Vuono, "National Strategy and the Army of the 1990s," 2–12.

135. Dubik, "On the Foundations of National Military Strategy," 21–46; Barnes, "The Diplomat Warrior," 55–63.

136. *Public Papers of the President of the United States: George Bush, 1992–1993*, 1:182–87.

137. Brown, *Kevlar Legions*, Kindle, 56; Millett and Maslowski, *For the Common Defense*, Kindle, locations 11861–73.

138. Shimko, *The Iraq Wars*, Kindle, locations 1591–1601.

139. Brown, *Kevlar Legions*, Kindle, 56–57.

140. Clancy and Franks, *Into the Storm*, Kindle, 271; Brown, *Kevlar Legions*, Kindle, 57; Millett and Maslowski, *For the Common Defense*, Kindle, locations 11895–11907, Shimko, *The Iraq Wars*, Kindle, locations 1622, 2381–401.

141. Kitfield, *Prodigal Soldiers*, Kindle, locations 5110–13; Shimko, *The Iraq Wars*, Kindle, locations 1634–39, 1658–63, 1670–85, 1688–93, 1697–1703, 1717–19.

142. Millett and Maslowski, *For the Common Defense*, Kindle, locations 11996–12008.

143. Shultz and Pfaltzgraff, *The Future of Air Power*, 157, 22, 31, 33; Shimko, *The Iraq Wars*, Kindle, locations 1652–55, 1775–79; Brown, *Kevlar Legions*, Kindle, 58–59.

144. Millett and Maslowski, *For the Common Defense*, Kindle, locations 12073–91; Brown, *Kevlar Legions*, Kindle, 59–60; Shimko, *The Iraq Wars*, Kindle, locations 1730–36, 2014–18.

145. Millett and Maslowski, *For the Common Defense*, Kindle, locations 12073–91, 12017–31.

146. Kitfield, *Prodigal Soldiers*, Kindle, locations 5936–40.

147. Hess, *Presidential Decisions for War*, 217–26.

148. Millett and Maslowski, *For the Common Defense*, Kindle, locations 11922–29; Shimko, *The Iraq Wars*, Kindle, locations 2994–98, 2973–91.

149. Center for Army Lessons Learned, "Desert Storm Lessons Learned," slides for a briefing to the chief of staff of the Army, Gen. Gordon R. Sullivan, Fort Leavenworth, KS, August 1991, Combined Arms Center Historical Archive, U.S. Army Combined Arms Center.

150. Vuono, "National Strategy and the Army of the 1990s," 2–12.

151. Kitfield, *Prodigal Soldiers*, Kindle, locations 5186–90; Center for Army Lessons Learned, "Desert Storm Lessons Learned," slides for a briefing to the chief of staff of the Army, Gen. Gordon R. Sullivan, Fort Leavenworth, KS, August 1991, Combined Arms Center Historical Archive, U.S. Army Combined Arms Center; Center for Army Lessons Learned, "Desert Shield/Storm Lessons."

152. Millett and Maslowski, *For the Common Defense*, Kindle, locations 11954–68; Summers, *New World Strategy*, 131; Center for Army Lessons Learned, "Desert Storm Lessons Learned."

153. Vuono, "National Strategy and the Army of the 1990s," 2–12.

154. Fitzgerald, *Learning to Forget*, Kindle, 88.

155. Center for Army Lessons Learned, "Desert Storm Lessons Learned," emphasis in the original.

156. Shimko, *The Iraq Wars*, Kindle, locations 2334–38, 175–77, 2697–703; Tom Clancy, Horner, and Koltz, *Every Man a Tiger*, Kindle, 359; Scales, *Certain Victory*, 355–84; Kitfield, *Prodigal Soldiers*, Kindle, locations 5568–72; Brown, *Kevlar Legions*, Kindle, 61.

157. Brown, *Kevlar Legions*, Kindle, 61; Shimko, *The Iraq Wars*, Kindle, locations 2738–40; Center for Army Lessons Learned, "Desert Storm Lessons Learned"; Center for Army Lessons Learned, "Desert Shield/Storm Lessons"; Shimko, *The Iraq Wars*, Kindle, locations 2218–24, 2707–8.

158. Shimko, *The Iraq Wars*, Kindle, locations 2738–40.

159. Center for Army Lessons Learned, "Desert Storm Lessons Learned"; Kitfield, *Prodigal Soldiers*, Kindle, locations 5491–95; Shimko, *The Iraq Wars*, Kindle, locations 2214–18.

160. Bush, "Radio Address."

161. "Operation Enduring Freedom: Fatalities by Year and Month," iCasualties.org, n.d., accessed December 4, 2017, http://icasualties.org/OEF/ByMonth.aspx; "U.S. Casualties in Iraq," Global Security, n.d., accessed December 4, 2017, https://www.globalsecurity.org/military/ops/iraq_casualties.htm.

162. U.S. Government Accounting Office, *Combating Terrorism*, 17–19.

163. "Text of Bush's Address," CNN, September 11, 2001, http://edition.cnn.com/2001/US/09/11/bush.speech.text/.

164. Associated Press, "Congress Reopens; Retaliation Is Urged," *Reading (PA) Eagle*, September 13, 2001.

165. "Transcript of President Bush's Address," CNN, September 21, 2001, http://www.cnn.com/2001/US/09/20/gen.bush.transcript/.

166. "Taleban Threatens to Retaliate," BBC, September 15, 2001, http://news.bbc.co.uk/2/hi/south_asia/1544957.stm; "Bin Laden: America 'Filled with Fear,'" CNN, October 7, 2001, http://archives.cnn.com/2001/WORLD/asiapcf/central/10/07/ret.binladen.transcript/index.html; "Retaliation: Second Night of Attacks," CNN, October 8, 2001, http://archives.cnn.com/2001/U.S./10/08/ret.retaliation.facts/index.html.

167. National Commission on Terrorist Attacks upon the United States, *The 9/11 Commission Report*, 330–34; Melissa Block, "Armitage Denies Making 'Stone Age' Threat," *All Things Considered*, NPR, September 22, 2006, http://www.npr.org/templates/story/story.php?storyId=6126088; "Retaliation: U.S., Britain Open Attack," CNN, October 7, 2001, http://archives.cnn.com/2001/U.S./10/07/ret.retaliation.facts/index.html; Christiane Amanpour, "Amanpour: Multiple Targets Hit in Afghanistan," CNN, October 7, 2001, http://archives.cnn.com/2001/WORLD/meast/10/07/amanpour.otsc/index.html.

168. Quinn-Judge, "How the Northern Alliance Plans to Win the War"; Amy Belasco, 'Troop Levels in the Afghan and Iraq Wars, FY2001–FY2012: Cost and Other Potential Issues," Congressional Research Service, July 2, 2009, http://www.fas.org/sgp/crs/natsec/R40682.pdf, 9.

169. Karon, "Can the Northern Alliance Control Kabul?"; Patrick Healy and Farah Stockman, "Taliban Flee Kandahar," *Boston Globe*, December 8, 2001, http://www.boston.com/news/packages/sept11/anniversary/timeline/stories/war_dec_07.htm.

170. Millett and Maslowski, *For the Common Defense*, Kindle, locations 12791–92.

171. Naylor, *Not a Good Day to Die*, 11–13; Fitzgerald, *Learning to Forget*, Kindle, 126.

172. Michael Isikoff and David Corn, *Hubris: The Inside Story of Spin, Scandal, and the Selling of the Iraq War*, 120–122; Philip Smucker, "How bin Laden Got Away," *Christian Science*

Monitor; Shimko, *The Iraq Wars*, Kindle, locations 3937–48. See also Rothstein, "America's Longest War."

173. Brown, *Kevlar Legions*, Kindle, 250; Millett and Maslowski, *For the Common Defense*, Kindle, locations 12933–40. See also Rothstein, "America's Longest War."

174. See Naylor, *Not a Good Day to Die*, 8–183; and Walling, *Enduring Freedom, Enduring Voices*.

175. Shimko, *The Iraq Wars*, Kindle, locations 3907–15; Khalil, "The Tangled History."

176. George W. Bush, "Text of President Bush's 2002 State of the Union Address," *Washington Post*, January 29, 2002, http://www.washingtonpost.com/wp-srv/onpolitics/transcripts/sou 012902.htm.

177. Norton, "The Loyal Jirga"; Smucker, "New Afghan Leader Faces a Rogues Gallery Government"; "Security Council Commends Afghan People on Successful Conduct of Emergency Loya Jirga," United Nations Security Council, June 26, 2002, https://www.un.org /press/en/2002/sc7435.doc.htm.

178. "Security Council Holds Iraq in 'Material Breach' of Disarmament Obligations, Offers Final Chance to Comply," United Nations Security Council, August 11, 2002, https://www .un.org/press/en/2002/SC7564.doc.htm.

179. Lydia Saad, "Top Ten Findings About Public Opinion and Iraq," Gallup, October 8, 2002, http://www.gallup.com/poll/6964/top-ten-findings-about-public-opinion-iraq.aspx; Michael Elliott, "Bush Isn't as Lonely as He Looks," *Time*, September 9, 2002, http://www.time.com /time/columnist/elliott/article/0,9565,349390,00.html; Authorization for Use of Military Force against Iraq Resolution of 2002, Pub. L. No. 107-243, October 16, 2002, https://www.govinfo .gov/content/pkg/PLAW-107publ243/pdf/PLAW-107publ243.pdf

180. MacLeod, "Arabs to Cheney: 'Curb Sharon before Saddam'"; MacLeod, "Time Exclusive: The Saudi Initiative Explained"; "Turkey Grants Overflight Rights to U.S.," CNN, March 23, 2003, http://edition.cnn.com/2003/WORLD/meast/03/23/sprj.irq.turkey .overflights/index.html.

181. Thalif Deen, "U.N. Agreed on Aid Role but Skirted Political Minefield," Inter Press Service, March 31, 2003, http://www.ipsnews.net/interna.asp?idnews=17188; Jim Lobe, "Poll Finds U.S. Public Rallying to Bush and Supporting U.N.," Inter Press Service, March 31, 2003, http://www.ipsnews.net/interna.asp?idnews=17187.

182. Millett and Maslowski, *For the Common Defense*, Kindle, locations 13018–22, 13022–23.

183. Millett and Maslowski, *For the Common Defense*, Kindle, locations 13039–53, 13068–81.

184. Fontenot, Degen, and Tohn, *On Point*, 80, 93–95, 100, 336; Ferguson, *No End in Sight*, 581–82; Millett and Maslowski, *For the Common Defense*, Kindle, locations 13091–97.

185. Millett and Maslowski, *For the Common Defense*, Kindle, locations 13097–109.

186. Shimko, *The Iraq Wars*, Kindle, locations 4284–89, 4290–96.

187. Fontenot, Degen, and Tohn, *On Point*, 80, 93–95, 100, 336; Ferguson, *No End in Sight*, 581–82.

188. Fontenot, Degen, and Tohn, *On Point*, 80, 93–95, 100, 336; Ferguson, *No End in Sight*, 581–82; Gordon and Trainor, *Cobra II*, Kindle, locations 9580–9616.

189. Shimko, *The Iraq Wars*, Kindle, location 5031.

190. Benson, "OIF Phase IV," 61–68.

191. Benson, "OIF Phase IV," 61–68; Shimko, *The Iraq Wars*, Kindle, locations 4349–56; Millett and Maslowski, *For the Common Defense*, Kindle, locations 13147–51.

192. Ricks, *Fiasco*, 161–66; Fontenot, Degen, and Tohn, *On Point*, 100; Shimko, *The Iraq Wars*, Kindle, location 5017; Amatzia Baram, *Who Are the Insurgents? Sunni Arab Rebels in Iraq*, Special Report 134 (Washington, DC: United States Institute of Peace, April 2005), https://www.usip.org/sites/default/files/sr134.pdf, 1–3.

193. Fitzgerald, *Learning to Forget*, Kindle, 135.

194. Associated Press, "Coalition Detaining about 200 Non-Iraqi Prisoners in Iraq," Fox News, May 8, 2003, accessed January 23, 2011, http://www.foxnews.com/story/0,2933,86398,00 .html; Associated Press, "Bush: Syria Must Not Harbor Iraqi Leaders," *USA Today*, April 14, 2003; New York Times, "U.S. Faces Foreign Fighters: Officials Say Resistance Appears to Be Organized," *Post and Courier* (Charleston, SC), June 22, 2003.

195. Cong. Rec., 107th Congress, 2nd Session., 2008, 148, pt. 1: 2250.

196. Fitzgerald, *Learning to Forget*, Kindle, 136; Fred Burton and Scott Stewart, "Jihadist Ideology and the Targeting of Humanitarian Aid Workers," Stratfor, October 22, 2008, http:// www.stratfor.com/weekly/20081022_jihadist_ideology_and_targeting_humanitarian_aid _workers; Associated Press, "UN Pulls Out of Baghdad after Bombings," *Sydney Morning Herald*, October 31, 2003, http://www.smh.com.au/articles/2003/10/31/1067233354087.html.

197. Ricks, *Fiasco*, 214–69; Dexter Filkins, "The Fall of the Warrior King," *New York Times*, October 23, 2005, http://www.nytimes.com/2005/10/23/magazine/23sassaman.html; "The Lost Year in Iraq," *Frontline*, PBS, October 17, 2006, http://www.pbs.org/wgbh/pages/fron tline/yeariniraq/etc/synopsis.html.

198. Ricks, *Fiasco*, 214–69; Filkins, "The Fall of the Warrior King"; "The Lost Year in Iraq"; Fitzgerald, *Learning to Forget*, Kindle, 139.

199. "The Toll of War in Iraq: U.S. Casualties and Civilian Deaths," NPR, n.d., accessed January 23, 2011, http://www.npr.org/news/specials/tollofwar/tollofwarmain.html; Steve Inskeep, "Bush: Saddam's Capture Ends Lawless Era," *All Things Considered*, NPR, December 14, 2003, http://www.npr.org/templates/story/story.php?storyId=1548212.

200. Associated Press, "Violence Strikes Iraq's Sunni Triangle," Fox News, March 31, 2004, http://www.foxnews.com/story/0,2933,115703,00.html; West, *No True Glory*, 100–126.

201. "Fallujah," Global Security, n.d., accessed April 9, 2006, http://www.globalsecurity .org/military/world/iraq/fallujah.htm; "Fighting in Fallujah," *PBS NewsHour*, PBS, April 30, 2004,https://www.pbs.org/newshour/show/fighting-in-fallujah; West, *No True Glory*, 256–62; Kaplan, "Five Days in Fallujah"; Millett and Maslowski, *For the Common Defense*, Kindle, locations 13290–301.

202. Ricks, *Fiasco*, 378–80.

203. Sebastian Usher, "Arab Media Question Abuse Trial," BBC, May 20, 2004, http://news .bbc.co.uk/2/hi/middle_east/3733585.stm; Rawe, "Iraq: The Sad Tale of Nick Berg."

204. Lisa Myers, "Who Are the Foreign Fighters in Iraq?," *NBC Nightly News*, June 20, 2005, http://www.nbcnews.com/id/8293410/ns/nbc_nightly_news_with_brian_williams-nbc_news _investigates/t/who-are-foreign-fighters-iraq/.

205. Mariam Karouny, "Iraq's SCIRI Party to Change Platform—Officials," Reuters, May 11, 2007, http://www.reuters.com/article/topNews/idUSYAT15330920070511; Al Marashi, "Iraq," 226; Edward Wong, "U.S. Begins First Major Assault on Iraqi Militia Led by Cleric," *New York Times*, May 5, 2004, http://www.nytimes.com/2004/05/05/international/middleeast /05CND-IRAQ.html; Luke Harding, "Sacred Shi'a Site Damaged as Tanks Move into Najaf," *Guardian*, May 15, 2004, http://www.guardian.co.uk/world/2004/may/15/iraq.lukeharding.

206. Fitzgerald, *Learning to Forget*, Kindle, 139; Mansoor, *Surge*, Kindle, 1–19.

207. Fitzgerald, *Learning to Forget*, Kindle, 144.

208. Bremer, *My Year in Iraq*, Kindle, 98; Millett and Maslowski, *For the Common Defense*, Kindle, locations 13280–90.

209. Shimko, *The Iraq Wars*, Kindle, location 5343; Melissa Block, "Attack on Shrine Is an Attack on Shiite Faith," *All Things Considered*, NPR, February 22, 2006, http://www.npr.org /templates/story/story.php?storyId=5228624.

210. Woodward, *The War Within*, 35–36. See also Dodge, "Grand Ambitions and Far-Reaching Failures."

211. Elhadj, *The Islamic Shield*, 183; Manzoor Alam, *War on Terrorism*, 222–26; Millett and Maslowski, *For the Common Defense*, Kindle, locations 13311–25,13221–23.

212. Elhadj, *The Islamic Shield*, 183; Alam, *War on Terrorism*, 222–26.

213. Millett and Maslowski, *For the Common Defense*, Kindle, locations 13248–50, 13188–221.

214. Fitzgerald, *Learning to Forget*, 143, 11; Millett and Maslowski, *For the Common Defense*, Kindle, locations 13327–31.

215. McGirk and Ware, "Losing Control?"

216. Millett and Maslowski, *For the Common Defense*, Kindle, locations 13392–95, 13419–22.

217. Millett and Maslowski, *For the Common Defense*, Kindle, locations 13419–22, 13436–49.

218. Waller, "Bush and Musharraf: Friends Again"; Baker, "Pakistan Braces for a Backlash after Taliban Raid"; Kilcullen, *Counterinsurgency*, 23.

219. Millett and Maslowski, *For the Common Defense*, Kindle, locations 13450–51.

220. "Abu Musab al-Zarqawi," *Telegraph*, June 9, 2006, http://www.telegraph.co.uk/news /obituaries/1520703/Abu-Musab-al-Zarqawi.html; Klein, "Why Bush Is (Still) Winning the War at Home"; "Bush Takes Blame for GOP Election Losses," CNN, November 8, 2006, http:// www.cnn.com/2006/POLITICS/11/08/election.bush/index.html.

221. Brown, *Kevlar Legions*, Kindle, 252, 253–254.

222. Brown, *Kevlar Legions*, Kindle, 256, 260; Millett and Maslowski, *For the Common Defense*, Kindle, locations 13250–55.

223. Brown, *Kevlar Legions*, Kindle, 320; Fitzgerald, *Learning to Forget*, Kindle, 152.

224. David H. Petraeus, "Battling for Iraq," *Washington Post*, September 26, 2004, quoted in Shimko, *The Iraq Wars*, Kindle, location 5373.

225. Shimko, *The Iraq Wars*, Kindle, locations 5072, 5221.

226. Shimko, *The Iraq Wars*, Kindle, location 5123.

227. Brown, *Kevlar Legions*, Kindle, 277.

228. Brown, *Kevlar Legions*, Kindle, 277, 323.

229. Fitzgerald, *Learning to Forget*, Kindle, 138, 150.

230. Fitzgerald, *Learning to Forget*, Kindle, 151.

231. Ricks, *Fiasco*, 420–24.

232. Fitzgerald, *Learning to Forget*, Kindle, 161.

233. Fitzgerald, *Learning to Forget*, Kindle, 161, Mansoor, *Surge*, Kindle, 137–46.

234. Fitzgerald, *Learning to Forget*, Kindle, 119.

235. Fitzgerald, *Learning to Forget*, Kindle, 119, 139.

236. Crane, *Cassandra in Oz*, Kindle, 43–60, 232.

237. Headquarters, U.S. Department of the Army, *Counterinsurgency*, FM 3-24, 2-2, 1-23.

238. Fitzgerald, *Learning to Forget*, Kindle, 166, 169; Headquarters, U.S. Department of the Army, FM 3-24, 1-10.

239. Headquarters, U.S. Department of the Army, FM 3-24, 5-1–5-2, 12-5–12-6.

240. Headquarters, U.S. Department of the Army, FM 3-24, 3-12.

241. Headquarters, U.S. Department of the Army, FM 3-24, 5-10.

242. David E. Sanger, "Bush Adds Troops in Bid to Secure Iraq," *New York Times*, January 11, 2007, http://www.nytimes.com/2007/01/11/world/middleeast/11prexy.html? sq=bush%20 baghdad%20anbar%20january%202007&st=cse&scp=10&pagewanted=all.

243. Associated Press, "Text of Gen. Petraeus' Letter to U.S. Forces," NBC News, September 7, 2007, http://www.nbcnews.com/id/20648010/ns/world_news-mideast_n_africa/t/text-gen -petraeus-letter-us-forces/; Duffy, "The Surge At Year One"; Fitzgerald, *Learning to Forget*, Kindle, 175.

244. Kagan, *The Surge*, 27–59.

245. Fitzgerald, *Learning to Forget*, Kindle, 178.

246. U.S. Department of Defense, *Measuring Stability and Security in Iraq*, ii–vii; "Iraq," Polling Report, n.d., accessed April 19, 2009, http://www.pollingreport.com/iraq.htm.

247. Sara Wood, "Petraeus Takes Command of Multinational Force Iraq," U.S. Department of Defense, February 12, 2007, http://archive.defense.gov/news/newsarticle.aspx?id=3032; "U.S. Casualties in Iraq"; "Operation Enduring Freedom."

248. "Bush Positive on U.S., Iraq Deal for U.S. Troop Withdrawal by 2012," Fox News, August 22, 2008, https://www.foxnews.com/story/bush-positive-on-u-s-iraq-deal-for-u-s -troop-withdrawal-by-2012.

249. "Biography: David Howell Petraeus," U.S. Central Command, n.d., accessed January 7, 2018, http://www.centcom.mil/ABOUT-US/LEADERSHIP/Bio-Article-View /Article/904777/david-howell-petraeus/; Abbaszadeh et al., "Provincial Reconstruction Teams," 4; "Provincial Reconstruction Teams in Iraq," U.S. Institute of Peace, March 20, 2013, https:// www.usip.org/publications/2013/03/provincial-reconstruction-teams-iraq; Proctor, *Task Force Patriot*, Kindle, locations 358, 367.

250. Naland, *Lessons from Embedded Provincial Reconstruction Teams in Iraq*, 1–6; Proctor, *Task Force Patriot*, Kindle, location 2989.

251. Miguel Marquez, "Finally! Iraqis Agree to Form Government Eight Months after Election," ABC News, November 11, 2010, http://abcnews.go.com/International /iraq-policiticians-finally-reach-deal-form-government/story?id=12119665&page=1.

252. Anthony Shadid, "Iraq's Last Patriot," *New York Times*, February 4, 2011, http://www .nytimes.com/2011/02/06/magazine/06ALLAWI-t.html; Marquez, "Finally!"

253. "Iraq Gripped by New Sunni-Led Insurgency," United Press International, November 8, 2010, http://www.upi.com/Top_News/Special/2010/11/08/Iraq-gripped-by-new-Sunni-led -insurgency/UPI-68131289234796/; "Last Man Standing: Saddam's Longtime Veep," United Press International, November 4, 2010, http://www.upi.com/Top_News/Special/2010/11/04 /Last-man-standing-Saddams-longtime-veep/UPI-11881288897426/; Regional Reachback Center Iraq Support Team, *Post-SOFA Motivations of Iraqi Insurgents*.

254. Sky, *The Unraveling*, Kindle, 322.

255. Tom Vanden Brook, "U.S. Formally Declares End of Iraq War," *USA Today*, December 15, 2011, http://usatoday30.usatoday.com/news/world/iraq/story/2011-12-15/Iraq -war/51945028/1.

256. Millett and Maslowski, *For the Common Defense*, Kindle, locations 13372–76.

257. Thompson, "Obama Weighs the Cost of an Afghan Surge"; Baker, "TIME's Interview with General Stanley McChrystal"; Barack Obama, "Remarks by the President in Address to the Nation on the Way Forward in Afghanistan and Pakistan," White House Office of the Press Secretary, December 1, 2009, https://obamawhitehouse.archives.gov

/the-press-office/remarks-president-address-nation-way-forward-afghanistan-and-pakistan; Fitzgerald, *Learning to Forget*, Kindle, 188.

258. Obama, "Remarks"; Hastings, "The Runaway General."

259. Lansford, *A Bitter Harvest*, 15–19.

260. Shimko, *The Iraq Wars*, Kindle, locations 3907–15; Rashid, *Descent into Chaos*, Kindle, 369–73.

261. Asad Hashim, "Pakistan Parliament Passes Landmark Tribal Areas Reform," Al Jazeera, May 24, 2018, https://www.aljazeera.com/news/2018/05/pakistan-parliament-passes -landmark-tribal-areas-reform-180524111258832.html; Rashid, *Descent into Chaos*, Kindle, 349–73; Saeed Shah, Adam Entous, and Gordon Lubold, "U.S. Threatens to Withhold Pakistan Aid," *Wall Street Journal*, August 21, 2015, http://www.wsj.com/articles/u-s -threatens-to-withhold-pakistan-aid-1440163925.

262. Kagan, *The Surge*, 27–59; Bacevich, *America's War for the Greater Middle East*, Kindle, locations 5865–911.

263. See Moyar, *Village Stability Operations and the Afghan Local Police*.

264. Steven Lee Myers and Elisabeth Bumiller, "Obama Calls World 'Safer' after Pakistan Raid," *New York Times*, May 2, 2011, http://www.nytimes.com/2011/05/03/world/asia/osama -bin-laden-dead.html?hp=&pagewanted=print; National Commission on Terrorist Attacks upon the United States, *The 9/11 Commission Report*, 330–34; Block, "Armitage Denies Making 'Stone Age' Threat"; "Retaliation: Second Night of Attacks"; Amanpour, "Multiple Targets Hit in Afghanistan."

265. Mark Landler, "U.S. Troops to Leave Afghanistan by End of 2016," *New York Times*, May 27, 2014, https://www.nytimes.com/2014/05/28/world/asia/us-to-complete-afghan-pullout-by -end-of-2016-obama-to-say.html; Baker, "TIME's Interview with General Stanley McChrystal"; Obama, "Remarks."

266. Eyder Peralta, "Obama Plans to Leave Residual Force of 9,800 in Afghanistan," NPR, May 27, 2014, http://www.npr.org/sections/thetwo-way/2014/05/27/316320567/obama-plans-to -leave-residual-force-of-9-800-in-afghanistan; VOA News, "Obama Won't Draw Down US Troops in Afghanistan in 2015," Voice of America, March 24, 2015, https://www.voanews.com /usa/obama-wont-draw-down-us-troops-afghanistan-2015; Matthew Rosenberg and Michael D. Shear, "In Reversal, Obama Says U.S. Soldiers Will Stay in Afghanistan to 2017," *New York Times*, October 15, 2015, http://www.nytimes.com/2015/10/16/world/asia/obama-troop-withdraw al-afghanistan.html?_r=0; Corey Dickstein, "Campbell Says US Troops Needed in Afghanistan for at Least 5 More Years," *Stars and Stripes*, February 2, 2016, http://www.stripes.com/news /campbell-says-us-troops-needed-in-afghanistan-for-at-least-5-more-years-1.391804.

267. McNerney, "Stabilization and Reconstruction in Afghanistan," 32–46. Admittedly, while similarly staffed principally by officials from other U.S. government agencies, PRTs in Afghanistan were led by the military.

268. Cordesman, *The Obama Strategy in Afghanistan*, 16–22.

269. Mark Landler, "The Afghan War and the Evolution of Obama," *New York Times*, January 1, 2017, https://www.nytimes.com/2017/01/01/world/asia/obama-afghanistan-war.html.

270. Zack Beauchamp, "18 Things about ISIS You Need to Know," *Vox*, November 17, 2015, https:// www.vox.com/world/2018/11/20/17995812/isis-islamic-state-18-things-you-need-to-know.

271. Tim Arango, "Iraqi Sunnis Frustrated as Awakening Loses Clout," *New York Times*, May 3, 2010, http://www.nytimes.com/2010/05/04/world/middleeast/04awakening.html?_r =1&scp=1&sq=sons%20of%20iraq&st=cse; "Has Sectarian Violence Returned to Iraq?" Al Jazeera, June 18, 2012, http://www.aljazeera.com/programmes/insidestory/2012/06/201 26186242422818.html.

272. Agence France-Presse, "'He Brought Fire to the Arab World': Tunisia Protest Icon's Mother Shares Her Son with the World," *Sydney Morning Herald*, March 5, 2011, http://www.smh.com.au/world/he-brought-fire-to-the-arab-world-tunisia-protest-icons-mother-shares-her-son-with-the-world-20110305-1biny.html; Jackie Northam, "Mubarak's Fall Spurs Calls To Rethink U.S. Policy," *Morning Edition*, NPR, February 15, 2011, http://www.npr.org/2011/02/15/133763952/mubaraks-fall-spurs-calls-for-u-s-policy-rethink.

273. "Syria's Chemical Weapons Stockpile," BBC, January 30, 2014, http://www.bbc.com/news/world-middle-east-22307705.

274. Remnick, "Telling the Truth About ISIS and Raqqa."

275. Remnick, "Telling the Truth About ISIS and Raqqa."

276. David Ignatius, "How ISIS Spread in the Middle East, And How to Stop It"; Tim Arango, "Uneasy Alliance Gives Insurgents an Edge in Iraq," *New York Times*, June 18, 2014, http://www.nytimes.com/2014/06/19/world/middleeast/former-loyalists-of-saddam-hussein-crucial-in-helping-isis.html?_r=0; "ISIS Executed Some of Its Former Baathist Allies in Iraq," *Business Insider*, April 7, 2015, http://mobile.businessinsider.com/isis-executed-some-of-its-former-baathist-allies-in-iraq-2015-4.

277. Associated Press, "U.S.- Supplied Equipment Abandoned by Iraqi Troops in Ramadi," CBS News, May 19, 2015, http://www.cbsnews.com/news/u-s-supplied-equipment-abandoned-by-iraqi-troops-in-ramadi/; Joshua Berlinger, "Who Are the Religious and Ethnic Groups under Threat from ISIS?" CNN, August 8, 2014, http://www.cnn.com/2014/08/08/world/meast/iraq-ethnic-groups-under-threat-isis/; Priyanka Boghani, "Can the Kurds Hold Out against ISIS?" *Frontline*, PBS, August 5, 2014, http://www.pbs.org/wgbh/frontline/article/can-the-kurds-hold-out-against-isis/.

278. Patrick Cockburn, "Iraq Crisis Exclusive: Us Rules Out Military Action until Prime Minister Nouri al-Maliki Stands Down," *Independent* (London), June 19, 2014, http://www.independent.co.uk/news/world/middle-east/iraq-crisis-exclusive-us-rules-out-military-action-until-pm-nouri-al-maliki-stands-down-9547311.html.

279. Michael R. Gordon and Julie Hirschfeld Davis, "In Shift, U.S. Will Send 450 Advisers to Help Iraq Fight ISIS," *New York Times*, June 10, 2015, http://www.nytimes.com/2015/06/11/world/middleeast/us-embracing-a-new-approach-on-battling-isis-in-iraq.html.

280. James Gordon Meek, Brian Ross, Rym Momtaz, and Alex Hosenball "'Dirty Brigades': US-Trained Iraqi Forces Investigated for War Crimes," ABC News, March 11, 2015, http://abcnews.go.com/International/dirty-brigades-us-trained-iraqi-forces-investigated-war/story?id=29193253.

281. Mahanta, "Fighting Terrorism in the Age of Trump."

282. David Jackson, "Trump Memorializes Navy SEAL Killed in Raid in Yemen at Dover Base," *USA Today*, February 1, 2017, http://www.usatoday.com/story/news/politics/2017/02/01/donald-trump-us-navy-seal-team-six/97351640/; Angela Dewan and Emily Smith, "What It's Like in the 7 Countries on Trump's Travel Ban List," CNN, January 30, 2017, http://www.cnn.com/2017/01/29/politics/trump-travel-ban-countries/.

283. Greg Myre, "Under Trump, U.S. Troops in War Zones Are on the Rise," NPR, December 1, 2017, https://www.npr.org/sections/parallels/2017/12/01/566798632/under-trump-u-s-troops-in-war-zones-are-on-the-rise.

2 SOMALIA, HAITI, AND FORCE XXI

WHEN GEN. GORDON R. SULLIVAN assumed the office of chief of staff, the U.S. Army was still basking in the glory of the Gulf War. Nevertheless, it also faced a number of looming issues. The Soviet Union, the focus of U.S. national security and foreign policy for the previous half century, had collapsed. The U.S. Congress was eager to reap the peace dividend to address America's burgeoning national debt. And the national security establishment was wondering on what, exactly, the Army should focus in the last decade of the millennium.

One would have thought that given Sullivan's background—as a veteran of the low-intensity conflict in Vietnam and an erstwhile supporter of more emphasis on low-intensity conflict doctrine—he would have taken a more balanced approach to building future Army capabilities. And the two major low-intensity conflicts that erupted during his tenure—in Somalia and Haiti—should have encouraged him to consider rebuilding the Army's capacity to engage in such conflicts.

Ultimately, however, Sullivan did not choose this course. While he made a few tentative moves in this direction—presiding over the development of some new low-intensity conflict doctrine and establishing the Peacekeeping Institute—he otherwise maintained the Army's laser focus on high-intensity conflict. He spearheaded the so-called modern Louisiana Maneuvers and a program of Advanced Warfighting Experiments that ushered in the era of Force XXI: an envisioned high-tech reshaping of the Army for an unequivocally high-intensity-conflict future.

General Gordon R. Sullivan

Gen. Gordon R. Sullivan was a 1959 graduate of Norwich University. He entered the Army as an officer in the armor branch, and he served two tours in Vietnam, but was particularly shaped by his first tour, 1962–63, during which he was wounded while serving as an adviser to the Civil Guard and

Self-Defense Corps—the precursors to the Regional Force and Popular Force, respectively, and both of which would be the focus of the Army of the Republic of Vietnam's (ARVN's) local security force training effort once the United Stated began its direct military intervention in the conflict.[1]

Sullivan had carried away from this experience a conviction that he and his fellow soldiers had been insufficiently prepared for the rigors of low-intensity conflict they faced in Vietnam.[2] As he would say many years later, based on his Vietnam experience as an adviser, "I don't think as an Army we prepared those guys—and I was one of them!" This experience would predispose Sullivan to take low-intensity conflict seriously when he arrived in 1987 at Fort Leavenworth, Kansas, as a major general to serve as the deputy commandant of the U.S. Army Command and General Staff College (CGSC).[3]

In his time as the deputy commandant at the CGSC, Major General Sullivan would preside over the Army's intellectual reaction to the Nunn-Cohen Amendment to the 1987 National Defense Authorization Act, which assigned primacy for low-intensity conflict to special operations forces (SOF), implicitly relieving the Army of this responsibility. Before Sullivan arrived as commandant, the Army had just taken its first tentative steps since the end of the Vietnam War toward greater doctrinal appreciation of low-intensity conflict. Fort Leavenworth had just produced the 1986 edition of Field Manual (FM) 100-5, *Operations*, which, for the first time since the Vietnam War, acknowledged that the Army would have to fight "on the unique battlefields of Low Intensity Conflict." The Department for Joint and Combined Operations (DJCO) at the CGSC had published the interim Field Circular (FC) 100-20, *Low-Intensity Conflict*.[4] While this circular was by no means perfect—most notably emphasizing the primacy of a host nation in the political dimensions of low-intensity conflict—it did acknowledge for the first time since the end of the Vietnam War that U.S. ground troops might be required to take a direct role in a conflict. And beyond Fort Leavenworth, in March 1986 the Joint Chiefs of Staff established the Army–Air Force Center for Low Intensity Conflict (A-AFCLIC) to study this form of warfare.[5]

In April 1987, a month after Sullivan arrived at Fort Leavenworth to assume his new duties, Congress passed the Nunn-Cohen Amendment as part of the 1987 National Defense Authorization Act. The law was primarily focused on establishing U.S. Special Operations Command with power analogous to the services to man, train, and equip SOF. But the law also implicitly transferred primacy for low-intensity conflict from the U.S. Army and U.S. Marine Corps to SOF.[6]

Yet Sullivan fought to force the Army to place more focus on low-intensity conflict. In May 1987 he invited the head of U.S. Southern Command, Gen. John R. Galvin—at the time overseeing the Army's support to insurgency operations in Nicaragua and counterinsurgency efforts in El Salvador—to speak to the student body. Exceeding the deference generally afforded to commanders-in-chief of regional commands, Sullivan also invited Galvin to bring key members of his Small Wars Operations Research Division, and, along with his team, to interact with CGSC students in smaller seminars. In May–June 1987 Sullivan brought together the DJCO and members of the U.S. Army School of Advanced Military Studies (SAMS), which had just added a block on low-intensity conflict to its curriculum, to discuss the new doctrine. Their conclusions challenged the Army orthodoxy that the tenets of AirLand Battle were universal and equally applicable to high- and low-intensity conflict.[7]

Sullivan's most radically move, however, was to suggest that low-intensity conflict be elevated to a coequal status with high-intensity conflict in doctrine, and that the upcoming FM 100-20, the manual that would describe such conflict, be coequal in precedence with the venerated FM 100-5, *Operations*, which enshrined AirLand Battle doctrine. Even before Nunn-Cohen relieved the Army of responsibility for low-intensity conflicts, this idea was anathema to an Army that had spent more than a decade ignoring the lessons of Vietnam and subsequent such conflicts in its pursuit of ever-greater capability to fight high-intensity conflicts. Many within the CGSC disagreed with this approach. In discussions between the DJCO and SAMS, significant disagreements arose on the role that the Army should play in low-intensity conflicts.[8]

Not surprisingly, the idea received a cool reception when Sullivan suggested it to General Carl Vuono in a briefing just before Vuono assumed office as chief of staff of the Army. Gen. Maxwell Thurman, who replaced Vuono as the commander of U.S. Army Training and Doctrine Command (TRADOC), also disagreed with many of Sullivan's ideas on low-intensity conflict. Vuono and Thurman blocked the effort to elevate low-intensity conflict doctrine to coequal status with high-intensity conflict doctrine. Still, Sullivan did convince Thurman to allow the DJCO to include in FM 100-20 "imperatives" for low-intensity conflict that were distinct from the tenets of AirLand Battle.[9]

This was not enough to placate Sullivan. At the end of July 1987, the CGSC held a tabletop wargame for the TRADOC leadership, Warfighting Seminar III, that was TRADOC's first ever low-intensity conflict exercise. In

December 1987 the CGSC's journal, the *Military Review*, published an article by Army colonel Richard Swain—who had developed the low-intensity
conflict curriculum for SAMS—to begin what was intended to be a yearlong
public debate on the idea of replacing the joint term *low-intensity conflict*
with an Army-specific term *operations short of war*, an idea that General
Thurman had already roundly rejected.[10]

Sullivan wasn't done. He bucked common Army wisdom that a force
optimized for high-intensity conflict would necessarily be adequate for the
lesser included task of low-intensity conflict; he added low-intensity conflict
capacity as a separate requirement in the Concept-Based Requirements
System, a bureaucratic mechanism that the Army used to determine its
future manning, organization, and training requirements.[11]

Sullivan's rebellion eventually began to infect the other senior Army
leaders at Fort Leavenworth. In a memorandum to the deputy commanding
general of the U.S. Army Combined Arms Center (CAC), Maj. Gen. Charles
Otstott, Sullivan praised Maj. Andrew Krepinevich's 1986 book *The Army
and Vietnam*. (This book—in diametric opposition to the Army orthodoxy
of Colonel Harry Summers's *On Strategy*—concluded that the Army had
lost in Vietnam because it neglected counterinsurgency in favor of fighting
a high-intensity conflict.) In March 1988 Sullivan convinced CAC commander Lt. Gen. Gerald Bartlett to endorse a request to General Thurman
for a half million dollars to study the development of a low-intensity conflict
capability within the Army.[12]

This appears to have been the last straw for the senior leaders of the Army.
General Thurman wrote a letter to General Vuono in which he expressed
concern with the direction that Sullivan and the DJCO were headed. He was
especially concerned that the new FM 100-20 might violate the spirit of the
Nixon Doctrine that had guided Army perceptions on low-intensity conflict
for nearly two decades. Thurman was adamant that "our military forces sent
to assist in counterinsurgency operations do not relieve the host nation of
their sovereign responsibility for resolving those issues which often prompt
the insurgency at its beginning."[13]

Chief of Staff Vuono directed Bartlett to stop work on FM 100-20 and
outsource its production to the newly formed A-AFCLIC, essentially taking
Fort Leavenworth out of the low-intensity conflict debate. Sullivan was
"kicked upstairs" to command the 1st Infantry Division at Fort Riley, Kansas,
in June 1988. Bartlett was replaced a month later as CAC commander by Lt.
Gen. Leonard P. Wishart III, who immediately set about bringing the Army's
low-intensity conflict doctrine back into line with the thinking of the senior

leadership. Wishart refused to endorse A-AFCLIC's version of FM 100-20, *Military Operations in Low Intensity Conflict*, until it more explicitly emphasized the "host nation responsibility" in low-intensity conflicts.[14] The chief of staff of the Army himself also directed A-AFCLIC to more forcefully emphasize the "indirect employment of US forces in LIC [low-intensity conflict]." During the manual's development, General Thurman likewise pressured A-AFCLIC to more explicitly emphasize the Nixon Doctrine in the manual. Thurman later held up publication because he wanted low-intensity conflict better tied to the tenets of AirLand Battle[15]—a decidedly high-intensity conflict doctrine. Senior Army leaders ultimately stalled publication of the manual until 1990.[16]

Today General Sullivan, now retired, says that Vuono and Thurman "didn't come down really hard" and there was not "any heavy-duty" chastisement over Sullivan's activities at the time. Yet he remembers that the "subtle climatics" made it abundantly clear that both men were "ambivalent" about the idea of elevating the importance of low-intensity conflict in Army doctrine. Sullivan remembers concluding at the time that Generals Vuono and Thurman didn't want any changes: "They wanted it to be what it had been." Vuono and Thurman preferred that high-intensity conflict as it was envisioned in the doctrine of AirLand Battle—and enshrined in FM 100-5—remain preeminent.[17]

Appropriately chastened at the time, Sullivan stopped talking about the topic of low-intensity conflict publicly or privately. That seemed to satisfy senior leaders of the Army and kept Sullivan's career on track; he was promoted twice, serving as the deputy chief of staff for operations and plans, and then the vice chief of staff of the Army before assuming the post of chief of staff of the Army on June 23, 1991.[18]

General Sullivan came to office in a time of turmoil for the U.S. Army and, in fact, the entire national security establishment. The Warsaw Pact had dissolved in November of the previous year and, two months after he took office, the Soviet Union passed into history.[19]

Congress was eager to reap a "peace dividend" from the end of the Cold War by slashing defense budgets. During the presidential campaign, President George H. W. Bush and Arkansas governor Bill Clinton sparred over who would slash defense budgets more deeply, with Bush promising $50 billion in cuts and Clinton promising $60 billion.[20]

Vuono, Sullivan's predecessor as chief of staff of the Army, had in many respects already set the course before Sullivan arrived. Faced with dwindling budgets and the demise of the Soviet Union, Vuono had to make a decision. On the one hand, he could stop the Army's modernization efforts. He could also slash junior manpower while retaining more senior soldiers, sustaining a cadre to serve as the foundation for an expensible Army, which had been the model before World War II. His third option was to cut whole units from the Army to reduce its size while continuing to modernize—building a leaner, smaller, force capable of smaller contingency operations. Cautious after the recent experience of Operation Just Cause in Panama and Operation Desert Storm in Iraq and Kuwait—which had each presented the Army with sudden, unanticipated, short-notice requirements for Army forces ready for deployment—Vuono and his planners chose this third option.[21]

Yet General Vuono's insistence on maintaining a continued high-intensity conflict capability was not only born of a fear of future small-scale, no-notice interventions but also a belief that a great power war was still a very real possibility. As he left office Vuono warned that war with the Russians still loomed: "History teaches us that the collapse of great empires seldom takes place without great upheaval . . . revolutionary changes in regimes, however benignly they may begin, often quickly dissolve into massive conflict." Even if a great power war were not to happen, Vuono still believed that the future demanded conventional deterrence: "Clearly, we have learned a key lesson of history—that poorly trained armies invite attack by enemies, incur casualties needlessly and ultimately suffer defeat."[22] Sullivan echoed Vuono's sentiments when he succeeded him in office.

The Gulf War played a prominent role in justifying continued force structure. The chairman of the Joint Chiefs of Staff, General Colin Powell, argued that the United States needed to maintain sufficient forces to execute a high-intensity conflict on the scale of the Gulf War, with additional manpower for extraneous peacekeeping or contingency missions. He argued for 1.6 million active duty personnel, but Congress only acquiesced to 1.4 million. This represented a 600,000 reduction from the levels available for the Gulf War.[23]

Ultimately, Congress cut much more deeply. And the Army felt the brunt of these reductions; its active force drew down from 770,000 in 1989 to under 495,000 in 1996. It reduced from sixteen to twelve active divisions and eliminated one of two corps in Germany.[24]

Not surprisingly given the ambiguity as to their purpose, the light divisions were the hardest hit in the Army's force reductions. They had initially been envisioned as low-intensity conflict forces, but their purpose gradually migrated toward operation in high-intensity conflict—a purpose for which they were woefully ill suited in comparison to the Army's heavy divisions.

Four of the Army's six light infantry divisions were cut, with deactivation beginning in 1992. Only the 10th Mountain Division and the 25th Infantry Division (Light) would survive the cuts.[25]

At the same time the Army was getting smaller, demands upon it were growing dramatically. Bill Clinton succeeded George H. W. Bush as president and came to office with a National Security Strategy of Engagement and Enlargement that would lead to a dramatic expansion of the Army's roles. President Clinton sought to exploit the opportunities created by the end of the Cold War to expand liberal democracy around the world.[26] In practice for the U.S. Department of Defense, this meant a much higher tempo of deployments to smaller contingency operations, many of which were low-intensity conflicts. Between 1989 and 1993, the U.S. military would participate in no fewer than forty-eight named operations—ranging in scale from the Gulf War and the conflict in Somalia to humanitarian relief after Hurricane Andrew and Operation Provide Comfort, which provided relief to Kurds in northern Iraq in 1991.[27]

Yet everything about Sullivan's outlook betrayed a fear that the Army would be caught unprepared for some impending, cataclysmic great power war. Throughout his tenure, the general kept two books on his desk: Col. Robert Doughty's *The Development of French Army Doctrine, 1919–1939* and *America's First Battles, 1776–1965*, edited by Lt. Col. Charles Heller and Brig. Gen. William Stofft.[28] The theme of both books was, of course, that failing to prepare for the next war or preparing for the wrong war brought dire consequences.

Speaking upon his assumption of duties as chief of staff on June 25, 1991, he explained that there would be no "time-out from readiness." Thus, Sullivan now clearly saw his task as executing the many contingencies that the Army might have to face while simultaneously continuing to modernize it toward ever-increasing readiness for Gulf War–style high-intensity conflicts. Yet in Sullivan's mind this did not just entail sustaining current capabilities. He believed that "the environment in which the Army operates is undergoing fundamental transformation" and that the Army had to change with it. His challenge, as he put it, was to "reshape the Army while maintaining readiness."[29]

Throughout the 1970s and 1980s, FM 100-5 had come to embody the Army's aspirations for its future capabilities. In the 1970s it had enshrined Active Defense. In the 1980s it had introduced the idea of AirLand Battle, which would find its full expression in the Gulf War. Thus it was natural that, to encapsulate his post–Cold War vision for the future, General Sullivan would seek to update FM 100-5 again. To lead this effort, he again looked to SAMS. A team led by the school's director, Army colonel James McDonough, set about rewriting the manual.[30]

Colonel McDonough and his team carried Sullivan's sense of alarm for the future into this writing effort. In an update on the writing of the manual in the pages of the *Military Review*, McDonough repeated the oft-cited caution from Sir Michael Howard against being "so wrong" in preparing in peacetime as to invite defeat when war finally comes. McDonough wrote that his team was mindful of the example of the "Russians in 1914 at Tannenberg, the French in 1940 at Seine and even the Americans in 1943 at the Kasserine Pass." The new FM 100-5 would reflect the sentiment that "by the first battle it is already too late."[31]

One new idea to emerge from McDonough and his team was the threat of low-tech adversaries acquiring high-tech weapons. In fact, this theme became a consistent refrain among senior Army leaders throughout the 1990s. As McDonough would write in 1991, "Even in Third World contingency operations, it will not be uncommon for our forces to face high-technology systems in the hands of an enemy." McDonough saw this problem as getting worse over time: "Even though we might have an initial advantage, the interim between the fielding of a new technology and its counter is rapidly diminishing."[32] This sense of impending crisis added urgency to the Army's need to transform itself.

Predicting the future of warfare was something of a booming industry in the early 1990s. The Institute for National Strategic Studies at the National Defense University (NDU) was only one of the agencies trying to predict what the future of warfare would look like. In its Project 2025 the institute focused on the "substantial military-technological advances" around the world and sought to "employ informed and disciplined imagination as its method, and plausibility as its criterion, to envisage events that could lead to a radically altered future." In illustrating the difficulty of trying to predict the future—but also betraying the high-intensity conflict bias of its outlook—the NDU asked the rhetorical question, "How predictable was Hitler in the early 1920s?"[33] Threats that this team was imagining were not radical at all. In fact, they were decidedly conventional and high-intensity in nature.

Project 2025 did acknowledge that "for the foreseeable future, we will not have a rival like the Soviet Union or the Axis powers—states gripped by totalitarian ideologies and equipped with large and sophisticated armed forces." But the project still found a reason for the U.S. military to continue to prepare for high-intensity conflict: "If the United States need not protect itself against a peer competitor in the immediate future, it will need military power to shape the calculations of possible peer challengers in decades to come." Moreover, if the United States did not continue to exploit the technologies of the revolution in military affairs (RMA), these "peer challengers" would be able to catch up with and surpass the current U.S. technological advantage. Project 2025 concluded, "As technologies developed by the civilian sector continue to outpace those developed by the military, other countries may be able to match us in certain key areas."[34]

In a paper Sullivan cowrote with Lt. Col. James Dubik that was presented to a February 1993 conference of the U.S. Army War College, the two explained their vision of the future. They warned that the primary adversary against which the Army needed to prepare was a theoretical future peer competitor who would acquire the same technological means that the United States had used to win the Gulf War: "Those who would consider threatening U.S. global interests are hard at work buying the hardware that they will need and learning their lessons from the Gulf War."[35]

For that reason, the Army had to continue to prepare for a future high-intensity conflict that would be dramatically altered by the RMA. Particularly, Sullivan and Dubik listed five "revolutions" that would change the face of modern land warfare: "lethality and dispersion; volume and precision of fire; integrative technology; mass and effects; and invisibility and detectability." Together these revolutions would make the Army faster, more precise, and more lethal. But these revolutions would also allow "smaller land forces [to] create decisive effects IF technology is used by high-quality, well-trained and well-led troops employing proper doctrine." Mindful of those who might see this as a reason to reap yet more peace dividends, they added, "there is a line below which technology can no longer compensate for cuts in force structure."[36]

To shape the doctrine, organization, and equipment to realize his vision of harnessing the RMA to stay ahead of hypothetical future competitors, Sullivan initiated a major experimentation effort. The biggest of these efforts, started in March 1992, was the Louisiana Maneuvers Task Force, organized under a brigadier general at Fort Monroe, Virginia and colocated with TRADOC.[37] The effort took its name from the massive exercise a half century earlier in which the Army trained and shaped itself in preparation for World War II.[38] But rather than tromping troops through the swamps of Louisiana, this task force organized a dizzying array of distributed and interconnected experiments—computer simulated and live—at posts across the Army. To protect the experimentation from institutional parochialism—and congressional politics—these experiments were overseen by a General Officer Working Group, a sort of board of directors chaired by the chief of staff of the Army and attended by the commanders of the Army's major commands—U.S. Army Forces Command (FORSCOM) and TRADOC chief among them.[39]

Perhaps the biggest initial insight to come out of the Louisiana Maneuvers—which made its way directly into FM 100-5—was the need for better deployability of the Army. The insight began as a lesson from the Gulf War, where the 82nd Airborne Division had sat precariously on the border between Kuwait and Saudi Arabia for an uncomfortably long time awaiting the arrival of follow-on forces.[40]

But this insight into the Army's need for greater deployability was also informed by the growing realization that the next crisis that required Army forces might occur suddenly and in an unexpected location. The Louisiana Maneuvers would address this problem with the concept of "power projection": U.S.-based forces rapidly deployable around the world through increased strategic mobility assets and pre-positioning of materiel configured to unit sets (POMCUS) in key locations. The Army sought the equivalent of the Cold War–era return of forces to Germany (REFORGER), but anywhere in the world and on a moment's notice.[41]

The Army set for itself the goal of being able to deploy a corps headquarters, one light division, and two heavy divisions anywhere in the world within thirty days and a full corps with five divisions within seventy-five days.[42] These were tall orders for a U.S. military that had only recently taken six months to build a similar force for the Gulf War.

Low-Intensity Conflict before Somalia

While most of the Army had remained focused on growing its capacity to deploy to and fight high-intensity conflicts, there was a nascent conversation still taking place about low-intensity conflict in the early 1990s. The two centers of this conversation within the Army were A-AFCLIC and the Low Intensity Conflict Proponency Office of the CAC at Fort Leavenworth.

During this period, a point of convergence between these low-intensity conflict observers and those advocating a continued focus on high-intensity conflict began to emerge; both camps believed that the future held more such conflicts. Two of the founding "assumptions" of A-AFCLIC at its inception in 1986 were that "conventional war, guerilla warfare, and terrorism will coexist in the future" and "the United States must possess a capability to confront such forms of conflict where U.S. interests are threatened." In May 1990 Lt. Col. Rudolph Barnes Jr. of the U.S. Army Reserve wrote that the collapse of the Soviet Union had created a "multipolar free-for-all in which former allies and adversaries now compete for world power and influence" and "the focus of competition among world economic powers is likely to be in the Third World." Looking to America's history since the end of World War II, Lt. Col. James Holt reached a similar conclusion. U.S. Air Force general George Butler, director of the Joint Staff J-5, cited Joseph Nye in the pages of *Parameters*, claiming that the collapse of the Soviet Union had released ethnic tensions suppressed by the bipolar international order and that, combined with the "failures in the human condition in the Third World," this would result in an increasing number of low-intensity conflicts.[43] At an April 1991 low-intensity conflict conference at Fort Leavenworth, Jerome Klingaman, a senior research fellow at the U.S. Air Force Center for Aerospace Doctrine, Research, and Education, argued that while anticommunism would no longer prompt intervention in low-intensity conflicts, "mounting instability in the developing nations will continue to generate insurgency" that would threaten U.S. or allied interests, prompting intervention.[44]

Most of those who believed that the Army should remain focused on high-intensity conflict also acknowledged the likelihood of future low-intensity conflicts. The NDU's Project 2025, which generally predicted the emergence of high-intensity, peer competitors to the United States, did still see a proliferation of low-intensity conflicts. The project acknowledged that the Army would continue to have to engage in these conflicts, and saw "a forcible change in regime by coup" in Third World countries as increasingly likely in a post–Cold War world. Members of the project also believed that, with the end of the Cold War, "a likely feature of the developing world in 2025" would be governments using subversion—supporting insurgencies in neighboring countries—as a cheap alternative to impose their will on neighbors.[45]

Even the chief of staff of the Army agreed that there would be an increase in the incidence of low-intensity conflicts in which the Army would be forced to participate.[46] The paper written by General Sullivan and Lieutenant Colonel Dubik noted that, with the collapse of the Soviet Union, the U.S. Army could expect to continue to be committed to "regional crises that require collective applications of military power in 'operations other than war.' These include humanitarian relief, peacekeeping, peace-enforcement, and peace-building."[47] Elsewhere Sullivan wrote, "Events around the world—in Somalia, Korea, Haiti, Kuwait, and the former Yugoslavia—amply demonstrate the post–Cold War expansion, in category, frequency and geographic dispersion, of missions we may be called upon to accomplish."[48]

Colonel McDonough, leader of the team writing the new FM 100-5, agreed. He wrote that the future held a "resurgence of ethnic animosities, national strife, contentious border disputes, aggressive religious fundamentalism and a growing number of regional instabilities."[49]

Where low-intensity conflict observers began to diverge with conventional Army thought was in their beliefs as to why such conflict would be prevalent in the future. Simply put, many of these observers believed that great power war was obsolete. Lieutenant Colonel Barnes wrote that "superpowers have . . . avoided direct military confrontation" in part because of the "awesome military destructive capabilities [of] weapons of mass destruction." He believed that the existence of nuclear weapons would preclude conventional war into the foreseeable future, but the great powers' pursuit of their interests would result in an increase in low-intensity conflicts. He concluded, "The decreasing likelihood of conventional war in Europe" had made "massive troop commitments . . . all but irrelevant." Holt seemed to agree with this assessment; in the epigraph to his March 1990 *Military Review* article, he quoted 1950s French counterinsurgency observer Colonel Roger Trinquier: "We still persist in studying a type of warfare that no longer exists and that we shall never fight again, while we pay only passing attention to the war we lost in Indochina and the one we are about to lose in Algeria." Admittedly, Project 2025, which predicted an array of high-tech challenges to the Army's dominance of high-intensity conflict, did acknowledge that "we live in an age in which wars between and within states outside the developing world have virtually ceased." The project claimed this was because of "fear of nuclear war, economic entanglement, [and] the desire to avoid another World War II."[50] But most who advocated greater preparation for high-intensity conflict did not come to this same conclusion.

Another area of divergence between low-intensity conflict observers and those who advocated a continuing focus on high-intensity conflict was the competing prescriptions from these two camps for the increasing likelihood of future such conflicts. For these observers, the likelihood of such conflicts in the future—and the increasing unlikelihood of future high-intensity conflicts, at least between great powers—necessitated that the Army develop a greater capacity to prosecute the former. General Fred Woerner, the commander of U.S. Southern Command, agreed that low-intensity conflicts were "high probability" in nature, writing, "Abrams tanks . . . have limited utility in the low-intensity/high-probability environment that characterizes [the Western] hemisphere." Instead, Woerner argued, the Army would need capabilities such as "security assistance, . . . civic action, psychological operations, engineer construction, . . . and infrastructure development."[51]

Many low-intensity conflict observers were almost apologetic as they insisted that the Army needed to develop more capacity to fight this type of war; they clearly saw the need for greater capacity in this area but had been professionalized in a culture that abhorred the suggestion. After a long article on the Army's lack of capacity to engage in low-intensity conflicts and an impassioned plea that this shortfall be corrected, Lieutenant Colonel Holt added that, of course, "there is no question that the US Army should maintain a credible force to meet its NATO [North Atlantic Treaty Organization] commitments—whatever they may turn out to be—in Central Europe."[52]

Other low-intensity conflict observers were reluctant to suggest that such conflict should be the U.S. Army's primary focus. While a SAMS student, Lt. Col. James Dubik insisted in his master's thesis that the Soviet Union was no longer a threat: "America's priority of effort and spending may have to go to counter the non-traditional, internal threats." But he also conceded that "armed forces correctly focus on the more traditional, external threats to national security. This focus must remain, for the world still contains powerful threats to the physical security of the United States and her citizens." Air Force general Butler acknowledged in the pages of *Parameters* that a "post-CFE [Conventional Armed Forces in Europe Treaty] Soviet aggression" was an "unlikely event," but still believed the United States should retain "the ability to reconstitute larger forces should the need arise."[53]

Some of those who advocated that the Army remain focused on high-intensity conflicts did concede that it needed to develop some capacity to fight low-intensity ones. McDonough acknowledged, "Future doctrine should be expanded to incorporate our evolving missions in areas such as stability operations, nation assistance and contraband flow" because the Army was finding itself "engaged around the world in a variety of missions that fall outside of" the scope of high-intensity conflict. McDonough added, "Doctrine should address nonconventional operations in operations short of war, during limited hostile action and in conditions of war and its aftermath" and that "security assistance, nation assistance, humanitarian assistance and disaster relief. . . . warrant doctrinal elaboration."[54]

Project 2025 recommended creating "countercoup forces, counterinsurgency teams, and diverse intelligence-gathering capabilities" as well as the ability to "help promote life-saving coups and support insurgents fighting for their freedom." Yet the project did still note that "in addition to planning for these special operation and low-intensity conflict scenarios, the United States will still need to keep a capability for short-notice, medium-scale military interventions."[55]

The vast majority of the Army, however, including virtually the entire senior leadership, remained exclusively focused on the problem of how to continue to modernize and prepare for high-intensity conflict in the face of the unwanted but unavoidable low-intensity conflict missions to which the Army would continue to be committed. In late 1991 General Sullivan convened a conference of the Army's senior leadership—five four-star generals and 133 other general officers—at Fort Leavenworth. A major theme of the conference's keynote address by Gen. (Ret.) John Vessey Jr.—who had been chairman of the Joint Chiefs of Staff in the early 1980s—was how to "structure our force to meet the threats spanning from insurgency to war" while "retain[ing] our technological superiority—preferrably [sic] pursuing 'leap-ahead' technology."[56]

Senior leaders dismissed calls to increase the Army's low-intensity conflict capacity by emphasizing the urgency of preparing for an imminent high-intensity conflict against an imagined future peer competitor. At the same conference at Fort Leavenworth, General Sullivan told attendees that there could be no "more Task Force Smiths." He repeated this theme—"No More Task Force Smiths"—in an interview with Soldiers Radio and Television and to a reporter from the Army's *Soldiers* magazine later that week. These were references to the Army task force initially sent to Korea in 1950 to meet the North Korean invasion of South Korea; the force was ill trained and ill equipped, and took grievous casualties while being routed from Seoul to the tiny Pusan perimeter on the southeast tip of the island.[57] But this idea also had another heritage: the turn by Generals William DePuy and Don Starry in the early 1970s to the Yom Kippur War for examples from which to shape the Army. If the Army was not prepared for the first battle of the next war, this line of reasoning concluded, it could be the United States' last.

Thus, FM 100-5—and, in fact, the entire Army—would remain unapologetically focused on high-intensity conflict. McDonough insisted, "Some things should not change," including "the conceptual ideas, tenets, imperatives and the battlefield framework" that had underpinned the Army's AirLand Battle doctrine.[58]

Embedded in McDonough's—and the Army's—strategy to remain focused on high-intensity conflict in the face of a growing number of low-intensity conflict missions was a dismissal that would often be repeated throughout the 1990s by senior Army leaders. McDonough wrote that the Army's focus on high-intensity conflict could continue because AirLand Battle concepts were equally applicable "across the operational continuum of peace, crisis and war."[59] This was a corollary to the notion that an Army unit that excelled at high-intensity conflict would automatically excel at low-intensity conflict—that is, that these conflict competencies were a lesser included subset of high-intensity ones.

A-AFCLIC and FM 100-20

While the Army had been focused on high-intensity conflict throughout the 1970s and 1980s, the conversation over low-intensity conflict had nonetheless continued. In the late 1980s, A-AFCLIC—charged with producing both Army and joint doctrine on low-intensity conflict, had become the dominant voice in the Army on the subject.

A-AFCLIC would eventually produce the 1990 edition of FM 100-20. Production of the manual would be heavily circumscribed by the senior leaders of the Army, who consistently pressed for less distinction between low- and high-intensity conflict doctrine and the application of the former (through such constructs as the tenets of AirLand Battle) to the latter.[60] The doctrine A-AFCLIC did finally produce still advanced the Army's understanding of low-intensity conflict.

Perhaps the 1990 edition of FM 100-20's most important contribution to such an understanding was its "low intensity conflict imperatives":

- Political dominance.
- Unity of effort.
- Adaptability.
- Legitimacy.
- Perseverance.[61]

Among the most interesting of these imperatives were the ideas of political dominance and legitimacy. Political dominance asserted the essentially political nature of low-intensity conflicts; this imperative challenged Army leaders to move beyond conventional, high-intensity approaches to the problems they faced in these conflicts and embrace "unorthodox" solutions "outside of traditional doctrine," where they were necessary to achieve the "political objectives" of the operation.[62]

Legitimacy, on the other hand, hid cognitive obstacles that would plague Army units fighting in low-intensity conflicts for the remainder of the decade. This imperative did rightly identify that a U.S. Army unit in a low-intensity conflict had to operate in and among a populace and that it was preferable to have this population's "willing acceptance." And FM 100-20 did correctly identify that elections in and of themselves do not confer legitimacy on a host nation government. But otherwise this definition was riddled with unspoken assumptions rooted in Western, liberal democratic notions of power deriving from the consent of the governed. This manual insisted that for authority to be "legitimate" it had to use "proper agencies for reasonable purposes." This definition also claimed that no party in a conflict could "create legitimacy for itself."[63] These wrongheaded notions missed the essential nature of low-intensity conflict as war; as Carl von Clausewitz explained, in war one uses "force to impose [one's] will" on others. In this context—and quite to the contrary of the assertions made in the FM 100-20—one could absolutely "create legitimacy for [one's] self" through force or the threat of force.[64]

With FM 100-20 complete, the center began work on the publication that would ultimately become the 1995 Joint Publication 3-07, *Joint Doctrine for Military Operations Other Than War*.

After the problematic publication in 1990 of FM 100-20,[65] the Army followed with the 1992 manual, FM 7-98, *Operations in a Low-Intensity Conflict*. The manual was generally a good summary of the tactical lessons from the low-intensity conflicts of the 1970s and 1980s but suffered from the general atrophy of low-intensity conflict doctrine over the period. And, on a deeper level, it treated the subject like an unavoidable but unpleasant chore in which the Army had to engage. For instance, the chapter on peacekeeping began with this epigraph:

> *"Peacekeeping isn't a soldier's job, but only a soldier can do it."*
> Anonymous Member, Peacekeeping Force[66]

Low-intensity conflict observers frequently highlighted the many ways in which the Army was unprepared to fight low-intensity conflicts. Lieutenant Colonel Holt warned that the Army was "training to fight the last war rather than the next one," adding that "to ignore operations short of war would be an abdication of responsibility and national trust." He believed that "the starting point is in leader education. The Army institutional school system must refocus to provide emphasis commensurate with the threat." Only then, Holt concluded, would "operations short of war share at least an equal billing with the improbable war in Central Europe."[67]

Holt was no doubt referring to the neglect of low-intensity conflict in the Army's formal education programs. At the CGSC, attended by nearly one thousand of the Army's top active duty majors each year, low-intensity conflict occupied forty-five hours of a curriculum spanning over five hundred hours in the 1991–92 academic year and only six electives of a menu of nearly one hundred were offered on the topic.[68]

Army captain Mark Rendina was much less diplomatic in his categorical condemnation of the Army's laser focus on high-intensity conflict in an October 1990 article in *Military Review*, noting, "We have a doctrine of AirLand Battle geared to fight a massed mechanized/armor battle on the open plains of Central Europe against a threat that is neither imminent nor probable. . . . We have made a tremendous investment in a string of weapons systems useful only on this imagined battlefield." Rendina warned that this focus on the wrong priorities carried the echoes of "the failures of other national armies as they prepared for war."[69]

Surprisingly, at least some mainstream Army officers agreed. An internal trip report of A-AFCLIC to a low-intensity conflict exercise held at the U.S. Army War College—a highly selective school for the top colonels in the Army—noted that at least "some of the students felt they needed more time for this block of instruction while others felt that regardless of the time available they could fill it with discussions about LIC." This report continued, "Others felt that some additional time could be well-spent if it was focused on efforts at original thought and innovative approaches to specific LIC challenges."[70]

Even Col. L. D. Holder acknowledged that the Army avoided thinking about low-intensity conflict. In an article about operational art for a September issue of *Military Review* he made the offhanded comment that "unconventional campaigns [were] a type of warfare for which there is adequate theory and example, but one about which most US professionals actively resist thinking." While years later as commanding general of the CAC at Fort Leavenworth he would do little to remedy this neglect, he added at the time that low-intensity conflicts "require the same attention and education that more conventional wars presently do. Many will argue that as the emergent dominant forms of war, they require more attention than any other type of war."[71]

Some observers disagreed that low-intensity conflict required special capabilities or training, arguing that a unit that was good at high-intensity conflict would, necessarily, be good at the lesser tasks required in a low-intensity conflict. This idea was even enshrined in Army doctrine. The otherwise generally helpful FM 7-98 offered a questionable insight: "No matter what parameters have been established for the use of force, a disciplined unit, with soldiers proficient at individual skills who are operating under a clear expression of the commander's intent, can perform successfully at the tactical level in [the low-intensity conflict] environment." It also offered a dubious assertion: "The tenets of AirLand Battle doctrine characterize successful conventional military operations and apply equally in LIC." Elsewhere the manual attempted a tortured application of the tenets of AirLand Battle to low-intensity conflict that did little to clarify either for the reader.[72]

Holt disagreed with those who believed that low-intensity conflict was simply a lesser included task of high-intensity conflict proficiency, writing, "The prevailing assumption appears to be that such operations can be handled as a simple planning responsibility and as an additional contingency for those forces prepared for commitment to a major European war." He objected to this thinking, adding, "This mind-set instilled in the leadership of the Army through its school system leads to an attempt to solve the LIC problem by large-scale capital items" and insisting that "the firepower, attrition oriented mind-set some feel is necessary in Europe to delay the decision to use nuclear weapons does not work in combating insurgency." The director for low-intensity conflict in the Office of the Assistant Secretary of Defense for Special Operations and Low-Intensity Conflict, William J. Olson, likewise attacked this notion, saying, "While regular forces are very well prepared to meet challenges at the mid-intensity level of conflict and above, their very preparations and the associated habits of mind do not make them equally prepared to cope with LIC."[73]

But while many acknowledged that the Army lacked the capability to successfully fight low-intensity conflicts, not all believed that it should even be employed in them. As early as 1988 Major Michael Symanski wrote in the pages of *Military Review* that the Army was simply incapable of fighting such conflicts. He argued, "The US military will be the likely loser in a LIC operation," adding, "The struggle will degrade the military institution, squander its combat power and alienate it from its parent society.[74]

Even those who believed the Army should engage in low-intensity conflicts occasionally fell into cognitive traps. Lieutenant Colonel Barnes echoed the Nunn-Cohen contention that SOF should have primacy in low-intensity conflict, writing, "special operations forces (SOF) such as psychological operations (PSYOP) and Special Forces (SF) share leading roles with CA [civil affairs] in LIC." He contended, "The one common denominator of the varied military missions of SOF in LIC . . . is that they must be conducted in politically sensitive environments for which conventional forces are not well suited."[75] Faulty thinking such as this relieved conventional Army units of the responsibility to prepare for low-intensity conflict or to engage in their political dimensions, a responsibility that, based on the Army's subsequent history, it would not be able to avoid when actually engaging in such conflicts.

Other low-intensity conflict observers saw this thinking for what it was: a cognitive trap that impeded learning. Air Force major J. C. Rhoades of A-AFCLIC would write in an internal memorandum, "[Low-intensity conflict] Involves All forces, not just Special Operations; LIC is an operational environment, Special Operations is a capability."[76]

As many observers arrived at the conclusion that the Army was not prepared for low-intensity conflict, they also highlighted specific capabilities the Army lacked. One of the key missing capabilities lay in assisting host nation militaries. Lieutenant Colonel Holt warned that the Army tended to make forces in its own image, overly focused on high-intensity conflict and high-tech weapon systems. This bred a "predilection for spending too much time in comfortable barracks" and not enough time out patrolling. Such a force was also apt to use "firepower that cannot distinguish between guerrillas and civilians," generating "political dissatisfaction and strengthen[ing] the insurgents' political position."[77]

FM 7-98 identified another capability that the Army would need in order to engage in low-intensity conflicts: integration with other U.S. government agencies, multinational partners, and intergovernmental organizations such as the United Nations (UN). In its chapter on peacekeeping, the manual warned, "The US may enter into PKOs [peacekeeping operations] under the auspice of an international organization in cooperation with other countries or unilaterally." The manual then provided detailed descriptions of how a peacekeeping mission might arise and be organized and executed, whether under or outside the auspices of the UN. FM 7-98 also acknowledged the need for the U.S. military to work with other agencies of the United States and other national governments. One of the "LIC imperatives" it enumerated was "unity of effort," the requirement to "integrate their efforts with both military and civilian organizations of the US and of countries we support" to achieve "interagency integration and cooperation."[78]

Many observers also noted that low-intensity conflict occurred in and among civilian populaces. Lieutenant Colonel Holt recounted the long history of urban guerillas in Latin America, fighting "where large groups of discontented people could be found." Lieutenant Colonel Barnes noted, "Government military forces that cannot protect the public from insurgent forces jeopardize the public support necessary for legitimacy and political control." But he also noted the requirement for restraint and the nuanced application of force, explaining that "military force must be constrained to avoid collateral damage and a loss of legitimacy."[79]

Low-intensity conflict observers also identified many key capabilities that the Army either had or needed to develop in order to engage foreign populaces. Lieutenant Colonel Barnes noted that Army units needed increased "language training and cultural orientation" before engaging in low-intensity conflicts. Likewise, FM 7-98 acknowledged, "The language, religious, and cultural differences between our society and those that soldiers may come in contact with pose additional challenges. The basic values and beliefs that are common to US soldiers are not universally embraced."[80]

FM 7-98 also broke new ground in acknowledging the role of media and PSYOP in low-intensity conflict, insisting, "Winning the information fight is often an overlooked aspect." This meant that "the ability to communicate with different agencies and the local populace," but it also meant that "any action can be exploited rapidly, by both friendly and enemy media and PSYOP efforts."[81]

Senior Army leaders—the chief of staff of the Army, Gen. Gordon Sullivan, chief among them—often missed the complexity introduced by civilians on the battlefield. In a report for other Army leaders on his 1991 trip to the Joint Readiness Training Center (JRTC) at Fort Chaffee, Arkansas, he wrote, "JRTC has done a good job of incorporating all aspects of potential fratricide play." Those factors included civilians and the host nation. Elsewhere Sullivan indicated that he saw these facets of the low-intensity conflict environment primarily as obstacles to "direct fire" and "indirect fire" rather than as people whose needs and desires had to be addressed in order to be successful. His goal was that a leader be "exposed" to these factors but he did not want "an overreaction" that might "take away aggressiveness."[82]

A few observers made the leap from understanding low-intensity conflict as occurring among the populace to a more difficult acknowledgment: the essentially political nature of such conflict. Lieutenant Colonel Barnes called for the Army to transform its leaders into "diplomat warriors."[83] He defined this as "a bastard stepchild born of the uneasy relationship between diplomacy and military operations. . . . To succeed, the diplomat warrior must understand the relationship of military operations to the political objectives of contemporary warfare." Barnes believed that advanced civil education was the key to achieving this balance.[84]

Even some more conventionally minded Army leaders recognized that there was a political dimension to low-intensity conflict and that even if an operation might begin with a high-intensity character, the conflict termination phase might metastasize into a low-intensity one. The principal author of FM 100-5, SAMS director McDonough, wrote in October 1991 that "large numbers of disoriented and often destitute prisoners of war" and "the rapid breakdown of civil order in the area of operations" might drive the conflict toward the lower end of the range of military operations. He noted that "conflict" resolution in both Panama and the Gulf War had required "operations after the cessation of hostilities" for which the Army was "without adequate doctrine."[85]

FM 7-98 included "LIC imperatives"—carried over from the 1990 FM 100-20[86]—that reflected an understanding of the political dimensions of low-intensity conflict, especially in the first imperative: "political dominance." Echoing FM 100-20, the manual explained, "Political objectives affect military operations in conventional war. In LIC operations, they drive military decisions at every level-from the strategic to the tactical." FM 7-98 also acknowledged the essentially political nature of insurgencies and counterinsurgency, stating, "The members of the insurgent force are organized along political lines to support political, economic, social, military, psychological, and covert operations." Elsewhere it added, "The initial goal of the insurgent movement is to replace the established government." Ultimately, counterinsurgency was a competition for political control of the country: "The amount of government control in an area directly affects the ability of the insurgent to operate. The more government control, the less successful are insurgent activities."[87]

FM 7-98 was particularly insightful in drawing connections between many capabilities required to operate in and among the populace and their effects on the political dimension of low-intensity conflict. The manual highlighted the connection between information and PSYOP and the political dimensions of low-intensity conflict, warning, "Low standards of living and desires for economic reforms may be popular causes of resentment toward the government's economic policies. . . . The insurgent seeks to exploit this situation through the use of PSYOP." FM 7-98 also tied adversaries' guerilla attacks to the political dimension of the conflict, writing that they add "to the perception that the government cannot or will not provide security for the population and its property. Elsewhere the manual added, "The safety of noncombatants and their property is vital to maintaining the legitimacy of the host government." It also highlighted the connection between rules of engagement (ROE) and the political dimension of low-intensity conflicts, explaining, "The unlimited use of firepower directed against civilians or their property may cause them to embrace the insurgent's cause."[88]

Yet FM 7-98 also included several cognitive obstacles to a full understanding of the political dimension of low-intensity conflict. First, the manual posited that a U.S. Army unit engaged in such a conflict could avoid the political dimensions of the conflict by limiting its mission. The chapter on peacekeeping stated, "A clear, restricted, and realistic mission must be given," and then proceeded to list the security tasks that it was acceptable to assign to an Army unit.[89] This, of course, presumed that some "other" agency would arrive to do the politics of the low-intensity conflict. In short, this limited list of missions was a recipe for forever war.

FM 7-98's answer to who would arrive to do the politics of low-intensity conflict reflected the Nixon Doctrine and the Reagan-era conception of counterinsurgency as primarily the responsibility of the host nation, explaining, "In counterinsurgency (COIN), the objective is for *the host government* to defeat an insurgency." The manual elsewhere argued, "The US government is responsible for influencing the host government's attitude toward democratic principles—it is not the commander's responsibility."[90] This exhortation is questionable on two levels: both the insistence that a host nation government be "democratic" and the U.S. Army's abdication of responsibility for the politics of low-intensity conflict to the host nation government.

Other observers also fell into this same cognitive pitfall, looking for others to engage in the political dimensions of low-intensity conflicts so that the Army didn't have to. As Lieutenant Colonel Barnes argued, "The primary mission of [U.S.] forces in LIC is advising and assisting indigenous military forces in accomplishing their politico-military objectives." Likewise, attendees at a April 30, 1991, low-intensity conflict conference at Fort Leavenworth concluded, "The U.S. Military role in nation assistance is supplemental to that of the Agency for International Development, which has primary responsibility for helping a friendly foreign country to develop its political, social, and economic capabilities." Lieutenant Colonel Holt wrote that while "the U.S. Army must be involved," still "the area of national interests, objectives and strategy is the principal province of the State Department."[91]

During this period even the most insightful observers were not ready to acknowledge the necessity for the Army to take sides and back a winner (designated by the president and administration) in order to bring a low-intensity conflict to a successful conclusion. FM 7-98, echoing the LIC imperatives from the 1990 FM 100-20, perpetuated the earlier manual's fallacious notion that "legitimacy" derives from a party's use of "proper agencies for good purposes." It likewise dismissed the notion that Army forces could create legitimacy for themselves through any means; instead, FM 7-98 contended, the Army had to endear itself to a population through its altruism. In the chapter on peacekeeping, the manual was more explicit on how Army forces maintained legitimacy, stating flatly, "The importance of the peacekeeping force being entirely neutral cannot be overstated." The chapter also enumerated principles for peacekeeping operations that included both "neutrality" and "balance"—the idea that the peacekeeping force should be comprised of a balance of nationalities that would make it appear more neutral.[92]

Panama Retrospectives

At the same time that the theoretical debate over definitions and LIC doctrine was taking place, a very real low-intensity conflict in Panama continued to offer up valuable lessons to the U.S. Army. Even though a duly elected political leadership was waiting to take power when the invasion occurred, the Army still struggled to achieve stability. The United States Military Support Group in Panama was activated in January 1990 but would not be deactivated until a year later because of challenges with stabilizing the country.[93] Many of the lessons of this ordeal were lost in the turmoil of the Gulf War, but lessons did continue to emerge after the Gulf War ended.

The continuing crisis in Panama lent ammunition to those who were already saying that the future held more low-intensity conflicts. Lt. Col. John Fishel of the Army Reserve contended in an article in *Military Review* that the low-intensity conflict that followed the invasion in Operation Just Cause had been virtually inevitable. He wrote, "The United States must commit itself solidly to assume responsibility for its actions," rebuilding countries it occupies.[94] Lawrence Yates, a historian for the Combat Studies Institute at Fort Leavenworth, noted that in addition to the low-intensity conflict that occurred after the invasion, the Army had been "continuously and, at times, deeply involved" in the less dramatic low-intensity conflict that preceded the invasion, beginning in mid-1987.[95]

The low-intensity conflict in Panama yielded the most lessons in the area of security force assistance—particularly in the establishment of the Panamanian National Police. The Panamanian Defense Force that had policed the country under Gen. Manuel Noriega had evaporated under the weight of American firepower during the invasion. Without a basic law enforcement infrastructure—even things as simple as functioning traffic signals—daily life in Panama City rapidly ground to a halt. Funds for military assistance to Panama had been frozen by the U.S. Congress before the invasion, when the country was still under the Noriega regime. After the invasion, there was a significant delay before the freeze was lifted, during which time security force assistance could not proceed. Police stations had no materials with which officers could write reports or issue citations. It was six months before police cars began arriving in Panama, and once they did arrive, fuel was limited to five gallons per car per day. Delays in developing a compensation system meant rural police often went months without pay. The Army's cash-for-weapons program had been incredibly efficient in taking weapons off the streets of Panama City and shipping them out of the country for demilitarization, but this efficiency had the unintended consequence of removing from the country all of the weapons the Army needed to arm a police force. There was a significant delay before this flow was reversed and the police were armed.[96]

In pursuing its security force assistance effort, the Army also learned lessons about working with other U.S. government agencies. In particular, it had to coordinate its efforts with the U.S. Justice Department's International Criminal Investigative Training Assistance Program (ICITAP). In a *Military Review* article in 1992, Lt. Col. John Fishel and Maj. Richard Downie noted the tension as the Army tried to convince the Justice Department to move beyond academy-based training—which would take eighteen months to produce a police force—and into precinct headquarters. ICITAP finally moved to "roll call training," brief classes that occurred during shift changes at the local police stations. Both the Army and the U.S. State Department were eager to hand off the entire mission of security force assistance to the Department of Justice, but ICITAP ultimately refused to take on the advisory role that the U.S. Army's military police (MPs) had assumed with the Panamanian National Police.[97]

The State Department—through its embassy in Panama—was theoretically in charge of the U.S. interagency effort in Panama after the invasion had ended. But it was inadequately staffed and oblivious to the prewar plan for reconstruction. Exacerbating these problems, the embassy continued to operate in peacetime mode, even granting ordinary leave to key personnel, at the height of the low-intensity conflict. The U.S. ambassador to Panama—ostensibly the head of the U.S. effort in the country—was replaced in the middle of the reconstruction effort, though there was a sizable gap between the departure of Ambassador Arthur Davis and the arrival of Ambassador

Deane Hinton. In the absence of an ambassador, the 7th Infantry Division commander—already pulling double duty as the leader of Joint Task Force (JTF) South during the low-intensity conflict—made unilateral decisions to keep the effort going. These tensions produced constant interagency squabbles between the Army, the Department of Justice, and the State Department over the allocation of resources and responsibilities.[98]

Low-intensity conflict observers also noted the challenges the Army faced as it operated in and among the populace in Panama. As Lawrence Yates noted, "Trained in a sterile 'force-on-force' environment, [Army forces] went through the mental agony of adapting: learning how to forgo combat in favor of psychological wargames, applying the ROE creatively to unique and unanticipated missions, . . . and learning how to operate in an environment in which friends (the majority of Panamanians) and enemies . . . were intermingled." Yates cited one Army brigade commander as saying he would happily trade one of his infantry rifle companies for an MP company "well trained in peacetime ROE."[99]

Yates also noted that the civil-military requirements in Panama rapidly overwhelmed the capacity of the limited number of civil affairs personnel on the ground. "As a result," he wrote, "combat units often found themselves performing civil affairs, constabulary, security patrols and other noncombat missions." He added that not all Army units were prepared for this transition, adding, "Initial civil affairs, civil-military, constabulary, stability and nation-assistance missions lacked adequate coordination."[100]

Even General Sullivan acknowledged these lessons. As the general would recall years later, "I was the G3 of the Army when we invaded Panama, and that went off like clockwork. . . . The next day . . . somebody called and said, 'Who is going to feed the Panamanians?'" We presumed they were going to feed themselves—wrong answer. We wound up with that on our plate. So the fact of the matter is, we had not prepared ourselves."[101] The building of the National Police in Panama also had a political dimension for which the U.S. Army was equally unprepared. The Panamanian government wrestled with whether to hire an all new police force, leaving thirteen thousand trained and disgruntled former police officers and soldiers in the populace or rehire these men, which would produce security more quickly but would also put the instruments of Noriega's tyranny back on the streets. The government chose the latter. To mitigate the risks of this choice, the government imposed measures to ensure that the leadership of the police was not tainted by its connection to the deposed Noriega regime. Many of the leaders that the new government placed in charge of the National Police had been jailed or exiled by Noriega. Others had barely survived the October 3, 1989, coup attempt against Noriega.[102]

Moreover, investigations of police officers—and prosecutions for those accused of crimes committed during the previous regime—continued after Noriega's ouster. Both the first police commander selected by the new government, Col. Roberto Armijo, and his deputy commander, Lt. Col. Aristides Valdonedo, were relieved and then indicted for crimes committed under Noriega.[103]

Also complicating the Army's attempt to build the police force were unauthorized acts of retaliation on the part of officials of the new government against the National Police. Officials contemptuous of the former police force under Noriega withheld funds and resources from the new police force out of petty revenge. As Fishel and Downie concluded, "There is always a cultural and political dimension that must be considered."[104]

Likewise, average Panamanian citizens, who had been victimized by the corrupt police force under the Noriega regime, were understandably concerned to see the same police officers returning to work in the same headquarters buildings. To address these concerns new Panamanian president Ricardo Arias Calderón established a new National Police leadership position—that of public zone commander—and made these commanders subordinate to the civilian governors. This move did little to improve civilian political control over the Panamanian National Police or alleviate the concerns of the Panamanian people. Aggressive human rights and ethics training was also required to transform Panamanian security officers who had once preyed on the populace into police officers sworn to protect and serve the people. Even the selection of new uniforms that were sufficiently dissimilar from those of the previous regime was intended to allay the public's fears. Despite these moves, the public understandably remained suspicious of the new police force.[105]

While low-intensity conflict observers did see the political dimension in Panama, they never came to see that the Army needed to back a winner in the conflict in order to reach a sustainable political settlement. In fact, most observers—in line with the doctrine of the day on legitimacy and neutrality—argued that the Army should *not* take sides. For instance, Fishel and Downie wrote that U.S. Army and State Department personnel needed to "avoid favoring any particular political party or group bias."[106] Ultimately, however, the conflict in Panama was concluded despite the Army's refusal to pick sides. President Calderón and his party were able to prevail and impose stability on Panama because Noriega's civilian, military, and paramilitary supporters were simply too weak after their losses in the initial invasion to continue to resist the weight of Panamanian public opinion and the new National Police Force. Thus, in a sense, the U.S. Army's success in very dramatically picking sides during the invasion—decimating Noriega's security apparatus—overcame the Army's later refusal to pick sides during the low-intensity conflict that followed. Had a suitable regime not already been in place, ready to take charge, the result might have been quite different.

May–June 1993
From Low-Intensity Conflict to Operations Other Than War

While low-intensity conflict observers struggled to capture the lessons of the conflict in Panama, the senior leaders of the Army strove to roll back their gains in Army doctrine. In June 1993 the Army finally published its new edition of FM 100-5, which did address low-intensity conflict to a greater degree than any previous edition. Today, Gen. (Ret.) Gordon Sullivan considers this a point of personal pride. He explains that while the discussion of low-intensity conflict in the 1993 edition of FM 100-5 was "not terribly long," at least "we got it in there."[107] Yet while low-intensity conflict was mentioned in the new FM 100-5, many of the most important points of understanding that were codified in FM 100-20 were absent from this later manual.

The 1993 edition of FM 100-5 purported to adapt the AirLand Battle doctrine of the 1982 edition to the reality that "the Cold War has ended," which changed "the nature of the threat [and] hence the strategy of the United States as well." Promisingly for those who might have hoped the Army would rebalance itself between low- and high-intensity conflict, this doctrine also promised that it "recognize[d] that Army forces operate across the range of military operations."[108] Yet while this new capstone Army doctrine did address low-intensity conflict, it also made a number of changes to extant doctrine that drove the Army decidedly downward on the ladder of understanding of such conflict.

Perhaps the most visible change that the 1993 edition of FM 100-5 made to the Army's existing doctrine on low-intensity conflict was to replace the term "low intensity conflict" with "operations other than war." This change in terminology was not just cosmetic. As Sullivan and Dubik explained in their 1993 paper, war was "the armies of one nation-state or alliance of nation-states fighting those of another" while "operations other than war" were "every other act of violence, use of force, or form of hostility" including "peacetime activities with very low levels of violence, crises, conflicts, war, and war termination activities." Sullivan and Dubik acknowledged

Clausewitz's conception of war and politics and Michael Howard's warning that these forms of conflict were "often indistinguishable from traditional war," but still, inexplicably, chose to classify everything besides combat between "the armies of . . . nation-states" as "other than war."[109]

Incredibly, the 1993 FM 100-5 chapter on operations other than war (OOTW) declared—in the very first sentence—that "the Army's primary focus is to fight and win the nation's wars." The manual unequivocally insisted, "Winning wars is the primary purpose of the doctrine in this manual." Operations other than war were activities outside the Army's "primary purpose" in which it was unfortunately sometimes forced to engage.[110]

FM 100-5 defined a "range of military operations" with some types of operations defined as "war" and others defined as "other than war."[111]

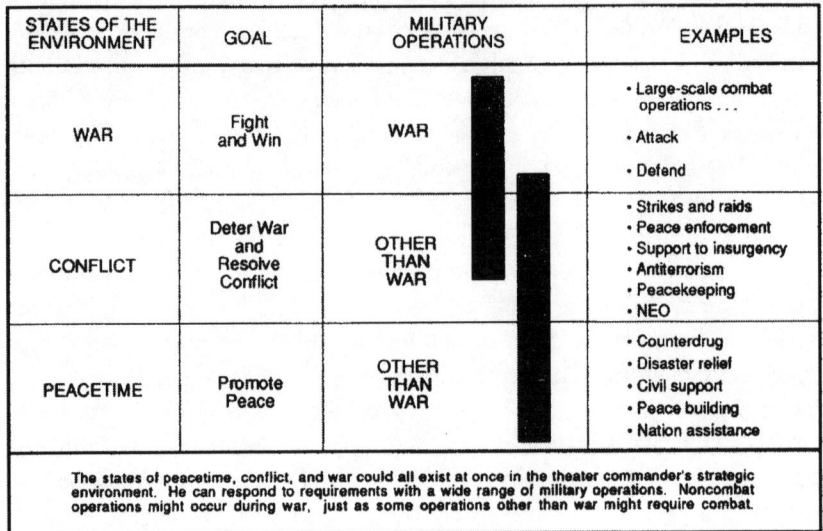

STATES OF THE ENVIRONMENT	GOAL	MILITARY OPERATIONS	EXAMPLES
WAR	Fight and Win	WAR	• Large-scale combat operations . . . • Attack • Defend
CONFLICT	Deter War and Resolve Conflict	OTHER THAN WAR	• Strikes and raids • Peace enforcement • Support to insurgency • Antiterrorism • Peacekeeping • NEO
PEACETIME	Promote Peace	OTHER THAN WAR	• Counterdrug • Disaster relief • Civil support • Peace building • Nation assistance

The states of peacetime, conflict, and war could all exist at once in the theater commander's strategic environment. He can respond to requirements with a wide range of military operations. Noncombat operations might occur during war, just as some operations other than war might require combat.

Figure 2.1. The range of military operations. *Source:* Headquarters, U.S. Department of the Army, *Operations*, FM 100-5 (1993), 2-1.

This segregation of environments and operations was illuminating. If the Army's purpose was to "win the nation's wars" and support to insurgency and peacekeeping were "operations other than war" and existed outside the state of "war,"[112] then these operations stood outside the Army's purpose.

In the 1993 edition of FM 100-5's diagram of the range of military operations, counterinsurgency— or even its post-Nunn-Cohen euphemism, "foreign internal defense"—does not appear.[113] Elsewhere the 1993 FM 100-5 reflected the Nunn-Cohen Amendment's delegation of counterinsurgency to SOF, explaining, "Due to their extensive unconventional warfare training, SOF are well-suited to provide this support [to foreign internal defense]." In fact, everywhere in the manual where foreign internal defense is mentioned, it is referred to as one of "five principal missions of special operations." Admittedly, the manual did later acknowledge that conventional forces might be required to engage in counterinsurgencies "when the scope of operations is so vast that conventional forces are required."[114] But this admission was little more than a footnote in the text.

The 1993 FM 100-5 also made other detrimental changes to low-intensity conflict doctrine. These changes are perhaps best illustrated by looking at an Army manual published in May 1993, FM 34-7, *Intelligence and Electronic Warfare Support to Low-Intensity Conflict Operations*, and comparing it to the 1993 FM 100-5.

Perhaps the most damaging change that the 1993 FM 100-5 made to low-intensity conflict doctrine was to excise "political dominance" from Army doctrine. FM 3-47 repeated the "imperatives" of low-intensity conflict enumerated in the 1990 FM 100-20, *Military Operations in Low-Intensity Conflict*: "political dominance," "unity of effort," "adaptability," "legitimacy," and "perseverance."[115] FM 3-47 warned readers that low-intensity conflicts presented "a constant flow of political-military missions" and suggested, "To deal with these, we must understand the driving political and socio-economic forces behind them." FM 3-47 also enumerated all of the ways in which civilian political sentiments could influence the low-intensity conflict environment. On the other hand, the 1993 FM 100-5's principles specific to OOTW—"objective," "unity of effort," "legitimacy," "perseverance," "restraint," and "security"[116]—excluded the all-important first "imperative" of low-intensity conflict from the 1990 FM 100-20: "political dominance."

In fact, the 1993 FM 100-5 argued that the Army should avoid the political dimensions of low-intensity conflict or hand off these responsibilities to someone else as soon as possible. Army primacy in the political dimension was to be the temporary exception rather than the rule: "The Army's presence and its ability to operate in crisis environments and under extreme conditions may give it the *de facto* lead in operations normally governed by other agencies." This manual insisted that the Army should only take the lead "until other US, international, interagency, or host nation agencies assume control." In the unlikely and unwanted circumstance that the Army was forced to temporarily assume primacy in the political dimension of a low-intensity conflict, Army leaders were told by FM 100-5 to "engage in joint planning with the State Department, relief agencies, and host nation

officials to prepare for a smooth and rapid transition to host country rule"
so that the host nation could take over the messy political dimension of
the conflict.[117]

The Army's doctrinal rejection of the political dimension of low-intensity
conflict would be sorely felt as the Army faced the first major such conflict
of General Sullivan's term as chief of staff.

Somalia, August 1992–March 1994: Operation Restore Hope

The regime of former major general Mohamed Siad Barre, which came to
power in Somalia in a bloodless coup in October 1969, survived decades
of intra- and intertribal wars, wars with neighboring Ethiopia, famines,
and subversion by Arab neighbors. Somalia and neighboring Ethiopia had
been pawns in the Cold War confrontation between the United States and
the Soviet Union throughout the 1980s, but with the collapse of the Soviet
Union, Somalia's diminished strategic importance translated into general
disinterest in the nation's plight.[118]

When the Barre regime did finally collapse in January 1991, Somalia
descended into anarchy and tribal warfare. The country fractured along political, ethnic, and tribal lines into competing factions that used financial and
food aid from the international community as weapons against one another
in their struggle to rule Somalia. As tribal violence became compounded
by the ravages of forced starvation and disease, the West was forced to take
notice. International relief agencies arrived to establish refugee camps and
distribute food aid, only to be threatened into paying "protection" to warlords and have their aid stolen by rival factions. Even the relief agencies' own
locally hired bodyguards got into the act, stealing supplies and threatening
the populace.[119]

The escalating response to this crisis eventually embroiled the United
States in a low-intensity conflict in Somalia. In April 1992, the UN authorized a small peacekeeping force—UNOSOM I—to enter Somalia, but this
force was quickly overwhelmed by the scale of the disaster. The United States
initially responded in August 1992 with Operation Provide Relief, providing
logistical support—primarily airlifting food aid into the interior of the
country for distribution by relief agencies. These supplies simply added fuel
to the growing fire, providing more ammunition with which rival factions
could starve their opponents and exert power over their own populations. In
December 1992 the United States was finally forced to deploy ground forces—
Operation Restore Hope—to ensure the food aid reached starving people
who needed it. State Department diplomats negotiated the acquiescence of

rival warlords—convincing them to withdraw their heavy weapons from the streets of Mogadishu—and 1,300 U.S. Marines arrived at the capital city's airport by helicopter while U.S. Navy SEALS emerged from the waves on the outskirts of the city, only to be greeted by eager journalists.[120]

The international military response of more than thirty-eight thousand troops, principally from France, Italy, Belgium, Morocco, Australia, Pakistan, Malaysia, and Canada, fell under the command of the Unified Task Force (UNITAF). Troops from twenty-three different nations and representatives from forty-nine different international relief agencies worked, frequently at cross-purposes, to feed the people of Somalia. The U.S. Army's contribution to UNITAF was the ten-thousand-soldier Task Force Mountain, with elements of the 10th Mountain Division from Fort Drum, New York, at its core. This Army contingent was led by the 10th Mountain Division's commander, Maj. Gen. Steven L. Arnold.[121]

UNITAF's efforts were complicated by the ongoing conflict between rival factions vying for control of Somalia. The Somali National Alliance (SNA) was led by a former general from the Siad Barre regime, Mohamed Farrah Aidid of the Habar Gidir subclan of the Hawiye clan. His principal opponent was Ali Mahdi Muhammad of the Abgaal subclan of the Hawiye clan, a former farmer and businessman who led a patchwork militia of fighters. The rivalry between these two factions had erupted into open warfare across Mogadishu from November 1991 to February 1992. U.S. ambassador Robert Oakley successfully negotiated a tenuous cease-fire to permit the entry of U.S. forces, but the constant threat of renewed fighting hung over Operation Restore Hope from its outset. Nonetheless, the momentary pause was enough to permit UNITAF to relieve the suffering. By the end of December, U.S. forces had spread across southern Somalia, the region hardest hit by famine and disease, and had distributed more than forty thousand tons of grain.[122]

But feeding hungry Somalis was not the same thing as healing Somalia's deep political divisions, and the UN mandate in Somalia held the seeds of future problems. UNITAF was a neutral party in Somalia, ostensibly favoring no particular faction in the conflict. And U.S. forces participating in the coalition effort believed that they had only been charged with relieving suffering, not forging an enduring political solution. The senior leaders of the U.S. contingent were resistant to any requests that they disarm the warring parties,[123] let alone involve themselves in the political dimension of the low-intensity conflict unfolding around them.

By early January 1993 Task Force Mountain was in position, providing for its own security and distributing relief supplies, but rigidly refusing to intervene while warring factions continued to victimize the Somali populace. An uneasy stability settled over Somalia, markets began to reopen, and starvation subsided. This relative calm was purchased at the price of eighteen UNITAF soldiers (ten who died in accidents) and nearly 400,000 dead Somalis (mostly children and the elderly).[124] Yet the country was no closer to a political settlement.

The UN took over the mission from the United States in May 1993 under the banner of UNOSOM II. Immediately upon taking over the mission, the UN began to pressure UNITAF to disarm the factions and force a political settlement. In response, Ambassador Oakley assembled the leaders of the warring factions at a conference in Ethiopia and convinced them to agree to a voluntary phased disarmament overseen by UNITAF. But actual compliance with this agreement was inconsistent. When groups of "technicals" (civilian trucks armed with heavy weapons) were stopped on the street, one would be surrendered while the others escaped.[125]

Moreover, Aidid's forces had played a key role in deposing Barre, and Aidid believed he had earned leadership of the country. When U.S. and UN forces, practicing a policy of strict neutrality, treated him like all of the other warlords, he perceived this as an attempt to readjudicate the outcome of the coup. In this mood of distrust, Aidid began hiding large numbers of tanks and artillery pieces out of the reach of international peacekeepers.[126]

With this expanded mission came dwindling numbers of American troops. One battalion of light infantry from the 10th Mountain Division and one U.S. Army Special Forces group remained in the country as the quick reaction force for a collection of infantry battalions from twenty other countries. Now the largest force in UNITAF was the four-thousand-strong Pakistani contingent. Sensing the opportunity and fearing disarmament by UNITAF, General Aidid unleashed his forces on UNITAF in a guerilla campaign across Mogadishu. In early June, Aidid's SNA attacked two Pakistani companies, killing twenty-four soldiers. UNOSOM II responded by declaring war on the SNA and employing attack helicopters and AC-130 Specter gunships to kill SNA fighters. The city rallied to Aidid's banner to repel the foreigners, causing the fighting to intensify.[127]

The United States responded to the growing violence by sending a SOF task force of U.S. Army Rangers, Delta Force operators, and SOF aviators as a second Quick Reaction Force (QRF) outside UNITAF command. This QRF launched a number of targeted, intelligence-driven raids to kill or capture General Aidid and his compatriots, culminating in the bloody Battle of Mogadishu on the night of October 3–4, 1993, immortalized in the

1999 novel and 2001 movie *Black Hawk Down*. During the night, eighteen American service members and as many as five hundred Somalis were killed; two UH-60 Black Hawk helicopters were destroyed. U.S. envoy to Somalia Admiral Jonathan Howe was recalled to the United States, Secretary of Defense Les Aspin resigned, and the United States completed an ignominious withdrawal from Somalia six months later.[128]

Operation Restore Hope seemed to confound conventionally minded senior Army leaders. The new FM 100-5 had only recently segregated Army operations into "war" (which the Army saw as its primary purpose) and "operations other than war" (which the Army saw as an undesirable but unavoidable tax on its time and resources). But as Major General Arnold would tell visitors to Somalia in January 1993 about the situation there, "It sure as hell ain't peace."[129]

While somewhat belatedly, General Sullivan and Lieutenant Colonel Dubik did eventually seem to grasp the full consequences of this revelation in their 1995 *Envisioning Future War*: "We will no longer be able to understand war simply as the armies of one nation-state or group of nation-states fighting one another. Somalia again demonstrates that this understanding is too narrow." They warned, "In not facing reality as it is, we could prepare the Army for the wrong war."[130] Yet despite this admission, the disastrous intervention in Somalia did nothing to change the course of the Army's march toward an ever-greater capacity to prosecute high-intensity conflicts.

The conflict in Somalia prompted many low-intensity conflict observers to warn that this conflict might be a harbinger of the future. Former foreign service officer Walter Clarke of the U.S. Army War College, writing in *Parameters*, warned, "The crisis in Somalia may be a paradigm of the New World Order. There are many more Somalias out there, especially in Africa, where debt, drought, disease, and politics threaten states with political implosion."[131]

In a 1993 article in *Military Review*, the former commander of the 10th Mountain division, Major General Arnold, warned that, like Somalia, "Bosnia and other potential hot spots" might force the United States to commit large numbers of troops to future low-intensity conflicts "regardless of desires to limit the commitment."[132] And Kenneth Allard of the NDU noted, "Far from ushering in an era of peace, our victory in the Cold War was quickly followed by combat in Operations Just Cause and Desert Storm." He added, "During UNOSOM II, for example, U.S. forces were also engaged in 12 other major operations requiring the formation of joint task forces—operations ranging from patrolling no-fly zones over Iraq to providing flood relief in the American Midwest."[133]

Though the term *asymmetry* was never used, the low-intensity conflict in Somalia also spurred observers to begin to note a mismatch between the high-tech, high-intensity conflict capabilities of the Army and the decidedly low-tech, low-intensity urban guerilla tactics that seemed to neutralize the Army's advantages. In a 1993 article Andrew Bacevich noted, "As a military commander, [Aidid] appears to have had one great insight: unlike Saddam, he knew that to play your enemy's game is the height of folly. On the other hand, to engage the opponent on terms that expose his weaknesses is to gain a priceless advantage." Bacevich added, "The lost battle for Mogadishu has shattered the dangerous illusion that the American military prowess displayed in the desert foretold an era of war without the shedding of American or civilian blood."[134]

Operations in Somalia also caused some low-intensity conflict observers to note the particular challenges presented by urban operations. Arnold wrote, "Operations in crowded streets, villages, and in urban environments of all types made daily life . . . dangerous," adding that "the populace seemed easily excitable and could be incited to participate in civil disturbances rather easily."[135] Unlike many of his contemporaries in the senior leadership of the Army, Arnold understood the essential character of urban environments: they were human environments full of people.

The logistical challenge of moving the Army to Somalia was massive. All told, the U.S. military would airlift more than 35,000 soldiers and 30,000 tons of cargo and supplies; somewhere around 280,000 tons of supplies and equipment and 14.38 million gallons of fuel would be sealifted. To sustain the operation, the military moved more than 1,192 containers of supplies.[136]

Army leaders and low-intensity conflict observers alike learned a number of lessons about deployability and sustainment from Somalia. Arnold noted that the Army's pre-positioned ships were unable to dock in the shallow ports of Mogadishu. Instead the division had to rely on aerial resupply. He also noted that his division had a critical shortage of contract specialists—"especially those with experience in local procurement"—for the division's logistical needs. Arnold mused that the sustainment problems reminded him of the challenges the Army had faced while deploying to Saudi Arabia for the Gulf War. Allard argued that the U.S. military was optimized to move forces to western Europe in support of NATO; Allard contended that low-intensity conflicts would probably occur in much more austere environments with much less developed ports and airfields.[137]

Low-intensity conflict observers examining Operation Restore Hope almost universally concluded that the Army was not trained to conduct the kind of operations it was called upon to execute in Somalia. The Center for Army Lessons Learned (CALL) published a "special edition" manual to provide units and individuals deploying to Somalia in early 1993 with tactics to deal with this unfamiliar environment. (Notably, a fellow at the Hoover Institution on War, Revolution, and Peace, Lt. Col. John Abizaid—who would go on to lead U.S. Central Command (CENTCOM) after the 2003 invasion of Iraq—was one of the authors of this interim manual.) CALL noted that soldiers in Somalia were faced with a number of situations— from requests for medical assistance from civilians, to the capture of civilian criminals, to confrontations with angry mobs of civilians for which they had never been trained; units had to create and train on new "battle drills" to deal with these unfamiliar situations.[138]

The 10th Mountain Division was also not correctly manned, equipped, or organized for the low-intensity conflict in Somalia. When it arrived, the division discovered that there was little use for the field artillery units and equipment it had brought. The artillery pieces were immediately shipped back to Fort Drum, and the division artillery headquarters was converted into a maneuver headquarters, placed in Humanitarian Relief Sector (HRS) Kismayu, and given control over U.S. Army Task Force 3-14 and the Belgian Army's 1st Parachute Battalion.[139] Major General Arnold observed that the 10th Mountain Division had a critical shortage of a number of capabilities that were desperately needed for operations in Somalia, including logisticians, MPs, and combat engineers. CALL likewise observed that the 10th Mountain Division—a light-infantry division that had supposedly been designed to fight in a low-intensity conflict environment—did not have enough or the right types of trucks and other tactical vehicles to move troops across the great distances that needed to be patrolled and secured. The distances across which Army units had to operate (infantry units commonly operated as far as fifty miles or more away from their parent headquarters) also highlighted the Army's lack of long-range communications equipment.[140]

The Army had also been unprepared conceptually for Somalia. Marine colonel F. M. Lorenze wrote in a 1994 *Parameters* article that the UNITAF commander and staff had to find "innovative" solutions to deal with the "nontraditional functions and activities" they were asked to provide in Somalia, and Major General Arnold wrote, "Battalion commanders and higher tend to be 'stretched' a little beyond conventional operations due to the complexities and the many 'players' involved in operations other than war."[141]

While some argued that Army units needed to be better manned, trained, equipped, and organized for low-intensity conflicts, others were not convinced that the Army should even be engaging in such conflict in the first place. Rafael Moreno and Juan Jose Vega described Somalia as a "'laboratory' of the new and revolutionary ideas of UN Secretary-General Boutros Boutros-Ghali about the imposition of peace and the reconstruction of nations in chaos" that left soldiers trapped in "a mission that neither they nor their superiors understood clearly." Moreno and Vega cautioned future leaders "to pause and reflect carefully on what can be reasonably expected of peacekeepers" before embarking on future Somalias.[142]

Others dismissed the need for special training or skills for low-intensity conflict. For instance, a CALL manual published after the Battle of Mogadishu echoed a warning that had often been repeated by Army chief of staff Sullivan: "Allowing OOTW requirements to distract Army training or equipment acquisition programs would degrade combat readiness and eventually lead to another 'Task Force Smith.'"[143]

Arnold argued that low-intensity conflict tasks were so similar to high-intensity conflict tasks that "well-trained, combat-ready, disciplined soldiers can easily adapt to peacekeeping or peace enforcement missions." He added, "Versatile units with flexible leaders (especially battalion commanders and up) are able to adjust to the complexities faced in operations other than war."[144] Allard dismissed the conceptual challenges Army leaders had in adjusting to low-intensity conflict, writing, "Planning for peace operations is much the same as planning for combat operations—except that peace operations are typically smaller and involve more fine tuning." Allard also went further than Arnold, claiming that while specialized training for low-intensity conflict might be a "good idea," still, "the most basic qualification of our Armed Forces to act as peacekeepers rests upon their credibility as warfighters. Their technical competence and physical prowess allow our soldiers . . . to prevail in any operational environmental." He added, "Success in peacekeeping operations depends directly upon small-unit tactical competence and the bedrock mastery of basic military skills."[145] These were all variations on the argument made by many Army senior leaders that low-intensity conflict competencies were a lesser included subset of high-intensity conflict.

These dismissals aside, low-intensity conflict observers identified many specific skills and capabilities that the Army lacked to operate in the conflict in Somalia. While he dismissed the need for specialized training, Allard did note that the Army had extreme difficulty interoperating with multinational partners. Simply sharing intelligence with international partners—even the commander of UNOSOM II, Turkish lieutenant general Çevik Bir—was complicated by arcane, Cold War–era rules on the handling of classified information. The Army solved the communications interoperability problems that had plagued it during the Gulf War, Allard wrote, by assigning liaison officers with satellite radios to each multinational contingent. He detailed how U.S. forces built a consensus among coalition partners by getting them to agree to ROE—with their own, minor, national caveats—in advance of joining the coalition. Ultimately, Allard—quoting Lt. Gen. Robert Johnston, commander of the Marine contingent in Somalia—concluded that "unity of command can be achieved when everyone signs up to the mission and to the command relationship." Clarke noted that one key lesson of Somalia was that multinational partners could only be relied upon if their contributing nations agreed "with the basic purpose and the goals for which the coalition was created." Without this agreement there could be no "unity of purpose in an operation" because each "coalition partners maintain[ed] direct links with their ministries of defense." Lorenze highlighted the delays in transition from U.S. to UN command as UNOSOM II stood up and individuals from dozens of contributing nations and the UN headquarters trickled into the country to take over the mission. Army major general Waldo D. Freeman, the chief of staff of CENTCOM, writing with Navy captain Robert Lambert and Army lieutenant colonel Jason Mims, highlighted the challenges created by the U.S. legal impediments to providing direct military support to multinational partners and the international political impediments to creating a consensus for action.[146]

Perhaps the most widely publicized lessons of Somalia came in the area of integration between conventional and SOF forces. During the tragic Battle of Mogadishu, there were significant delays in getting UNITAF forces in motion to rescue members of Task Force Ranger because the SOF task force had not coordinated its activities—or even notified the 10th Mountain Division of its activities—until it needed help. This was emblematic of the separate chain of command under which Task Force Ranger operated. General Sullivan's successor as Army chief of staff, Gen. Dennis Reimer, would later blame the debacle in Mogadishu on this arrangement and also muse that the Rangers had gotten somewhat complacent. Yet he also claimed the mission had been a misuse of SOF forces for routine operations in the first place.[147]

Few lessons about security force assistance emerged from Somalia. Walter Clarke did note that UNITAF attempted to establish a police force, enrolling five thousand former police officers to help establish security in Mogadishu.[148] But this effort never took root because of the level of violence in Somalia and the absence of a national government, and few lessons were recorded.

In a clear sign that the Army had not institutionalized the lessons of Panama, Army units in Operation Restore Hope struggled to conduct operations in and among civilian populations. Allard noted that Army leaders lacked education and experience in dealing with foreign populations, highlighting the "importance of knowing the country, the culture, the ground, and the language as a pre-condition for military operations." He concluded that the Army needed specialized training in, among other things, "local culture" before engaging in a low-intensity conflict operation. Arnold was blunter in a December 1993 article in *Military Review*, admitting, "We did not know a great deal about Somalia." To fill this gap, Arnold continued, "UNITAF spent a great deal of time learning more about the clans, the subclans, the political factions, the warlords, the local clan elders, sultans and history."[149]

CALL noted that Army units were unfamiliar with the techniques required to gather intelligence from civilian populations in a low-intensity conflict, writing, "The most basic intelligence in a low-intensity conflict scenario is invariably provided by humans, the best and most important HUMINT [human intelligence] source always being the soldier or marine in the field." Allard added that intelligence in such conflict was not always about the enemy and that "the definition of who or what the enemy is in a peace operation is not always clear." Even civil-military operations centers (CMOCs) that dealt exclusively with local leaders and nongovernmental organizations (NGOs) became sources of intelligence in such an environment.[150]

Civil-military operations were a particular focus of lessons learned in Somalia. CALL noted that Army units, unprepared for civil affairs operations, were forced to "raid [their] dispensaries and mess halls for medicine and food to distribute." CALL's 1993 special manual on Somalia included an entire appendix on NGOs operating in Somalia and how to leverage their capabilities to assist Somalis and understand the complex clan relationships within the country.[151] For Freeman, Lambert, and Mims, "humanitarian relief organizations, various UN agencies and the Somali population" were simply a "challenge" to be overcome in achieving the military mission of providing security. Allard disagreed, writing, "The real 'peacekeepers' in a peace operation are the humanitarian relief organizations (HROs) that provide both aid for the present and hope for the future." Both Major General Arnold and Colonel Lorenze recounted the challenges of disarming the local security forces that protected humanitarian relief organizations (HROs) by day and used their weapons to victimize the populace by night. Lorenze highlighted the establishment of CMOCs at the battalion level as very successful in coordinating military and relief efforts. Arnold highlighted the intelligence and cultural context that individual NGO workers provided to military forces, gained while delivering relief to the Somali people.[152]

Low-intensity conflict observers also noted that the Army lacked other key capabilities to operate effectively in and among the populace of Somalia. Arnold wrote that there was a lack of sufficient medical specialists to allow his division to provide medical services to the vast number of civilians in need of care. His unit also lacked sufficient civil affairs personnel to effectively coordinate division efforts with local leaders and NGOs. He also acknowledged that this latter task of coordinating with local leaders and NGOs was so vast that it could not be relegated solely to civil affairs personnel. Instead conventional units had to "fall back on the doctrinal manuals" and fill this need themselves.[153]

ROE were a particular area of interest for observers trying to draw lessons from Somalia. Major General Arnold wrote that he lacked sufficient lawyers to help his division fight within the ROE, noting that this problem was compounded by the challenging environment in Somalia: "Our soldiers were required to shift rapidly from assisting the Somali people to conducting combat operations, sometimes within seconds." He added, "Extensive training, individual restraint and appropriate rules of engagement" were required to be successful in this environment. Colonel Lorenze noted that combatants in Somalia "used women and children as active participants, with a mix of carefully coordinated infantry tactics," complicating the application of "international humanitarian law."[154]

Allard provided a number of interesting observations about ROE and the nuanced application of force. He wrote that soldiers on the ground learned that while applying ROE was essential, they only worked in combination with "the use of water bottles and smiles as basic negotiating tools." He also noted the connection between the nuanced application of force and the political objectives of the conflict, writing, "ROE are not only life and death decisions but also critical elements in determining the success or failure of a peace operation; that means that the determination of ROE is a command decision."[155]

Allard also noted that the pressures of ROE in operations among the populace drove some to seek less lethal means to exert force in Somalia. Throughout Operation Restore Hope there was a "constant search for more accurate, less deadly munitions . . . to adapt military power to those situations where the line between combat and non-combat is difficult to draw." Marine colonel Lorenze noted that the use of nonlethal weapons like tent stakes, barbed wire on the outside of vehicles, and cayenne pepper spray were effective in deterring looting from moving convoys.[156]

Many observers highlighted lessons learned in Somalia about the need to influence local opinion in order to effectively operate in and among the population. Lorenze highlighted that UNITAF violence against women and children who were used as combatants by warring parties in Somalia created a backlash among the populace even though it was a legal use of force. He added that UNITFAF PSYOP personnel countered this press with announcements in Somali-language papers and local radio broadcasts to publicize atrocities by warring factions. Arnold also highlighted the ways in which PSYOP teams used "newspaper, radio, leaflet drops, [and] personal contacts" to counter "wild claims by warlords" of U.S. atrocities. In fact, this aspect of the conflict was so important that the 10th Mountain Division designated it as a "battlefield operating system" on par with intelligence and maneuver. Still, Arnold complained that his unit lacked sufficient PSYOP capabilities to adequately influence civilian populations.[157]

Arnold likewise complained that his unit lacked sufficient public affairs officers. According to Allard, U.S. forces within UNOSOM II had no public affairs personnel at all. Still, Army units learned important lessons about influencing U.S. and international public opinion about Operation Restore Hope. Allard noted that "an efficient means of dealing with visitors, including not only the media but congressional leaders and other public figures," was critical in shaping international opinion about the low-intensity conflict in Somalia and building a consensus within the coalition. Lorenze noted two incidents in which the use of deadly force created negative press for U.S. forces both at home and internationally,[158] incidents that might have been mitigated with sufficient public affairs personnel.

An unhealthy preoccupation with the metrics of population security prevented some observers from moving beyond the acknowledgment of low-intensity conflict's nature as occurring among the populace to an understanding of the political dimensions of the conflict in Somalia. For instance, Allard argued, "The best measures of success may well be those that signal reductions in the level of violence." He added, "Other important indicators may be expressed in terms of the numbers of children being fed, gallons of potable water being pumped, or weapons being turned in." These measures, he told readers, could help in "answering one basic question: 'How will we know when we have won?'"[159] This preoccupation with measuring success was dangerous; it created the illusion of progress while obscuring the true but inherently qualitative and unquantifiable goal in Somalia: a political solution to the conflict.

Some observers did at least come to acknowledge the political dimensions of the low-intensity conflict in Somalia. As Arnold wrote, "Our battalion commanders became quite knowledgeable on the history and *the current political power structure* of their assigned areas. We used this collective knowledge and lessons learned to help guide our daily operations."[160]

CALL also tacitly acknowledged the political dimension of the conflict in its 1993 special edition manual on tactics; it began with a detailed treatise on the history, culture, clans, and political factions and the ways that these elements impacted the low-intensity conflict environment in Somalia. The manual also provided a list of cultural and political "dos and don'ts" to help soldiers operating in the environment to avoid aggravating the delicate political situation in the country.[161]

For observers uncomfortable or unfamiliar with the political dimensions of low-intensity conflict, "negotiations" became a euphemism for engaging in the politics of Somalia. Arnold wrote that his leaders "conducted direct negotiations with warring faction leaders within each HRS and attempted to establish a dialogue between the factions, as well as establish a process for disarmament." Allard went further, writing that peacekeepers conducted "delicate negotiations with clan warlords to assure the security of relief supplies." He posited that "negotiating skills and techniques were essential to mission accomplishment," adding that whether they wanted to admit it or not, U.S. Army units disarming Somali factions were engaged in politics. Allard explained that disarmament was a "bright line" after which U.S. forces were engaged in the political dimensions of the conflict because "in societies where peacekeeping may be needed, the distribution of arms reflects internal power structures (political, cultural, ethnic, or even tribal) that can be expected to fight to maintain their position."[162]

Other observers went even further, acknowledging that the Army was a player in the political drama unfolding in Somalia. Former foreign service officer Walter Clarke wrote, "The introduction of a substantial international force into Mogadishu and southern Somalia in December 1992 directly affected the internal lines of communication and balance of political forces of local leaders who had been at war with one another for nearly two years." He added that engaging these political forces was essential to ending the conflict: "From the outset, it was clear that the success of the Unified Task Force

(UNITAF) would be judged not by how many people it helped to feed, but by the political situation it left behind." For Clarke, starvation was a symptom of the political problem—"The starving of Somalia were primarily minority clans or refugees. . . . away from their home areas and rejected by the warlord leaders of opposing clans"—and, while UNITAF had succeeded in feeding the starving, "what was missing was a strategic vision for Somalia, one that could have integrated political goals with the missions assigned to the military." Clarke concluded that UNOSOM I had set the stage for the later failure of UNOSOM II by "letting the appointees of Ali Mahdi and Aideed [sic] provide political, police, and judicial liaison with UNOSOM and UNITAF," which had "proved to be a very ineffective expedient" because it reduced the "likelihood that effective political processes would be established by Somalis not associated with them." But, more fundamentally, Clarke insisted, simply by establishing security and feeding the populace UNITAF was engaging in a political act. He noted that "by making it harder for some of the gangs of thugs to do business, UNITAF, and by extension the United Nations, earned the enmity of the warlords so affected."[163]

Similarly, historian Caleb Carr wrote just after the Battle of Mogadishu that Somalia was forcing military and civilian leaders to acknowledge "the long overdue recognition that military intervention, by definition, cannot be nonpolitical: if we send American and U.N. forces abroad because of a "humanitarian" crisis, we will necessarily come into conflict with those indigenous political leaders who are not capably addressing—or, as in the case of Somalia, are actively abetting—that crisis."[164] Carr added, "There is no middle road in such a conflict . . . : if we enter the fray we become political players, whereas any attempt to portray a political conflict as a humanitarian crisis is simply sidestepping the terrible choice before us." The Army could not succeed in such an environment, Carr concluded, with "half-measures aimed at anything less than a thorough reordering of the . . . tribal, ethnic, and religious factions."[165]

Even Major General Arnold, commander of the 10th Mountain Division in Somalia, seemed to have reached this realization. In his December 1993 article in *Military Review*, he applied the Clausewitzian idea of the "center of gravity" to low-intensity conflict in a fascinating way that would not be seen again until the reemergence of counterinsurgency doctrine in the Iraq War more than a decade later. He wrote, "In my view, the center of gravity of the operation in Somalia is the erosion of the independent power of the warlords." More important, he saw a role for military force in facilitating this erosion: "This can be accomplished voluntarily, as warlords join in a coalition, confederation or some form of national government, or it can be done involuntarily as military situations arise." Arnold did concede that Somalia would not be "on the road to full recovery" until it reached a political solution either with or despite the warlords. Yet, overly optimistically, he concluded, "We have come very close to establishing the right environment to enable the Somalis to arrive at a '*Somali solution*.'"[166]

The idea that the Army had a duty to engage in the political dimension of the low-intensity conflict in Somalia was very contentious. Many—both low-intensity conflict observers and conventional Army thinkers—argued that the Army shouldn't or couldn't engage in the politics of a host nation. For instance, Lorenze argued that in a foreign culture, "Alliances are complex and probably cannot be understood by someone who is not native."[167]

Others sought to limit the mission of the Army in Somalia so that it might avoid having to engage in the politics of the low-intensity conflict. This tendency sprang from the Powell Doctrine's principle that U.S. military forces should not be committed to combat without a clearly defined mission and end state. For instance, Allard argued that UN mandates needed to be more "precise and fully reflect a clear understanding of a given situation and its military implications." Freeman, Lambert, and Mims argued that U.S. forces needed an "early and clear definition of an achievable mission" to avoid having to engage in the political dimension of low-intensity conflicts. Rafael Moreno and Juan Jose Vega insisted, "Do not begin participation until there is a clear end-point established," and added, "One of the biggest flaws of peacekeeping operations is that no one knows when they will finish." Moreno and Vega also invoked two other principles from Secretaries of Defense Colin Powell and Caspar Weinberger, insisting that the United States must "clearly define why the mission is necessary in the light of national interests and be sure that public opinion and government understand and share the definition."[168]

Allard insisted that UNOSOM II had failed because Army forces had strayed from their CENTCOM-assigned mission, which he claimed was to "provide military assistance in support of emergency humanitarian relief to Kenya and Somalia." For Allard, the UN insistence that Army forces should disarm the warlords to force a political settlement had doomed the effort. He wrote, "To commit military forces to the mission of forcibly disarming a populace is to commit those forces to a combat situation that may thereafter involve them as an active belligerent"—that is, a political participant in the conflict. Allard did admit that the Army's mission in Somalia implied the responsibility to "restore order in southern Somalia," but used the CENTCOM mission statement to argue that once the Army had established "a secure environment for uninterrupted relief operations," the operation should have been terminated and transferred to UN peacekeeping forces.[169]

This idea that the Army could avoid the political dimension of low-intensity conflicts like the one in Somalia by dumping these aspects of the mission on the UN pervaded discussions about the war. For instance, Freeman, Lambert, and Mims argued that dumping the mission on the UN—"effecting a smooth transition to UN forces as soon as practicable"—*was* the U.S. political objective in Somalia because "successful peacemaking and humanitarian operations in Somalia under direct UN control would further enhance the UN's leadership role in resolving future conflicts."[170] They made this incredible claim despite the fact that neither President George H. W. Bush nor President Bill Clinton had ever stated that legitimizing the UN was a goal of Operation Restore Hope.

Freeman, Lambert, and Mims faulted NGOs and the UN for being reluctant to assume the mission despite assurances from the United States that it would remain committed to the peacekeeping operation—assurances that ultimately rang hollow as America completed an ignominious withdrawal. Likewise, Allard accused UN secretary-general Boutros Boutros-Ghali of delaying the transition in hopes that a delay would give the U.S. forces more time to disarm the warlords. Allard also faulted Boutros-Ghali for advocating rebuilding "the country's fragmented institutions 'from the top down'—an exercise akin to nation-building,"[171] implying that nation building was an illegitimate mission for military forces.

On the other hand, Clarke saw this relegation of politics to the UN itself as the root of problems in Somalia. This he thought problematic because, "unable to look beyond the warlords, [the UN] failed to develop coherent political goals for the entire population of Somalia." Meanwhile, he added, "UNITAF forces generally followed a policy of nonconfrontation with [the warlords]. In the end, this policy was interpreted by some of the warlords and their allies as weakness." As a result, Clarke concluded, "the UN was unable to fill the political vacuum that existed in the UNITAF area of operations," setting the stage for the disastrous Battle of Mogadishu.[172]

Carr similarly decried these attempts to avoid engaging in the political dimension of the low-intensity conflict in Somalia: "First and foremost, the United States and the United Nations entered the affair believing that they could place combat troops in a foreign country and then direct them to ignore existing political realities and pursue extra-military—which is to say extra-political—ends." While he acknowledged that "many in the West are uncomfortable with such an approach," Carr insisted that engaging in the politics of Somalia was essential to achieving an enduring solution to the conflict.[173]

This debate exposed a fundamental flaw in the Army's doctrine for low-intensity conflict. Born of the supposed lessons of Vietnam, the Nixon Doctrine, and the Reagan-era model for low-intensity conflict, the core tenet of Army low-intensity conflict doctrine held that the responsibility for the political dimension of such conflict rested with the host nation government. But this tenet was based on the unspoken assumption that there *was* a host nation government and, unfortunately, Somalia had no government to assume this responsibility. As Lorenze wrote, "There was little precedent for Operation Restore Hope, in large part because there was no sovereign nation." He added, "The term 'host nation support' had no real application in Somalia." Arnold lamented, "Neither a national government nor regional governments existed; only self-appointed local leaders bent, for the most part, on extortion and abuse of power." He continued, "Nonexistent were the police, justice system, schools, public water, public electricity and transportation system."[174]

With the UN unreliable and the host nation nonexistent, those desperate to see the Army avoid the political dimension of the low-intensity conflict in Somalia searched for other places to dump this responsibility. Allard argued, "The real responsibility for nation-building must be carried out by the civilian agencies of the government better able to specialize in such long-term humanitarian efforts." The State Department was the leading candidate to take over the politics of Somalia. Freeman, Lambert, and Mims wrote that the State Department was relied upon to "provide a linkage between the United States, UN agencies and the various political factions in Somalia."[175]

The Army might have been desperate to have the State Department insulate it from the politics of Somalia, but the State Department completely lacked the capacity to deal with a disaster on the scale of Somalia. As Arnold wrote, "There were not adequate State Department or other diplomatic personnel in-theater to help in each HRS, so our leaders . . . became involved in negotiations and disarmament talks with warlords and faction leaders."[176]

Other cognitive obstacles also blocked Army units from completely engaging in the political dimension of the low-intensity conflict in Somalia. Perhaps the biggest of these obstacles was the preoccupation of U.S. Army units in Somalia with force protection. Along with influencing the population, the 10th Mountain Division also elevated force protection to a "battlefield operating system" on par in importance with intelligence, mobility, or sustainment. CALL warned Army forces against becoming familiar with the populace due to force protection concerns, advising that commanders "impose substantial limitations on off-post travel."[177] While this move might have helped keep soldiers safe, it also made it harder for them to interact with the population and engage in the political dimension of the conflict.

Notably, a few low-intensity conflict observers moved beyond a basic understanding of the political dimensions of the conflict and began to understand that the international community had, in fact, picked sides in the conflict. Clarke pointed out that UNOSOM I did unintentionally take sides; he wrote that UNOSOM I accorded General Aidid "virtual chief of state status by various diplomatic and business delegations" and added that Aidid "was permitted by UNOSOM I to determine the membership of police and judicial committees, some of which became extensions of his broad-based criminal organization."[178]

Many—both well-meaning low-intensity conflict observers and more mainstream Army thinkers—instinctively rejected the idea of taking sides in the conflict in Somalia. Lorenze wrote that UNITAF—having limited itself to providing security—was successful in its mission. He did acknowledge that UNOSOM II had the expanded mission of nation building but confusingly insisted that it should do this without taking sides: "The challenge for UNOSOM II is to accomplish the expanded mission without becoming embroiled in the factional fighting to the point of backing one faction against the others."[179] Allard noted, "During operations where a government does not exist, peacekeepers must avoid actions that would effectively confer legitimacy on one individual or organization at the expense of another." He fully understood the difficulty of maintaining neutrality in a conflict, writing, "Because every military move will inevitably affect the local political situation, peacekeepers must learn how to conduct operations without appearing to take sides in internal disputes between competing factions." Echoing contemporary doctrine on OOTW, Allard still believed, however, that this neutrality was key to the "legitimacy" of the peacekeeping force itself. He also cited Ambassador Robert Oakley as pointing out how ROE could assist "peacekeepers who wish to avoid becoming active belligerents." CALL insisted in its 1993 special edition manual that soldiers must "be impartial" and not "discuss or comment on the opposing forces except in the performance of duty" to maintain their legitimacy as peacekeepers. Freeman, Lambert, and Mims wrote that CENTCOM established guidelines for U.S. forces in

Somalia to ensure that "all coalition forces were treating the many diverse Somali factions in essentially the same manner."[180] By not picking a winner, the international community was condemning Somalia to war without end.

A more insidious obstacle to accepting the necessity to pick sides in the low-intensity conflict in Somalia was the desire to see an inclusive, Western-style democracy in the country that respected the rights of all of its citizens. Clarke, who otherwise had a clear-eyed understanding of the politics of the conflict, fell prey to this cognitive trap, writing, "The ability to provide the people of Somalia an opportunity to act politically without coercion would constitute a success for the various international groups in Somalia." Clarke believed "the organization of a national conference should become a funda-mental component of every United Nations Chapter VII operation."[181] His solution, while noble, missed the essential fact that empowering the power-less citizens of Somalia was essentially picking the side that had already lost the conflict before the peacekeeping force ever arrived; they had lost to all of the factions that victimized them—the factions that vied for power through the force of arms, inflicting violence and starvation on each other and the helpless populace outside their respective camps.

Notably, Allard criticized the commander of the Italian contingent for en-gaging in exactly the activities in which U.S. Army forces desperately needed to engage to end the conflict; the Italians had picked sides, opened nego-tiations with General Aidid, and attempted to build a political settlement. The Americans objected, and cajoled the UN command into demanding the Italian commander's relief and removal. Neither occurred and, as Allard put it, "life went on."[182] One can only speculate as to how differently Operation Restore Hope might have unfolded if the U.S. Army had spearheaded this same effort.

A very few observers did finally reach the understanding that the Army did, in fact, *have* to pick sides in Somalia—or any low-intensity conflict, for that matter—to solve the political problems that precipitated the conflict in the first place. Carr wrote that a key lesson of Somalia was that "we must determine before going in the legitimacy of those leaders, as well as (should they be at war with each other) the validity of their various positions in the conflict." The inescapable consequence of this determination, he added, was that "we must be prepared, should we choose military intervention, to arrest [illegitimate] leaders, disarm their followers, and create a U.N. protectorate whose term may be far longer even than that in Somalia." Ironically, Moreno and Vega, who questioned the wisdom of engaging in peacekeeping missions, agreed with Carr, noting that U.S. forces should "never enter in the middle of a civil war." Instead, the United States should "choose a side and then try to achieve peace."[183]

Army units and interested observers did learn lessons from Somalia, but the Army refused to institutionalize them. Instead it engaged in assigning blame elsewhere. The Army's failure in Somalia, which had led to the death of eighteen special operators and countless Somalis, was not, Army senior leaders and pundits concluded, a result of a faulty approach or the Army's stubborn refusal to engage in the politics required to produce peace. Rather, this faction argued, it was the result of "mission creep"—a gradual changing of the mission assigned to military forces over the course of a low-intensity conflict—imposed on U.S. forces in Somalia by the UN or by the Clinton administration's unclear strategic objectives. In this manner the Army summarily dismissed the chorus of observers arguing that Somalia was the shape of wars to come and that the Army needed to reform itself to better fight low-intensity conflicts, thereby closing the book on the entire episode and resuming its headlong race toward ever-greater capacity to fight high-intensity conflicts.

The UN bore the brunt of criticism over Somalia. Allard noted that the increasing number of peacekeeping operations had "strained the ability of the United Nations to manage them effectively," adding, "Because the United Nations also lacks standard doctrine, tactics, and equipment, command and control is a problem for all but small operations in generally peaceful environments." He also highlighted problems with the manning of the UN staff at UNOSOM II, blaming the UN's "foot-dragging" in assuming the mission from U.S. forces. He also blamed the other contributing nations for failing to send all of the pledged forces or adhere to their "terms of reference" with the UN once the environment became less permissive. Allard concluded, "No soldiers of any nationality should be expected to serve under the U.N. command structure in any combat setting until . . . reforms . . . have been put in place."[184]

Others blamed mission creep. Arnold wrote, "Our mission was to secure relief operations in our assigned Humanitarian Relief Sectors (HRS) and break the cycle of starvation." "Disarmament," he observed, "was one of those missions that was added to the plate." Somewhat diplomatically, he added, "Some 'mission creep' is inevitable in operations other than war. . . . Some nation assistance at the grass-roots level became necessary in Somalia, although it was not an assigned mission." Moreno and Vega were blunter, writing that a key lesson of Somalia was, "Do not change the mission's objectives in mid-undertaking."[185]

Many blamed the UN for this mission creep. Lorenze noted that "the UNITAF Commander had stated on numerous occasions that disarming Somalia was not his mission. He maintained this position despite several statements to the contrary by the UN Secretary General." Yet, Lorenze added, the UNITAF commanders were eventually compelled to begin confiscating weapons.[186] Freeman, Lambert, and Mims also blamed the UN for mission creep, insisting, "Forces assigned to Restore Hope were tasked to provide a safe environment for humanitarian operations while operating under a UN mandate." While CENTCOM crafted "a clear, achievable mission statement for the operational commander in Somalia," they claimed, it "received numerous requests for coalition forces to perform additional tasks." The UN "pressed UNITAF to begin disarming factional militia forces . . . [and] sought to involve UNITAF in reestablishing a national police force for Somalia and assisting in the repatriation of Somali refugees." These tasks, they concluded, "represented a new phenomenon labeled 'mission creep.'"[187]

While he cited the UN as his primary culprit for mission creep, Freeman, Lambert, and Mims also took a thinly veiled swipe at the other agencies of the U.S. government operating in Somalia: "The phenomenon also points out the difficulty of achieving consensus when other agencies with key roles in the operation have differing views of the desired end state." Only the "persistence, patience and pragmatism" of the commander in chief of CENTCOM, they claimed, thwarted these efforts by other government agencies to expand the mission.[188]

Not everyone, however, was ready to excuse Army forces for the failure in Somalia. Allard argued that Army units in Somalia lacked a clear understanding of their mission, and laid the blame for this on the military leaders themselves: "The U.S. mission to support UNOSOM II . . . was considerably more open-ended, although this fact may not have been well appreciated when the operation began." He added, "The real issues were the lack of agreement between the United States and the UN about the conditions at the time of the transition and the military capabilities required to carry out the expanded mandate of UNOSOM II." Allard did make vague recommendations about using metrics to limit missions and clearly define exit strategies but stopped short of claiming U.S. forces had exceeded the parameters of their mission. For Allard, the problem was not mission creep but mission clarity; he concluded, "Clear U.N. mandates are critical to the planning of the mission because they shape the basic political guidance given to U.S. forces by our National Command Authorities."[189]

Clarke was even more explicit in rejecting charges of mission creep, writing, "Complaints about a change in the mission are unjustified. By its very nature, Operation Restore Hope was always more than a simple humanitarian operation." He insisted that the mission clearly created "a mandate for an extended period of 'nation-building' in Somalia" and blamed the U.S. forces' initial foot-dragging on disarming the warring factions—artificially and willfully limiting their mission—for dooming UNOSOM II. In a direct challenge to those who argued that the UNITAF mission had been a success and UNOSOM II had lost the war, Clarke wrote, "The strongest criticism leveled at the UNITAF intervention in Somalia is not that it did so much, but that it did so little. UNITAF did not disarm the warlords or establish law and order." He added, "A narrow mandate can be pursued in future such operations, but in the end someone must pay the price of earlier short-term successes." He concluded that this refusal to engage in the messy political dimension of the low-intensity conflict in Somalia, "combined with the lack of a clear political agenda for Somalia, greatly reduced the likelihood that UNOSOM II could ever have attained its political and nation-building objectives."[190]

Carr was equally explicit in rejecting the idea of mission creep, writing, "Obviously, the general Somali population could not be helped . . . unless the feuding clans were disarmed; but the U.N. forces declared that the task of disarmament exceeded the terms of their mandate." Citing Clausewitz, he concluded, "The notion of a nonpolitical military intervention is therefore worse than chimerical: it is oxymoronic, carrying with it from the start the seeds of its own eventual frustration."[191]

The Battle of Mogadishu did not cause the U.S. failure in Somalia; rather, this disastrous battle was the culmination of a failure caused by the Army's insistence on artificially limiting its mission to avoid the political dimension of the conflict and its refusal to identify and then back a winner in the conflict. Yet, many—Gen. Dennis Reimer chief among them—instead cited the rejection by Secretary of Defense Les Aspin of the request by Maj. Gen. Thomas Montgomery, the commander of U.S. Forces Somalia and deputy commander of UNOSOM II, for the deployment of armor to Somalia as the reason for America's failure there. Secretary of Defense Aspin ultimately resigned in large part over these accusations. It is true that without tanks or armored personnel carriers, the QRF from the 10th Mountain Division sent to rescue Task Force Ranger on the night of October 3–4, 1993, during the

Battle of Mogadishu had to rely on Pakistani armor for protection and mo-bility.[192] But it was the lack of will, not the lack of tanks, that had prevented the Army from forging a sustainable political settlement in Somalia in the months that preceded this battle.

Others—implicitly or explicitly—condemned the Bush and Clinton administrations for their decision to intervene in Somalia in the first place. Moreno and Vega suggested that, in light of the UN's ineptitude in managing peacekeeping missions, the United States should "return to the cautious peacekeeping practices of the past; to be more realistic in the expectations of what can be accomplished and the objective possibilities of each of the peace missions." They conceded, "European and American public opinion . . . displays very low levels of tolerance to suffering" and warned, "a few deaths from their own side generate[s] a sense of panic."[193]

The Clinton administration and the U.S. Department of Defense were clearly chastened by their experience in Somalia. The administration's 1994 National Security Strategy reflected many of the objections that critics of low-intensity conflict interventions had expressed after the Battle of Mogadishu. Alluding to the Powell Doctrine, the strategy promised to "employ rigorous criteria" before the United States "consider[ed] contributing U.S. forces to a UN peace operation."[194]

The Army embraced the excuses pundits had constructed to absolve the Army of its guilt for the failure of Operation Restore Hope in Somalia and actively avoided learning lessons from this low-intensity conflict. The frank, unfavorable after-action review from Maj. Gen. David Meade, commander of the 10th Mountain Division, was poorly received by the senior Army leadership. Army chief of staff General Gordon Sullivan instead dispatched the commander of U.S. Army Forces Command, Gen. Dennis Reimer, to Somalia. Reimer returned and provided a verbal report to the chief of staff but fastidiously avoided putting his observations in writing. While he later claimed that this was done to protect "sensitive operational methods," this strains credibility given both that he derides Meade's report in the same breath and that the Department of the Army headquarters in the Pentagon is constantly awash in secret documents about "sensitive operational meth-ods" that are routinely safeguarded against release.[195] It is much more likely that both Reimer and Sullivan were afraid that his report might be used as ammunition by those outside the senior Army leadership to force the Army to reform itself.

Rather than institutionalize the lessons of Somalia, the Army institution-alized a visceral aversion to peace operations. FM 100-23-1, *Multiservice*

Procedures for Humanitarian Assistance Operations, published in 1994, a year after the Battle of Mogadishu, made it clear that the Army should not be used to intervene in low-intensity conflicts in the absence of a strong host nation government.

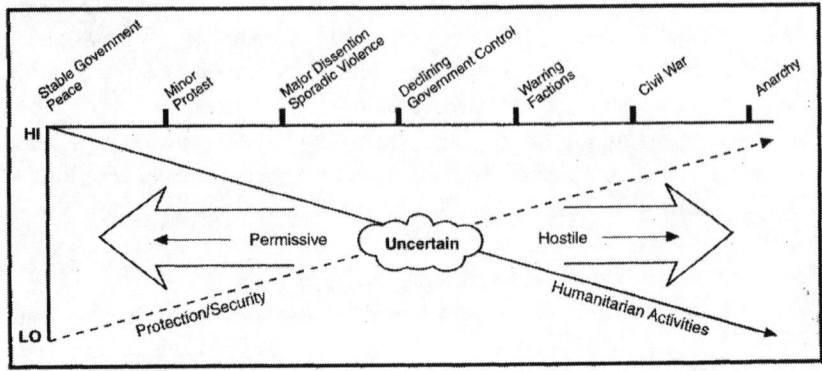

Figure 2.2. Humanitarian operations after Somalia. *Source:* U.S. Army Training and Doctrine Command, *Multiservice Procedures for Humanitarian Assistance Operations*, FM 100-23-1 (1994), 1-7.

The implication of this diagram could not be clearer. As the environment of an operation moved from major dissention and sporadic violence toward declining government control, the requirements for protection and security went up and the ability to execute humanitarian activities went down. The outcome of low-intensity conflict operations in an environment at or beyond declining government control became uncertain.[196] This was a clear reassertion of the Nixon Doctrine and the Reagan-era model for low-intensity conflict: all such conflicts should be prosecuted through a host nation government.

FM 100-23-1 also spelled out conditions, based on America's experience in Somalia, that had to be in place before PKOs could even be contemplated. For instance, the manual insisted that for UN PKO to be successful, "hostile parties must be separated and substantially disarmed." The manual was silent on what other military force would come into the country and take care of disarming the belligerents before the U.S. Army arrived. The manual also set other impossibly high bars for Army participation in a peacekeeping operation, writing that such an operation will have "an even higher probability of failure" if "the mandate is ambiguous or unclear," "weapons are readily available," or "the UN PKO chain of command is poorly disciplined."[197] This was a virtual checklist of conditions that were almost certain to be present in any peacekeeping operation—a backhanded way of saying that the Army should not be committed to low-intensity conflict.

The manual also enshrined the idea of mission creep in Army doctrine. The passage on this topic was a blatant renunciation of the Army's guilt for failing to engage the politics of the low-intensity conflict in Somalia, instead laying the blame on the UN.[198] In part this passage read,

> Military forces will undoubtedly receive numerous requests to perform additional tasks, as was the case in Somalia. . . . The UN also pressed the force to begin disarming factional militia.
>
> These tasks represented the phenomenon labeled *mission creep.* . . . Due to political agendas, key participants in the operation sought to expand the unified task force (UNITAF) activities and AOs beyond the initial, carefully limited scope of securing the environment for humanitarian relief operations.[199]

This passage is nearly a word-for-word recitation of Freeman, Lambert, and Mims's complaint from their *Military Review* article published the previous year.[200]

Missing from all of these analyses was the fact that it was the United States that expanded the mission in Somalia. As Allard rightly points out, after the June 5, 1993, attack that killed twenty-four Pakistani soldiers, the United States was the lead country in the drafting of UN Security Council Resolution 837, which set UNOSOM II to the task of finding Mohamed Farrah Aidid.[201]

Regardless of the facts, the Army codified its dismissals of the lessons of the low-intensity conflict in Somalia, closed the book on the entire episode, and resumed its headlong march toward ever-greater capacity to fight high-intensity conflicts.

Force XXI and Haiti

At the same time the tragedy was unfolding in Somalia, the Department of Defense was in the depths of a reexamination of the Army's purpose in the post–Cold War world. As President Clinton assumed office, his new secretary of defense, Les Aspin, began a Bottom-Up Review (BUR) of U.S. defense capabilities. The Office of the Secretary of Defense, the Joint Staff, and each service's staff formed standing committees to address different topics within the review. Despite the fact that the BUR unfolded in the aftermath of the Battle of Mogadishu, it still concluded that the services should remain focused on high-intensity conflict. The U.S. military would shape itself to simultaneously fight two "major regional conflicts" (high-intensity conflicts), presumably in Iran or Iraq and Korea. The BUR process was wracked with service parochialism and defense contractor mischief and further delegitimized by the resignation of Secretary Aspin,[202] but preparation for two major regional conflicts would still remain in the U.S. military's strategy for the rest of the century.

Even as it concluded that it should remain focused on high-intensity conflict, the Army found itself grappling with the inconvenient truth that it was heavily engaged in low-intensity conflicts around the world. By June 1994 the Army had soldiers deployed in more than sixty countries. Five hundred soldiers in Skopje, Macedonia, helped an international coalition enforce an embargo against Serbia. In Incirlik, Turkey, U.S. soldiers participated in the effort to feed Kurdish refugees in northern Iraq. More than twenty-one thousand soldiers were deployed worldwide, a more than 300 percent increase in operational tempo since spring 1990. Even after the last forces left Somalia, nearly sixteen thousand soldiers were still deployed abroad.[203]

Over this same period—between spring 1990 and June 1994—the drawdown of the Army continued. The Army's budget had been cut by 40 percent. In response to the BUR, the chairman of the Joint Chiefs of Staff, Gen. Colin Powell, proposed a "base force" that would reduce the number of personnel in the U.S. military by 25 percent. The Army's share of this manpower reduction was a cut from two million to 1.5 million soldiers in active, reserve, and National Guard forces, a greater reduction in total personnel than the other three services combined saw in the same period. The posture of the Army was also changing. The Department of Defense closed sixty-two installations across the United States and 380 overseas, and more than half of those were Army installations. More than 100,000 soldiers had been redeployed from overseas bases back to the United States, leaving 125,000 soldiers still forward stationed at bases in Europe, Korea, and Panama.[204] While its overseas presence had been reduced, the Army was still forward deployed and preparing to fight a great power war against a nonexistent peer competitor.

Gen. Gordon Sullivan's Louisiana Maneuvers were well underway, refining the required capabilities, envisioning and designing the force to fight this theoretical future war, and developing the strategy to realize this vision. On March 8, 1994, Sullivan announced that the future force he envisioned would be called Force XXI.[205] In their April 1994 *Military Review* article, Sullivan and Dubik announced that the strategy to realize this vision would be called "transformation." Force XXI would be optimized to fight in "the conditions under which America will use its Army," including "coalitions, sometimes ad hoc; interagency operations; precise rules of engagement, executed under the eye of near-instantaneous, global media; [and] perhaps unreasonable expectations concerning casualties."[206] At first glance, these conditions—most of which were present in Somalia—seemed to indicate that Sullivan had taken to heart the lessons of Operation Restore Hope and was rebalancing the Army to deal with both low- and high-intensity conflict environments.

But then Sullivan and Dubik described the type of force they envisioned to succeed in this environment, and it became clear that they were still laser focused on dominating imagined future peer competitors in a high-intensity conflict. Force XXI would prevail in the conditions they described by harnessing information age tools—"speed, customization and precision," which had "already arrived on the battlefield." Sullivan and Dubik expected this force to face "information age peers" and defeat them through superior knowledge and speed of decision and action. They wrote that this future force would "know where the enemies are and are not"; would know where its own forces were, "much more accurately than before"; and "this enemy and friendly information" would be "distributed among the forces . . . to create a common perception of the battlefield among the commanders and staffs." Sullivan and Dubik explained, "Speed and precision result from maneuver platforms, fire support and sustainment systems and command and control platforms that are linked digitally." The article even provided a graphic to illustrate the accelerating speed of warfare. The figure was built around the "observe–orient–decide–act" (OODA) loop framework developed by Air Force colonel John Boyd in the 1950s and popularized in the 1970s and 1980s by RMA enthusiasts.[207]

In an extended version of the article published as a white paper by the Strategic Studies Institute (SSI) of the U.S. Army War College, Sullivan and Dubik expanded on this idea of accelerating decision making, writing that transformation would "allow our task forces to observe, decide, and act faster and more precisely than before." They promised that "in the future, soldiers will be able to mass the effects of fire support or maneuver forces from dispersed locations and do so nearly simultaneously." They summarized, "Speed and precision, lethality and versatility—these are the capabilities that the Army is building into its forces today."[208]

As the Army expanded on these ideas in the following months, it began to conceptualize not just how Force XXI would fight but how it would get to the battlefield and sustain itself during the fight. An October 1994 Army

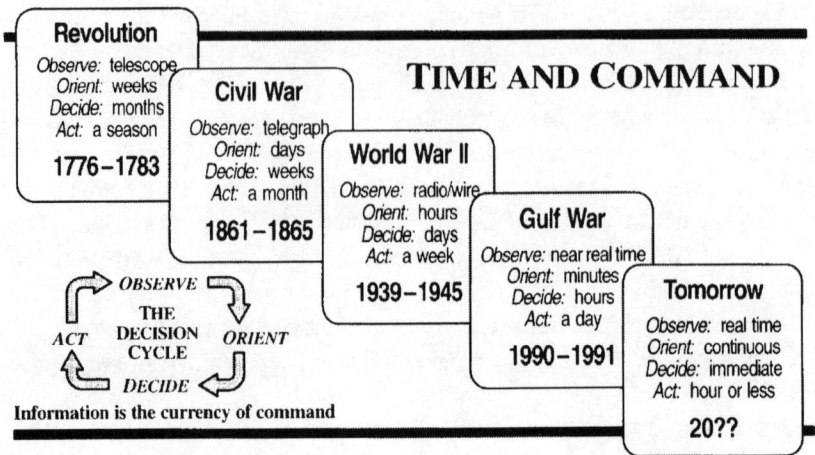

Figure 2.3. Transformation and the speed of decisions. *Source:* Gordon Sullivan and James M. Dubik, "War in the Information Age," *Military Review* 74, no. 4 (April 1994).

white paper explained, "The Army must be capable of dominating maneuver, conducting precision strikes, winning the battlefield information war, [and] protecting the joint force." But the Army also wished to address the deployability problems that had been so visible in the build up to the Gulf War. This white paper explained that the Army had to be capable of rapidly "projecting and sustaining combat power."[209]

All of these requirements and concepts were codified in the August 1994 publication by the U.S. Army Training and Doctrine Command (TRADOC), in TRADOC Pamphlet (PAM) 525-5, *Force XXI Operations*. The core concepts of this aspirational document that would define the shape of Force XXI as "*doctrinal flexibility, strategic mobility, tailorability and modularity, joint and multinational connectivity,* and the *versatility to function in War and OOTW.*" The first three concepts described how the Army would fight a future high-intensity conflict against an imagined peer competitor. Facilitated by "quality leaders and soldiers," the principle of doctrinal flexibility would allow the Army to deal with an environment that would be "varied and multifaceted and have a great potential for surprise across the operational spectrum." Strategic mobility would allow the Army to overcome the deployability issues it had consistently experienced both in the Gulf War and in the multiple low-intensity conflicts with which it had been confronted since the end of the Cold War—all at unexpected times and locations. Tailorability and modularity would allow the Army to rightsize itself for the conflicts with which it was faced, providing it with the ability to leave behind capabilities it didn't need and add capabilities it did need for a specific conflict. But the last concept was a euphemism to allow the Army to ignore low-intensity conflict and continue to focus on a supposedly imminent future high-intensity conflict. Versatility in War and OOTW provided an Army optimized to meet the "requirement to be trained and ready—to win the land battle," which "remain[ed] the absolute priority" while still being able to passably engage in low-intensity conflicts. Well-trained and disciplined units, PAM 525-5 explained, "provided with sufficient time and resources to train, can transition to OOTW missions as required."[210]

With the path for transformation set, the Army's regimen of experimentation accelerated. Progress on developing and fielding technologies was monitored by a new organization created in July 1994, the Army Digitization Office. The Louisiana Maneuvers Task Force moved out of Fort Monroe, and into the Pentagon so that General Sullivan could exert more direct control over the transformation strategy they were developing: the Force XXI Campaign Plan. Technologies that showed promise in simulation exercises were tested in the field with units in "advanced technology demonstrations" and "advanced concept technology demonstrations" against criteria jointly developed by the Army and defense contractors. As units were built with these new technologies, the units were tested against the desired capabilities for Force XXI in advanced warfighting demonstrations and, finally, advanced warfighting experiments (AWEs).[211]

The first AWE, called Desert Hammer VI, took place in April 1994 during Rotation 94-07 at the National Training Center (NTC) at Fort Irwin, California. A brigade-size force from the 24th Infantry Division (Mechanized) was outfitted with digital equipment that provided position location and communication between combat vehicles. This brigade was pitted against the NTC's highly trained Opposing Force in a simulated high-intensity conflict. This would be followed in November 1995 by Warrior Focus, an AWE with Col. James Dubik's light infantry 2nd Brigade, 10th Mountain Division at JRTC Rotation 96-02. This experiment focused on the digitization of light forces and the implementation of additional night vision capabilities. While the results of these experiments were inconclusive and highlighted problems with fielding digital equipment and training soldiers to use it, they were nevertheless successful enough to prompt the Army to publish a schedule of AWEs culminating in a 1997 test of a digitized brigade, the experimental force.[212]

In October 1994, to capture General Sullivan's vision for Force XXI, the Army produced a white paper titled *Decisive Victory: America's Power Projection Army* that acknowledged the warnings from low-intensity conflict observers that the Army would continue to face more such conflicts, its authors noting, "The Army's challenge . . . will be to respond to threats ranging from regional war, through lesser conflicts, to peace operations." But whereas such observers argued that this meant that the Army needed some specialized low-intensity conflict capabilities, the paper went on to argue, "The nation . . . needs an Army that can respond to a range of missions: from humanitarian assistance through regional war to general war." The Army could not, therefore, afford to specialize in low-intensity conflict because it had to "be prepared, if necessary, to confront any future global threat." Like General Sullivan, the paper argued, "The Army exists to fight and win the nation's wars." The main purpose of transformation was to build an Army capable of "victory over an information-based state" that would "entail not only sufficient destruction of the armed forces and physical warmaking capability, but also dominance of its information system."[213]

For General Sullivan, bent on transformation toward Force XXI, low-intensity conflicts were unavoidable, but largely a distraction from transformation. Sullivan agreed with low-intensity conflict observers that the Army would be forced to engage in more such conflicts, but unlike these observers, Sullivan believed that he had to minimize such conflicts' impact on the Army so as not to impede the progress of transformation. He wrote, "We understand . . . that there is no 'time-out' from our requirements to be trained and ready, to succeed at whatever the nation asks of us." But while the Army dealt with these nuisance low-intensity conflicts, it was "moving out . . . digitizing the battlefield right now" to prepare for the imagined coming great power war against a nonexistent peer competitor.[214]

The euphemism that the Army began to use to describe its nonresponse to the demands of low-intensity conflict was "versatility." As *Decisive Victory* insisted, "The Army does not design units nor do units train against a precise mission, threat, or set of employment conditions"—an ironic assertion given that the Army had been designed and trained against a war with the Soviet Union in Western Europe since the end of World War II.[215]

Nevertheless, this white paper explained, "they are designed to be highly flexible—to be able to be employed in a wide variety of situations." It concluded, "It is this strategic and operational *versatility* that, for example, enables American combat infantry battalions to be not only the best infantry in the world, but also, in the words of both MFO [Multinational Force and

Observers] and UN Commanders, the best peacekeepers in the world."[216] This passage highlighted the foundational assumption of the Army's response to the demands of low-intensity conflict; the "best infantry in the world" were clearly also "the best peacekeepers in the world," because low-intensity conflict proficiency was a lesser included subset of high-intensity conflict proficiency.

One might rightly wonder how Army leaders could justify to themselves the risk of remaining focused on preparing for a highly unlikely great power war while refusing to prepare for what they themselves acknowledged were highly likely future low-intensity conflicts. A *Parameters* article written by Sullivan and Lt. Col. Andrew Twomey in autumn 1994 provides a clue. As Sullivan frequently did, he began by citing historian Michael Howard's charge to military and civilian leaders on preparation for war, that it is "the task of military science in an age of peace to prevent the doctrines from being too badly wrong." In that context, Sullivan and Twomey explained, it was best to prepare for the war for which the risks of failure were greatest: "We must . . . build a force with the capability to win in the most important contingencies, while retaining the versatility, flexibility, and residual force to win across the range of uncertainty inherent in our forecasts of the future." For Sullivan and Twomey, high-intensity conflicts were "the most important contingencies" and the Army would accept risk in all other types of conflict, relying on "versatility, flexibility, and residual force" to muddle along until the next great power war.[217]

While Army transformers generally dismissed the lessons of Somalia, it did affect the thinking of low-intensity conflict observers, and some of this thinking did make it into Army doctrine. In addition to the October 1994 FM 100-23-1, *Multiservice Procedures for Humanitarian Assistance Operations*, the Army also published FM 100-23, *Peace Operations*, in December 1994. While FM 100-23-1 discouraged interventions in low-intensity conflict in the absence of a host nation and enshrined the idea of mission creep, both manuals asserted that the U.S. military *did* have a role to play in low-intensity conflict, a role only it could fill. FM 100-23-1 explained that the Army was "uniquely qualified to plan and execute" humanitarian assistance missions because it had "the organizational structure, educated and trained personnel, essential equipment, rapid worldwide deployability, and ability to operate in austere physical environments."[218] Given the number of leaders who believed low-intensity conflict was a nuisance at best and an illegitimate use of the Army at worst, this was a stark admission.

Not surprisingly, many within the mainstream of Army thought disagreed. In a press conference about the BUR, the chairman of the Joint Chiefs of Staff, Gen. Colin Powell, told reporters that despite the proliferation of low-intensity conflicts, the Army's mission remained "to fight and win the Nation's wars." Col. (Ret.) Harry G. Summers Jr., author of the 1980s book *On Strategy*—which codified the Army's rejection of such conflicts as a legitimate mission for Army forces—was one of the louder voices of opposition to the use of the Army in low-intensity conflicts. In congressional testimony in late 1993, Summers decried U.S. military intervention in low-intensity conflicts as "growing out of civilian academic conceits that one can change the world with the tools of social science."[219] He claimed that military leaders were concerned that "an overemphasis on peacekeeping and other nonmilitary operations would erode the military's sense of its core missions and responsibilities." The labeling of peacekeeping and other low-intensity conflicts as "nonmilitary operations" was a subtle but unambiguous rejection of the legitimacy of interventions in these conflicts as proper missions for the Army. Summers also argued that these "peripheral missions" would erode the Army's ability to fight high-intensity conflicts, writing, "That rifle company [on peacekeeping duty] in Somalia is degrading its wartime skills. . . . We need to understand that, as we saw in the Gulf War, the training of maneuver units today is a full-time job."[220] This was a different line from the one most Army transformers took. Most argued that low-intensity conflict tasks were a lesser included subset of high-intensity conflict tasks and thus required no special training; Summers's argument at least implicitly accepted that low-intensity conflict tasks were somehow different.

Moreover, this rejection of the legitimacy of low-intensity conflict missions was a much harder line than General Sullivan was prepared to take. In their autumn 1994 article in *Parameters*, Sullivan and Twomey wrote that Army forces could be used effectively to implement peace agreements, contain conflicts, or even "impose peace through the forceful disarming of a hostile movement." This could be read as a modest acknowledgment that Operation Restore Hope and other low-intensity conflicts in which the Army was engaged had been legitimate uses of the Army.[221]

Just after the Battle of Mogadishu, the Army gave another signal of its acceptance, at least of the requirement to engage in low-intensity conflicts—and at least a tacit acknowledgment that Army units required at least some additional specialized training to engage in these types of conflicts. In November 1993, the Army conducted a training rotation at the JRTC—recently moved to Fort Polk, Louisiana—that tacitly focused on low-intensity conflict. According to Sullivan and Twomey, the scenario included "ethnic conflict," and the training area "was dotted with soldiers, civilians, and representatives from the same nongovernmental organizations that we have seen in Somalia and Bosnia. Representatives from the International Red Cross, Save the Children, the United Nations Department of Humanitarian Affairs, a United Nations Disaster Assistance Relief Team, CARE, World Vision, [and] media representatives."[222]

Even more promising, the Army seemed to be coming to the realization that war—whether high-intensity or low-intensity conflict—was an activity that occurred in and among civilian populaces. The October 1994 white paper *Decisive Victory* explained, "Decisive victory will most likely require the establishment of control, the domination of land and populations through the occupation of terrain." The new doctrine published during this period went even further, contending that the Army had to do even more than dominate or control populations; to be successful in low-intensity conflicts, these manuals insisted, the Army had to influence, secure, and provide aid to civilian populations. The October 1994 humanitarian assistance manual identified that PSYOP could play a role in "overcom[ing] hostile attitudes of the local populace" and "provid[ing] a commander with real-time analysis of the perceptions and attitudes of the civilian population and the effectiveness of the information being disseminated."[223] This was a dramatic departure from Cold War Army doctrine, in which the sole role of PSYOP was to influence enemy military forces.

Unfortunately, in addition to an aversion to low-intensity conflicts in all but the most ideal situations, and codifying mission creep, this new doctrine also codified certain cognitive obstacles to understanding low-intensity conflict. FM 100-23-1 advocated applying statistical measures of effectiveness (MOEs) to the inherently qualitative process of fighting a low-intensity conflict in and among the populace. The manual insisted, "Commanders need some means to evaluate operations." It did caution that "MOEs cannot cover every aspect of a mission; therefore, commanders should resist heavy reliance on them," but it still insisted that a unit needed MOEs to address the "difficult challenge of determining whether or not the force is meeting mission objectives." The manual then went into great detail on how to develop MOEs that were "*appropriate, mission-related, measurable, reasonable in number, sensitive,* and *useful.*"[224]

Still, both manuals—FM 100-23 and FM 100-23-1—did roll back some of the less helpful changes to low-intensity conflict doctrine made in the 1993 edition of FM 100-5, *Operations*. For instance, while FM 100-23-1 did not restore "political dominance" as a principle of OOTW, it did highlight the political dimension of low-intensity conflicts, explaining, "As compared to war, MOOTW [military operations other than war] are more sensitive to political considerations because of the overriding objective to limit potential hostilities." Similarly, FM 100-23 explained of low-intensity conflict that, "because of the potential linkages between combatants and noncombatants, the political and cultural dimensions of the battlefield become more critical to the conflict."[225]

While these manuals accepted the role of the Army in the political dimension of low-intensity conflicts, the chief of staff of the Army, Gen. Gordon Sullivan, seemed to reject this responsibility. In the 1994 *Parameters* article, Sullivan and Twomey explained that in the case of such conflicts, "We can expect long-term solutions to be found primarily through political, not military, means. Military means may well be [required] to assist in the resolution of these conflicts, but we should expect the use of force to be tightly linked and coordinated with other forms of national power."[226] This segregation of activities in a low-intensity conflict into "military" and "political" means was subtle but significant. By using "military" rather than "violent" or "combat," Sullivan and Twomey were signaling that "political" activities were not in the realm of the "military" (i.e., the Army). The "other forms of national power" were presumably other agencies of the U.S. government—such as the U.S. State Department—to which the Army should subordinate itself in a low-intensity conflict, letting them take the lead in the political dimension of the conflict.

Sullivan and Twomey were even more explicit in rejecting the Army's role in nation building: "The Army should not take the lead in organizing or supporting the formation of democratic institutions in other nations." They allowed that the Army could "provide security" and "provide medical treatment; build roads, buildings, and ports; and deliver a variety of supplies." Yet they insisted, "Nation-building is not an Army issue, but the Army is prepared to support those agencies of the government which are directly concerned with that task."[227]

FM 100-23-1 echoed General Sullivan's belief that the Army should outsource the political dimension of low-intensity conflicts to other agencies. The manual insisted, "Mission success depends on the US military turnover of HA [humanitarian assistance] responsibilities, including security, to the host nation or relief organization." This exhortation was followed by a very detailed explanation of how to coordinate the handover of the conflict.[228]

The manual also reiterated the fallacious principle that the Army should avoid taking sides in a conflict: "In most short-term, foreign [humanitarian assistance] operations, neutrality is an important aspect." Notably, FM 100-23 was more circumspect about neutrality, explaining that peacekeeping operations, normally under Chapter VI of the UN Charter, required "strict impartiality," but peace enforcement, which normally occurred under Chapter VII, might be characterized by "questionable impartiality." Given the normal instinctive insistence even among low-intensity conflict observers upon neutrality in such conflicts, this was a dramatic departure. Yet the manual did still insist that humanitarian assistance *should* be impartial. Quoting the directors of the Humanitarian and War Project, the 1994 edition of FM 100-23 insisted that a principle of humanitarian assistance was nonpartisanship. The manual explained that Army forces were "not to advance political, sectarian, or other agendas," and insisted flatly that the Army "should not take sides in conflicts."[229]

In 1993, in perhaps the most significant move toward institutionalizing the lessons being learned about low-intensity conflict since the creation of A-AFCLIC in the 1980s, General Sullivan established the Peacekeeping Institute (PKI) just next door to the U.S. Army War College at Carlisle Barracks, Pennsylvania. Today, General (Ret.) Sullivan recalls that the selection of the site for the PKI was made because "Carlisle worked for me"; the U.S. Army War College reported directly to the chief of staff of the Army.[230]

Sullivan would later say that the establishment of the PKI was a response to his observations from the aftermath of the invasion of Panama and the first days of Operation Restore Hope in Somalia, when "the Army [was] not fully prepared to deal effectively with the complexities of the humanitarian aid process." He added that civil-military operations and interactions with NGOs was a particular shortfall. Today General Sullivan particularly remembers that, on a trip to Somalia in 1992, he was "disappointed" to see that units were "learning on the fly how to get along with NGOs." He recalls believing at the time, "We had to do something."[231]

There were other reasons that the PKI was established. Sullivan would elsewhere say that his decision to establish it was also driven by his conviction that "military organizations perform better if they have a doctrine and people have thought about it . . . and [are] trained to do it." He concluded, "That's why PKI was developed." Today he acknowledges that the establishment of the PKI was also intended to signal to political leaders that the Army was "not a 'reluctant dragon' in the missions we were being given."[232]

U.S. Army colonel Karl Farris, who had previously worked with the UN mission participating in the low-intensity conflict in Cambodia, was tapped as the PKI's first director. Colonel Farris would serve as the director of the institute for three years, overseeing its growth to a meager staff of eight— civilians and military service members.[233]

Farris developed the purpose for the PKI in cooperation with his commander, the commandant of the U.S. Army War College, Maj. Gen. William Stoft. The institute had three initial functions: it would provide peacekeeping instruction to students at the U.S. Army War College (the top lieutenant colonels and colonels in the Army, as well as officers from partner nations and sister services); provide liaisons from the Army to the UN Department of Peacekeeping Operations and other international organizations and NGOs; and provide educational resources, training, or even personnel to the geographical combatant commands for peacekeeping operations. It was in this latter capacity that the PKI produced an after-action review in cooperation with U.S. European Command on Operation Support Hope, which was conducted by about 3,600 service members in Rwanda, Uganda, and Zaire. Later, Colonel Farris personally participated in the 10th Mountain Division's planning for Haiti.[234]

Ultimately, however, the PKI was too isolated from the rest of the Army to truly institutionalize low-intensity conflict lessons within such a vast organization. While Sullivan had hoped that the organization would begin to institutionalize low-intensity conflict by capturing "lessons learned, writing doctrine, conducting training and educating our leaders,"[235] in practice the PKI only captured—but seldom published—lessons learned. Moreover, while Sullivan had created the mechanism to capture these lessons, he had not demanded that the Army institutionalize them by integrating them into training or creating any incentives for Army officers to find and study those lessons.

Father Jean-Bertrand Aristide became Haiti's first democratically elected president in supervised elections on December 16, 1990. In September 1991, only a few months after Aristide took office, he was deposed in a military coup by Lt. Gen. Raoul Cédras. Aristide fled to the United States and began canvassing the U.S. government for support.[236]

In mid-1993 the UN and the Organization of American States brokered the Governor's Island Accord, which was to permit the return of President Aristide to Haiti by October 30, 1993. Yet when the USS *Harlan County*, loaded with peacekeeping troops, attempted to dock in Port-au-Prince a few weeks ahead of the implementation of the accord, it was turned away at the port by a gun-toting mob. As the ship approached, the mob reportedly chanted "Somalia! Somalia!" The U.S. Navy began a blockade of the island nation, but this, too, failed to dislodge the junta.[237]

By mid-1994 it was clear that Lieutenant General Cédras would not willingly comply with the will of the international community. The 10th Mountain Division and 82nd Airborne Divisions were briefed on Operation Uphold Democracy and directed to begin planning and rehearsing for either a forcible or an unopposed entry into Haiti. This was followed days later by UN Resolution 940, which called for the "application of all necessary means" to restore Aristide to power.[238] By September 1994, elements of the 10th Mountain Division were aboard the aircraft carrier USS *Eisenhower*, departing from Norfolk, Virginia, for Haiti.

On September 17, 1994, as U.S. forces closed on Haiti, Jimmy Carter, Senator Sam Nunn, and the chairman of the Joint Chiefs of Staff, Gen. Colin Powell, arrived in Haiti in a last-ditch effort to avert war. The next day, with the 82nd Airborne Division loaded in aircraft and en route from Fort Bragg, North Carolina, to execute an airborne assault on the island nation, Lieutenant General Cédras and his puppet president Émile Jonaissant agreed to step aside and permit the implementation of the Governor's Island Accord. Beginning on the morning of September 19, 1994, the 10th Mountain Division began a helicopter-borne air movement of its forces from the USS *Eisenhower* to seize the Port-au-Prince airfield and port.[239]

In the first days of Operation Uphold Democracy, the U.S. contingent was under the command of Lt. Gen. Hugh Shelton, commander of the XVIII Airborne Corps, designated as Combined Joint Task Force (CJTF) 180. After October 24, 1994, the XVIII Airborne Corps departed and command fell to Maj. Gen. David Meade, commander of the 10th Mountain Division, designated CJTF 190. Throughout the operation, the senior U.S. commander in the operation was also double-tasked as the Multinational Force (MNF) Haiti commander, leading both the U.S. contingent and the coalition partners participating in the operation.[240]

By mid-October 1994, nearly twenty-one thousand American service members—mostly from the Army and the Marines—were operating inside Haiti. Reflecting lessons from Somalia, the Army deployed eight MP companies and nearly four dozen armored vehicles. The core of the U.S. Army force in Haiti was the 1st and 2nd Brigade Combat Teams (BCTs) of the 10th Mountain Division. They were joined by almost three hundred international soldiers gathered into an infantry battalion under a Caribbean Command, nearly six hundred other coalition troops (mostly from Bangladesh), and more than eight hundred International Police Monitors (IPMs). Overall, twenty countries contributed more than two thousand people to MNF Haiti. This force was arrayed across base camps in the principal cities of Port-au-Prince and Cap-Haïtien.[241]

The occupation of Haiti came largely without violence against the coalition troops. This was in no small measure due to the presence and confidence of the soldiers of the 10th Mountain Division, many of whom had served in the much meaner streets of Somalia only a few years earlier. But it was also the result of the Army artificially circumscribing its mission in Haiti. Zealously protecting itself from mission creep, the 10th Mountain Division initially refused any humanitarian or reconstruction requests as nation building. Instead Army forces focused on a weapons buyback program, building the national police force, and joint patrolling—with both the police and the former regime's army forces—to maintain security.[242]

While this approach prevented the Army from becoming embroiled in the politics of Haiti, it also allowed the tragedy of September 20, 1994, to occur, when Haitian police killed two unarmed protesters. Haitian security force violence continued to escalate, culminating in a firefight on September 24 between U.S. Marines at Cap-Haïtien and the Forces Armées d'Haïti (FAdH, the Haitian Armed Forces) that left one marine wounded and ten Haitian soldiers dead. In response to these setbacks, the Army did finally expand its mission to include raids on weapons caches and the capture of "wanted people."[243]

The Army did achieve the goals that it set for itself in Operation Uphold Democracy. Over the following month, President Aristide returned to Haiti and Lt. Gen. Raoul Cédras and his ally, Brig. Gen. Philippe Biamby, departed. The 10th Mountain Division would continue to execute low-intensity conflict operations across the country until January 1995, when they turned over the mission to the 25th Infantry Division, which rotated into Haiti from Hawaii. The 25th Infantry Division would remain in Haiti for three months before departing and transferring the mission to the UN Mission in Haiti (UNMIH).[244]

Yet ultimately the United States did not achieve its strategic goal of stability in Haiti. Because the Army artificially circumscribed its mission, it did not solve any of the deep-rooted political or economic problems in Haiti. Despite the efforts of literally hundreds of NGOs and millions of dollars in financial and food aid, political strife and environmental degradation caused Haiti to remain among the poorest countries in the world. And despite the efforts of both the United States and the UN to "uphold democracy" in Haiti, President Aristide turned out to be just as authoritarian and brutal as the Haitian military leaders who deposed him. After reelection in 2000, he was forcibly removed from power on February 28, 2004, by the Haitian and American militaries and sent into exile in South Africa.[245]

The UN anticipated withdrawing forces from Haiti in March 1996, per the terms of UN Resolution 940. This mission did finally end, three months late, in June of that year. But the mission was immediately followed by a succession of other UN peacekeeping missions in Haiti that similarly failed to create a political solution for the country. As of this writing, the UN Mission for Justice Support in Haiti is just the latest incarnation of the continuous UN military presence in Haiti since 2004.[246]

The entire senior leadership of the 1990s Army had a front-row seat for Operation Uphold Democracy. The 2nd BCT commander was Col. James Dubik, former special assistant to the chief of staff of the Army and future deputy commanding general for transformation of TRADOC. Because Haiti was not far from Washington, DC, there were frequent visits from senior Army leaders; General Sullivan visited multiple times. U.S. forces in Haiti also received visits from the chairman of the Joint Chiefs of Staff, Gen. John Shalikashvili; the vice chief of staff of the Army, Gen. Dennis Reimer; the deputy commander in chief of U.S. Atlantic Command (and future TRADOC commander), Lt. Gen. William Hartzog; future FORSCOM commander Maj. Gen. John Hendrix; and future Army vice chief of staff Maj. Gen. Jack Keane.[247]

The U.S. Army did learn lessons from Operation Uphold Democracy, beginning with those about deployability. The proximity of Haiti to the United States and the use of light infantry forces permitted easy movement to the island. Even so, the use of Army helicopters launched from a Navy aircraft carrier was innovative. And, more important, despite the fact that the 10th Mountain Division was a light infantry division, the Army discovered that it was still very heavy in terms of strategic lift requirements. Deploying the U.S. military to Haiti required 539 sorties by fifty-one helicopters, an airlift of 970 aircraft flights, forty-six separate vehicle convoys, 238 railcars, fourteen ships, and fifteen barges. While this was significantly less than the deployment to Saudi Arabia for Operation Desert Storm, it was still significant. The deployment was further complicated by the fact that Cap-Haïtien's airfields were incapable of receiving aircraft larger than the Air Force's C-130 propeller-driven airplane.[248]

The Army did learn quite a bit about sustainment—specifically the necessity of contractors—from Operation Uphold Democracy. The Brown and Root Corporation eventually took over all logistics for ground operations in Haiti on November 14, 1994, only a few months after the operation began. By contracting their logistical support, the U.S. force was able to draw down to six thousand people—mostly combat troops rather than logistics personnel—by the end of the year. But this solution did not come without its own complications. Under the contracts, 542 U.S. civilians were joined by 541 international subcontract workers and 731 local contract workers. The contracts not only flooded the local economy with destabilizing sums of money but also brought the Caribbean region $7 million,[249] creating a disincentive for Caribbean Command partner forces to help bring the conflict to a conclusion.

The Army relearned many lessons about the travails of long-term residence in base camps. As phone companies added phone banks to allow soldiers to call home, the Army–Air Force Exchange System added post exchanges, and the United Service Organizations added recreation centers, Army units took on an inertia that made moving out into the cities to operate in and among the populace that much harder. This negative inertia was compounded by excessive concerns over force protection, which created further disincentives against leaving base camps or interacting with the populace.[250]

The Army also got a good deal of experience in rotating forces. The 25th Infantry Division executed a phased relief in place of the 10th Mountain Division beginning on December 28, 1994, with the deployment of advanced party personnel. After a handover of operations, the units conducted a transfer of authority for the mission on January 15, 1995.[251]

In after-action reviews, the 10th Mountain Division staff rejected the contention that it—or the larger Army—was not manned, trained, or equipped to engage in low-intensity conflicts. The 10th Mountain Division staff observed that while the Army continued to classify operations such as those being conducted in Haiti as OOTW, the conflict in Haiti sat somewhere between peacetime noncombat operations and wartime combat operations. They argued that for this reason the tasks they were asked to accomplish in Haiti had a close "correlation with [the division's] wartime mission essential tasks." Thus they argued that Operation Uphold Democracy had been "good training" on their high-intensity conflict tasks.[252]

Of course, this was a self-assessment, and was no doubt shaded by the 10th Mountain Division's relative familiarity with low-intensity conflict tasks, having engaged in Operation Restore Hope only two years earlier.

But this may also have reflected the belief—often expressed by Army transformers—that low-intensity conflict tasks were a lesser included subtask of high-intensity conflict tasks. This misperception pervaded the 10th Mountain Division's after-action reviews of Operation Uphold Democracy. One division staff officer wrote, "Even though many missions were nonstandard, infantry doctrine still provides the appropriate foundation for the tasks and individual skills required to perform peace operations." Another wrote, "It may be OOTW from a distance, but for the soldiers on the ground it looks and feels like war." And yet another wrote, "Haiti was excellent training and excellent for readiness." This after-action review concluded, "The combat readiness of the division was enhanced by our involvement in Operation Uphold Democracy."[253]

While they may have felt that low-intensity conflict tasks were a subset of their "wartime mission essential tasks," these staff officers still conducted extensive predeployment training on many tasks that were not within the normal scope of high-intensity conflict competencies. The division's predeployment training included such collective unit tasks as securing NGOs, securing the U.S. embassy, port security operations, and rules of engagement training. The staff acknowledged that once they were notified of their impending deployment, they felt it "very important to get the training changed right away." They also admitted that it was hard to adjust the training away from typical high-intensity conflict tasks because of the "enormous inertia" toward that type of training. In their after-action review, the 10th Mountain Division staff noted that the "division master training plan must change based upon the assigned mission" and the division's "modified training plan prior to deployment contributed considerably to mission success."[254] Given the scale of the effort required to recalibrate the division's training prior to deployment, the staff's exhortations that low-intensity conflict tasks were simply a lesser included subset of high-intensity conflict tasks and required no additional training ring somewhat hollow.

Even with predeployment training, Army units in Haiti found themselves ill prepared for the challenges they faced. Engineer units trained to breach obstacles in the plains of northern Europe were ill equipped to rehabilitate schools, dig wells, and conduct power plant assessments. The 10th Mountain Division's engineer officer observed, "Much of our current joint engineer doctrine is based on the environment of the Cold War." Likewise, the division G/J2 (intelligence officer) observed that his staff was trained to collect and analyze signal intelligence, but in Haiti "HUMINT quickly became the paramount collector." The intelligence staff soon became as consumed with tracking crime statistics in cooperation with the MPs and infantry BCTs as it had previously been with finding threatening individuals in the country.[255]

The 10th Mountain Division lacked more than the required training; it was also missing many key capabilities and not correctly organized for low-intensity conflict. The division had to be augmented by, among other capabilities, two MP brigades, an engineer brigade, a PSYOP group, a civil affairs task force, and a public affairs detachment.[256] In order to operate as a JTF headquarters—CJTF 190, coordinating and synchronizing the activities of Air Force, Navy, and Marine elements as well as those of international coalition partners in a low-intensity conflict environment—the three-hundred-man 10th Mountain Division staff had to be augmented with as many as five hundred additional people. The division had learned from its experience in Somalia that its considerable complement of field artillery was of little use in a low-intensity conflict environment; the division's howitzers were not even deployed to Haiti. The 10th Mountain Division's artillery headquarters was reorganized into a third maneuver brigade headquarters, Task Force Mountain. The staff of one of the field artillery battalions was reorganized to act as the staff for the IPMs. Other elements of the two artillery battalions in Haiti conducted convoy security or guarded key facilities.[257]

Perhaps no element of the U.S. Army mission in Haiti was so stretched beyond its normal high-intensity conflict missions as were the MPs, who operated as site security, protected key leaders, and seized weapon caches. They executed customs inspections and enforced traffic laws and responded to looting and civil disturbances. Perhaps their most important role was in training and conducting joint patrols with the Haitian Interim Public Security Force (IPSF).[258]

Of course, many believed that Army units shouldn't even be engaged in low-intensity conflict operations, let alone be training and organizing to better execute them. General Reimer would say of his visit to Haiti in late 1994, when he was still the FORSCOM commander, "I was convinced that we needed to get out of Haiti." He added that the operation was deteriorating the unit's high-intensity conflict proficiency.[259]

This was a common refrain, even from inside the staff of the 10th Mountain Division, who especially noted the deterioration of "aviation readiness" and the atrophy of infantry skills. To combat this perceived problem, the 10th Mountain Division diverted vast resources from its low-intensity conflict mission in Haiti to build and train on a multipurpose range complex (MPRC) inside Haiti and maintain "readiness." This was not just a distraction; it had an immediate negative impact on the division's ability to succeed in the low-intensity conflict in which it was engaged. Civilians had to be moved from their homes, and the civilian government had to be coerced into providing land for the facility. Civil affairs direct support teams and tactical PSYOP teams that might otherwise have been engaging the populace of Haiti or working on reconstruction tasks were instead employed talking to local villages about staying away from the range complex while live-fire training was being executed. OH-58C helicopters were equipped with loud-speakers and flown across the countryside to warn villagers of the dangers of interfering with the training. Instead of rebuilding Haiti, two engineer battalions were employed nearly full-time in constructing the facility. Fire support officers and soldiers who might otherwise have been coordinating and executing civil affairs tasks were instead employed in scheduling and managing the range. Where money might have been spent on rebuilding Haiti, it was instead spent on real estate leases and construction contracts to build the MPRC. Shipping resources might have been used to bring humanitarian aid to the Haitian people, but instead, "state-of-the-art radio controlled, solar-powered targets were shipped and installed by the Army Training Support Center, Fort Eustis, VA." The entire endeavor would have been comical had the costs to the Haitian people of not solving the deep-rooted political and economic problems of Haiti not been so dire.[260]

The 10th Mountain Division was sacrificing success in the war in which it was very much presently engaged for the sake of readiness for an imagined future great power war. This point seemed to be lost on Lt. Col. David Stahl and the division staff. Stahl wrote, "Units needed a means to maintain readiness in warfighting skills. A multi-purpose range complex (MPRC) would allow units to train on their warfighting METL [mission essential task list]."[261] The 10th Mountain Division's impression of its mission in Haiti could not be clearer: Operation Uphold Democracy was not warfighting, but a distraction from the division's true mission: readiness for the next great power war.

Whether or not it hurt its high-intensity conflict proficiency, the 10th Mountain Division did still learn lessons about the unique capabilities that the Army required to operate in a low-intensity conflict. Perhaps the most lessons were learned while rebuilding Haitian security forces. The FAdH had been the source of the coup that had deposed President Aristide, and the day after Aristide returned to Haiti there was no coherent police force in the capital of Port-au-Prince. The division participated at the most senior levels in building the Haitian security forces by interacting with generals from the FAdH and officials from the Ministries of Defense, Justice, and the Interior.

Much of the Army's direct interaction with Haitian police was done by MPs. But conventional U.S. Army forces in Haiti also coordinated their efforts with SOF in rural areas and IPMs.[262]

The IPMs were augmented by the 1st Battalion, 7th Field Artillery staff, which provided planning and administrative support. Together with the MPs and the U.S. Department of Justice's ICITAP, the IPMs did a great deal to professionalize and increase the confidence and competence of the fledgling IPSF,[263] but the goal of a full-fledged national police force remained elusive.

One major obstacle was finding suitable manpower. A major effort within the rebuilding of both the FAdH and the IPSF centered on "vetting" current employees and new recruits for loyalty to the "legitimate" government and ensuring that they had not engaged in human rights abuses. After vetting, ICITAP provided a formalized, six-day training program for the approved recruits before they were put on the street in joint patrols with U.S. Army MPs or IPMs. Despite this herculean effort, between October 24, and December 17, 1994, slightly fewer than three thousand police had matriculated through the entire process.[264]

The 10th Mountain Division also learned a number of lessons while working through the challenges of coordinating and interoperating with international coalition partners. Things as simple as telephone communications and—in the age before the internet—facsimile were complicated by the inability to share classified information with partners. Providing spare parts for the variety of vehicles used by coalition partners also proved a challenge. Even the act of moving coalition partners by air into and out of Haiti, which required U.S. visas for moving through U.S. airports, presented challenges.[265]

The U.S. Army units in Haiti might also have learned some negative lessons about operating as part of a multinational force. Chastened by its experience operating within a UN command as part of UNOSOM II, the 10th Mountain Division was much happier being in the lead as the CJTF headquarters. Lt. Col. David Stahl of the 10th Mountain Division wrote that, in Operation Uphold Democracy, "command and control of coalition forces was a big success," adding, "There never was any question about who they worked for or to whom they reported." According to Stahl, the only friction occurred when "40 French soldiers who worked for neither MNF nor the UN arrived in Haiti. They had a separate agreement with the GOH [government of Haiti] and attempted to move into some police stations to train Haitian police." This friction was only resolved, he added, once they stopped operating independently and were integrated into the international effort.[266] While leading the international coalition certainly made the operation easier for the 10th Mountain Division, this expedient may have prevented it from learning more lessons about how to interoperate with coalition partners.

The 10th Mountain Division had a great deal more integration with SOF than it was able to achieve in Somalia. SOF was stationed in twenty-seven locations across the country, integrating with local police forces. These operations were coordinated and integrated with platoon- to company-sized conventional infantry operations by the 10th Mountain Division to seize weapons caches. Like coalition forces, however, SOF units were subordinate to the 10th Mountain Division commander,[267] which may likewise have prevented the division from learning more lessons about interoperability.

Army units in Haiti learned a number of new lessons about operating in and among the populace and in an urban environment. First, the 10th Mountain Division again learned the value of Army linguists, who were a critical and scarce resource in Haiti both in dealing with the populace and coordinating life support from local contractors.[268]

The division also learned how to employ PSYOP to influence populations. Lieutenant Colonel Stahl wrote, "During all phases of the operation, psychological operations (PSYOP) were particularly effective in setting the conditions for the introduction of US/coalition forces and in the protection of nonbelligerents." To synchronize the effects of PSYOP, civil affairs, and security force assistance efforts, the division established a "targeting" process that coordinated Army, Air Force, and Marine intelligence and assets to achieve the division's objectives, not just with regard to the amorphous "enemy" but also in respect to the people of Haiti. One division staff officer noted that PSYOP was only one piece of a suite of capabilities required to operate in and among the populace. He observed, "Infantry units were habitually task organized with Military Intelligence, PSYOPS, Civil Affairs and linguists," adding that "every operation required detailed coordination" of these assets. PSYOP was especially effective when used in conjunction with MPs and light helicopters to communicate with or disperse mobs.[269]

Just as had the 10th Mountain Division commander in Somalia, the division staff in Haiti came to see the ability to influence the population as a key capability. Echoing Major General Arnold's comments after Somalia, the division staff said in an after-action review briefing, "PSYOP is almost a separate battlefield operating system in this kind of environment." In military parlance, this meant that, in the low-intensity conflict in Haiti, influencing the population was as important as intelligence, artillery and aviation fires, or infantry maneuver. Building on lessons from Somalia, the 10th Mountain Division staff discovered that not only the method of delivery, but also a consistent message was important. The staff added that PSYOP teams "were critical in Haiti at the strategic level right on down through the tactical level." They explained that PSYOP was "often the weapon of choice."[270]

The division staff also came to appreciate the value of influencing local and international audiences through the media. They understood the importance of international perceptions of their actions, writing, "The J2 [intelligence section] and the J3 [operations section] constantly monitored CNN for information about the JOA [joint operational area]." The staff also found that their reception of and communications with senior military and civilian government visitors and media figures had a dramatic impact on the perception of their operations among international audiences: "Visitors, especially statesman / diplomatic leaders as well as media personnel are critical to maintaining public support for OOTW operations." The 10th Mountain Division also coordinated with the U.S. Information Service in "targeting audiences" inside Haiti and the surrounding region "with specific messages and themes."[271]

The U.S. Army in Haiti also pioneered the use of embedded media in Operation Uphold Democracy. The XVIII Airborne Corps commander, Lt. Gen. Hugh Shelton, penned an article with Lt. Col. Timothy Vane upon his return from Haiti touting embedding as the best means for reporters to "understand the military operational context and background in order for success on the battlefield or in OOTW to be properly understood by the public," and added, "Commanders or senior military officials can no longer get away with a 'no comment' answer regarding American troop use in trouble spots worldwide." Shelton and Vane enumerated ten "principles of information for news media covering DOD [Department of Defense] operations" that rejected the pooling of reporters in favor of getting them as close to Army units during operations as possible.[272] These principles were a direct rejection of the normal practice of tightly controlling media coverage. But they were also a cultural change away from the hostility toward journalists that had pervaded in the Army since the Vietnam War. It is, of course, no coincidence that Lieutenant General Shelton had been a brigade commander in the 82nd Airborne Division but remained at Fort Bragg while the rest of his division departed for Operation Urgent Fury, the invasion of Grenada.[273] He almost certainly watched press coverage of the invasion as it unfolded, tinged with journalists' bitter recriminations over the tight control of the press during the operation.

Both CJTF 180 (XVIII Airborne Corps) and later CJTF 190 (the 10th Mountain Division) maintained a Joint Information Bureau of public affairs service members to manage this program. In addition to holding regular press briefings, these service members also facilitated "linkups of media with operational units." The public affairs staff observed in their after-action review that this new "open approach to the media worked at all levels." They added, "In about a ten-week period, nearly 1300 media personnel were accredited. . . . Many roamed the city and countryside, stopping to cover events and units as they encountered them." This idea of trusting "commanders and soldiers who understand the importance of telling their stories honestly and openly" was a dramatic departure from the prior public affairs practice of trying to control the message and what reporters saw.[274]

Building on its experiences in Somalia, the 10th Mountain Division also developed several new capabilities for civil-military operations in Haiti. The division was augmented by civil affairs soldiers who manned a humanitarian assistance coordination center, which operated at the U.S. Agency for International Development building in downtown Port-au-Prince, coordinating the division's efforts with the more than four hundred NGOs and other international organizations operating inside Haiti. Army civil affairs officers—many of whom worked in similar fields in their civilian employment—also led ministerial adviser teams that provided advice to key ministries of the Haitian government. They worked under the direction of the ambassador, but technically belonged to the division. As the division's civil affairs personnel concluded in their after-action review, "One of the keys to success in Uphold Democracy and any other operation of its type is the close coordination of military efforts with the efforts of other agencies and organizations, both U.S. and international."[275] The MNF also provided support to other U.S. government agencies and NGOs, moving over 100,000 tons of humanitarian supplies for NGOs in Haiti.[276]

The infantry BCTs' fire support elements—normally charged with planning and coordinating artillery fires and the employment of attack aircraft—were instead employed as brigade CMOCs. The 2nd BCT commander, Colonel Dubik, observed that the CMOC was "critical" to his organization's success in Haiti. At the battalion and company level, fire support officers and soldiers also coordinated and executed civil-military operations.[277]

Despite the MNF's insistence that it was not in Haiti to conduct nation building, the civil-military effort there moved beyond working through other U.S. government agencies and NGOs to directly assisting the government and population of Haiti. In one example—Operation Light Switch—the 10th Mountain Division provided more than 108,000 gallons of fuel to power Haitian power plants. The MNF also restored one-third of Port-au-Prince's total capacity to produce potable water and brought drinking water to seventeen other cities around Haiti. Working with UNICEF and the Haitian government, the MNF inoculated as many as 2.1 million Haitian women and children.[278]

The MNF still faced difficulties with civil-military operations, however. Army lawyers struggled with antiquated laws that prevented the 10th Mountain Division from using its resources to address urgent needs for the people of Haiti. Rules allowed Army units to spend money on much needed reconstruction projects for the Haitian people only if they also helped U.S. forces. Very early in the operation, everything from schools and wells to street cleanup and port repairs began to be justified as somehow helping U.S. forces. Yet with the Haitian economy in collapse due to a UN embargo and political turmoil, the small local contracts of $2,500 or less for local goods that Army units were allowed to establish quickly destabilized the economy and aggravated the already chronic scarcity of supplies for the civilian populace. Likewise, the weapons buyback program intended to get weapons off of the streets made recipients of the cash rewards targets for mugging and poured nearly $2 million of destabilizing cash into the Haitian economy.[279]

The 10th Mountain Division learned a number of new lessons about implementing rules of engagement. Building on lessons from Somalia, it began to distribute and train on the rules of engagement as early as possible. Yet the division had initially expected to be part of a forcible entry into Haiti. When this did not materialize, division staff struggled to rewrite and redistribute more restrictive rules of engagement before soldiers arrived in the country. Army lawyers also discovered that the rules that were the best from a legal perspective were not necessarily the most easily understood or implemented on the ground. Numerous questions arose every day of the operation, requiring lawyers to be present to provide answers.[280]

With the new skills that the Army learned to operate in and among the populace also came cognitive obstacles to moving beyond seeing people as objects to be acted upon to seeing them—and their attitudes—as the political objective of the low-intensity conflict. Chief among these cognitive obstacles was the unhealthy obsession with measuring the effectiveness of operations in Haiti. And chief among these measures was the division's metrics on crime. Soon after Operation Uphold Democracy began, the 10th Mountain Division's G2 (intelligence) section began tracking incidence of crime by Haitians. Soon this metric became the chief measure of the progress toward a "stable and secure environment," the division's "most important" mission objective.[281] While the division G2 did eventually begin segregating the statistics into a measure of which faction was perpetrating the violence and what segment of the population was a victim of it, much of the violence—and the resultant deaths—remained classified as "indiscriminate violence." Senior U.S. military and civilian visitors were frequently dismayed by the increasing incidence of violence in the capital, which rose sharply after the return of Aristide. The division staff allayed these concerns by explaining that although "there was [actually] a marked increase in the reporting and investigating of crime," in any case, "crime in PAP [Port-au-Prince] was much less than that of any city of comparable size in the US."[282]

The division also tracked other statistics, including "detainees that were being acquired by the MNF, operations at the police stations throughout the country, the ability of NGOs/PVOs [private volunteer organizations] to do their jobs, the ability of the Aristide government to go about its business, the return of commerce in local markets and many other indicators."[283] This obsession with quantifying progress obscured the United States' true objectives in Haiti. The amassed statistics created the illusion of progress, while the division actually drew no closer to a real solution to the deep political problems in Haiti—an inherently qualitative, unquantifiable objective.

The 10th Mountain Division was keenly aware of the experience it had gained in Somalia and how its experiences in Haiti had differed. In their after-action review, the division staff identified several factors that were different in Haiti, including "a functioning Haitian government" and a "legitimate president and government."[284] By way of contrast, the division staff identified several factors unique to Somalia:

- No function in government.
- No head of state or single recognized leader of the country.
- Aideed [sic] & his clan were against US/UN and willing to fight.[285]

Notably, these were all political factors. The division staff was tacitly acknowledging the primacy of the political dimension of this low-intensity conflict, a level of understanding they struggled to reach during Operation Restore Hope.

In fact, however, in Operation Uphold Democracy it was impossible for the 10th Mountain Division to avoid the political dimensions of the conflict. At the strategic level, the United States had decided to depose the junta of Lieutenant General Cédras and return President Aristide to power, a clear intervention in the politics of Haiti. At the tactical level this required the U.S. military to engage in the political dimension of the conflict and, in certain cases, compel the transfer of political power from former regime elements to the new government. The commander of the 2nd BCT, Col. James Dubik, and his battalion commanders found themselves negotiating with mayors, American civilian organizations, and town and village councils and leaders in Cap-Haïtien. These negotiations had to be coordinated with the negotiations by SOF with local leaders, citizens, and FAdH leaders in rural areas. Dubik, his subordinates, and their staffs learned the art of preparation and rehearsal for negotiations,[286] but also how to navigate the political interests of the different factions in Haiti. The division learned to sort the political leaders with whom it engaged by echelon into spheres of influence, with more senior U.S. military leaders engaging more senior Haitian political and military leaders.[287]

Chastened by their experiences in Somalia, the leaders of the 10th Mountain Division in Haiti began to understand that they could not relegate the political dimensions of the conflict to other U.S. government agencies such as the State Department or Department of Justice. The ambassador did accompany senior U.S. Army leaders when they met with Cédras and his compatriots before their departure. And the Department of Justice did administer ICITAP. But the 10th Mountain Division staff found that many agencies lacked personnel, were too concerned with security to interact with the populace of Haiti, or simply "had not developed a plan to follow on after a successful military operation." As a result, the 10th Mountain Division commander, according to the division staff, "talked directly to the . . . Prime Minister, Minister of Defense, and the Minister of Justice on a regular basis." The staff added, "Other members of the MNF coordinated with other GOH ministries as did the Ministerial Advisory Teams." One of the division's deputy commanding generals, Brig. Gen. George Close Jr., regularly engaged with the mayor of Cap-Haïtien. Commanders at both the division and brigade levels also engaged members of the Haitian parliament on a regular basis. More junior commanders engaged local leaders in their areas. State Department representatives almost never attended any of these meetings.[288]

The 10th Mountain Division's experience in Somalia was not entirely beneficial to its performance in Haiti. According to Robert Bauman of the Combat Studies Institute, the division's experiences in Somalia caused it to maintain "a posture of maximum vigilance while assuming minimum risk to its personnel." He added, "Most . . . personnel in Port-au-Prince remained permanently locked down in well-bunkered compounds." As a result, he noted, "Direct engagement of the populace was minimal, at least within the limits of the Haitian capital."[289] This overconcern for force protection—presumably born of the American public's reaction to the casualties from the Battle of Mogadishu—had a clear negative impact on the division's ability to engage in the political dimension of the low-intensity conflict in Haiti.

Moreover, too often the 10th Mountain Division staff conflated the establishment of security and the return of President Aristide with engaging in the political dimensions of the conflict. Two core elements of the division's own mission statement were to "establish and maintain a stable and secure environment" and to "facilitate the return and proper functioning of the GOH." Not addressed were the political conditions that the 10th Mountain Division would have to set before and after the return of President Aristide to solidify his power. The division staff seems to have honestly believed that providing security was the only necessary condition required. The first key task of the MNF commander's intent was to "establish a safe and secure environment." In one after-action review slide, the 10th Mountain Division Staff insisted, "All success is based on a stable and secure environment." Elsewhere the staff wrote, "There were four main mission points—the most important was to establish and maintain a safe, secure and stable environment."[290]

A key to this conflation of security with politics was the constant use of "stable and secure" together when describing the desired environment—as if stability, created by engaging in the political dimensions of the conflict, and physical security, imposed by force or the threat of force, were the same thing. The 10th Mountain Division staff, at one point in their after-action review, did acknowledge that these were different.[291] Yet nowhere in their mission statement, the division commander's intent, or the division's actual operations did they actually direct or execute actions that directly contributed to improving "political and economic" stability. They seemed genuinely to think that imposing security on the population, physically removing Cédras and his cronies, and returning Aristide to the presidential palace would, in and of itself, create this stability. The history of Haiti since the departure of the U.S. Army has proved otherwise.

Frankly, SOF operating in rural areas did much of the work to create the political and economic stability in Haiti that conventional forces should have been working to effect. As Lt. Col. David Stahl unabashedly admitted, "While the conventional forces focused on providing security, presence and higher level governmental support, the Special Forces focused on shoring up the Haitian population from the bottom up. . . . Encouraging towns and villages to organize and help themselves; holding town meetings explaining democracy, rights and the Haitian constitution; organizing and mentoring newly founded Haitian security forces, local governments and their leaders; and interfacing with International Police Monitors and foreign security forces were just some of these SOF efforts."[292] Conventional Army units did engage in negotiations with leaders at the local through national levels, but these contacts were not nearly as extensive as the effort expended by SOF in the rural areas. One can only wonder how differently Operation Uphold Democracy might have turned out if conventional forces had mounted similar, integrated political efforts in Port-au-Prince and Cap-Haïtien.

Rather, the U.S. Army in Haiti deliberately circumscribed its political engagement, intentionally limiting its role. Throughout the operation, the 10th Mountain Division insisted that it was "not tasked to disarm Haitian forces or to undertake significant 'nation building.'" At the outset, the division even tried to avoid "maintaining public order" but was forced to take a direct role when they "were embarrassed by egregious incidents of Haitian-on-Haitian violence."[293] The division never sought—nor saw it as its responsibility—to find a sustainable political solution to the conflict.

Instead, reflecting contemporary Army doctrine, the division sought to dump the political dimension of the low-intensity conflict in Haiti on the UN as quickly as possible. The division staff wrote in their after-action review, "From the beginning we prepared to hand over operations to the United Nations." Thus, every threat was judged on the degree to which it could "complicate UNMIH transition" rather than how it supported or threatened the resolution of the political competition between Aristide's and Cédras' supporters.[294]

On some level it is shocking that the Army was not more successful in forging a political solution in Haiti. After all, in accordance with both U.S. government and UN policy, Operation Uphold Democracy began with the Army decidedly taking sides in the conflict and backing a winner: the elected president of Haiti, Father Jean-Bertrand Aristide. Throughout after-action reviews of the operation, the 10th Mountain Division staff consistently referred to his administration as "the legitimately elected government." The BCTs conducted raids that successfully detained the MNF's "most wanted" individuals, remnants of the former regime. The MNF rounded up those it deemed a threat to the force or to Aristide. At one time or another, more than 361 people were detained. On one day in the first two weeks of October, in a raid on the far-right-wing Front pour l'Avancement et le Progrès Haitien (FRAPH, the Front for the Advancement and Progress of Haiti) party headquarters, the MNF rounded up forty Cédras supporters.[295] While these missions were military in nature, they clearly had profound political implications in favor of one faction in the conflict: the Aristide administration.

But just as the MNF began to build momentum toward bringing the low-intensity conflict in Haiti to a successful conclusion, the U.S. Army confusingly stopped taking sides in the conflict. The MNF engaged with the ousted regime and its remnants. Even as it called the Cédras regime "illegitimate," both the XVIII Airborne Corps and the 10th Mountain Division commanders—as well as the U.S. ambassador to Haiti—engaged in discussions with Cédras and his compatriots. After Aristide's return, the president was coerced into meeting with the FAdH commanders who had deposed him.[296]

The 10th Mountain Division seemed particularly tone-deaf in refusing to take sides in building a police force and army in Haiti. The division staff did acknowledge that "the FAdH was a center of anti-Aristide and anti-democratic power," but from the outset, the MNF still doggedly "tried to keep the FAdH involved in the security apparatus." When former regime elements within the FAdH used their position within the military to wield "excessive force . . . against the populace" that supported Aristide, they received nothing more than remedial training on human rights from coalition forces. The MNF also pressured the Aristide government to integrate remnants of the former regime's police force into the IPSF. Adding fuel to the simmering political tensions within the Haitian security forces, the MNF then transformed 967 Haitian refugees at Guantanamo Bay, Cuba—most of them Aristide supporters who had fled after Cédras's coup—into police officers and returned them to Haiti to be integrated into the IPSF.[297] Unsupervised, these police frequently detained people hostile to Aristide's rule, only to have those detainees released by U.S. Army MPs at the detention facility because they were being "held without cause." Yet, despite having just witnessed all of the above firsthand, the 10th Mountain Division was surprised that politics was interfering with the formation of a police force. One exasperated staff officer wrote, "It appeared to many that the new police force was going to be very politicized."[298]

The division staff seemed genuinely proud that they had upset both sides in the conflict with their neutrality. They wrote in an after-action review briefing that Aristide and his party, Lavalas, were "disappointed when we did not immediately destroy the army as an institution and kill or arrest all those they thought were responsible for their long-term suppression and misery." At the same time, they bragged that Cédras's supporters in the FAdH, FRAPH, and a criminal organization called the Attachés were "disappointed when we collected their heavy weapons and crew-served weapons from their weapons company first and then the police stations and army barracks around the country." The division staff added, "They were disappointed we did not do more to control the Lavalas elements." The 10th Mountain Division conducted raids on weapons caches, and while "most of the targeted weapons stockpiles belonged to anti-Aristide and right wing individuals, or groups such as the FRAPH," the MNF also rounded up weapons belonging to supporters of Aristide.[299]

Because it did not understand the political dimension of building the Haitian security forces, the leadership of the 10th Mountain Division was confounded by the Aristide government's actions to undermine or abolish the FAdH, a political arm still controlled by Aristide's political opponents. Worse, the U.S. Army in Haiti worked to undermine Aristide's efforts to solidify his power—the power that the Army had just restored to him by expending the money, time, and effort to depose Cédras and then return Aristide to Haiti. On December 12, 1994, the Ministry of Defense ordered the FAdH to demobilize, but the MNF commander and U.S. ambassador pressured Aristide to rescind the order. When Aristide's prime minister put the FAdH on leave and cut their numbers to 1,500 on December 19, the MNF commander and the ambassador tried to pressure the prime minister again. The Ministry of Defense again responded by disbanding the entire FAdH on December 23. Concerned by the angry crowds of Haitian soldiers who were gathering at the FAdH headquarters, the MNF commander paid them from U.S. funds. When this failed to quell the tensions, shooting broke out at the FAdH headquarters. The U.S. Army's 1st Battalion, 22nd Infantry Regiment was dispatched to the FAdH headquarters, where it returned fire. In the incident seven Haitians were wounded and three were killed. The 10th Mountain Division took eighty-three prisoners (most of whom were released soon thereafter) and occupied the headquarters.[300] Two days later the Aristide government excised the IPSF—mostly remnants of Cédras's police force—from the FAdH, transferring them in toto to service in the Ministry of Justice.[301]

The 10th Mountain Division staff seemed oblivious to the political dimension of these actions by Aristide's administration, writing, "These uncoordinated actions did not come from the palace, [but] rather from ministers and others in the GOH." The 10th Mountain Division staff explained in its after-action review that it resisted these efforts by the Aristide administration because the "precipitous, uncoordinated action by GOH thwarts progress & moment[um] and threatens the stable and secure environment."[302]

The MNF had become fixated on maintaining a "stable and secure environment" rather than helping the Aristide administration displace Cédras's rule, the route to a sustainable political settlement to the war. In talking points developed for the division commander to engage President Aristide during the crisis, the 10th Mountain Division staff protested Aristide's actions because they "placed[d] MNF in the middle," forcing them to take sides in the conflict. These actions, they protested, "complicate[d] transition to UNMIH."[303] The Army was not interested in reaching a political settlement in Haiti; they simply wanted to hand the whole mess off to the UN as quickly as possible.

Initially, the chief of staff of the Army, Gen. Gordon Sullivan, declared operations in Haiti one of the Army's "significant military successes," mentioning it alongside the Gulf War in the 1995 white paper he cowrote with Lt. Col. Anthony Coroalles.[304] But with the benefit of hindsight, even Sullivan was forced to admit that the Army fumbled at the goal line in Haiti. In interviews conducted with him between 2002 and 2004, he admitted, "Frankly, we had a win and I think we walked away from it." He elaborated, saying he believed that "the Haiti operation was very successful," but he regretted that the Army "left the way that we did." Still, despite this reflectiveness, Sullivan refused to blame the ultimate loss in Haiti on the Army's urgency to dump the mission on the UN or the failure to back the chosen winner in the conflict and establish a stable government. Instead he blamed the loss on the American public and the national leadership. He concluded, "The rest of this country could not get behind Haiti, the other elements of power could not get behind it."[305]

In its 1995 strategic assessment—which attempted to predict the near-future strategic environment—the U.S. Army War College's SSI acknowledged, "U.S. forces will continue to be used in Military Operations Other Than War." With an eye on a continuing conflict in the former Yugoslavia, the SSI added, "Ethnicity and religion have supplanted ideology as social forces most likely to promote violence" and warned, "The United States could find itself under pressure to increase its military role in the Balkans, almost certainly through expanded air strikes and then, possibly, with ground forces."[306] In short, the SSI predicted that the future held more low-intensity conflicts.

For low-intensity conflict observers, the proper response to this reality was obvious: the Army needed to build greater capacity to fight and win in these conflicts. In a paper presented to the 1995 Annual Strategy Conference at the U.S. Army War College, Jeffrey Record asked the question, "Ready for What and Modernized against Whom?" He criticized the results of the Defense Department's BUR, writing that it "ignores the impact of Haiti- and Somali-like operations," and questioned the efficacy of the strategy of preparation for "two nearly simultaneous major regional contingencies," asking, "How likely is it that we would be drawn into two major wars at the same time? What are the opportunity costs of preparing for such a prospect?" Record noted that the only time the United States had actually had to fight two simultaneous wars was during World War II. He insisted—in an assertion that approached heresy at the Army War College—that the "chances for another world war . . . disappeared with the Soviet Union's demise." Record joined Andrew Bacevich in arguing that "the age of conventional military practice is drawing to a close." Record insisted instead that "we are entering an era in which the predominant form of conflict will be smaller and less conventional wars waged mostly within recognized national borders." In a particularly prescient passage, he warned that "the likely spread of politically radical Islam" was among the factors that "portend a host of politically and militarily messy conflicts." Record concluded, "Our present strategy portends an excessive readiness for the familiar and comfortable at the expense of preparation for the more likely and less pleasant"[307]

While he never used the word, Record also joined Andrew Bacevich in warning that enemies would increasingly threaten the U.S. Army with *asymmetry*: the use of low-tech means to neutralize the Army's high-tech, high-intensity conflict capabilities. He argued, "Desert Storm's very success will encourage our adversaries to side-step head-on collisions with U.S. conventional military power, in favor of strategies and tactics against which that power is poorly suited to respond." He added, "Our military failures and humiliations for the most part have been at the hands of opponents having little or nothing in the way of sea and air power, or even ground force other than light infantry. Most of them could not hope to prevail over U.S. forces conventionally. But they did prevail because they employed . . . unconventional strategy and tactics. . . . U.S. military power was . . . embarrassed in . . . Somalia by opponents who succeeded in denying to U.S. forces the kind of targets most vulnerable to overwhelming firepower." In light of this threat, he concluded, "Stuffing money into the defense budget readiness accounts prepares us for conventional warfare but not for much else."[308]

Record was not alone in worrying that the Army might be preparing for the wrong war. Even the project director of Force XXI, Army major general Michael Garret, admitted that the 1995 TRADOC PAM 525-5, *Force XXI Operations*, which described the desired capabilities for this future force, "lacked a clear conceptual path to meet challenges of MOOTW [Military Operations Other than War]." He added that the document "failed to address the complexities of achieving MOOTW desired end states." Garret recommended "a separate effort to better define its role in MOOTW."[309]

Yet the response of the pamphlet itself to such objections was that low-intensity conflict was a lesser included task of high-intensity conflict competency. It insisted that well-trained and disciplined units, "provided with sufficient time and resources to train, can transition to OOTW missions as required." This assertion was reinforced in a TRADOC conference report on Force XXI that cited "front-line troops well versed in war-fighting skills" as "almost unanimously" believing that "it makes most sense to conduct military operations other than war with existing forces, and "forces should not be earmarked for peace operations nor should new forces be created."[310]

The Army also rejected the requirement to educate its leaders in the lessons that it was learning in low-intensity conflicts in places such as Somalia and Haiti. During the period that the Army was engaged in these conflicts, the amount of time dedicated to low-intensity conflict in the CGSC curriculum actually decreased. In the 1993–94 academic year, the total number of hours dedicated to low-intensity conflict in the five-hundred-hour core curriculum dropped from forty-five to thirty-six (and this even with six electives still offered). Even after the disastrous Battle of Mogadishu and the ignominious withdrawal of U.S. forces from Somalia, the curriculum would remain at these reduced levels for the 1994–95 academic year.[311]

While the Army transformers dismissed the need for a focus on such conflicts, the low-intensity conflict observers continued to emphasize the importance of the political dimension of the conflicts. In the pages of *Parameters*, John W. Jandora, a U.S. Marine Corps Reserve colonel and supervisory intelligence analyst with the U.S. Army Special Operations Command at Fort Bragg, examined the nature of the threats the United States was facing in these new conflicts. He described the latest variety of low-intensity conflicts, including those in Somalia and the emerging one in Bosnia-Herzegovina, a "multidimensional factional conflict" characterized by "numerous armed groups which align and realign in ever-changing alliances." Jandora noted that the Army was ill-prepared to deal with these types of conflicts in which a "simple dichotomy no longer suffices as a formula," and that U.S. Army forces operating in Somalia had found it necessary to include "nontraditional categories" in their intelligence planning that were critical to "the success or failure of [the] operation." He added, "Military planning and intelligence analysis must move beyond the Cold War mind-set and its preoccupation with standing, conventional forces" to a consideration of the political motivations of multiple classes of threats and other parties in a low-intensity conflict.[312]

By this time many in Congress were worried that the rapidly transforming, high-intensity conflict Army might not be headed in the right direction to address the threats of the twenty-first century. To investigate this question, Congress directed the Department of Defense to form the Commission on Roles and Missions to review U.S. military effectiveness. The commission included civilian defense leaders like its chairman, future deputy secretary of defense John P. White, and a retired senior general or flag officer from each service. The commission found that the U.S. military needed to "improve [its] capabilities to deal with new challenges of the post–Cold War world." The report went on to identify four "emerging mission areas that demand immediate attention." Two of these four—"Peace Operations" and "OOTW"—represented a direct challenge to the Army's laser focus on preparing for a high-intensity conflict against a future peer competitor.[313]

Dismissing Low-Intensity Conflict

Yet the recommendations of this commission had no impact on the direction of the Army's transformation. The Joint Staff was only marginally involved in the commission, and the U.S. military establishment largely ignored its findings,[314] continuing its headlong march toward ever-greater capacity to fight high-intensity conflicts.

Admittedly, some of the Army's transformation also benefited the Army's ability to execute operations in a low-intensity conflict. By 1995 the Army had five war reserve stockpiles pre-positioned around the world. One was in the United States, and another in Europe. A third was afloat, generally, in the Indian Ocean, while the fourth and fifth were in Korea and Southwest Asia, respectively. This forward positioning of equipment mirrored the increasingly expeditionary nature of the Army. In the 1995 fiscal year, more than twenty thousand soldiers were deployed to around eighty different countries—this over and above those troops forward stationed in overseas bases around the world.[315]

But other innovations were clearly focused on an anticipated high-intensity conflict against an imaginary peer competitor and had little or no impact on the Army's ability to succeed in the low-intensity conflicts it was actually fighting. In 1995, General Sullivan proclaimed that the information age was ushering in a "Third Wave" of warfare.[316] And in the white paper that Sullivan and Coroalles published in March 1995, just a few months before the end of Sullivan's tenure as chief of staff of the Army, it was clear that he was still convinced that these low-intensity conflicts were a distraction from preparation for the real threat. They wrote that the "defining characteristic of war" was "simultaneity, the simultaneous employment of overwhelming combat power throughout the breadth and depth of the operational area to paralyze the enemy." While Sullivan himself claimed that this principle also applied to low-intensity military operations other than war, his focus was clearly on high-intensity conflict, the Army's preferred milieu of warfare since at least World War II. Achieving this capacity was at the heart of Sullivan's vision: "a new force for a new century-Force XXI."[317]

Oblivious to the irony, Sullivan and Coroalles warned, "The reason that many armies have failed to change with the conditions is that armies are by nature conservative institutions, generally resistant to change." Yet it was precisely because Sullivan and the institution he led were "conservative" and "generally resistant to change" that he saw no reason to change the course of his Army's transformation away from ever-greater capacity to fight high-intensity conflicts. Despite a string of dismal outings in low-intensity conflicts since the end of the Gulf War—and every indication that many more would follow—he remained stubbornly focused on fighting a type of warfare that was rapidly becoming obsolete. For Sullivan, despite the low-intensity conflicts in which it might be engaged, "the Army's fundamental character [was] to fight and win the nation's wars."[318]

Conclusion

Two important points of convergence between low-intensity conflict observers and Army transformers emerged during Sullivan's tenure as chief of staff of the Army. Both acknowledged that the future held many more low-intensity conflicts. And while they were not yet using the word *asymmetry*, low-intensity conflict observers began to warn that future adversaries would find low-tech, low-intensity conflict means to challenge the Army in asymmetric ways that negated its high-tech, high-intensity conflict advantages. Observers and transformers differed sharply on the consequences of these insights, however, with the observers arguing that the Army needed to improve its capacity to wage low-intensity conflicts and the transformers seeing these conflicts as a dangerous distraction from preparation for a theoretical future high-intensity conflict against a nonpeer competitor.

The coincidental participation of the 10th Mountain Division in the low-intensity conflicts in both Somalia and Haiti clearly demonstrated the potential value of institutionalizing the lessons the Army was learning from its experiences in such conflicts. While the division did not reach the highest levels of understanding—accepting its responsibility to pick sides and back a winner in the conflict in Haiti—it did make dramatic improvements in its capacity to operate in such environments between Operation Restore Hope and Operation Uphold Democracy. The division was much more effective in using PSYOP and public affairs to effectively operate in and among the populace. In Haiti the 10th Mountain Division, from the platoon to the division level, also began to employ its leaders in negotiations with the civilian and military leaders, from the local to the national level, and thus beginning to engage the political dimension of the conflict. These improvements that the division made in its operations in Haiti were a direct result of the lessons it learned in Somalia.

Yet the U.S. Army writ large refused to institutionalize these lessons. While joint doctrine at the end of General Sullivan's tenure reasserted many of the lessons that low-intensity conflict observers had drawn from Somalia and Haiti, Army doctrine generally regressed during this period, replacing the term "low intensity conflict" with "operations other than war," excising "political dominance" from low-intensity conflict doctrine, institutionalizing the idea of mission creep, and questioning the legitimacy of committing Army units to low-intensity conflicts where no host nation was present.

In fairness, General Sullivan did establish the PKI during his tenure and continued to mention "operations other than war" in his writings on the future of the Army. But the PKI turned out to have very little impact on the Army, its doctrine, or its prosecution of low-intensity conflicts. And while Sullivan may have kept these operations in mind, the transformation that was taking shape under his supervision deliberately and explicitly ignored the necessity to prepare for or optimize the Army to engage in these operations.

Unfortunately, this trend would continue—even accelerate—under General Sullivan's successor as chief of staff of the Army, Gen. Dennis Reimer. Unlike Sullivan, Reimer had never demonstrated a passion for institutionalizing or elevating the precedence of low-intensity conflict. In fact, as chief of staff of the Army, Reimer would act to roll back many of the gains observers had made in institutionalizing the lessons of low-intensity conflict. Reimer presided over the destruction of A-AFCLIC and nearly dismantled the PKI. And as the Army engaged in a new low-intensity conflict in Bosnia-Herzegovina, Reimer would actively insulate both the Army and its transformation from the lessons that units were beginning to learn there.

NOTES

1. Sullivan and Coroalles, "The Army in the Information Age," vi; Gordon R. Sullivan, interviews with John R. Dabrowski, February 14, 2002–November 4, 2008, transcript, Senior Officer Oral History Collection, U.S. Army Military History Institute, U.S. Army Heritage and Education Center, i–v; U.S. Combined Arms Center, *1987 Annual Historical Review*, 99–108; Townsend, "Combined Action Platoons in the Vietnam War," 31.

2. U.S. Combined Arms Center, *1987 Annual Historical Review*, 99–108.

3. Gordon R. Sullivan, telephone interview with the author, June 15, 2018; U.S. Combined Arms Center, *1987 Annual Historical Review*, 37.

4. Headquarters, U.S. Department of the Army, *Operations*, FM 100-5 (1986), 1; U.S. Combined Arms Center, *1987 Annual Historical Review*, 99–108.

5. Fitzgerald, *Learning to Forget*, Kindle, 68, 73; Armed Forces News Service, "The Center for Low Intensity Conflict Closes after 10 Years," Federation of American Scientists, June 26, 1996, https://fas.org/irp/news/1996/n19960626_960615.html.

6. National Defense Authorization Act for Fiscal Year 1987, Pub. L. No. 99–661, November 14, 1986, Title XIII, Part B.

7. U.S. Combined Arms Center, *1987 Annual Historical Review*, 67–70, 99–108.

8. U.S. Combined Arms Center, *1987 Annual Historical Review*, 99–108.

9. U.S. Combined Arms Center, *1987 Annual Historical Review*, 99–108.

10. U.S. Combined Arms Center, *1987 Annual Historical Review*, 99–108.

11. U.S. Combined Arms Center, *1987 Annual Historical Review*, 99–108.

12. Gordon R. Sullivan, "Subject: Close Combat," memorandum for deputy commanding general Charles Ostott, U.S. Combined Arms Center, January 29, 1988, on low-intensity conflict, frame 590, reel 47670, microfilm collection, Records of the Army–Air Force Center for Low Intensity Conflict, U.S. Air Force Historical Research Agency; Gordon R. Sullivan, "Subject: Low-Intensity Conflict Capabilities Requirements Study," memorandum for commander Maxwell R. Thurman, U.S. Army Training and Doctrine Command, n.d. [February– March 1988], on low-intensity conflict, frame 595, reel 47670, microfilm collection, Records of the Army–Air Force Center for Low Intensity Conflict, U.S. Air Force Historical Research Agency.

13. Fitzgerald, *Learning to Forget*, Kindle, 78.

14. Carl E. Vuono to Gerald T. Bartlett, March 13, 1987, detailing creation of Army–Air Force Center for Low Intensity Conflict and directing that they lead effort to create FC 100-20, frame 326, reel 47670, microfilm collection, Records of the Army–Air Force Center for Low Intensity Conflict, U.S. Air Force Historical Research Agency; U.S. Combined Arms Center, *1987 Annual Historical Review*, 99–108; Sullivan, interviews with Dabrowski, i–v; Gerald B. Thompson, "Subject: Revisions to FM 100-20 / AFM 2-20," memorandum for deputy commandant, U.S. Army Command and General Staff College, March 20, 1989, detailing proposed changes to FM 100-20 concerning counterinsurgency and the response of the U.S. Combined Arms Center, frames 378–80, reel 47671, microfilm collection, Records of the Army–Air Force Center for Low Intensity Conflict, U.S. Air Force Historical Research Agency.

15. Albert M. Barnes, "FM 100-20 / AFM 2-XY Read Ahead," memorandum for commanding general Maxwell Thurman, U.S. Army Training and Doctrine Command, May 19, 1988, on comments on FC 100-20, frames 726–27, reel 47670, microfilm collection, Records of the Army–Air Force Center for Low Intensity Conflict, U.S. Air Force Historical Research Agency;

Army–Air Force Center for Low Intensity Conflict, "BACK CHANNEL," memorandum drafted for commanding general Maxwell Thurman, U.S. Army Training and Doctrine Command,
n.d. [October 27, 1988], to send to the chief of staff of the Army, Gen. Carl Vuono, about obstacles to publication of FM 100-20, frames 1109–10, reel 47670, microfilm collection, Records
of the Army–Air Force Center for Low Intensity Conflict, U.S. Air Force Historical Research
Agency; William F. Furr, "FM 100-20 / AFM 2-20 Chronology of Events," Army–Air Force
Center for Low Intensity Conflict, October 27, 1988, frames 1185–92, reel 47670, microfilm
collection, Records of the Army–Air Force Center for Low Intensity Conflict, U.S. Air Force
Historical Research Agency.

16. Albert M. Barnes, "Trip Report," memorandum on meeting with commanding general
Maxwell Thurman, U.S. Army Training and Doctrine Command and Gen. (Ret.) Paul Gorman
about obstacles to publication of FM 100-20, October 27, 1988, frames 1107–8, reel 47670,
microfilm collection, Records of the Army–Air Force Center for Low Intensity Conflict, U.S.
Air Force Historical Research Agency; Army–Air Force Center for Low Intensity Conflict,
"A-AF CLIC Weekly Update, 4–8 December 1989," n.d. [December 1989], frames 877–79,
reel 47671, microfilm collection, Records of the Army–Air Force Center for Low Intensity
Conflict, U.S. Air Force Historical Research Agency; Hunt, "OOTW: A Concept in Flux," 3–10.

17. Sullivan, interview with the author.

18. Yarrison, *The Modern Louisiana Maneuvers*, 1–15.

19. Millett and Maslowski, *For the Common Defense*, Kindle, locations 11791–804.

20. *Public Papers of the President of the United States: George Bush, 1992–1993*, 1:182–87;
Millett and Maslowski, *For the Common Defense*, Kindle, locations 12112–19.

21. Brown, *Kevlar Legions*, Kindle, 73–74, 85.

22. Carl S. Vuono, quoted in Brown, *Kevlar Legions*, Kindle, 45; Holt, "LIC in Central
America," 2–15.

23. Crane, "Peace Dividends and Benevolent Interventions."

24. Crane, "Peace Dividends and Benevolent Interventions"; Millett and Maslowski, *For the
Common Defense*, Kindle, locations 12157–72.

25. Fitzgerald, *Learning to Forget*, Kindle, 83.

26. White House, *A National Security Strategy of Engagement and Enlargement*, i–ii.

27. Fitzgerald, *Learning to Forget*, Kindle, 92.

28. Brown, *Kevlar Legions*, Kindle, 85.

29. Brown, "Defense Transformation Redux," 23–26; Sullivan, *The Collected Works of the
Thirty-Second Chief of Staff, United States Army*, 4, 98.

30. Headquarters, U.S. Department of the Army, *Operations*, FM 100-5 (1976), 1-1, emphasis in the original; Shimko, *The Iraq Wars*, Kindle, locations 1265–73; Chapman et al., *Prepare
the Army for War*, 51.

31. McDonough, "Building the New FM 100-5," 2–12.

32. McDonough, "Building the New FM 100-5," 2–12.

33. Institute for National Strategic Studies, *Project 2025*, 3.

34. Institute for National Strategic Studies, *Project 2025*, 61–63.

35. Sullivan and Dubik, "Land Warfare in the 21st Century," viii.

36. Sullivan and Dubik, "Land Warfare in the 21st Century," iii, xi, xv, xxiv, xxv, emphasis
in the original.

37. Chapman et al., *Prepare the Army for War*, 31–32; Yarrison, *The Modern Louisiana
Maneuvers*, vi.

38. Chapman et al., *Prepare the Army for War*, 31–32.

39. Brown, *Kevlar Legions*, Kindle, 91–92.

40. Headquarters, U.S. Department of the Army, *Operations*, FM 100-5 (1993), 3-1–3-6.

41. Butler, "Adjusting to Post–Cold War Strategic Realities," 2–9; Brown, *Kevlar Legions*, Kindle, 93.

42. Brown, *Kevlar Legions*, Kindle, 126.

43. Army–Air Force Center for Low Intensity Conflict, "Army/Air Force Center for Low Intensity Conflict (CLIC) Activation Plan, Annex A, Terms of Reference," January 29, 1986, frame 548, reel 47670, microfilm collection, Records of the Army–Air Force Center for Low Intensity Conflict, U.S. Air Force Historical Research Agency; Barnes, "The Diplomat Warrior," 55–63; Holt, "LIC in Central America," 2–15; Butler, "Adjusting to Post–Cold War Strategic Realities," 2–9.

44. Tinder, "Low Intensity Conflict," 23; John B. Hunt, ed., *LIC Conversations: An After-Action Report*, Regional Military Studies Office, Foreign Military Studies Office, Fort Leavenworth, KS, n.d. [1991], Combined Arms Center Historical Archive, U.S. Army Combined Arms Center.

45. Institute for National Strategic Studies, *Project 2025*, 4, 7.

46. Sullivan and Dubik, "Land Warfare in the 21st Century," iii.

47. Sullivan and Dubik, "Land Warfare in the 21st Century," iii.

48. Sullivan, "A Vision for the Future," 6, quoted in Shimko, *The Iraq Wars*, Kindle, locations 3009–3013.

49. McDonough, "Building the New FM 100-5," 2–12.

50. Barnes, "The Diplomat Warrior," 55–63; Roger Trinquier, *Modern Warfare*, quoted in Holt, "LIC in Central America," 2–15; Institute for National Strategic Studies, *Project 2025*, 7.

51. Fitzgerald, *Learning to Forget*, Kindle, 79.

52. Holt, "LIC in Central America," 2–15.

53. Dubik, "On the Foundations of National Military Strategy," 21–46; Butler, "Adjusting to Post–Cold War Strategic Realities," 2–9.

54. McDonough, "Building the New FM 100-5," 2–12.

55. Institute for National Strategic Studies, *Project 2025*, 25.

56. Gordon R. Sullivan, "Subject: Trip Report—Joint Readiness Training Center, Ft. Chaffee, AR, Ft. Bliss, TX, and Ft. Leavenworth, KS, 17–21 November 1991," memorandum to the chief of staff, U.S. Army, November 27, 1991, Ike Skelton Combined Arms Research Library, 10–11.

57. Sullivan, "Subject: Trip Report," 10, 12; Hastings, *The Korean War*, 15–98.

58. McDonough, "Building the New FM 100-5," 2–12.

59. McDonough, "Building the New FM 100-5," 2–12.

60. Army–Air Force Center for Low Intensity Conflict, "A-AF CLIC Weekly Update, 4–8 December 1989"; Hunt, "OOTW: A Concept in Flux," 3–10; Army–Air Force Center for Low Intensity Conflict, "Back Channel"; Furr, "FM 100-20 / AFM 2-20 Chronology of Events."

61. Headquarters, U.S. Departments of the Army and the Air Force, *Military Operations in Low Intensity Conflict*, FM 100-20, 1-5.

62. Headquarters, U.S. Departments of the Army and the Air Force, FM 100-20, 1-5.

63. Headquarters, U.S. Departments of the Army and the Air Force, FM 100-20, 1-6.

64. Clausewitz, *On War*, Kindle, 75; Headquarters, U.S. Departments of the Army and the Air Force, FM 100-20, 1-6.

65. Army–Air Force Center for Low Intensity Conflict, "A-AF CLIC Weekly Update, 4–8 December 1989"; Hunt, "OOTW: A Concept in Flux," 3–10.

66. Headquarters, U.S. Department of the Army, *Operations in a Low-Intensity Conflict* FM 7-98, 4-1–4-9. Rather than originating with an "anonymous member" of a "peacekeeping

force," this quote has been attributed in various sources to former UN secretary-general Dag Hammarskjöld or military sociologist Charles Moskos.

67. Holt, "LIC in Central America," 2–15.

68. U.S. Army Command and General Staff College, *United States Army Command and General Staff College Catalog, Academic Year 1991–1992*, CGSC Circular No. 351-1, July 1991 Command and General Staff College Papers, Ike Skelton Combined Arms Research Library, 33, 42–43, 45, 49–51, 59–60, 62, 64–65, 68.

69. Rendina, "An Officer Corps for the 1990s," 64–73.

70. H. Lee Dixon, "Trip Report," report on trip to U.S. Army War College in Carlisle Barracks, PA, to observe a low-intensity conflict exercise conducted by War College seminars, May 7–11, 1990, May 16, 1990, frames 1417–25, reel 47671, microfilm collection, Records of the Army–Air Force Center for Low Intensity Conflict, U.S. Air Force Historical Research Agency.

71. Holder, "Educating and Training for Theater Warfare," 85–99.

72. Headquarters, U.S. Department of the Army, FM 7-98, 1-1-1-4, 2-1-2-21.

73. Holt, "LIC in Central America," 2–15; William J. Olson, quoted in Holt, "LIC in Central America," 2–15.

74. Fitzgerald, *Learning to Forget*, Kindle, 71.

75. Barnes, "The Diplomat Warrior," 55–63.

76. J. C. Rhoades, "Information Paper, Subject: Low Intensity Conflict (LIC)," Army–Air Force Center for Low Intensity Conflict, Langley Air Force Base, VA, February 7, 1990, frames 1459–60, reel 47671, microfilm collection, Records of the Army–Air Force Center for Low Intensity Conflict, U.S. Air Force Historical Research Agency.

77. Holt, "LIC in Central America," 2–15.

78. Headquarters, U.S. Department of the Army, FM 7-98, 4-1-4-9, 1-1-1-4.

79. Holt, "LIC in Central America," 2–15; Barnes, "The Diplomat Warrior," 55–63.

80. Barnes, "The Diplomat Warrior," 55–63; Headquarters, U.S. Department of the Army, FM 7-98, 1-1-1-4.

81. Headquarters, U.S. Department of the Army, FM 7-98, 1-1-1-4.

82. Sullivan, "Subject: Trip Report," 4–5.

83. Barnes, "The Diplomat Warrior," 55–63.

84. Barnes, "The Diplomat Warrior," 55–63.

85. McDonough, "Building the New FM 100-5," 2–12.

86. Headquarters, U.S. Departments of the Army and the Air Force, FM 100-20, 1–5.

87. Headquarters, U.S. Department of the Army, FM 7-98, 1-1-1-4, 2-1-2-21.

88. Headquarters, U.S. Department of the Army, FM 7-98, 2-1-2-21.

89. Headquarters, U.S. Department of the Army, FM 7-98, 4-1-4-9.

90. Headquarters, U.S. Department of the Army, FM 7-98, 1-1-1-4, 2-1-2-21, emphasis added.

91. Barnes, "The Diplomat Warrior," 55–63; Hunt, *LIC Conversations*; Holt, "LIC in Central America," 2–15.

92. Headquarters, U.S. Department of the Army, FM 7-98, 1-1-1-4, 4-1-4-9.

93. Crane, "Peace Dividends and Benevolent Interventions."

94. Fishel and Downie, "Taking Responsibility for Our Actions?," 66–72.

95. Yates, "Joint Task Force Panama," 58–71.

96. Fishel and Downie, "Taking Responsibility for Our Actions?," 66–72.

97. Fishel and Downie, "Taking Responsibility for Our Actions?," 66–72.

98. Fishel and Downie, "Taking Responsibility for Our Actions?," 66–72.

99. Yates, "Joint Task Force Panama," 58–71.

100. Yates, "Joint Task Force Panama," 58–71.

101. Gordon Sullivan, comments at the Association of the United States Army, Washington, DC, November 25, 2013, quoted in C. J. Restemayer, "PKSOI Anniversary Gathering: Celebrating 1993–2013," *Peace & Stability Operations Journal Online*, December 2013, http://pksoi.armywarcollege.edu/default/assets/File/Peace_Stability_Journal_Special_Edition.pdf.

102. Fishel and Downie, "Taking Responsibility for Our Actions?," 66–72.

103. Fishel and Downie, "Taking Responsibility for Our Actions?," 66–72.

104. Fishel and Downie, "Taking Responsibility for Our Actions?," 66–72.

105. Fishel and Downie, "Taking Responsibility for Our Actions?," 66–72.

106. Fishel and Downie, "Taking Responsibility for Our Actions?," 66–72.

107. Sullivan, interview with the author.

108. Headquarters, U.S. Department of the Army, FM 100-5 (1993), v–vi.

109. Sullivan and Dubik, "Land Warfare in the 21st Century," xii, xiii–xiv.

110. Headquarters, U.S. Department of the Army, FM 100-5 (1993), v, 13-0-13-8.

111. Headquarters, U.S. Department of the Army, FM 100-5 (1993), 2-0-2-1.

112. Headquarters, U.S. Department of the Army, FM 100-5 (1993), vi, 2-0-2-1.

113. National Defense Authorization Act for Fiscal Year 1987, Pub. L. No. 99–661, November 14, 1986.

114. Headquarters, U.S. Department of the Army, FM 100-5 (1993), 2-20, 13-8.

115. Headquarters, U.S. Department of the Army, *Intelligence and Electronic Warfare Support to Low-Intensity Conflict Operations*, FM 34-7, 1-1-1-5.

116. Headquarters, U.S. Department of the Army, FM 34-7, 1-1-1-5; Headquarters, U.S. Department of the Army, FM 100-5 (1993), 13-0-13-8.

117. Headquarters, U.S. Department of the Army, FM 100-5 (1993), 3-11-3-12, 13-0-13-8.

118. Center for Army Lessons Learned, *Somalia Operations Other Than War*.

119. Center for Army Lessons Learned, *Somalia Operations Other Than War*; Stewart, *The United States Army in Somalia, 1992–1994*, 5–8.

120. Stewart, *The United States Army in Somalia*, 5–9.

121. Stewart, *The United States Army in Somalia*, 10–11.

122. Stewart, *The United States Army in Somalia*, 10–11.

123. Stewart, *The United States Army in Somalia*, 10–11.

124. Stewart, *The United States Army in Somalia*, 10–11; Millett and Maslowski, *For the Common Defense*, Kindle, locations 12260–98.

125. Stewart, *The United States Army in Somalia*, 12–14; Curtiss, "In Somalia, the Goal Must Be 'Do No Harm,'" 38.

126. Curtiss, "In Somalia, the Goal Must Be 'Do No Harm,'" 38.

127. Millett and Maslowski, *For the Common Defense*, Kindle, locations 12260–98.

128. Bowden, *Black Hawk Down*, 332–33; Millett and Maslowski, *For the Common Defense*, Kindle, locations 12260–98.

129. Arnold, "Somalia," 26–35.

130. Sullivan and Dubik, *Envisioning Future War*, 35, 52, quoted in Fitzgerald, *Learning to Forget*, 97.

131. Clarke, "Testing the World's Resolve in Somalia," 42–58.

132. Arnold, "Somalia," 26–35.

133. Allard, *Somalia Operations*, xv–xviii.

134. Bacevich, "Learning from Aidid," 32, 34, quoted in Shimko, *The Iraq Wars*, Kindle, locations 2946–49.

135. Arnold, "Somalia," 26–35.

136. Freeman, Lambert, and Mims, "Operation Restore Hope," 61–72; Allard, *Somalia Operations*, xv–xviii.

137. Allard, *Somalia Operations*, xv–xviii.

138. Center for Army Lessons Learned, *Somalia Operations Other Than War*.

139. Arnold, "Somalia," 26–35.

140. Allard, *Somalia Operations*, xv–xviii; Center for Army Lessons Learned, *Somalia Operations Other Than War*.

141. Lorenz, "Law and Anarchy in Somalia," 27–41; Arnold, "Somalia," 26–35.

142. Moreno and Vega, "Lessons from Somalia," 11.

143. Center for Army Lessons Learned, *US Army Operations in Support of UNOSOM II: Operations Other Than War*, I-1–I-11, quoted in Fitzgerald, *Learning to Forget*, 103.

144. Arnold, "Somalia," 26–35.

145. Allard, *Somalia Operations*, 1–86.

146. Allard, *Somalia Operations*, 72–73; Clarke, "Testing the World's Resolve in Somalia," 42–58; Lorenz, "Law and Anarchy Somalia," 27–41; Freeman, Lambert, and Mims, "Operation Restore Hope," 61–72.

147. Dennis J. Reimer, interview with Lewis Sorley, 2000, transcript, Senior Officer Oral History Project, U.S. Army Military History Institute, U.S. Army Heritage and Education Center, 180–95; Allard, *Somalia Operations*, xv–xviii.

148. Clarke, "Testing the World's Resolve in Somalia," 42–58.

149. Allard, *Somalia Operations*, xv–xviii; Arnold, "Somalia," 26–35.

150. Allard, *Somalia Operations*, xv–xviii.

151. Center for Army Lessons Learned, *US Army Operations in Support of UNOSOM II*, I-1–I-11; Center for Army Lessons Learned, *Somalia Operations Other Than War*.

152. Freeman, Lambert, and Mims, "Operation Restore Hope," 61–72; Allard, *Somalia Operations*, 60; Lorenz, "Law and Anarchy in Somalia," 27–41; Arnold, "Somalia," 26–35.

153. Allard, *Somalia Operations*, xv–xviii; Arnold, "Somalia," 26–35.

154. Arnold, "Somalia," 26–35; Lorenz, "Law and Anarchy in Somalia," 27–41.

155. Allard, *Somalia Operations*, xv–xviii.

156. Allard, *Somalia Operations*, xv–xviii; Lorenz, "Law and Anarchy in Somalia," 27–41.

157. Lorenz, "Law and Anarchy in Somalia," 27–41; Arnold, "Somalia," 26–35; Allard, *Somalia Operations*, xv–xviii.

158. Allard, *Somalia Operations*, xv–xviii, 70–90; Lorenz, "Law and Anarchy in Somalia," 27–41.

159. Allard, *Somalia Operations*, 33–34.

160. Arnold, "Somalia," 26–35, emphasis added.

161. Center for Army Lessons Learned, *Somalia Operations Other Than War*.

162. Arnold, "Somalia," 26–35; Allard, *Somalia Operations*, xv–xviii.

163. Clarke, "Testing the World's Resolve in Somalia," 42–58.

164. Carr, "The Consequences of Somalia," 1.

165. Carr, "The Consequences of Somalia," 4.

166. Arnold, "Somalia," 26–35.

167. Lorenz, "Law and Anarchy in Somalia," 27–41.

168. Allard, *Somalia Operations*, xv–xviii; Freeman, Lambert, and Mims, "Operation Restore Hope," 61–72; Moreno and Vega, "Lessons from Somalia," 11–12.

169. Allard, *Somalia Operations*, xv–xviii, 12–15.

170. Freeman, Lambert, and Mims, "Operation Restore Hope," 61–72.

171. Freeman, Lambert, and Mims, "Operation Restore Hope," 61–72; Allard, *Somalia Operations*, xv–xviii.

172. Clarke, "Testing the World's Resolve in Somalia," 42–58.

173. Carr, "The Consequences of Somalia," 1–4.

174. Lorenz, "Law and Anarchy in Somalia," 27–41; Arnold, "Somalia," 26–35.

175. Allard, *Somalia Operations*, 80–81; Freeman, Lambert, and Mims, "Operation Restore Hope," 61–72.

176. Arnold, "Somalia," 26–35.

177. Arnold, "Somalia," 26–35; Center for Army Lessons Learned, *Somalia Operations Other Than War*.

178. Clarke, "Testing the World's Resolve in Somalia," 42–58.

179. Lorenz, "Law and Anarchy in Somalia," 27–41.

180. Allard, *Somalia Operations*, xv–xviii; Center for Army Lessons Learned, *Somalia Operations Other Than War*; Freeman, Lambert, and Mims, "Operation Restore Hope," 61–72.

181. Clarke, "Testing the World's Resolve in Somalia," 42–58.

182. Allard, *Somalia Operations*, 49–53.

183. Carr, "The Consequences of Somalia," 1–4; Moreno and Vega, "Lessons from Somalia," 11–12.

184. Allard, *Somalia Operations*, 1–86.

185. Allard, *Somalia Operations*, xv–xviii, 24–28, 29–32, 79–83; Arnold, "Somalia," 26–35; Moreno and Vega, "Lessons from Somalia," 11–12.

186. Lorenz, "Law and Anarchy in Somalia," 27–41.

187. Freeman, Lambert, and Mims, "Operation Restore Hope," 61–72.

188. Freeman, Lambert, and Mims, "Operation Restore Hope," 61–72.

189. Allard, *Somalia Operations*, xv–xviii.

190. Clarke, "Testing the World's Resolve in Somalia," 42–58.

191. Carr, "The Consequences of Somalia," 1–4.

192. Reimer interview, 180–92; Allard, *Somalia Operations*, xv–xviii; Bowden, *Black Hawk Down*, 331; Allard, *Somalia Operations*, xv–xviii.

193. Moreno and Vega, "Lessons from Somalia," 11–12.

194. White House, *A National Security Strategy of Engagement and Enlargement*, 13.

195. Reimer interview, 180–92.

196. U.S. Army Training and Doctrine Command, *Multiservice Procedures for Humanitarian Assistance Operations*, FM 100-23-1, 1-7.

197. U.S. Army Training and Doctrine Command, FM 100-23-1, 2-9.

198. U.S. Army Training and Doctrine Command, FM 100-23-1, 1-8.

199. U.S. Army Training and Doctrine Command, FM 100-23-1, 1-8, emphasis in the original.

200. Freeman, Lambert, and Mims, "Operation Restore Hope," 61–72.

201. Allard, *Somalia Operations*, xv–xviii.

202. Aspin, *Report of the Bottom-Up Review*, 19; Brown, *Kevlar Legions*, Kindle, 162.

203. U.S. Department of the Army, *Decisive Victory*, 15–16, 10; Sullivan and Twomey, "The Challenges of Peace," 4–17; Sullivan and Dubik, *War in the Information Age*, 13.

204. U.S. Department of the Army, *Decisive Victory*, 15–16; Sullivan and Dubik, *War in the Information Age*, 13; Sullivan and Twomey, "The Challenges of Peace," 4–17. See also Stockton, "When the Bear Leaves the Woods."

205. Brown, *Kevlar Legions*, Kindle, 92.

206. Sullivan and Dubik, "War in the Information Age," 46–62.

207. Sullivan and Dubik, "War in the Information Age," 46–62; John R. Boyd, "Organic Design for Command and Control," slide presentation, Defense and the National Interest, Air Power Australia, February 2005, https://www.ausairpower.net/JRB/organic_design.pdf.

208. Sullivan and Dubik, *War in the Information Age*, 17–18.

209. U.S. Department of the Army, *Decisive Victory*, 8–9.

210. U.S. Army Training and Doctrine Command, *Force XXI Operations*, PAM 525-5, 3-1-3-2, emphasis in the original.

211. Brown, *Kevlar Legions*, Kindle, 91, 124.

212. 10th Mountain Division, "10th Mountain Briefs," in *Operation Uphold Democracy: US Forces in Haiti*, CD-ROM, Norfolk, VA: U.S. Atlantic Command, September 1997; Chapman et al., *Prepare the Army for War*, 31–32.

213. U.S. Department of the Army, *Decisive Victory*, 1, 2; Sullivan and Dubik, "War in the Information Age," 46–62.

214. Sullivan and Dubik, "War in the Information Age," 46–62.

215. Reimer interview, 324–27.

216. U.S. Department of the Army, *Decisive Victory*, 11, emphasis added.

217. Sullivan and Twomey, "The Challenges of Peace," 4–17.

218. U.S. Army Training and Doctrine Command, FM 100-23-1, 1-5–1-6.

219. Fitzgerald, *Learning to Forget*, Kindle, 89, 100.

220. Haskin, *Bosnia and Beyond*, 202; Fitzgerald, *Learning to Forget*, Kindle, 100.

221. Sullivan and Twomey, "The Challenges of Peace," 4–17.

222. Sullivan and Twomey, "The Challenges of Peace," 4–17.

223. U.S. Department of the Army, *Decisive Victory*, 7; U.S. Army Training and Doctrine Command, FM 100-23-1, 3-7.

224. U.S. Army Training and Doctrine Command, FM 100-23-1, 4-5–4-6, emphasis in the original.

225. U.S. Army Training and Doctrine Command, FM 100-23-1, 1-5–1-6, 1-8–1-9; Headquarters, U.S. Department of the Army, *Peace Operations*, FM 100-23, 1.

226. Sullivan and Twomey, "The Challenges of Peace," 4–17.

227. Sullivan and Twomey, "The Challenges of Peace," 4–17.

228. U.S. Army Training and Doctrine Command, FM 100-23-1, 4-15–4-18.

229. U.S. Army Training and Doctrine Command, FM 100-23-1, 1-5–1-6, 1-2, 28.

230. Michael Bruno and Jason Kring, "The History of PKI/PKSOI: The Early Years: A History of the U.S. Army Peacekeeping Institute and a Snapshot of the Past 20 Years," *Peace & Stability Operations Journal Online*, December 2013, http://pksoi.armywarcollege.edu/default/assets/File/Peace_Stability_Journal_Special_Edition.pdf; Sullivan, interview with the author.

231. Sullivan, comments at the Association of the United States Army; Sullivan, interview with the author.

232. Sullivan, comments at the Association of the United States Army; Sullivan, interview with the author.

233. Bruno and Kring, "The History of PKI/PKSOI."

234. Karl Farris, "The History of PKI," *Peace & Stability Operations Journal Online*, December 2013, http://pksoi.armywarcollege.edu/default/assets/File/Peace_Stability_Journal_Special_Edition.pdf; "PKI/PKSOI: Key Events of the Past 20 Years," *Peace & Stability Operations Journal Online*, December 2013, http://pksoi.armywarcollege.edu/default/assets/File/Peace_Stability_Journal_Special_Edition.pdf; David T. Stahl, ed., "10th Mountain Division, Operation Uphold Democracy, Operations in Haiti," in *Operation Uphold Democracy: US Forces in Haiti*, CD-ROM, Norfolk, VA: U.S. Atlantic Command, September 1997.

235. Sullivan, comments at the Association of the United States Army.

236. Brown, *Kevlar Legions*, Kindle, 114; Dupuy, *The Prophet and Power*, 62–83; 10th Mountain Division, "10th Mountain Briefs"; Crane, "Peace Dividends and Benevolent Interventions."

237. 10th Mountain Division, "10th Mountain Briefs"; Haley, *Strategies of Dominance*, 96; Von Einsiedel and Malone, "Haiti," 471–73.

238. Stahl, "10th Mountain Division"; 10th Mountain Division, "10th Mountain Briefs."

239. Brown, *Kevlar Legions*, Kindle, 114; 10th Mountain Division, "10th Mountain Briefs."

240. 10th Mountain Division, "10th Mountain Briefs."

241. 10th Mountain Division, "10th Mountain Briefs"; Stahl, "10th Mountain Division."

242. Stahl, "10th Mountain Division"; Crane, "Peace Dividends and Benevolent Interventions." See also Bailey, Maguire, and Pouliot, "Haiti: Military-Police Partnership for Public Security."

243. Crane, "Peace Dividends and Benevolent Interventions"; Stahl, "10th Mountain Division."

244. 10th Mountain Division, "10th Mountain Briefs"; Millett and Maslowski, *For the Common Defense*, Kindle, locations 12299–309.

245. Ann Crawford-Roberts, "A History of United States Policy towards Haiti," Brown University Library Center for Digital Scholarship, n.d., accessed March 10, 2018, https://library. brown.edu/create/modernlatinamerica/chapters/chapter-14-the-united-states-and-latin -america/moments-in-u-s-latin-american-relations/a-history-of-united-states-policy -towards-haiti/; Heather Nauert, "UN Mission for Justice Support in Haiti (MINUJUSTH)," press release, U.S. Department of State, October 16, 2017, https://www.state.gov/un-mission -for-justice-support-in-haiti-minujusth/.

246. 10th Mountain Division, "10th Mountain Briefs"; "Haiti—UNMIH," United Nations Department of Public Information, September 1996, accessed March 10, 2018, http://www .un.org/Depts/DPKO/Missions/unmih_b.htm; Ann M. Simmons, "U.N. Peacekeepers Are Leaving after More Than Two Decades, but Where Does That Leave Haiti?" *Los Angeles Times*, April 17, 2017, http://www.latimes.com/world/global-development/la-fg-un-haiti-mission -20170417-story.html; Crawford-Roberts, "A History of United States Policy Towards Haiti"; Nauert, "UN Mission for Justice Support in Haiti."

247. James Dubik, interview with Robert M. Mages, 2008, transcript, Senior Leader Debriefing Program, U.S. Army Military History Institute, U.S. Army Heritage and Education Center, iii–iv; 10th Mountain Division, "10th Mountain Briefs."

248. 10th Mountain Division, "10th Mountain Briefs"; Stahl, "10th Mountain Division."

249. 10th Mountain Division, "10th Mountain Briefs"; Stahl, "10th Mountain Division."

250. Stahl, "10th Mountain Division."

251. Stahl, "10th Mountain Division"; 10th Mountain Division, "10th Mountain Briefs."

252. 10th Mountain Division, "10th Mountain Briefs."

253. 10th Mountain Division, "10th Mountain Briefs"; Stahl, "10th Mountain Division."

254. 10th Mountain Division, "10th Mountain Briefs"; Stahl, "10th Mountain Division."

255. 10th Mountain Division, "10th Mountain Briefs"; Stahl, "10th Mountain Division."

256. 10th Mountain Division, "10th Mountain Briefs."

257. Stahl, "10th Mountain Division"; 10th Mountain Division, "10th Mountain Briefs."

258. Stahl, "10th Mountain Division."

259. Reimer interview, 246.

260. Stahl, "10th Mountain Division."

261. Stahl, "10th Mountain Division."

262. Stahl, "10th Mountain Division."
263. 10th Mountain Division, "10th Mountain Briefs."
264. Stahl, "10th Mountain Division."
265. Stahl, "10th Mountain Division."
266. Stahl, "10th Mountain Division."
267. Stahl, "10th Mountain Division"; 10th Mountain Division, "10th Mountain Briefs."
268. Stahl, "10th Mountain Division."
269. Stahl, "10th Mountain Division."
270. 10th Mountain Division, "10th Mountain Briefs."
271. Stahl, "10th Mountain Division."
272. Shelton and Vane, "Winning the Information War in Haiti," 3–9.
273. H. Hugh Shelton, interview with Russell Riley, transcript, Presidential Oral Histories, University of Virginia Miller Center, May 29, 2007, https://millercenter.org/the-presidency/presidential-oral-histories/henry-hugh-shelton-oral-history-chairman-joint-chiefs.
274. Stahl, "10th Mountain Division."
275. 10th Mountain Division, "10th Mountain Briefs"; Stahl, "10th Mountain Division."
276. Stahl, "10th Mountain Division."
277. 10th Mountain Division, "10th Mountain Briefs."
278. Stahl, "10th Mountain Division."
279. Stahl, "10th Mountain Division."
280. Stahl, "10th Mountain Division."
281. Stahl, "10th Mountain Division."
282. Stahl, "10th Mountain Division"; 10th Mountain Division, "10th Mountain Briefs."
283. Stahl, "10th Mountain Division."
284. 10th Mountain Division, "10th Mountain Briefs."
285. 10th Mountain Division, "10th Mountain Briefs."
286. Stahl, "10th Mountain Division."
287. 10th Mountain Division, "10th Mountain Briefs."
288. Stahl, "10th Mountain Division"; 10th Mountain Division, "10th Mountain Briefs."
289. Baumann, Yates, and Washington, "Introduction: The Meaning of Somalia," 3.
290. Stahl, "10th Mountain Division"; 10th Mountain Division, "10th Mountain Briefs."
291. Stahl, "10th Mountain Division."
292. Stahl, "10th Mountain Division."
293. Brown, *Kevlar Legions*, Kindle, 114.
294. 10th Mountain Division, "10th Mountain Briefs"; Stahl, "10th Mountain Division."
295. Stahl, "10th Mountain Division"; 10th Mountain Division, "10th Mountain Briefs."
296. Stahl, "10th Mountain Division."
297. Stahl, "10th Mountain Division"; 10th Mountain Division, "10th Mountain Briefs."
298. Stahl, "10th Mountain Division"; 10th Mountain Division, "10th Mountain Briefs."
299. 10th Mountain Division, "10th Mountain Briefs"; Stahl, "10th Mountain Division."
300. 10th Mountain Division, "10th Mountain Briefs."
301. 10th Mountain Division, "10th Mountain Briefs"; Stahl, "10th Mountain Division."
302. 10th Mountain Division, "10th Mountain Briefs."
303. 10th Mountain Division, "10th Mountain Briefs."
304. Sullivan and Coroalles, *The Army in the Information Age*, 18.
305. Sullivan interviews with Dabrowski, 300–301.
306. Earl H. Tilford Jr., "Introduction," in Tilford, ed., *World View*, 1–3.

307. Aspin, *Report of the Bottom-Up Review*, 19; Record, "Ready for What and Modernized against Whom?," vi–vii, ix–x, v.

308. Record, "Ready for What and Modernized against Whom?," ix–xii.

309. Michael Garret, quoted in Fitzgerald, *Learning to Forget*, Kindle, 97.

310. U.S. Army Training and Doctrine Command, PAM 525-5, quoted in Fitzgerald, *Learning to Forget*, 97; Institute for Foreign Policy Analysis, *Summary Report of a Conference on Operations Other Than War, sponsored by HQ TRADOC*, April 11, 1995, 7, quoted in Fitzgerald, *Learning to Forget*, 97.

311. U.S. Army Command and General Staff College, *United States Army Command and General Staff College Catalog, Academic Year 1993–1994*, CGSC Circular No. 351-1, July 1993, Command and General Staff College Papers, Ike Skelton Combined Arms Research Library, 46, 49, 52–53, 63–64, 67, 70–71; U.S. Army Command and General Staff College, *United States Army Command and General Staff College Catalog, Academic Year 1991–1992*, CGSC Circular No. 351-1, July 1991, Command and General Staff College Papers, Ike Skelton Combined Arms Research Library, 42–43, 45, 49–51, 59–60, 62, 64–65, 68; U.S. Army Command and General Staff College, *United States Army Command and General Staff College Catalog, Academic Year 1994–1995*, CGSC Circular No. 351-1, July 1994, Command and General Staff College Papers, Ike Skelton Combined Arms Research Library, 65.

312. Jandora, "Threat Parameters for Operations Other Than War," 55–67.

313. Brown, *Kevlar Legions*, Kindle, 80–97; Commission on Roles and Missions of the Armed Forces, *Directions for Defense*, ES-2, ES-4.

314. Brown, *Kevlar Legions*, Kindle, 80–97.

315. Brown, *Kevlar Legions*, Kindle, 97–98, 106.

316. Sullivan, "A Vision for the Future," 6, quoted in Shimko, *The Iraq Wars*, Kindle, locations 3014–3018.

317. Sullivan and Coroalles, "The Army in the Information Age," 18, 12, Sullivan, "A Vision for the Future," 6.

318. Sullivan and Coroalles, "The Army in the Information Age," 18, 12; Sullivan, "A Vision for the Future," 6.

3 BOSNIA-HERZEGOVINA
AND THE ARMY AFTER NEXT

AS CHIEF OF STAFF of the U.S. Army, Gen. Dennis Reimer would oversee an aggressive suppression of the lessons that the Army had been learning from the low-intensity conflicts in which it was engaged around the world. Reimer administered the effort to purge the remnants of low-intensity conflict doctrine from the Army's flagship manual. The Army–Air Force Center for Low Intensity Conflict (A-AFCLIC) was disbanded and, in what was perhaps General Reimer's biggest contribution to transformation, the Army After Next would set the Army on the path to "leap ahead" of the current rash of low-intensity conflicts to envision a hypothetical future of high-intensity conflicts waged against peer competitors.

While Reimer and other Army transformers imagined this high-tech future, the transformation under Force XXI, begun by Reimer's predecessor, Gen. Gordon Sullivan, continued at a brisk pace, providing the Army with a dizzying array of new network- and computer-enabled capabilities and ever-greater capacity to engage in high-intensity conflicts. General Reimer also continued the Advanced Warfighting Experiments.

This retrenchment of transformation against the efforts of low-intensity conflict observers came despite the commitment of the Army to the biggest low-intensity conflict it had faced since the Vietnam War: the North Atlantic Treaty Organization (NATO) intervention to end the civil war in Bosnia-Herzegovina. Starting in December 1996 the Army deployed tens of thousands of soldiers to a peacekeeping operation to end the sectarian carnage in the Balkan States, a move that would precipitate an avalanche of lessons that filled the pages of Army journals. Throughout the mission, General Reimer ruthlessly protected his high-intensity conflict transformation from what he saw as the distraction of the low-intensity conflict in Bosnia-Herzegovina. As a result, lessons learned in this conflict were not institutionalized and therefore had no impact on the Army's organization, equipment, or training outside those units directly involved in it.

General Dennis Reimer

Gen. Dennis Reimer was commissioned as a field artillery lieutenant at the U.S. Military Academy at West Point, New York, in 1962. He served two tours in Vietnam, his first as a first lieutenant, serving on an advisory team working with an Army of the Republic of Vietnam battalion. This was Reimer's only direct experience as a participant in a low-intensity conflict during his entire Army career. He would leave Vietnam in July 1965, just ahead of President Lyndon Johnson's announcement that he was sending a sizable ground force to directly engage the Viet Cong and North Vietnamese Army in South Vietnam.[1]

Reimer returned to Vietnam as a major. In September 1968, he served as the executive officer and S3 (operations officer) for the 2nd Battalion, 4th Artillery of the 9th Infantry Division. The battalion conducted decidedly high-intensity conflict operations, providing tens of thousands of rounds of artillery fire in support of the 3rd Brigade, 9th Infantry Division in Long An Province southwest of Saigon.[2]

Reimer's path to chief of staff of the Army passed through the operational—rather than the more theoretical doctrine and force design—elements of the Army. After commanding the 4th Infantry Division, he served in the Pentagon for three years—between May 1990 and March 1993—as the deputy chief of staff for operations and plans (DCSOPS, pronounced "dess-ops") and as the vice chief of staff of the Army. He served more than two years as the commander of U.S. Army Forces Command (FORSCOM). As a long-term member of General Sullivan's four-star "board of directors," Reimer was intimately familiar with Sullivan's transformation when he assumed office as the thirty-third chief of staff of the Army on June 20, 1995.[3]

As chief of staff of the Army, to a much greater degree than any of his immediate predecessors, General Reimer deferred to the informal, extrainstitutional leadership of the former four-star leaders of the Army. General Sullivan had assembled his "board of directors" from the senior leaders of the Army—the U.S. Army Training and Doctrine Command (TRADOC) and FORSCOM commanders chief among them. Continuing a practice he had begun as vice chief of staff of the Army and continued as FORSCOM commander, General Reimer instead established an annual "Army leadership seminar" attended by a constellation of retired four-star Army generals that he used "as sort of a Board Directors." Reimer would later say that he considered it "like taking your orals" and "as a good event to get an azimuth check."[4] While this practice might have been reassuring for General Reimer, it reinforced, rather than challenged, underlying assumptions and made innovative solutions harder to find; Reimer's Army leadership seminars reinforced the power of the Army's organizational culture and yearly applied an old lens to new problems—problems like the increasing demands on and consistently poor performance of the Army in low-intensity conflicts.

In addition to the Army's continued challenges with executing low-intensity conflicts, its fiscal future was also precarious. The downsizing of the Army was largely complete when General Reimer assumed office, but national defense budgets would remain at reduced levels for the foreseeable future.[5]

This reduced funding did not, however, translate into reduced demands on the Army. In the year before Reimer took office, 22,200 soldiers were deployed in seventy countries around the world. Throughout the first year of his tenure, an average of 21,500 soldiers were deployed around the world each day. These numbers were in addition to the 125 thousand soldiers still stationed in bases outside the continental United States.[6]

By the time Reimer took office, transformation had gained considerable momentum. Since 1993, TRADOC had envisioned establishing an experimental force (EXFOR) that could be used as a permanent test unit for transformation technologies. In March 1995, just before Reimer assumed his duties as chief of staff of the Army, the 2nd Armored Division at Fort Hood, Texas, was designated as that EXFOR. (The 2nd Armored Division would be redesignated as the 4th Infantry Division in January 1996 but would remain the EXFOR.)[7]

When General Reimer assumed the role of chief of staff of the Army in June 1995, he immediately endorsed the approach of his predecessor. Reimer would continue to build Force XXI with an unrelenting focus on high-intensity conflict. He, too, insisted that the Army's purpose was to "fight and win our Nation's wars."[8]

The framework under which Reimer would execute transformation was the Joint Staff's *Joint Vision 2010*, published in 1996, which was a tacit response to the final report of the congressionally mandated Commission on Roles and Missions convened in 1995. Whereas the commission's final report had called for greater attention to low-intensity conflict,[9] *Joint Vision 2010* defiantly reaffirmed the revolution in military affairs (RMA) and the U.S. military's focus on preparation for high-intensity conflict.

The vision trumpeted its goal of "achieving dominance across the range of military operations" and acknowledged that, while it sought to describe a force that could "fight and win against any adversary at any level of conflict," the Army "must also be able to employ these forces in operations other than war to assist in the pursuit of other important interests." Nonetheless, each of the vision's "four operational concepts"—"dominant maneuver, precision engagement, full dimensional protection, and focused logistics"—was decidedly focused on high-intensity conflict.[10] The manual's depiction of "dominant maneuver" made it clear that *Joint Vision 2010* was a vision of a future high-intensity conflict-focused force.

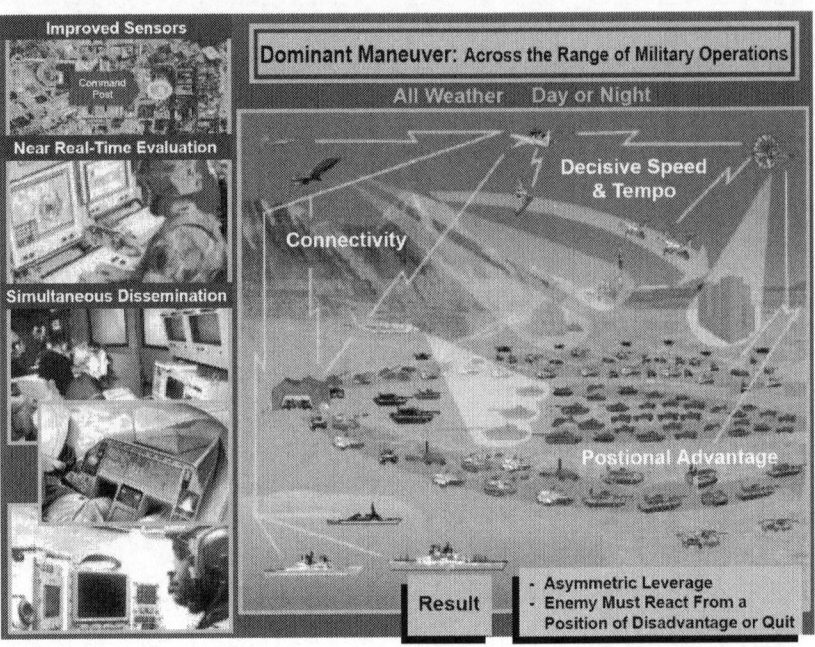

Figure 3.1. The high-intensity conflict focus of dominant maneuver. *Source:* Joint Staff, *Joint Vision 2010*, 20.

Joint Vision 2010 was even more explicit than the Army XXI vision in arguing that "operations other than war" were a lesser included activity of high-intensity conflict. The vision explained that its four operational concepts would "enable us to dominate the full range of military operations from humanitarian assistance, through peace operations, up to and [including] the highest intensity conflict." Yet throughout the document, "full spectrum dominance" is used synonymously with "dominance" in a high-intensity conflict environment. *Joint Vision 2010* justified this conflation by explaining, "Other operations, from humanitarian assistance in peacetime through peace operations in a near hostile environment, have proved to be possible using forces optimized for wartime effectiveness."[11]

To optimize itself for high-intensity conflict, *Joint Vision 2010* posited that the U.S. military had to acquire greater technology to compensate for its diminished numbers: "Faced with flat budgets and increasingly more costly readiness and modernization, we should not expect a return to the larger active forces of the Cold War period." It also noted, "Technologically superior equipment has been critical to the success of our forces in combat," claiming that through harnessing the power of the RMA, the U.S. military would manage to do more with less. Doing more with less was not just a response to the downsizing of the U.S. military; it was also a response to the deployability issues that had troubled Army leaders since Operation Desert Storm. It would "reduc[e] the need for time-consuming and risky massing of people and equipment."[12]

To create a sense of urgency for the transformation it proposed, *Joint Vision 2010* posited three ideas that would remain pillars of Army transformation thought until the war in Iraq. First, it contended that future adversaries would transform in the same way that the U.S. military was transforming: toward ever higher technology and ever more integrated networks. Second, future adversaries would seek "asymmetry" by using the "information technology" of the RMA to negate the advantages of the U.S. military rather than duplicate them capability for capability. And, finally, potential adversaries would be able to transform "very rapidly," "outrunning" the U.S. military's ability to react. This final idea created a sense of urgency to beat theoretical future adversaries to "dominant battlespace awareness" that would render the "battlespace considerably more transparent" for the first force to acquire these technologies.[13]

The idea of asymmetry was particularly critical to transformation thought throughout the rest of the 1990s, and it was a point of convergence with low-intensity conflict thought during the same period. While they never used the word *asymmetry*, Andrew Bacevich and Jeffrey Record began—after the U.S. Army's disastrous showing in Somalia—to highlight the way in which adversaries were using low-tech, low-intensity conflict means to negate advantages afforded the Army by its dominant high-tech, high-intensity conflict capabilities. *Joint Vision 2010* did use the word *asymmetry*, but did so in a different way. Like many low-intensity conflict observers, the vision worried that a "future adversary" might develop "military capabilities that provide asymmetrical counters to US military strengths, including information technologies." But transformation enthusiasts believed that, rather than finding low-tech, low-intensity conflict means to neutralize the Army's dominant capabilities, adversaries would "use technology to make rapid improvements."[14] This difference is significant; whereas the low-intensity conflict observers' vision of asymmetry might have driven the Army to develop greater capacity to fight such conflicts, the dominant, transformation vision of asymmetry instead drove the U.S. military to chase high-tech, high-intensity conflict capabilities at a breakneck pace.

Notably, though *Joint Vision 2010* was dominated by a high-intensity conflict vision of the future, it did contain a passage that seemed to acknowledge the limits of technology and highlight the capabilities uniquely required to succeed in low-intensity conflicts: "We cannot assume that all new concepts will be equally valuable in all operations. . . . Extensive physical presence . . . may be required to fully neutralize enemy forces [or] deal with prisoners and potentially hostile populations."[15] This passage represented a stark admission of the limits of the RMA for transformers bent on optimizing the U.S. military for high-intensity conflict. Yet some important caveats moderate the impact of this passage. First, the emphasis on "physical presence" reveals the intent of this statement to be more a justification for not cutting the number of soldiers in the Army than an admission of the unique demands of low-intensity conflict. Second, this is only one paragraph in a thirty-five-page document touting the promise of technology to reshape the U.S. military for the next great power war.

While General Reimer's approach to the Army's future clearly reflected the high-intensity conflict focus of *Joint Vision 2010*, it also reflected deeper underlying assumptions within the Army's culture. In an exit interview after stepping down as chief of staff of the Army, Reimer explicitly rejected those who suggested that the era of high-intensity conflict might be drawing to a close. Instead Reimer countered, "Wishful thinking has it that we have [seen] . . . an end to war—but history tells us that is not the case." To prove his point, he harked back to a charge Gen. Douglas MacArthur had given a group of West Point cadets—almost certainly including one Dennis Reimer—more than thirty years earlier: "Through all this welter of change your mission is fixed, determined, inviolable. It is to win our nation's wars." Reimer also cited MacArthur's warning, "If you lose, the nation will be destroyed" and admitted, "It's engrained on my mind."[16]

General Reimer did understand that the U.S. Army he inherited struggled in the many low-intensity conflicts it had faced and would continue to face during his tenure. He would say after his departure as chief of staff of the Army, "In the post–Cold War world we had . . . to be able to stabilize countries and help people and that required a more sophisticated approach than just figuring out how to defeat the Soviets on the plains if Europe."[17] But his goal was to meet these demands "while executing the most fundamental restructuring since the end of World War II" and remaining "trained and ready for the uncertain world we were facing."[18]

Always foremost in General Reimer's mind was an imminent high-intensity conflict: "I did not know when we would be tested again; I only knew we would and we needed to be ready." Just as had General Sullivan, General Reimer saw the demands of low-intensity conflicts as a dangerous distraction. This sense of urgency drove many of the decisions Reimer made during his tenure overseeing the Army's transformation.[19]

General Sullivan's vision—Force XXI—would move the Army from its Cold War, threat-based focus on the Soviet Army to a capabilities-based focus on a broader range of possible threats and environments. But whereas Sullivan had focused on the conceptual foundations of this vision, it would fall to Reimer to actually develop and implement the technologies to realize such a vision.[20]

Under the day-to-day direction of TRADOC commander Gen. William Hartzog, the Army sought to give its forces "dominant knowledge," a common understanding across units of friendly and enemy locations. The Army would test, adopt, and integrate a dizzying array of digital systems—both custom and off-the-shelf—into the Army's existing combat platforms. Overall, the building of the EXFOR would require the acquisition, installation, and testing of nearly five thousand pieces of equipment—1,200 of them Appliqué computers—on more than nine hundred vehicles. This was the path by which the U.S. Army would harness the promise of the RMA.[21]

To test and field all of this new equipment, the regimen of Advanced Warfighting Experiments (AWEs) begun by Sullivan continued under Reimer, supervised by TRADOC and General Hartzog.

The Army before Bosnia-Herzegovina

The state of low-intensity conflict thought at the beginning of Reimer's tenure as chief of staff of the Army was perhaps best summarized in Joint Publication (JP) 3-07, *Joint Doctrine for Military Operations Other Than War*, written by A-AFCLIC. The manual was by no means perfect; it echoed the 1993 edition of Field Manual (FM) 100-5, *Operations*, regarding the principles of operations other than war—"objective, unity of effort, security, restraint, perseverance, and legitimacy" and, like FM 100-5, omitted the principle of "political dominance" that had been listed in the 1990 FM 100-20, *Military Operations in Low Intensity Conflict*. It also reflected the Nixon

Doctrine and Reagan-era philosophy of low-intensity conflict: supporting a host nation government while it did the heavy lifting of prosecuting such a conflict.[22]

Yet, on balance, JP 3-07 was a clear reassertion of the lessons of recent low-intensity conflicts that had been diluted by the 1993 FM 100-5. JP 3-07 codified the continuing agreement between low-intensity conflict observers and more conventional military thinkers that the future held many more such conflicts. The manual acknowledged, "It is expected that Armed Forces of the United States will increasingly participate in [operations other than war]."[23]

Low-intensity conflict observer Steven Metz of the U.S. Army War College's Strategic Studies Institute (SSI) made a very similar observation, citing Robert Kaplan in warning that "the trend toward anarchy will eventually win out" and the governments of the Third World would wither away. In a particularly prescient passage, he predicted, "Reactionary insurgency, in which a religious-based group attempts to seize power from a secular, modernizing government—as the Iranians did in 1979—may be common." Elaborating, he warned that this new breed of low-tech adversary "will largely be urban with an emphasis on terrorism rather than on rural guerrilla war." In another note, eerily suggestive of the Taliban in Afghanistan, Metz added, "This type of insurgency will be most dangerous if it again becomes a technique of interstate conflict, with external sponsors using insurgency to weaken an opponent." He could have just as easily been describing al-Qaeda or the Islamic State in Iraq and Syria when he wrote, "Future American engagement in counterinsurgency might also provoke domestic terrorism. With easy global transportation [and] the existence of a variety of emigre communities in the United States . . . , insurgents could open an 'American front.'" Metz concluded, "As our limited experience with 'holy terrorists' in the Middle East shows, we are ill-equipped to deal with the root causes of religion-driven violence."[24]

But transformation advocates differed sharply from low-intensity conflict observers on how to respond to the likelihood of more such conflicts. In one of its less helpful contributions to low-intensity conflict doctrine, JP 3-07 codified this transformation view. For instance, it acknowledged that "*US forces need to be prepared*." But the manual insisted that operations other than war were a less important task than war: "Commanders must remember that their primary mission will always be to prepare for, fight and win America's wars. This is the US military's most rigorous task and requires nothing less than top priority when training and equipping our forces."[25]

This relative diminution of low-intensity conflict was also reflected in the U.S. Army's professional education. In the 1995–96 academic year, the number of hours dedicated to low-intensity conflict at the Command and General Staff College at Fort Leavenworth, Kansas, did return to its levels from the beginning of the 1990s, forty-five hours—this up from a low of thirty-six hours in the 1993–94 academic year. But this still represented a tiny fraction of the total five-hundred-hour core curriculum.[26]

Many low-intensity conflict observers complained that the Army was not manned, trained, or equipped to fight such conflicts. Even the 1995 JP 3-07 acknowledged that the U.S. military was not trained and ready for low-intensity conflicts, but "for most types of MOOTW [military operations other than war], *military personnel adapt their warfighting skills to the situation.*" On a positive note, the manual did acknowledge, however, that "for some MOOTW (for example, humanitarian assistance and peacekeeping operations) *warfighting skills are not always appropriate.*" In these situations, the manual advocated education of officers and noncommissioned officers (NCOs) and training for "individuals, staffs, and units." Elsewhere, however, it seemed to eschew training for MOOTW and claim that education of officers and NCOs was enough: "The lack of opportunity to train for a specific operation is in large part overcome by military leaders who have a solid foundation of MOOTW provided through the military education system," additionally noting, "*A well-trained force can adapt to MOOTW under the leadership of officers and NCOs educated in the principles and types of MOOTW.*"[27] The manual was, however, silent on the fact that over the course of their careers, Army officers received very little if any education on low-intensity conflict from the institutional Army's educational system.

Metz, on the other hand, concluded that the time had come for a radical reshaping of the U.S. military. Citing retired Army colonel Rod Paschall—a former Special Forces officer and at the time the director of the U.S. Military History Institute at Carlisle Barracks, Pennsylvania—Metz made a radical recommendation: "The United States may need a new organization to confront new forms of insurgency. . . . Post–Cold War forms will be far removed from the Army's traditional areas of expertise and will be more police functions than military ones. The Army should thus encourage the formation of a permanent civil-military cadre of experts with a strong emphasis on law enforcement and intelligence collection and analysis."[28] Though he thought this new force should be civilianized, separate for from the Army, Metz was suggesting that the U.S. military needed a force specialized for low-intensity conflicts.

The SSI's Earl H. Tilford Jr. took on those Army transformers who argued that low-intensity conflict was a lesser included task of high-intensity conflict and, thus, required no special preparation, writing, "In 1962, at the start of the U.S. commitment to the war in Vietnam, Chairman of the Joint Chiefs of Staff, Army General Lyman L. Lemnitzer claimed that forces constituted for war in Europe could just as easily fight and win against guerrillas in Indochina." He added, "There was a general acceptance of the notion . . . if American forces could fight and prevail over Soviet or Chinese forces in conventional or nuclear war, they could certainly win any lesser order conflict quickly and with less application of more or less the same kind of force. In Vietnam that notion proved tragically flawed." Tilford also joined Jeffrey Record and earlier low-intensity conflict observers in warning that low-tech, low-intensity conflict adversaries could neutralize the advantages of an Army optimized for high-tech, high-intensity conflict. But Tilford offered a new twist on this warning, noting that the information technologies that the Army was pursuing would not protect it from this asymmetry: "Given the inevitability that this will occur, any strategy that may be developed during the RMA that does not anticipate and plan for these counterstrategies will not serve the nation well."[29]

Other low-intensity conflict observers identified the capabilities and skills that the Army lacked for operating in such a conflict. The 1995 JP 3-07 highlighted that Army units faced challenges in operating in a multinational environment due to "political considerations," "language barriers," "cultural backgrounds," "military capabilities and training," "equipment interoperability," and "logistic support system coordination." The manual also suggested that liaison's to multinational partners could ease these challenges and—reflecting the Nunn-Cohen Amendment to the 1987 National Defense Authorization Act's relegation of low-intensity conflict to special operations forces (SOF)—that SOF elements were ideal for this purpose: "SOF-unique capabilities in language and cross-cultural training, their regional orientation and forward deployment, and focus on independent small unit actions *make them one of the principal forces of choice to complement and support multinational operations objectives.*"[30]

By far the most important contribution that the 1995 edition of JP 3-07 made was to restore the political dimension—excised from low-intensity conflict doctrine in the 1993 edition of FM 100-5—to military doctrine. JP 3-07 declared, "*Political objectives drive MOOTW at every level from strategic to tactical.*" The manual explicitly rejected those—both well-meaning low-intensity conflict observers and Army transformers—who hoped to avoid the political dimension by avoiding "mission creep," writing, "*Commanders should remain aware of changes not only in the operational situation, but also to changes in political objectives that may warrant a change in military operations.*" Failure to do so, the manual warned could, over time, "lead to disconnects between political objectives and military operations."[31]

Like FM 100-5, JP 3-07 separated military activities into "war" and "military operations other than war." But unlike in FM 100-5, in JP 3-07 it was the relative weight of the political dimension in an operation that distinguished between the two. War, the manual explained, was the "conduct [of] large-scale, sustained combat operations to achieve national objectives or protect national interests." On the other hand, low-intensity conflict was dominated by the political dimension: "All military operations are driven by political considerations," but "in MOOTW, political considerations *permeate all levels* and the military may not be the primary player."[32]

While many low-intensity conflict observers continued to argue the essentially political nature of such conflicts, contemporary Army and joint doctrine continued to reflect the Nixon Doctrine and the Reagan-era model for such conflicts: the host nation government bore the largest share of the responsibility for engaging in the political dimensions of the conflict. The 1995 JP 3-07 used the euphemisms "nation assistance" and "support to counterinsurgency" (rather than just "counterinsurgency"), defining the two as "*civil or military assistance . . . rendered to a nation by US forces* within that nation's territory during peacetime, crises or emergencies, or war." The manual also dumped some responsibility on the U.S. Department of State, adding, "All nation assistance actions are integrated through the US Ambassador's Country Plan." Elsewhere, the JP 3-07 used the euphemism "foreign internal defense" (FID) for counterinsurgency, defining it as "encompass[ing] the total political, economic, informational, and military support provided to another nation to assist its fight against subversion and insurgency." But in addition to reflecting the Nixon Doctrine and the Reagan-era model for counterinsurgency, this passage also reflected the Nunn-Cohen Amendment, which gave primacy for counterinsurgency to SOF.[33] JP 3-07 insisted, "FID is a principal special operations mission."[34]

The SSI's Steven Metz was not convinced that the Army could so easily avoid the political dimension of a low-intensity conflict, as JP 3-07 asserted, noting that "El Salvador is thought to have proven the correctness of our strategy and doctrine" but adding, "Future counterinsurgency may not emulate the past; the similarities between Vietnam and El Salvador may be much greater than those between El Salvador and what comes after it."[35]

Despite its effort to relegate the political dimension of low-intensity conflict to agencies other than the Army, the 1995 JP 3-07 had still returned the political dimension to official U.S. Department of Defense doctrine on low-intensity conflict after its two-year absence.

Unfortunately, this modest acknowledgment of the Army's role in the political dimension of such conflicts would not be enough to permit it to successfully conclude its next low-intensity conflict, this time in Bosnia-Herzegovina.

Bosnia-Herzegovina, Part One: December 20, 1995 to 2004

In the aftermath of the Cold War, the former Yugoslavia shattered along ethnic and religious lines into five separate countries—Bosnia-Herzegovina, Croatia, Macedonia, Slovenia, and Serbia (which continued to consider itself, and the break-away provinces, Yugoslavia). Serbs had been the preeminent religious and cultural group in the former Yugoslavia, dominating both the Yugoslav Communist Party and the national armed forces. The capital of Yugoslavia had been the Serbian city of Belgrade. Serbians saw themselves as the defenders of Slavic culture and the only true patriots, having resisted both the Turks and the Germans in earlier wars.[36]

The postcollapse war between Croatia and Serbia quickly concluded, but the civil war inside Bosnia-Herzegovina resisted settlement. Ninety-nine percent of Bosnians identified themselves as Slavs, and nearly all Bosnians spoke Serbo-Croatian. The Bosnian population of about 4.3 million was split among Muslim Bosniaks (40 percent), Eastern Orthodox Bosnian Serbs (31 percent), Roman Catholic Bosnian Croats (15 percent), and other religions (the remaining 14 percent).[37]

In and around the country of Bosnia-Herzegovina, militia forces and criminal gangs—armed with everything from small arms to armored vehicles from the former Yugoslav Army—engaged in brutal acts of sectarian violence against one another and in murder and "ethnic cleansing" against civilian populations that killed as many as 250 thousand people and eventually rendered more than half the country's population refugees or internally displaced people. By 1995 there were at least six different armies— the Muslim-dominated Bosnian Army, the Bosnian Croatian Defense Council, the Bosnian Serb militia, the Croatian Army, a militia of Croatians from the self-proclaimed autonomous state of Serbian Krajina, and the Yugoslav Army from Serbia—fighting inside Bosnia-Herzegovina.[38]

As the fighting grew, so did concern in European capitals that it might spread to neighboring states in the Balkans. In February 1992, in an effort to halt the fighting, the United Nations established the UN Protection Force (UNPROFOR), a multinational force that eventually included thirty-eight thousand troops from thirty-seven different countries spread across more than seven thousand bases in the former Yugoslavia. It was the largest peace-keeping mission in UN history.[39]

On the ground, however, this UN force was largely a spectator to the violence rather than a peace enforcer. Though it would emphatically argue otherwise, by committing to safeguarding Bosnia-Herzegovina's 350 thousand internally displaced people—most of them Bosnian Croats and Muslim Bosniaks—UNPROFOR had unintentionally chosen to support the losing side in the conflict. To make matters worse, Muslims—including armed militia fighters—flooded into UN "safe zones," making them a target for Serbian violence. UNPROFOR's weak mandate, its lack of cohesion, and the huge number of national caveats from the contributing nations rendered UNPROFOR impotent in trying to stop the violence.[40]

In Bosnia-Herzegovina, no religious group was innocent in the conflict; all engaged in violence against civilians and ethnic cleansing. In fact, Bosnian Croats and Muslim Bosniaks were responsible for most of the UNPROFOR casualties. Additionally, Bosniaks were occasionally guilty of firing on their own people to provoke NATO and UNPROFOR retaliation against the Bosnian Serbs. But Bosnian Serbs were guilty of some of the worst atrocities of the war. In April 1992, Serbs massacred around one hundred Bosniaks in Bijeljina, initiating the flight of hundreds of thousands from northeast Bosnia-Herzegovina. Perhaps the Bosnian Serbs' worst atrocity came in July 1995, with the murder of between seven and eight thousand Bosniaks in the "safe" city of Srebrenica—in full view of Dutch peacekeepers, one hundred of which were taken prisoner. Soon after, in August 1995, a mortar round landed in a marketplace in Sarajevo, killing thirty-seven and wounding scores more; most blamed the Serbians for the incident.[41]

In 1994 NATO and the United States began gradually to escalate military pressure—primarily through air strikes—on the Bosnian Serbs. The United States also sent a mechanized infantry battalion to Macedonia in a symbolic show of force. Bombing gradually turned the tide and put the Bosnian Serb militias and the Yugoslav Army on the run. In December 1995 the warring parties signed the Dayton Accords, ending the fighting and delineating lines between the warring parties in Bosnia-Herzegovina.[42]

A provision of the Dayton Accords was an international Implementation Force (IFOR) that would, among other things, establish and enforce a zone of separation, protect the civilian populace, and create the conditions for the reestablishment of civil governance. Its mission was called Operation Joint Endeavor.[43]

The conditions at the beginning of Operation Joint Endeavor favored success. First, the war was over. The Croatian Army was the dominant military force in Bosnia-Herzegovina; the country had essentially become a Croat protectorate. Only the northeast, the Bosnian Serb region that had been declared Republika Srpska, remained outside Croatian control. Moreover, all of the parties had been exhausted by the conflict; the factions had lost the will to fight. And those who wished to continue the conflict faced the prospects of international isolation.[44]

Second, IFOR had a much more robust mandate and many fewer national caveats than UNPROFOR, making IFOR a much more effective force; it could compel compliance from each faction. The core of the U.S. contingent of IFOR, Task Force Eagle, was the 1st Armored Division with the division headquarters, two armored brigades, an aviation brigade, and attached enablers such as engineers, field artillery, military intelligence, and military police. Altogether, the U.S. contribution to the 60 thousand-man IFOR was 17,500 troops.[45]

Today, the shape of Operation Joint Endeavor looks eerily similar to the stability phase of Operation Iraqi Freedom. Bosnia-Herzegovina was divided into three Multi-National Divisions (MNDs). American forces assumed control of MND-North and assumed varying degrees of authority over forces from countries including Russia, Turkey, Poland, and Denmark. Prior to deployment, U.S. forces went through rigorous training, including a "mission readiness exercise" at the Combat Maneuver Training Center (CMTC) at Hohenfels, Germany.[46] Once on the ground in Bosnia-Herzegovina, Task Force Eagle executed operations and logistics from forward operating bases. Units tried to balance force protection with the need to interact with the population; the tactic that emerged to travel around the country was the four-vehicle convoy. Intelligence personnel and linguists were always in short supply. Operations were unusually decentralized in comparison to the Army's contemporary high-intensity conflict-focused doctrine and training; specific patrols of towns and villages, riot control operations, and even negotiations with local leaders were frequently initiated and executed at the battalion, company, or even platoon level.[47] While the level of violence was considerably lower and mines replaced improvised explosive devices as the top threat to foot and mounted patrols, in virtually every other way that mattered Operation Joint Endeavor was a scale model of Operation Iraqi Freedom.

Bosnia-Herzegovina revealed the depths of the Army's unpreparedness to conduct low-intensity conflict operations. Despite promises before the deployment that Operation Joint Endeavor would only last a year, U.S. Army forces in Bosnia-Herzegovina seemed to have no idea how to bring the low-intensity conflict to an end. The 1st Infantry Division replaced the 1st Armored Division in November 1996, the mandate for IFOR was extended, IFOR became the Stabilization Force (SFOR), and the mission was renamed Operation Joint Guard. The conditions inside the country had improved sufficiently that the size of the international force was cut in half, from sixty thousand to thirty thousand; a full ten thousand of the troops that remained were from the U.S. Army. The U.S. contingent of SFOR was also a lighter force than the U.S. contingent in IFOR had been; much of the armor from the 1st Armored Division returned to Germany and was not replaced.[48]

Many of the men who would lead the Army into the twenty-first century had a front row seat for the low-intensity conflict in Bosnia-Herzegovina. V Corps and future TRADOC commander Lt. Gen. John Abrams commanded U.S. Army Europe (Forward), or USAREUR (Forward), in Bosnia. His lead planner was Army lieutenant colonel Peter Schifferle, who would go on to become the deputy director of the School of Advanced Military Science (SAMS) at Fort Leavenworth. Col. John Batiste, who would go on to command the 1st Infantry Division in the early days of Operation Iraqi Freedom, commanded the 2nd Brigade Combat Team (BCT), 1st Armored Division during the initial deployment of Task Force Eagle to Bosnia-Herzegovina. The chief of staff of the 1st Armored Division during its first deployment to Bosnia-Herzegovina was Col. John Sloan Brown, who would go on to be a brigadier general and run the transition team for the next chief of staff of the Army. In fact, the first three senior military commanders of the multinational, postinvasion effort to stabilize Iraq—Lt. Gen. Ricardo Sanchez, Gen. George Casey, Gen. David Petraeus, and Gen. Ray Odierno—all participated in the low-intensity conflict operations in the Balkans.[49]

As early as 1996, a number of agencies throughout the Army began sending observers to Bosnia-Herzegovina to collect lessons on low-intensity conflict. Among these agencies was the Army's Peacekeeping Institute, which also convened two conferences—in May 1996 and April 1997 at the U.S. Army War College at Carlisle, Pennsylvania—to gather together U.S. and foreign military officers and civilians to discuss these lessons. More than one hundred individuals from fifty different agencies in the United States and around the world gathered at these two conferences; the participants arrived at recommendations that were captured in an article in the War

College's journal, *Parameters*. As with previous low-intensity conflicts, Operation Joint Endeavor prompted many to warn that the future held more such conflicts. As Dickinson College professor and low-intensity conflict observer Max Manwaring warned in the article about this conference, "Many have concluded . . . that conflict of the sort encountered in the Balkans may well be a harbinger of future US military operations." In his conclusion, Manwaring insisted that the Army needed to reshape itself for "political partnership" between itself and "civilian organizations" by instituting "doctrinal and structural change" to optimize itself for "complex humanitarian relief or peace support operations." He indicted the Army's senior leadership for failing "to understand and to behave as though the Cold War is over," insisting that it was a failure of "leader judgement."[50]

Despite all of this effort to capture the lessons of Operation Joint Endeavor, the transformers chose not to address the Army's unpreparedness to fight a low-intensity conflict, preferring instead to focus narrowly on what the operation revealed about continued problems with deployability—namely, that the Army was still too heavy. Deployment of the 1st Armored Division began on December 16, 1995. Moving more than nine thousand people and twenty thousand short tons of U.S. equipment into Bosnia-Herzegovina had required nearly four hundred trains with more than seven thousand railcars, over 1,400 sorties of cargo aircraft, over four hundred buses, over two hundred commercial truck convoys, and forty-two military convoys. The deployment was further complicated by the flooding of the Sava River, which overflowed its banks on December 28, 1995. A massive engineering effort, supported by CH-47 helicopters from the division's aviation brigade, was required to cross the river. Even with this massive effort, the first 140 vehicles did not cross into Bosnia-Herzegovina until December 31. The full deployment of more than twenty-eight thousand people, eleven thousand vehicles and pieces of equipment, and 145 helicopters was not complete until February 15, 1996, nearly two and a half months after it had begun.[51]

The Army also grappled with other deployability challenges that it had faced in previous low-intensity conflicts. Because much of the engineer, public affairs, and civil affairs capability that the Army needed for such conflicts resided in its reserve components, there was a significant delay in activating, mobilizing, and deploying these soldiers before they could be employed. As the assistant division commander for maneuver of the 1st Armored Division, Brig. Gen. Stanley Cherrie, wrote in a *Military Review* article after his return from Bosnia-Herzegovina, "We had to get our civil affairs and media operations apparatus into high gear early. . . . Early enhancements to the Tuzla airfield were an absolute necessity if we were to establish an effective deployment and sustainment 'air bridge.'" Yet these soldiers were the last to arrive in Bosnia-Herzegovina.[52]

In Bosnia-Herzegovina the Army also learned lessons about sustaining such a large international force in an austere environment. The remoteness of the country and the demand to move supplies to it from bases deep in Central Europe over air and ground put a huge demand on the Army's logistics capabilities. This was further complicated by the vastness of the task to be accomplished; IFOR was deployed across twenty-four base camps across the zone of separation. Sustaining this force required three convoys and twelve air sorties providing seventy-five thousand meals, 192 thousand gallons of water, 130 thousand gallons of fuel, and 133 short tons of other supplies every single day.[53]

The lessons to come from the rotation of forces were mixed. Low-intensity conflict observers argued that the constant rotation of forces hurt the Army's effectiveness in such conflict; as soon as a unit finally understood its area of operations and how to do all of tasks required of it, it rotated out to be replaced by the next unit. Just the simple act of building relationships with local leaders was complicated by the rotation of Army leaders every six months to a year. On the other hand, Army transformers—General Reimer chief among them—believed that rotation helped preserve the Army's readiness for an imminent high-intensity conflict against an imagined future peer competitor.[54]

While the Army focused on lessons surrounding the deploying, sustaining, and rotating of units in Bosnia-Herzegovina, it was woefully unprepared to actually engage in the low-intensity conflict in the country. USAREUR tacitly acknowledged this fact when it embarked on two massive training efforts—Mountain Eagle I in September–November 1995 at the CMTC, and Mountain Eagle II in December 1995 in Schweinfurt, Germany—to prepare the forces of the 1st Armored Division for deployment. The 1st Armored Division and the forces that followed it as Task Force Eagle spent weeks prior to deployment training on patrolling, crowd control, establishing and maintaining checkpoints, mine awareness and clearing operations, civil-military operations, and an array of other tasks. As a measure of how alien low-intensity conflict tasks were to the forces of USAREUR, Brigadier General Cherrie would later call this training "the toughest, most-concentrated training our Army has ever done" and adding, "This preparation was the most strenuous training regimen I had been exposed to in my then 31-year career."[55]

Because the Army had not institutionalized the lessons of Somalia and Haiti, the leadership of the division likewise lacked an understanding of the principles of low-intensity conflict. To catch up, Army leaders received intensive education during this predeployment training period, learning about the history of Bosnia-Herzegovina and hearing accounts of the conflict from people who had served in the region during the first half of the 1990s. They also talked to leading media figures about the international media environment in which they would be executing Operation Joint Endeavor. But there was also a great deal of education on the basic tenets of low-intensity conflict, of which most Army leaders were ignorant.[56]

The Army also found itself having to train its junior officers to deal with the unfamiliar environment of decentralized operations necessary to conduct the low-intensity conflict in Bosnia-Herzegovina. Leaders had to learn to respond with minimal guidance in urgent circumstances in the cases of negotiations, crowd control, and settling disputes. While Army leaders could look to recent doctrine such as FM 100-23-1, *Multiservice Procedures for Humanitarian Assistance Operations,* for some help, they more frequently had to look to Nordic or UN manuals for guidance on operating in these unfamiliar conditions.[57]

The U.S. forces designated for Operation Joint Endeavor were also not properly equipped for low-intensity conflict; the division began a herculean effort to reequip itself only months before entering Bosnia-Herzegovina. In addition to new tactical satellite communications gear to span the vast distances over which the Army would be forced to operate, the division also had to rapidly field more survivable "up-armored" high-mobility, multipurpose wheeled vehicles (HMMWVs).[58]

The Army also was not properly organized for the operation. It struggled with a persistent shortage of engineers and civil affairs personnel even after reserve component soldiers arrived in Bosnia-Herzegovina. As with previous low-intensity conflicts, the Army's intelligence capabilities—geared toward signals and image intelligence collection against a peer competitor—were ill suited to the human intelligence requirements that are essential in such conflicts. Additionally, there were never enough linguists on the ground to support daily operations in the country. As in Somalia and Haiti, in Bosnia-Herzegovina the U.S. Army found little use for field artillery; the 1st Armored Division artillery was instead employed in a cantonment of the rival factions' combat vehicles, investigating mine strikes, and inspecting weapon sites— unfamiliar tasks for which the division artillery had to develop its own ad hoc procedures.[59]

As in Haiti, military police were in high demand and asked to execute unfamiliar tasks such as operating as maneuver forces, interacting with international law enforcement agencies, and executing law enforcement directly within the populace of Bosnia-Herzegovina. The U.S. Army once more found itself training local police forces as the International Police Task Force, composed of a mélange of law enforcement professionals from countries around the world, proved incapable of effectively training the police or conducting law enforcement itself; the force was neither armed to protect itself and the populace nor culturally and linguistically equipped to investigate crimes in the foreign environment of Bosnia-Herzegovina.[60]

The Army also committed to the training and equipping of the nascent Army of Bosnia-Herzegovina. The goal was that with U.S. assistance, the new Army of Bosnia-Herzegovina would reach rough parity with the armies of Croatia and Serbia as a hedge against future aggression from either country.[61] In the end, this proved too high a bar; the latter armies were equipped with the collective arsenal of the former Yugoslavia. As of this writing, the Armed Forces of Bosnia and Herzegovina are still woefully less capable than other militaries in the region.

The low-intensity conflict in Bosnia-Herzegovina was inherently multinational. NATO's Allied Rapid Reaction Corps—commanded by British lieutenant general Sir Michael Walker—directed the operation from Sarajevo as the headquarters of IFOR. The three MNDs were led by France, the United Kingdom, and the United States. When NATO conducted its transfer of authority with UNPROFOR on December 20, 1995, some seventeen thousand international troops from that force were absorbed into IFOR across all three MNDs. Altogether, the IFOR consisted of fifty-five thousand troops from thirty-five different countries.[62] Even within the U.S. sector in MND-North, there was a hodgepodge of around five thousand troops from other countries; in addition to the U.S. 1st Armored Division at Tuzla, there was a brigade of troops from Nordic countries and a Polish parachute battalion at Doboj, a reinforced Turkish mechanized battalion at Zenica, and two Russian parachute battalions posted near Srebrenica.

Multinational operations presented the Army with special challenges for which it was simply not prepared. Max Manwaring noted what he called a "systemic disconnect": the Army insisted on operating outside the NATO or UN chain of command, which might have better integrated the Army with its international partners in the country. He advocated four measures of cooperation for the Army to successfully integrate itself into multinational operations. First, he wrote, "The primary peacekeeping parties must be in general agreement with regard to the objectives of a political vision and the associated set of operations." Second, the Army had to participate in "an executive-level management structure that can and will ensure continuous cooperative planning and execution of policy among and between the relevant US civilian agencies and armed forces." Third, the Army needed to "ensure clarity, unity, and effectiveness by integrating coalition military, international organization, and nongovernmental organization processes with US political-military planning and implementing processes." Fourth, and more important than the other three measures, Manwaring stressed, "Unity of effort requires education as well as organizational solutions." In a thinly veiled attack on contemporary military education, he added, "Unity of effort ultimately entails the type of professional military education and leader development that leads to effective diplomacy, as well as to military competence."[63]

While most acknowledged that the Army was not prepared for the low-intensity conflict in Bosnia-Herzegovina, not everyone agreed that the Army *should* be prepared for—let alone employed in—such conflicts. The army chief of staff, Gen. Dennis Reimer, complained to fellow generals that the demand of low-intensity conflicts was overtaxing the Army to a "red line."[64]

Yet the supposed exhaustion of the Army was not nearly as great a problem as General Reimer contended. Purportedly to highlight the extent of this problem, Reimer instituted a system by which units would report their deployment tempo (DEPTEMP). In a single year of reporting, only 8.6 percent of Army units exceeded 120 days deployed, and only 3.7 percent exceeded 180 days. Moreover, if the Army had institutionalized the lessons of its low-intensity conflict experiences, their predeployment training—a component of the DEPTEMP—would have been much shorter. And, of course, much of the DEPTEMP was not the result of training for or executing low-intensity conflicts. Instead it was incurred after redeployment from low-intensity conflicts, in gunnery and maneuver training to rebuild "eroded" skills in high-intensity conflict tasks[65]—returning units to readiness for a great power war against a peer competitor.

It is much more likely that Reimer made such warnings and instituted such reporting procedures to discourage civilian leaders from committing Army forces to future low-intensity conflicts. In his mind, Bosnia-Herzegovina was not a war—the Army's purpose for existence. Instead it was a dangerous distraction from the Army's transformation—its quest for ever-greater readiness to fight high-intensity conflicts.

Yet even Reimer is on record as saying that despite his misgivings, engaging in low-intensity conflict deployments helped the Army retain talented soldiers. In an interview after his tenure as chief of staff, he admitted as much: "Quite frankly, I think many of these people who deployed to Bosnia felt like they'd made a contribution." He added, "They were able to work on their own education while there and things like that." Altogether, he concluded, "We retained better than we predicted. I think part of the reason was due to the facts; there was some additional money associated with these deployments and people felt good about what they were doing."[66] This was Armyspeak for an acknowledgment that soldiers who had deployed to Bosnia-Herzegovina reenlisted at higher rates than soldiers who had not.

Brig. Gen. Stanley Cherrie stopped short of calling Operation Joint Endeavor a distraction from preparedness for a future high-intensity conflict. Still, he hurried to note that while the 1st Armored Division artillery did execute low-intensity conflict tasks on behalf of the division, it "always stayed trained and ready to fire." He also said that the division had to do extra work, "continuing to train in conventional ways to the extent possible as we did with our tanks, Bradleys and dismounted rifle squads in Hungary, and our small arms in-country" in order to "reduce conventional warfighting 'decay.'"[67]

But Cherrie ultimately acknowledged that "combat forces [were] the right forces for PE [peace enforcement] missions. . . . The materials of war [are] a powerful motivator of compliance." He provided a powerful anecdote to prove his point: "This opinion was shared by at least one Bosnian corps commander who, when pointing to 4,500 of his troops in formation, stated: 'All my men out there are fighters, not yet soldiers. You Americans are soldiers. You all dress alike, you all have discipline, you have clean weapons at the ready, you always travel in four vehicle convoys, even your helicopters fly in formation. Soldiers do that and we notice it.'"[68] Cherrie's observation goes to the very heart of the reason why only the U.S. Army could achieve U.S. political ends in a low-intensity conflict. If—as advocates of the Nixon Doctrine and the Reagan-era model for counterinsurgency contended—host nation soldiers were enough to do the job, the introduction of Army troops would not have been necessary in the first place. If the U.S. State Department or UN diplomats could solve the problem, then the problem was not a low-intensity conflict; it was normal, peaceful, political competition. The heart of the reason why disciplined, capable, and appropriately trained and equipped soldiers are necessary in low-intensity conflict is simple: such conflict is war, and war is prosecuted by soldiers. In this environment, only an Army capable of using calibrated force or threat of force and willing to engage in the political dimension of the conflict could win.

That said, the performance of the Army in operations among the populace in Bosnia-Herzegovina made it clear that it had not institutionalized the lessons of Somalia and Haiti. Max Manwaring noted that the Army desperately needed "relevant doctrine at the conceptual level for multilateral peace and stability operations" that recognized "the real locus of power, e.g., the civil population, in a given operational area."[69]

The Army relearned many lessons about exercising nuanced application of force under strict rules of engagement. Young platoon leaders and company commanders were called upon to balance intimidation and negotiations as they tried to quell demonstrations, settle disputes, and dismantle illegal militia checkpoints. Col. Gregory Fontenot, commander of the 1st BCT, 1st Armored Division, explained to his soldiers that firing their weapons constituted a tactical defeat, and killing a civilian might well be a strategic defeat.[70]

Some believed that the Army's ability to apply nuanced force in Bosnia-Herzegovina was hindered rather than helped by its preparation for Operation Joint Endeavor. IFOR had been designed and prepared to "compel compliance" with the Dayton Accords through "overwhelming force." Thus, in predeployment training, soldiers were trained in many combat tasks that were not only unneeded but also possibly conditioned soldiers to use more force than was required. Maj. Walter Piatt—who would go on to command a BCT of the 25th Infantry Division in Iraq in 2009 and later, as a major general in 2018, command the entire 10th Infantry Division in Iraq—told interviewers that his training for Bosnia-Herzegovina had been predominantly in combat tasks. He added that this training had caused his unit to be overly forceful in interactions with refugees and nongovernmental organizations (NGOs).[71]

The chief of staff of the Army, General Reimer, disagreed; he told fellow generals that he was heartened to find, when inspecting training in Germany, that "the training was tougher than the real thing." Reimer cited Command Sgt. Maj. Timothy Beck of V Corps as comparing this training "to swinging a heavy bat in practice" so that the real burden during execution would be easier.[72]

In another indication that the Army had not institutionalized the lessons of Somalia and Haiti, Task Force Eagle lacked the nonlethal capabilities to apply force without severely injuring or killing civilians. Consisting of two armored BCTs with more than one hundred tanks each, the 1st Armored

Division was chronically ill equipped for riot control and crowd dispersal. Thus handicapped, soldiers were forced to innovate, just as they had in previous low-intensity conflicts. In one example in Čelić in early 1997, Capt. Mike Slocum found himself facing an angry mob attempting to stop his convoy en route to conduct a weapons inspection. To disperse the crowd, he called in a helicopter to use its rotor wash—the strong winds created by the beating of its rotors against the air.[73]

Intense concern for force protection and avoiding casualties served as a cognitive obstacle to Task Force Eagle's ability to move beyond operating in and among the populace to actually engaging the political dimension of the low-intensity conflict in Bosnia-Herzegovina. With memories of the fallout from the battle of Mogadishu still fresh in their minds, the members of the administration of President Bill Clinton put intense pressure on military leaders to avoid casualties; "force protection" even appeared in the mission statements of SFOR and Task Force Eagle. As Brigadier General Cherrie would write, "Paramount in everything we planned and accomplished was a concentration on force protection." Thus, American forces traveled around the country in armored vehicles and body armor while their counterparts from other countries strolled through villages in only their uniforms with soft caps and sipped coffee at cafés. These individual requirements were compounded by the requirement that soldiers travel in four-vehicle convoys for mutual protection—a requirement that made journeys outside heavily fortified base camps rare.[74]

Up to 50 percent of the U.S. Army's combat forces in Bosnia-Herzegovina were consumed in force protection tasks. As Manwaring complained, "In Operation Joint Endeavor, the US force protection effort has taken on a higher degree of importance than the peace and stability mission in Bosnia itself."[75]

Lt. Col. Mark Viney, commander of the 1st Squadron, 4th Cavalry Regiment in Bosnia-Herzegovina, was unforgiving in his condemnation of the Army's obsession with force protection in Operation Joint Endeavor, writing that it created a "zero-defects culture." According to Viney, "Our senior military leaders understood that minimizing casualties was a measure of success, like it or not." He quipped, "Whether force protection was prudence or just covering one's ass depended largely on the rank of the observer."[76]

Admittedly, the Army did gradually relax its posture as the operation moved into its second and third years. The four-vehicle convoy requirement was relaxed for psychological operations (PSYOP) and civil affairs soldiers in 1997.[77] But most other force protection measures continued well beyond 1997.

Civil-military operations in particular suffered as a result of excessive concern for force protection. By the time IFOR was established, more than three hundred registered NGOs were in Bosnia-Herzegovina. But because of force protection restrictions on travel—four-vehicle convoys, in particular—small teams of soldiers, such as civil affairs teams, were frequently not able to travel outside base camps. Civil affairs teams were not truly effective in engaging these organizations or coordinating U.S. relief activities with NGO activities until 1997, when restrictions were relaxed. But even after these activities began, the Army struggled to meet the needs of the populace. As Manwaring put it, "The urgency of developing mature multilateral civil-military doctrine for contemporary peace and security requirements is clear."[78]

Because the Army had failed to institutionalize the lessons of Somalia and Haiti, in Bosnia-Herzegovina it once again struggled to repurpose its PSYOP capabilities—designed to persuade enemy soldiers in a high-intensity conflict to lay down their arms—to influence the civilian populace. A Defense Science Board study on the topic concluded that the Army was unprepared to plan, resource, or support the demands of PSYOP in a postconflict environment. Force protection concerns did make it difficult for PSYOP soldiers to interact with the populace, because they operated in teams too small to form four-vehicle convoys. But much of the Army's challenges were the result of a poor understanding of cultures and attitudes; some particularly disastrous PSYOP products included quotes from Yugoslav Serbian strongman Marshal Tito—hated by most Bosnians but especially by non-Serbs—or quotes from western European Enlightenment figures—of whom most Bosnians were completely ignorant.[79]

Other failures resulted from trying to achieve very difficult political ends solely through the meager means provided by PSYOP—for instance, trying to convince Bosnian Serbs not to leave their homes in other factional regions for the safety of Republika Srpska. PSYOP soldiers drawing on lessons from the decidedly Third World countries of Somalia and Haiti were also ill prepared for the challenges of producing PSYOP products for television and radio, which were ubiquitous in Bosnia-Herzegovina. They finally achieved success in these media by hiring former BBC producer Karen Holman to help them prepare their messages for broadcast.[80]

Another capability that the Army had to develop in the midst of Operation Joint Endeavor was the public affairs capacity to engage international media. Initially the Army maintained an information center at the IFOR headquarters. But the efforts of this element were not coordinated with the media engagement efforts conducted by the British armored division and the U.S.

1st Armored Division in their own sectors. Eventually public affairs efforts were consolidated in Sarajevo under a single Coalition Press Information Center, and the quality of media engagements began to improve.[81]

But the Army also repeated some of its success from the earlier low-intensity conflict in Haiti. There is no direct documentary evidence as to whether this was a lesson learned from that conflict or whether public affairs soldiers in Bosnia-Herzegovina simply arrived at the same solution independently, but "embedding" of journalists in U.S. Army units across the country was commonplace in both Operation Joint Endeavor and Operation Joint Guard. The positive reception to this practice among the media was magnified by the fact that the British rejected this practice and exercised much tighter control on the press in their own sector.[82]

One obstacle to moving beyond an understanding of the requirement to operate in and among the people to a deeper understanding—that is, an acknowledgment of the political dimension—reemerged in the low-intensity conflict in Bosnia-Herzegovina: the Army's obsession with quantifying its activities and elements of the environment, of chasing statistics as a measure of winning rather than actually seeking a political solution to the conflict. In his article on the 1st Armored Division's operations in Bosnia-Herzegovina, Cherrie explained that there was an "absolute necessity" for the division "to develop a system capable to track hundreds of bits of critical information and display them in a rapid, easily understood manner." The zone of separation was divided into minute blocks, and each block was rated with an "an arrow-like, color-coded (red, amber or green) indicator . . . that gave a quick visual reference of the operation[']s status in that particular block." Everything from the sentiment of the populace to living conditions to security was measured, totaling "thousands of information bytes that needed to be monitored."[83]

While they had engineered an impressive feat of statistical wizardry, it was apparently lost on Cherrie and the staff of the division that, even if every block was "green" in every category, the conflict would not be over; that required a political settlement, something inherently qualitative in a way that defied quantification. Every time a group from one sect tried to assert its freedom of travel through or reclaim property in another sect's area, violence would spike,[84] blocks would turn from green to red, and the U.S. Army would be dispatched and turn the block back to green. But without a political settlement, this cycle was all but meaningless and the low-intensity conflict in Bosnia-Herzegovina would remain a forever war.

Officially, the U.S. Army tried to avoid the political dimension of the low-intensity conflict in Bosnia-Herzegovina by artificially limiting its mission there. USAREUR narrowly defined the "military tasks" that it derived from the Dayton Accords:

- Ensure that IFOR could defend itself and move freely.
- Supervise selective marking of boundaries and ZOSs [zones of separation] between the parties.
- Monitor—and if necessary enforce—the withdrawal of forces to their respective territories. . . .
- Provide a safe environment for civil peace-implementation functions (assisting the UNHCR [UN High Commissioner for Refugees] and other international organizations engaged in humanitarian work in Bosnia and Herzegovina).
- Help observe and protect the civilian population, refugees, and displaced persons.
- Help monitor the clearance of minefields and other obstacles. [85]

The Army did also acquiesce to implementing the Dayton Accords' provision for disarming "armed civilian groups" and conducted periodic inspections of weapon sites. But none of these narrow tasks addressed the political dimension of the conflict, and all were complete by June 1996,[86] yet the low-intensity conflict in Bosnia-Herzegovina was no closer to a conclusion.

Once on the ground, the Army tried equally hard to avoid the political dimension of the conflict. In one example in March 1996, sixty-six Bosnian Serb thugs began terrorizing Bosnian Serb civilians in Sarajevo in order to "encourage" them to move to Republika Srpska. In the process, the mob burned a number of buildings, including the UNHCR's warehouse. IFOR fastidiously refused to intervene. It was only after Secretary of State Warren Christopher became personally involved that the Army moved to arrest sixteen of the instigators. By then, of course, this episode of ethnic cleansing was a fait accompli.[87]

While the Army may have tried to avoid the political dimension of the low-intensity conflict in Bosnia-Herzegovina, at the tactical level engaging in politics was unavoidable. Junior Army leaders faced the demand of establishing civil governance and practicing the "art of street diplomacy." Officers grappled with a complex web of history, family ties, and ethnic and religious conflicts to weave together political, economic, and social solutions in their areas of responsibility.[88]

In fact, soldiers on the ground found it nearly impossible to sidestep this facet of the conflict. All over the country, from the squad to the division level, Army leaders were confronted with the political dimension of the conflict. For example, in early 1996, Col. John Batiste, the commander of the 2nd BCT, 1st Armored Division, tried to secure the heavy weapons inside the Bosnian Serb enclave at Mount Zep. He was met by Bosnian Serb general Ratko Mladić, who refused to allow an inspection of the weapons. A "rent-a-mob" of drunken Serbs was bussed in and began shoving and spitting on Batiste and his party. The IFOR commander, U.S. Navy admiral Leighton Smith Jr., eventually had to intervene, bringing in the president of the Bosnian Serb enclave to break the impasse. Meanwhile, the international media noted that this was one of dozens of meetings between Colonel Batiste and General Mladić, an indicted war criminal; yet the Army had not bothered to arrest him.[89]

As the IFOR mission gave way to the SFOR mission, Operation Joint Endeavor became Operation Joint Guard, and it became clear that other agencies—the U.S. State Department, North Atlantic Council (NAC) diplomats, or the UN—were not going to achieve a political solution to the low-intensity conflict in Bosnia-Herzegovina, so the international community put more pressure on the U.S.-led coalition force to try to do so. The new mandate for SFOR—UN Security Council Resolution 1088—gave military forces in Bosnia-Herzegovina a new mission. In addition to its tasks to "prevent the resumption of hostilities" and "promote a climate conducive to pushing the peace process forward," the Army was now also directed to "provide selective support to civilian organizations within its capabilities." The Army added to its mission a goal to "assist international organizations to set the conditions for civilian implementation of the GFAP [General Framework for the Agreement of Peace, aka the Dayton Accords] in order to transition the area of operations to a stable environment." The Army began to help organize elections, establish local governments and police forces, return displaced persons, and provide humanitarian aid. But the cost of the Army's procrastination was that it would now have fewer forces with which to complete its considerably expanded mission; with the transition from IFOR to SFOR, the number of troops dropped from sixty thousand to thirty-two thousand.[90]

Low-intensity conflict observers recorded the lessons the Army was learning about the political dimension of the conflict in Bosnia-Herzegovina. As Manwaring noted, "The political complexity of contemporary peace operations stems from the fact that *intrastate* conflicts such as those in the former

Yugoslavia are the result of careful political consideration and strong political motivation." The consequences of this fact were twofold. "First," Manwaring explained, "in *intrastate* conflict, confrontation is transformed from the level of military violence to the level of a political-psychological struggle for the proverbial 'hearts and minds' of a people." In this environment, he added, "The blunt force of military formations supported by tanks and aircraft could be irrelevant or even counterproductive." Instead, Manwaring insisted that the Army had to look to " 'soft' political, economic, psychological, and moral power—supported by information operations, careful intelligence work, and surgical precision at the more direct military or police level." But there was also, he explained, a "second leadership and cooperation level of political dominance" in low-intensity conflicts, and this level required "the greatest civil-military and military-military diplomacy, cooperation, and coordination." He advocated increased education of Army officers in low-intensity conflict and especially civil-military and multinational operations.[91]

Manwaring placed the blame for the Army's challenges with its failure to institutionalize low-intensity conflict—and particularly its political dimensions—in its doctrine and training. As he put it, "The need for more mature peace and stability operations doctrine is made clear when every civil and military organization involved in missions such as those in the former Yugoslavia operates under its own procedures or doctrine," adding, "Extant doctrine is generally designed to provide conventional military solutions to traditional military problems." At the root of the shortfall in Army doctrine was, in Manwaring's opinion, that in Army doctrine, "there is little or no doctrinal recognition of the fact that peace and stability operations are primarily multinational, political, and psychological in nature."[92]

For his part, Gen. Dennis Reimer seemed oblivious to the necessity to engage the political dimension of the low-intensity conflict in order to find a sustainable political solution in Bosnia-Herzegovina. After leaving office, he would comment that the Joint Chiefs of Staff "always wanted to focus on end state and nobody else really did." He added, "We should have held people's feet to the fire, that 'one-year,' but we didn't do it." Reimer concluded, "Once you get involved, then you're stuck with it. I don't really see . . . any good end state or any day certain that we're going to be out of Bosnia." For Reimer, the Army had been a victim of an ill-advised intervention, powerless to determine the date of its own exit absent permission from the president to leave. The general appears to have genuinely seen no role for the Army in creating the political conditions in Bosnia-Herzegovina that would have made its exit possible.[93]

Because of such sentiments, IFOR pressed for early elections per the timetable established in Dayton. The decision was perhaps also driven by a naïveté about the power of democratic processes to replace violent political expression. In any case, it was primarily driven by the United States' urgent desire to conclude the IFOR mission on the president's timetable and bring its troops home. But the elections did more harm than good. Sectarian firebrands played on the fears of the public that U.S. and international military peacekeepers would stick to their pledge and leave after one year. Average citizens, fearing fighting might later resume, backed these sectarians over those who might be more likely to back a political accord with the other factions in the conflict; votes fell strictly along sectarian lines.[94] The result solidified the power of precisely the hard-liners who were the greatest obstacles to a negotiated political settlement to the low-intensity conflict in Bosnia-Herzegovina.

Arriving at a sustainable political solution there would require more than solving the small, local political problems in each town or city. It would require the Army to back a winner to bolster the U.S. government's strategic decision to support one side in the conflict. Insistence on neutrality had already doomed UNPROFOR; because they wished to remain neutral, rather than use force to stop human rights abuses, UN forces stood by and watched as each faction ethnically cleansed their respective regions.[95]

But the impulse to remain neutral in the conflict was deeply engrained in the Army's culture, from the top down. General Reimer said, after leaving office as chief of staff of the Army, that he and the rest of the Joint Chiefs of Staff had been "concerned about . . . the idea of training and equipping the Muslim forces" in Bosnia-Herzegovina. "We wanted to stay out of that. If we were going to be a peacekeeping force, we thought we had to be neutral." According to Reimer, the Joint Chiefs believed that training the Bosniaks would " require us to give up our neutrality." To mitigate this risk, he explained, the mission of training the Bosniaks was instead given to military contractors from Military Professional Resources, Inc.[96]

Chastened by its experience in Somalia in hunting the warlord Mohamed Farrah Aidid, the Army was particularly resistant to tracking down indicted war criminals (designated as PIFWCs, personnel indicted for war crimes), the vast majority of whom were Bosnian Serbs. The chairman of the Joint Chiefs of Staff, Gen. John Shalikashvili, would later say of his discussions with administration officials about the matter that he made it clear that "we weren't going to take them unless we stumbled upon them." He added that this insistence was a reflection of the U.S. military's experience in Somalia. In fact, the Joint Chiefs had refused to concur with the Dayton Accords until any language even implying that the military would capture war criminals was expunged from the agreement.[97]

The Army was right to fear that hunting war criminals would jeopardize its hopes of remaining neutral in the conflict. In July 1997, after the British Army in its own sector captured one Bosnian Serb indicted for war crimes and killed another, U.S. forces in MND-North suffered an unusual escalation in violence, including a bomb and an antitank rocket attack.[98] But the simple fact was that to end the low-intensity conflict in Bosnia-Herzegovina, the Army was going to have to round up the Bosnian Serb instigators who were the biggest obstacles to a political settlement there.

Instead, the Army scrupulously refused to take sides. Brigadier General Cherrie wrote that IFOR carefully selected base camp locations so that no one faction's area would have more foreign troops than any other. He seemed oblivious to the fact that IFOR had, in the process, placed foreign troops in the areas of the factions with which they had the most affinity. The Turkish headquarters was placed in the Federation territory dominated by Muslims; the Russians were stationed in the Bosnian Serb Republika Srpska.[99] These countries clearly *had* picked sides in the conflict.

This wasteful insistence on neutrality was largely lost on the populace of Bosnia-Herzegovina. Opinion sampling from across all of the sects in the country consistently showed that Bosnian Serbs believed that international forces favored the other factions at their expense. This was a reflection of the fact that the Bosniaks and Bosnian Croats largely supported the Dayton Accords—and were rewarded for their compliance—while Bosnian Serbs largely opposed the accords—and were penalized for resistance.[100] The energy expended trying not to take sides in the low-intensity conflict was more than just wasteful; this neutrality was a recipe for a forever war in Bosnia-Herzegovina.

The Army After Next: 1996–1997

While low-intensity conflict observers grappled with the lessons that the Army was learning in Bosnia-Herzegovina, the rest of the Army continued on its headlong march toward transformation: the ever-greater capacity to fight high-intensity conflicts. The next major milestone in this journey was another revision of FM 100-5, *Operations*. Planned for release in the summer of 1998, work began on it in June 1996 within SAMS. The update was to reflect the recent experimentation and thinking on the future Army and capture the aspirations of *Joint Vision 2010*.[101]

But the manual was also intended to place low-intensity conflict—which would now be called "stability and support operations"—safely within the category of additional but nonessential tasks the Army was also occasionally required to do. Thus, when he initiated the effort to revise the Army's flagship doctrinal manual, one of TRADOC commander Gen. William Hartzog's first directives was the "doctrinal integration of peace operations, humanitarian assistance operations, and other military operations short of general war into the body of operational doctrine." General Hartzog sought to erase all distinctions between war, "the Army's primary focus," and operations other than war. The initial draft of the manual was completed in October 1996 and it was distributed for Army-wide review beginning in mid-1997.[102]

The 1993 edition of FM 100-5 had likewise tried to reign in low-intensity conflict observers. It had reclassified low-intensity conflict as operations other than war and excised the principle of political dominance from the doctrine that had been described in the 1990 edition of FM 100-20, *Military Operations in Low Intensity Conflict*. But A-AFCLIC had undone some of this regression in low-intensity conflict thinking with its publication of a joint manual (which nominally trumped any Army doctrine), the 1995 JP 3-07, Joint Doctrine for Military Operations Other Than War. This joint doctrine had reasserted the unequalled importance of the political dimension to low-intensity conflict.[103]

Presumably to protect the 1998 edition of FM 100-5 from suffering the same fate, General Reimer summarily closed A-AFCLIC by withdrawing Army support from the organization, and it officially closed its doors on June 28, 1996. The curt official justification provided by Army transformers for the closure of the important organization was that the Army had sufficiently captured the lessons of low-intensity conflicts and "understood their principles well." Later in 1996, General Reimer further justified this decision with the incredible claim, "We've done the Somalias, the Bosnias, the Haitis . . . and they've all been done well."[104] The chief of staff of the Army was either genuinely oblivious or willfully ignoring the fact that low-intensity conflicts continued, unabated, in all three of these countries—with or without the presence of U.S. troops.

Ironically, the 1998 edition of FM 100-5 would never be published. A new version of the manual would not be published until June 2001, and it would be designated FM 3-0, *Operations*, to align it with the new, joint doctrinal publication numbering convention.[105]

Gen. Gordon Sullivan had called the publication of the TRADOC Pamphlet 525-5, *Force XXI Operations*, the "the first step of our doctrinal journey into the future." In the spring of 1996, General Reimer took the next step, initiating an effort to look beyond Force XXI to the more distant future of the Army. General Reimer and his TRADOC commander, General Hartzog, envisioned an "Army After Next"—an Army after Force XXI—that would be even more optimized to fight a high-intensity conflict against an imagined peer competitor that they claimed would emerge fifteen to thirty years into the future. Reimer would later explain, "*Army XXI* will maintain and improve America's warfighting edge." Yet, he continued, "The AAN [Army After Next] will provide the nation with overmatch capabilities . . . to protect the nation's interests against any peer competitor."[106]

At its heart, the Army After Next was a scheme to justify new platforms and weapons for the Army; General Reimer insisted that, from 2010 to 2015, the Army's current major weapons platforms—the Big Five systems, including the M1 Abrams Tank and the M2 Bradley Infantry Fighting Vehicle—would "approach technological obsolescence." This idea would appear again and again in writings about the Army After Next. To solve this problem, Reimer hoped, "The critical technological advancements of the next 20 years will be incorporated into new operating systems and weapon platforms." But the Army After Next also sought to find an elusive combination of technology and military doctrine that would allow it to surpass imagined peers, as Army transformers believed the German Army had with blitzkrieg technology and tactics before World War II. As Reimer would later write, he wanted to eschew "'creep-ahead' technologies" in favor of "'leap-ahead' technologies."[107]

To define the shape and requirements of this new force, in February 1996 Generals Reimer and Hartzog initiated the Army After Next project—an experimentation regimen enabled by computer simulation and tabletop wargaming exercises. A cycle of midyear concept papers followed by winter wargames and experiments at the Army's new wargaming facility, the Center for Strategic Leadership at the U.S. Army War College, would incrementally define the shape of this future Army. The project itself was massive, rapidly growing in personnel to two hundred permanent planners and representatives from the joint services. The first winter wargame, from late January to early February 1997, included more than four hundred participants—Army officers, other U.S. government officials, foreign military officers, and academic advisers. These experiments would be overseen by Maj. Gen. Robert

Scales, commandant of the U.S. Army War College from August 1997 until his retirement in November 2000.[108]

Before the Army could embark on finding the Army After Next, however, it had some housekeeping to do. The Joint Staff had published the *Joint Vision 2010*, and the Army needed to publish its own vision to communicate how it was nested within this joint concept. The result was *Army Vision 2010*, published in November 1996. Much of the document was a tortured attempt to graft the six things that the Army wanted to do ("Project the Force," "Gain Information Dominance," "Shape the Battlespace," conduct "Decisive Operations," "Protect the Force," and "Sustain the Force") onto the four core concepts of *Joint Vision 2010* ("Full Dimension Protection," "Dominant Maneuver," "Focused Logistics," and "Precision Engagement").[109] Nonetheless, *Army Vision 2010* was still useful as a glimpse into the Army's conception of the future of warfare.

To "Project the Force" the Army would have to be more deployable, configuring itself into "Modular Organization[s]" and using pre-positioned equipment and "Army War Reserve Prepositioned Stocks" to rapidly deploy "Joint, Lethal, Early Entry Forces . . . Directly to Combat." Going forward the Army would launch invasions of another country directly from the United States, with no painful, six-month-long buildup of forces as had preceded the Gulf War. To achieve this, according to the explanation of the concept of "Decisive Operations," the Army would have to "Mass Effects, Not Forces," substituting "Information Dominance," "Lethality at Extended Ranges," "Precision Systems & Munitions," and "Mobility, Speed, [and] Agility" for numbers. It also would have to "Shape the Battlespace" by "Dominat[ing an] Expanded Multidimensional Battlespace" with "Simultaneity" and "Precision Systems and Munitions." Because the Army would have a much-diminished force, it would have to "Protect [that] Force" with "Speed, Agility, Long Range Weapons," and "Real Time Intelligence" that would allow it to "Avoid Detection [and] Prevent Acquisition." The Army After Next would also have "Improved Ballistic Protection" that would give it "Early Warning" to "Avert Hits." Key to all of these concepts was to "Gain Information Dominance" through "Wireless Communications" and "Advanced Network Technology," which would provide the Army with "Linked Strategic, Operational, and Tactical Sensors and [Command, Control, Communications, Computers, and Intelligence]."[110] In short, the Army After Next was the translation of the RMA to high-intensity conflict ground operations.

Army Vision 2010 did acknowledge some role for the Army in military operations other than war, but low-intensity conflict operations were little more than a footnote in this Army transformation vision of the future. Ultimately, these activities were simply other things that the Army did beside its primary function: war.[111]

The 1997 QDR

Army Vision 2010 came just in time to prepare the Army for the interservice scrum that was the Quadrennial Defense Review (QDR) process. Dissatisfied with the pace of defense adaptation to the post–Cold War world, Congress mandated that the Department of Defense would, every four years, review its capabilities, ostensibly to ensure that they were appropriate for the threats and strategic environment of the future. In reality, this review was a delicate ballet between Congress, the Joint Chiefs of Staff and their respective service staffs, and the secretary of defense and his office—with the chairman of the Joint Chiefs of Staff and the Joint Staff trying to mediate the conflict. The Army was fighting to retain numbers and defend its own conception of transformation against those RMA enthusiasts who believed that advanced sensors and precision munitions connected by computer networks could make ground warfare obsolete.[112]

In many respects, the result of the QDR was a forgone conclusion; it took as its starting point the Bottom-up Review's sizing mechanism—two "major regional conflicts"—and then, in experimentation, envisioned high-intensity conflicts beginning within the next fifteen years against threats up to a "Peer Competitor." The result was a negotiated settlement that preserved the interests of the individual services while addressing the interests of the administration as represented by the secretary of defense. That settlement heavily favored the Army's current and envisioned future ground combat capabilities, to be realized through transformation.[113]

The final report of the 1997 QDR tried desperately to reconcile the reality that "the prospect of a horrific, global war has receded" with the end of the Cold War, while—to justify continued exorbitant spending on national security—insisting that "new threats and dangers—harder to define and more difficult to track—have gathered on the horizon." The QDR report dismissed those who argued that "America's military establishment and forces are trapped hopelessly in the past, still structured and struggling to fight yesterday's wars." Yet, as of 1997, 15 percent of the federal budget and 3.2 percent of the country's gross national product—$250 billion—was still being spent on defense, despite the absence of an existential threat to the country's survival. While the Department of Defense may have insisted that it was not "trapped hopelessly in the past" and that it could "maintain the capability to *respond* to the full spectrum of threats," $44 billion of the total defense budget was still being spent on the acquisition of weapon systems that were very much designed to fight "a horrific, global war."[114]

Hence, while the QDR report acknowledged that the threat of global war had receded and Russia and the former Warsaw Pact countries were cooperating with the United States in missions around the world, the report still saw lurking high-intensity conflict adversaries around every corner, citing "the threat of coercion and large-scale, cross-border aggression against U.S. allies and friends in key regions by hostile states with significant military power." Chief among these threats was what President George W. Bush would later call the axis of evil: Iran, Iraq, and North Korea.[115]

Low-intensity conflict was an afterthought in the QDR report. The threat of terrorism against "Americans at home in the years to come" received only a single sentence of acknowledgment. The danger that "failed or failing states may create instability, internal conflict, and humanitarian crises" was similarly only briefly mentioned. Despite the fact that interventions to combat this danger had consumed the Army since the end of the Cold War, they were hardly mentioned in the QDR report amid warnings of the proliferation among potential future high-intensity conflict adversaries of "nuclear, biological, and chemical (NBC) weapons and their means of delivery; information warfare capabilities; advanced conventional weapons; stealth capabilities; unmanned aerial vehicles; and capabilities to access, or deny access to, space."[116]

To the extent that low-intensity conflict was addressed at all, it was treated as a risk to preparedness for "fighting and winning two major theater wars" nearly simultaneously. The report explained that because the U.S. military was being excessively engaged in low-intensity conflicts—"substantial levels of peacetime engagement overseas as well as multiple concurrent smaller-scale contingency operations"—it faced the additional challenge of being able "to transition to fighting major theater wars from a posture of global engagement"—*engagement* being precisely the word that the Clinton administration used to justify committing the Army to such conflicts. The QDR report warned ominously that such engagement forced the United States to be "extremely selective" in where it committed forces, at least implying that worrying about how it might address low-intensity conflicts put at risk the U.S. ability to respond to high-intensity conflicts. To underline the magnitude of the risk, the report warned, "Failure to halt an enemy invasion rapidly can make the subsequent campaign to evict enemy forces from captured territory much more difficult, lengthy, and costly."[117] In other words, low-intensity conflicts were a dangerous distraction from the U.S. military's primary responsibility: fighting and winning America's wars.

The report went on to state, "In the period beyond 2015, there is the possibility that a regional great power or global peer competitor may emerge. Russia and China are seen by some as having the potential to be such competitors." The report briefly (in fact, in a single word) acknowledged that this possibility was "unlikely," but then insisted that, if a peer threat did emerge in "the period from 1997 to 2015" it "would have highly negative consequences that would be very expensive to counter." Thus, the QDR report argued, "in an uncertain, resource-constrained environment," the sensible and "relatively inexpensive way to manage the risk of being unprepared to meet a new threat, developing the wrong capabilities, or producing a capability too early and having it become obsolete by the time it is needed" was an inexorable, paced development of technologies that would give the U.S. military ever-greater capacity to face this mythical future peer competitor in high-intensity conflict.[118]

While the QDR report did acknowledge of both China and Russia that "their respective futures are quite uncertain," the prospect of a potential successor to the Soviet Union as the nemesis to the United States was clearly intended to justify continued excessive spending on defense. The report dismissed those who argued "that because we no longer face the challenge of a global peer competitor like the Soviet Union, we would be best served as a nation by focusing our energies at home and only committing military forces when our nation's survival is at stake." Instead it claimed that such people were guilty of "a 19th century view of the world, which ignores the impact of global events on our nation, the growing interdependence of the world economy, and the acceleration of the information technology revolution." The report likewise painted low-intensity conflict observers who argued that the U.S. military should improve its capacity to engage in these conflicts as advocating "relieving human suffering wherever it exists, and promoting a better way of life, not only for our own citizens but for others as well." The report derided this view as advocating that the U.S. military be "world policemen," instead advocating a middle path that would maintain for the United States "unparalleled military capabilities" while spending aggressively to realize the future promise of the RMA.[119]

In describing this promise, the QDR report proclaimed, "The information revolution is creating a Revolution in Military Affairs that will fundamentally change the way U.S. forces fight." The report endorsed *Joint Vision 2010* as a "template for seizing on these technologies and ensuring military dominance," but it also proposed "three alternative paths"—two straw men and one middle way—to envision the future. The report rejected a path that "focus[ed] more on current dangers and opportunities" and saw "today's threats demanding more attention and tomorrow's threats far enough away to give us ample time to respond" (which, of course, would have driven the Defense Department to increase its capacity for low-intensity conflicts, very much the current threat facing the Army). But the QDR report also insisted that the Department of Defense was not focusing only on "future dangers," "devoting more resources to building the future force" at the cost of "significant reductions in the current force" and a sharp reduction in the U.S. military's "ability to shape the international environment" (again, precisely the language the Clinton administration used to describe interventions in low-intensity conflicts). Instead, the report assured its readers, it would "strike a balance between the present and the future, recognizing that our interests and responsibilities in the world do not permit us to choose between the two."[120]

The result of this strategy would be a U.S. military that would use a "*shape-respond-prepare*" strategy and "trim current forces—primarily in the 'tail' (support structure) and modestly in the 'tooth' (combat power)" in order to harvest "funding for the next generation of systems—such as information systems, strike systems, mobility forces, and missile defense systems—that will ensure our domination of the battlespace in 2010 and beyond." Rather than meeting the many—and growing—present demands to engage in low-intensity conflicts around the world, the U.S. military would continue to focus on being "capable of fighting and winning two major theater wars nearly simultaneously" in some hypothetical future.[121]

The Department of Defense, with the full concurrence of the Army, was sacrificing the Army's capacity to execute low-intensity conflicts—which were manpower intensive, especially in the "tail" that the QDR report proposed to cut—in order to increase its capacity to execute some future high-intensity conflict, which required grasping for the extravagantly expensive brass ring of RMA-enabled technology. The Army would cut fifteen thousand soldiers from the active force, primarily from combat support and service support units. At the same time, within the reserve component, the Army would "accelerate conversion of some units from combat to combat support and combat service support roles."[122] The term "combat support" was military jargon for capabilities such as engineering, military policing, and managing civil affairs—precisely the capabilities that were in highest demand and shortest supply in low-intensity conflicts.

This was an overt rejection of the consistent lessons of the Army's low-intensity conflict interventions since the end of the Cold War—especially in Somalia, Haiti, and Bosnia-Herzegovina. In each of these conflicts participants and low-intensity conflict observers noted that too many of these essential capabilities resided in the reserve component and were too slow to mobilize and deploy. As a result of the QDR, the Army was moving even more of these capabilities into the reserve component, aggravating rather than solving this chronic problem of the Army's capacity to effectively engage in low-intensity conflicts.

At the conclusion of the QDR, General Reimer endorsed its findings by repeating its major points in the pages of the *Military Review*. In reality, though, Reimer saw the results of the QDR as a negotiated settlement in which the Army had gotten two-thirds of what it wanted. He told an interviewer after leaving office that the strategy to emerge from the QDR review was "based upon three pillars," two of which fell directly in line with Reimer's desires as expressed in *Army Vision 2010*. The Army had "to be able to respond to a crisis anywhere in the world"—with Reimer apparently interpreting "crisis" to mean only a high-intensity conflict—and "prepare for an uncertain future"—which Reimer appeared to understand as one with an emergent peer competitor. But these pillars had come at a cost to the Army, as it would also have to continue to "shape the environment that we live in." As Reimer explained, "That meant stability operations as far as I was concerned."[123]

Apparently to mitigate the risk that this burden might derail transformation, General Reimer doubled down on the warnings of the QDR report. Whereas the report warned that the threat of a future peer competitor in a high-intensity conflict was "unlikely" but dangerous, Reimer warned that history proved that the emergence of a peer competitor was not just likely but inevitable. He recounted the history of the Army's supposed unpreparedness for both World War I and World War II before going on to recount President Harry Truman's Cold War–era exhortation that "we must be prepared to pay the price for peace, or assuredly we will pay the price of war" and bemoan the Army's subsequent unpreparedness for the war in Korea.[124] The implication was clear: every time the Army let its guard drop, it was unpleasantly surprised by the emergence of a new peer competitor.

For Reimer, "the Soviet Union's disintegration" and "the breakup of nation-states" had "promoted new and potentially destabilizing trends" that had forced the U.S. Army to intervene. But these challenges, Reimer believed, could not be allowed to take Army leaders away from their focus on the real danger: "At least 56 countries already are capable of engaging in mid- intensity conflict, each having military forces that include at least 700 tanks or armored personnel carriers, 100 combat aircraft, 500 artillery pieces and more than 100,000 soldiers."[125] The general further warned that "precision-guided munitions and high-technology weapon's proliferation among developing nations will make future battlefields, even in the developing world, high-risk environments." He likewise softened the caveats of the QDR report, conceding only that "the emergence of a global competitor against the United States in the next quarter-century *seems* unlikely." Reimer would only concede that "there will be no global peer competitor between now and 2010," but was quick to add, "Regional powers armed with modern weapons certainly will be an international security environment feature, and great powers such as Russia and China may well assert their will in areas they deem within their sphere of influence, thereby challenging US national interests."[126]

In this conception of the immediate future for the Army, the post–Cold War era was not an end to the threat of global war and an opportunity to consolidate the Free World's gains from the collapse of the Soviet Union. Instead, Reimer wrote, "The Cold War's end provides the United States a strategic window of opportunity. . . . There is no global competitor to challenge US worldwide interests. This strategic situation provides us with the opportunity to think imaginatively about the future as we try to divine what capabilities will be possible and *necessary* in the decades ahead."[127] In this conception of the strategic environment, a future high-intensity conflict against a peer competitor was all but inevitable. The Army could either use this brief respite to prepare or suffer the consequences of its failure to innovate.

Notably, Reimer also recognized that "the notion of nationalism based on 'ethnic purity' contains the seeds of endless and intractable conflicts. Some fundamentalist religious movements advocate violence and vengeance." Yet neither of these dangers prompted him to warn that the Army should be equally prepared for low-intensity conflict threats. Instead he continued to insist, "The Army's fundamental purpose is to fight and win our nation's wars."[128]

Breaking the Phalanx

Through the QDR, Force XXI and the AWEs continued to march along at a steady clip. In March 1997, a massive AWE at the National Training Center in the deserts of Fort Irwin, California, tested new organizations, equipment, and doctrine to speed communication and improve situational awareness across Army units. The exercise involved an entire digitized brigade of the 4th Infantry Division from Fort Hood—the EXFOR—testing eighty-seven different systems of the proposed Force XXI.[129]

In the meantime, a key focus of the Army After Next would be its deployability. In his *Military Review* article, Reimer described the Army After Next as "a logistically unencumbered force with greater lethality, versatility and strategic and operational mobility." He defined the term "logistically unencumbered" to mean " 'just-in-time,' rather than 'just-in-case.' "[130] While this did little to decipher the technojargon surrounding the Army After Next, it was at least clear that Army transformers envisioned a future force that would be pared down to its bare essentials and much lighter than the Army of the day.

Within the mainstream of transformation thought, the principal critic of the Army's course was Col. Douglas Macgregor, author of the 1997 book *Breaking the Phalanx*. While superficially he disagreed with the vision of Force XXI and the Army After Next, his principal critique was not that the Army was preparing to fight the wrong type of war but that it was too focused on technology and not going far enough in reimagining the Army's organization and doctrine to truly embrace the promise of the RMA and transform the face of warfare: "Because it is fashionable to speak of the decisive role technology plays in the 'revolution in military affairs' (RMA), much less attention is paid in military circles to the complex set of relationships that actually link technology's military potential to strategy and organization for combat (doctrine)."[131] However rebellious Colonel Macgregor may have felt while he was writing these words, this critique was well within the pale of contemporary military thought in the 1990s, and the debate he raised was one that senior leaders of the Army were very comfortable having. It was certainly much more palatable to them than was a debate over whether the Army should focus on getting better at the low-intensity conflicts in which it was currently very much engaged or whether it should instead continue to optimize itself for some—admittedly unlikely—hypothetical future high-intensity conflict against a peer competitor. Not surprisingly, General Reimer distributed Macgregor's book to his staff and senior leaders and directed that they all read it.[132]

Bosnia-Herzegovina, Part Two: The 1997 QDR and Beyond

Just as the IFOR mission had been promised to only last a year, the SFOR mission had been promised to only last eighteen months. This too proved to be wishful thinking.[133] As hard as the Army tried to ignore the low-intensity conflict in Bosnia-Herzegovina or its costly lessons, the war continued with no end in sight.

In 1997 the German Army sent a contingent to support SFOR, the first deployment of German troops outside Germany since World War II. This dramatic move was driven by Germany's desire to repatriate Bosnian refugees temporarily living in Germany. That same year, the 1st Armored Division again assumed Task Force Eagle. In 1998 the mission was renamed Operation Joint Forge and the 1st Armored Division was replaced by the 1st Cavalry Division and then the 10th Mountain Division before returning for a third rotation. The mission would continue, with no end in sight, until well into the Iraq War. The mission in Bosnia-Herzegovina did, however, steadily decrease in size: by 1999 the U.S. contingent in SFOR was only four thousand troops.[134]

As the low-intensity conflict in Bosnia-Herzegovina dragged on, more future leaders of the twenty-first-century Army would be exposed to this unfamiliar form of warfare. Brig. Gen. James Dubik, who would go on to serve as the deputy commanding general for transformation at TRADOC, served as the assistant division commander for support for the 1st Cavalry Division when it assumed the role of Task Force Eagle in Bosnia-Herzegovina from 1998 to 1999. Lt. Col. Tony Cucolo would go on to be the commander of the 3rd Infantry Division in Iraq and then the commandant of the U.S. Army War College, but in 1997 he commanded the 3rd Battalion, 5th Cavalry Regiment in Bosnia-Herzegovina. Lt. Col. Kevin Benson served as a battalion commander with the 1st Cavalry Division during its deployment to Bosnia-Herzegovina;[135] he would go on to become director of plans for the Third Army during the planning for the 2003 invasion of Iraq and, later, the director of SAMS.

Leadership changes at the senior levels of the international effort in Bosnia-Herzegovina would have a dramatic effect on the low-intensity conflict in subsequent years. In 1997 USAREUR commander and future chief of staff of the Army, Gen. Eric Shinseki, took over as the commander of SFOR. At the same time, Gen. Wesley Clark—a longtime friend of President Bill Clinton—took over as Supreme Allied Commander Europe. Back in Washington, DC, Madeleine Albright succeeded Warren Christopher as secretary of state.[136]

With this changing of the guard came a new, more aggressive attitude toward engaging the political dimension of the low-intensity conflict in Bosnia-Herzegovina. In a private communication, Clark explained to Shinseki that he wanted to "split the Serbs," reducing the influence of Belgrade on Bosnian Serbs.[137] Clark also wanted to accelerate the return of large numbers of refugees and internally displaced people, which would clearly upset the tenuous stability that IFOR and SFOR had produced in the country. In addition, General Clark insisted that U.S. Army forces in Bosnia-Herzegovina begin rounding up indicted war criminals.

As he assumed command of SFOR, General Shinseki made changes of his own. The Army embarked on an effort to restructure the Bosnian police forces and reduce the sectarian abuses being committed by the Interior Ministry's Special Police. Shinseki also pursued Clark's goal of driving a wedge between the Bosnian Serbs and Belgrade. In September 1997—with the approval of the NAC and the High Representative for Bosnia and Herzegovina—SFOR seized five Bosnian Serb television stations, ostensibly for speaking out against the Dayton Accords.[138]

As SFOR raised the pressure for a political settlement, sectarian tensions predictably rose. In late 1997, violent riots and clashes broke out in Brčko and Bijeljina in the U.S. sector. In part, these riots were about repatriation. But they were also driven by a competition between two rival civilian leaders inside Republika Srpska. In April 1998, Bosnian Croats began a series of riots and protests in response to the return of Bosnian Serb refugees to the town of Drvar in the Canadian sector of MND-Southwest. Tensions rose to the point that the Canadians considered the operation a nascent counterinsurgency—but without any direct engagement between the insurgents and the conventional forces of SFOR. Tensions ultimately resulted in the burning by Bosnian Croats of fifty homes owned by Bosnian Serbs.[139]

Despite this increased pressure, SFOR never found a political solution to the low-intensity conflict in Bosnia-Herzegovina. In December 2004 the SFOR mission finally ended, but NATO troops were simply replaced by European Union peacekeeping troops. As of this writing, those peacekeepers are still stationed in Bosnia-Herzegovina, a hedge against the persistent threat of a resumption of communal violence.[140]

By the late 1990s there was almost universal agreement—among both low-intensity conflict observers and Army transformers—that the future held more such conflicts. As Army colonel Tony Cucolo noted, "The great majority of armed conflicts in the world are *intra*national." He added, "There

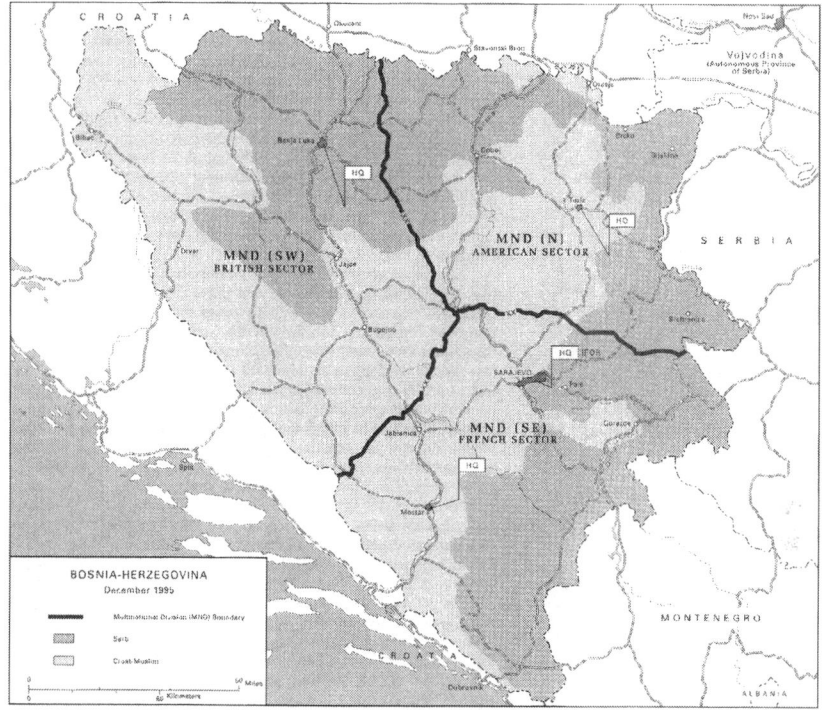

Figure 3.2. Operation Joint Endeavor. *Source:* U.S. Army Center for Military History, *Bosnia-Herzegovina: The U.S. Army's Role in Peace Enforcement Operations 1995–2004*, 15.

is little argument that the United States will continue to intervene selectively in intranational conflicts. . . . What is debated is how, not whether, the US military will participate in peace operations."[141]

As it became clear that the low-intensity conflict in Bosnia-Herzegovina would not be ending anytime soon and as the Army continued to rotate units through the operation, it settled into a routine cycle of preparing units for deployment. The Army established standardized predeployment low-intensity conflict training at one of its combat training centers, the Combat Maneuver Training Center (CMTC) in Hohenfels, Germany. CMTC observer/controllers (O/Cs) educated themselves in peace operations by studying writings by low-intensity conflict observers and through trips to Bosnia-Herzegovina to observe U.S. Army operations there. During training exercises for units deploying to Operation Joint Guard and, later, Operation Joint Forge, the CMTC employed soldiers who had previously served as opposing forces in high-intensity conflict training as civilian role-players, pretending to be

Bosnian civilians. The CMTC also employed linguists to train leaders to interact with the population through interpreters.[142] As it turned out, all of these measures were another eerie preview of the war in Iraq.

Despite this dramatic shift in the training focus at the CMTC, the broader U.S. Army still refused to institutionalize the lessons of low-intensity conflict. For instance, describing in the *Military Review* the 1st Cavalry Division's predeployment training for its rotation to Bosnia-Herzegovina, the commander of the 3rd BCT, 1st Cavalry Division, Col. Ben Freakley (who would go on to command the 10th Mountain Division and Combined Joint Task Force 76 in Afghanistan in 2005), acknowledged, "Conduct of PSO [peace and stability operations] is a dramatic shift in the 1st [Cavalry Division's] METL [mission essential task list]," adding, "We all had to learn and master new and different tasks."[143] This admission was not surprising, since the heavy 1st Cavalry Division had been carefully insulated from the lessons of low-intensity conflict throughout the 1990s, protected as it prepared for an imminent high-intensity conflict against an imagined future peer competitor.

Yet when faced with the challenge of preparing this force for a very real, truly imminent low-intensity conflict in Bosnia-Herzegovina, the U.S. Army refused to change the training at either of its combat training centers in the continental United States: the National Training Center at Fort Irwin, California, or the Joint Readiness Training Center (JRTC) at Fort Polk, Louisiana. Instead it sent the leaders of the 1st Cavalry Division to the CMTC to "right seat rid[e]" with the O/Cs as they trained USAREUR units for their own deployments. The 1st Cavalry Division then had to train itself for Operation Joint Forge the best it could.[144] This was done presumably to minimize the impact of the low-intensity conflict in Bosnia-Herzegovina, preventing it from becoming a larger distraction to Army transformation.

As operations in Bosnia-Herzegovina ground on and the mission expanded, the Army became ever more aware of its unpreparedness to operate in and among civilian populaces, an inescapable facet of low-intensity conflicts. The contributing militaries of SFOR—the U.S. Army chief among them— had been reluctant to enforce law and order for fear of jeopardizing their neutrality. The International Police Task Force, in turn, had proven incapable of building a suitable local police force that would not use its power in a sectarian fashion. As riots began to break out across the country in response to repatriation and the capture of indicted war criminals, soldiers struggled to apply rules of engagement in the face of angry crowds. In late 1997, during the violent riots in Brčko and Bijeljina, soldiers were remarkably restrained in their response to civilian violence. Five American soldiers received Purple Heart medals for wounds received from a demonstrator with a rifle, but no civilians were shot by American soldiers in response.[145]

What the Army lacked—despite having operated in the country for over two years—was nonlethal means to compel angry civilians to cease their violence. The Army, unwilling to venture into tasks it considered civilian responsibilities, lacked basic riot control gear such as shields, batons, and face masks to quell civil disturbances. Faced with the dilemma of growing civil unrest, Task Force Eagle finally relented, implementing a training program and distributing foam batons, pellets and adapters for rifles, foam grenades, and dye markers. Yet because of national caveats SFOR did not implement these changes with other contributing nations across Bosnia-Herzegovina. For instance, Canadian law strictly prohibited its military from quelling civil disturbances—in Canada or any other country. Thus, the Canadians failed to implement these measures in their own sector and were similarly unprepared when riots broke out in Drvar.[146]

Rising tensions in Drvar forced the NAC—exasperated with SFOR's inability to provide adequate public safety in Bosnia-Herzegovina—to create a paramilitary gendarmerie called the Multinational Security Unit and consisting of international police units—the Italian Carabinieri and Spanish Civil Guard chief among them. This new force, of around eight hundred police officers, finally filled the gap between reluctant international military forces and sectarian local police and began to quash the escalating sectarian tensions in Bosnia-Herzegovina.[147] While effective, this move had the unintended consequence of relieving the Army of the responsibility for maintaining civil order in Bosnia-Herzegovina, retarding further learning in this area.

As the crises of 1997 and 1998 abated, the Army began to contemplate its continuing problems with balancing the need to interact with the populace against excessive concern for force protection. During his tenure as SFOR commander from October 1998 to 1999, General Montgomery "Monty" Meigs encouraged division commanders to reduce force protection measures—allowing their soldiers to wear soft caps instead of helmets when outside base camps but inside vehicles. Division commanders resisted, still concerned about avoiding casualties. Notably, while Meigs wanted his division commanders to assume this risk during his tenure as SFOR commander, when Meigs was commander of the 3rd Infantry Division in Bosnia-Herzegovina, he instituted none of these reduced force protection measures.[148]

Col. Tony Cucolo derided the Army's overconcern with force protection: "Being overly concerned about taking casualties could lead to a 'bunker mentality.' . . . A bunker mentality . . . sacrifices the opportunity to engage leaders from all sides, which could eventually deprive the intervention force of credibility, essentially rendering those 'in the bunker' . . . mere spectators to peacebuilding efforts."[149] For Cucolo, this risk aversion jeopardized both the Army's understanding of the environment and its ability to engage in the political dimension of a low-intensity conflict.

This aversion to the risk of casualties among the Army's senior leaders stemmed from a lack of understanding of the importance of interacting with the populace and—by extension—a lack of understanding of the political dimension of low-intensity conflict. In the July–August 1997 edition of *Military Review*, Lt. Col. Walter Kretchik railed against these force protection issues, comparing the much-reduced posture of foreign troops to the heightened force protection posture of U.S. soldiers. But he failed to explain that the cost of this heightened force protection was reduced interaction with the populace, something that was essential if the U.S. Army was going to forge a political settlement in Bosnia-Herzegovina and extricate itself from the conflict.[150] In this light, it is not surprising that division commanders refused to reduce the force protection posture of their forces; why risk casualties if you don't see or understand a cost to excessive force protection?

Colonel Cucolo did understand the essentially political dimension of the low-intensity conflict in Bosnia-Herzegovina. Generalizing from this conflict to a wide range of "intrastate conflicts," he explained in a spring 1999 article in *Parameters* that conflict was

> . . . . embedded in long-standing relationships among people who live in close geographic proximity. They act on deep-seated (often cross-generational) fears, passions, and hatred. In these conflicts, people seek security in and identify with something close to their control, perhaps a region, a religion, or an ethnic alignment.[151]

Cucolo noted that the "prevailing attitude among some senior leaders" was that solving political problems in Bosnia-Herzegovina "was 'out of [Task Force Eagle's] lane.'" He added, "In-depth preparation of junior leaders in [negotiation] skills was very low on the training priority list" prior to the deployment. Admittedly, battalion commanders and higher Army leaders did get some rudimentary "political-military training" that included "negotiation techniques" and the culture of the different sects in Bosnia before their deployment.[152] But more junior officers and NCOs did not receive this training.

Moreover, this rudimentary training proved inadequate as the political dimension pervaded at every echelon of the conflict. As Cucolo wrote, "Acting as [negotiators] was a sustained role throughout the battalion's deployment. . . . What was surprising was the extent to which soldiers and leaders at platoon, company, and battalion level became involved as third-party actors."[153] Without the willingness and training to engage in the political dimension of low-intensity conflict, Task Force Eagle couldn't solve the political problems that threatened renewed violence.

Surprisingly, Gen. Dennis Reimer, who generally seemed oblivious to the political dimension of low-intensity conflict or the necessity to engage this dimension to end such conflicts, was pleased by Cucolo's success in Brčko, noting that Cucolo "did a wonderful job." Reimer was particularly pleased with Cucolo's technique of hiring multisectarian teams to do work and only paying them if they worked together.[154] While he seemed to miss the political foundations of this technique, he also didn't argue with success.

Colonel Cucolo was correct in claiming that senior leaders did not believe the Army should be engaged in the political dimensions of the conflict. In an unpublished "thought piece" written to fellow general officer Lt. Gen. Thomas Burnette, Brig. Gen. James Dubik confronted the problem of how to "reduce the time our military forces . . . have to be involved or the size of the military force required after initial intervention" in low-intensity conflicts. His insights were based on his experiences in Haiti as a brigade commander and in Bosnia-Herzegovina—from where he wrote the document—as an assistant division commander. Confronting "the difficulties with 'getting out' of an intervention like Haiti and [Bosnia]," Dubik believed the obstacle was "that even after we 'take down' the military and overt para-military [forces], the governments in question retain power and frustrate the civil capability from developing." Under present Army doctrine, Dubik believed that the options were either to "end up being 'trapped' into expanding our mission in order to make it work, or just 'declare victory and leave.'"[155]

Discussing the extended time that Army forces had been required to remain in Haiti and Bosnia-Herzegovina, Dubik suggested that the initial entry force in such operations be followed by some hypothetical "National Judicial Force" (NJF) with some law enforcement capability to deal with "covert para-military [and] criminal element[s]." But this force would also have what Dubik considered nonmilitary capabilities, such as dealing with "police, judicial, and governmental process[es]" to wrest the nonmilitary, illegal levers of power from the leaders that the United States wished to supplant. Brigadier General Dubik envisioned this force being a completely American equivalent to the "local International Police Monitors (in Haiti) and International Police Task Force (in Bosnia)" consisting of both judges and police. This would allow "BOTH the military to leave, AND the civil capability to have the cover it needs." He explicitly stated that an NJF would resolve the problems of "mission creep" and relieve the military of the responsibility to secure a "stability force" of other U.S. and international civilian agencies that he believed had responsibility—rather than the Army—for achieving political settlements in low-intensity conflicts.[156] This solution was an attempt to relieve the Army of the responsibility to grapple with both the multinational *and* the political dimensions of low-intensity conflict.

Yet at the core of this was wrong-headed thinking about low-intensity conflict as "operations other than war." Dubik believed that once an initial intervention force took "away the government's ability to use its military or para-military" and the "government loses control and cannot resist," the Army didn't "need force any more" and could "start sending . . . military forces home, and begin the shift from force to civil capability."[157] His core misconception was that force or the threat of force is only useful against the organized army of a nation and has no utility against armed civilians, militia forces, criminal gangs, or insurgents. Or, perhaps, his core misconception was that fighting such elements was beneath the dignity of the U.S. Army.

Dubik had been a direct participant in Operation Uphold Democracy in Haiti and Operation Joint Forge in Bosnia-Herzegovina. He should have known better. He wrote in this "thought piece" that, in both Haiti and Bosnia, the U.S. Army was able to "take away the government's ability to use its military and paramilitary . . . very quickly." He also acknowledged that "each government merely shifted to [a] different form of force to remain in power. These forms are: covert para-military; the criminal element, and its police, judicial and governmental processes."[158] But Dubik seemed oblivious to the possibility that the Army might change its methods of operation to use force or the threat thereof to neutralize these lesser threats.

Equally revealing are the reasons underlying Brigadier General Dubik's assertion that an NJF was more suitable for these postconflict low-intensity conflict tasks. In a clear reference to the requirement to hunt down indicted war criminals in Bosnia-Herzegovina that had recently been placed on U.S. forces, Dubik explained that, after the initial, high-intensity conflict phase, "those that have been coerced into signing an agreement . . . remain in power." He explained, "Military forces intervening are not able to take them out of power in any legally legitimate and acceptable way."[159] Dubik's NJF was nonmilitary and, thus, suitable to execute a task that was somehow inappropriate for Army forces: "By turning intelligence into evidence the NJF could get at just that strata of corruption that is both outside of the military's area of expertise and frustrating to the overall strategy objective we seek to attain."[160] It is telling that Dubik's solution to the problem was the arrival of some other force to assume the duty of navigating the political dimensions of the low-intensity conflict. Apparently it never occurred to him that the Army—the force already present in the conflict zone, immersed in the problem, and with the military capability to compel compliance—might be capable of taking on this task without help from some other agency, let alone that it might in fact be a component of the Army's self-proclaimed highest purpose: to fight and win the nation's wars.

Colonel Cucolo explicitly rejected such thinking. He noted, "Current thinking on the matter, reinforced in military doctrine, supports participation of the armed forces to create an environment in which other organizations do the peacebuilding once there is a peace on which to build." He also acknowledged the reasons Army transformers frequently gave for the Army to avoid involvement in low-intensity conflicts: "Such involvement has nothing to do with their warfighting skills, erodes their readiness, and is done better by organizations that specialize in humanitarian and developmental work." But, Cucolo insisted, "Intrastate conflict resolution and peace operations in such an environment require military involvement as a third party for mission success." He continued, "Notwithstanding the theorists and their rhetorical exercises, on the ground there are no sharp divisions among peace enforcement, peacekeeping, and peacebuilding," adding that "military units . . . dispersed among the warring factions with a mandate to implement the Dayton Peace Agreement in the face of pervasive ethnic hatred" were better positioned than any outside agency to engage the political dimension of the low-intensity conflict in Bosnia-Herzegovina. Only "military units committed to the intervention" could compel warring parties to reach a settlement in such an environment. For Cucolo, the consequences of this fact were clear: "Therefore, the relevant awareness and skills should be added to the military's warfighting ability."[161]

Yet even Colonel Cucolo, who fully grasped the consequences of the political dimension of low-intensity conflict, failed to reach the highest level of understanding: an acknowledgment that the Army must back a winner in the conflict in order to bring it to an acceptable conclusion. He did acknowledge that, when acting as an intermediary between warring parties in Bosnia-Herzegovina, Army leaders needed to be "quasi-mediator[s]," and "not neutral," but instead with their "own interests and point of view." While facilitating negotiations, he added, Army negotiators needed to be "an advocate for [their] side." This promising start notwithstanding, Cucolo fell into the same cognitive trap as many of his contemporaries, contending that the Army's only "side" was, in fact, "absolute impartiality." Because "parties will watch closely for any reason, real or perceived, to question the fairness of implementation," he explained, the Army's imperative in facilitating negotiations needed to be the "'No Winner' factor." He made this exhortation even though he acknowledged, "The distinction was lost on Serbs in and out of uniform; what they perceived was the US Army in Bosnia training their Bosniac and Croat adversaries."[162]

This passage illustrates just how engrained the principle of neutrality was in the Army's thinking about low-intensity conflict; even someone like Col. Tony Cucolo, who bucked the Army's traditional deference to hierarchy to contradict senior Army leaders and instead contend that the political dimension of low-intensity conflict *was* the domain of the U.S. Army, still couldn't see past the cognitive obstacle of neutrality to a full understanding of how to fight and win such conflicts.

Low-Intensity Conflict Thought through 1996 and Beyond

While soldiers and leaders struggled to bring the low-intensity conflict in Bosnia-Herzegovina to a close, the Army was busy insulating its leaders from the lessons being learned in Operations Joint Guard and Joint Forge. The Army's formal education gradually phased out dedicated classes on low-intensity conflict. In the 1997–98 academic year at the Command and General Staff College (CGSC), the number of hours dedicated to low-intensity conflict actually went up, from forty-five to fifty-five hours of the five-hundred-hour curriculum. But in the 1998–99 academic year, the specialized block of instruction on low-intensity conflict was eliminated and the MOOTW instruction was rolled into a 128-hour course that also covered major theater war. In the CGSC course guide that year, MOOTW topics only received one sentence of treatment in a page-long description of this course on Army operations.[163] Despite this attempt to suppress the lessons of the low-intensity conflicts in which the Army was engaged, conflict observers continued to provide insights in the pages of the Army's academic journals without interruption.

Both Army transformers and low-intensity conflict observers continued to predict that the future held many more such conflicts. In 1996 Steven

Metz and several of his colleagues at the SSI penned the U.S. Army War College's annual predictions for the future strategic environment. They predicted an increased demand for "military activities other than war such as peace operations," warning, "The Army will have to decide whether warfighting is *the* function for which it exists or simply one function (albeit an important one) among several." In a radical departure from Army transformation orthodoxy, they added that "warfighting with armor-heavy divisions . . . will probably become no more than the co-equal of other tasks, and may eventually become a secondary mission if enemies like Iraq and North Korea reform or collapse." Metz and his colleagues added that it was a "myth" that "the United States can disengage from the conflict prone parts of the world, thereby obviating the need for direct involvement." Admittedly, they also believed that the idea that "the world will see no more conventional wars" was also a myth. Elsewhere, however, Metz emphasized that in most conceivable futures the possibility of "traditional state-on-state warfare is insignificant."[164]

Army transformers agreed that the future held more low-intensity conflicts. Even *Army Vision 2010* warned that the future strategic environment would continue to produce a "near-term increased demand for operations on the lower end of the spectrum of crisis," including "humanitarian relief, peacekeeping, peacemaking, etc." The 1997 QDR report likewise warned that low-intensity conflicts, especially peace operations, would dominate military operations for the immediate future, noting that "U.S. participation in smaller-scale contingency operations" would "still likely pose the most frequent challenge for U.S. forces through 2015 and may require significant commitments of forces, both active and Reserve." Among the specific types of operations in which the Army could expect to continue to engage, the QDR report explained, were "noncombatant evacuation operations . . . peace enforcement . . . peacekeeping, humanitarian assistance, and disaster relief." Even the Army After Next project noted that researchers believed that because of "ethnic rivalries, nationalism, religion-based antagonisms, and competition for scarce economic resources, including water," low-intensity conflicts would continue "and perhaps intensify" into 2020 and beyond.[165]

While both low-intensity conflict observers and Army transformers agreed that the future held more such conflicts, their prescriptions for dealing with this reality varied greatly. In a 1996 *Military Review* article, John B. Hunt wrote, "The Army is developing impressive new technical means to speed information acquisition, storage, retrieval and distribution." He warned, however, that they would be of little use in low-intensity conflicts of the future. Additionally he noted, "Any attempt to conduct such an operation by conventional AirLand Battle methods has little chance for success. It might be worthwhile to reconsider the Somalia experience in this light."[166]

For the chief of staff of the Army, Gen. Dennis Reimer, the continued prevalence of low-intensity conflict, "for which the Army does much of the 'heavy lifting,'" was an unwanted but unavoidable distraction from readiness for the next great power war. As late as 1999 Reimer continued to warn, "The Army can't take a time out from readiness" because, he lamented, "Time is not necessarily on our side."[167] For Army transformers the imminent high-intensity conflict against an imagined future peer competitor always loomed just over the horizon.

Many observers continued to argue that the Army was not manned, trained, equipped, or organized for low-intensity conflicts. As Ralph Peters wrote in an article for *Parameters*, "The deployment of a reinforced brigade cripples multiple corps" yet the Army refused to reorganize itself for low-intensity conflict. As a result, Peters added, for each low-intensity conflict "the cost to our military establishment, already slimmed to fragility and poorly structured for missions short of war, can be exorbitant." For Peters, the conclusion was simple: the Army needed to begin "reforming [its] thought, doctrine, and training to better reflect the world as it is and will be."[168]

The Army seemed to concede the point that it was not trained in such conflict when it executed a low-intensity conflict rotation for the Warrior Brigade (2nd BCT) of the 25th Infantry Division from Hawaii at the JRTC in 1994. The scenario was a peace enforcement mission in which U.S. forces controlled a buffer zone between belligerents, conducted "search and attack operations" to find and destroy hostile insurgents and dealt with UN representatives. The tasks were sufficiently unfamiliar that, once it was over, the exercise warranted an article in *Military Review*. But ultimately the event failed to train the key elements of low-intensity conflicts: operating in and among the populace and engaging the political dimension of the conflict. There were only a handful of role-players acting as civilians on the battlefield, and the civilian government was barely replicated. Moreover, the "peace enforcement" phase of the exercise was preceded by a decidedly high-intensity helicopter-borne air assault into the training area and followed by an equally high-intensity "conventional military" operation to repel an attack by and launch a counterattack against an armored opposing force.[169] While it purported to be a low-intensity conflict, in reality this JRTC rotation differed little in substance from most other JRTC rotations; the Army was paying lip service to training for low-intensity conflict while continuing to focus on increasing its high-intensity conflict capacity.

But some still drew lessons from this rotation. Writing about it, Col. Charles H. Swannack Jr. conceded that the Army was not properly trained or educated for low-intensity conflict. While he believed that it needed to keep its "primary focus on warfighting capabilities," Swannack also believed that at least those Army units designated to conduct specific peace enforcement operations needed to "make certain intellectual adjustments to adapt to this convoluted military environment" and required additional "mental preparation and practical experience necessary to perform future PE [peace enforcement] operations." Specifically, Swannack believed that Army leaders lacked "negotiation skills, knowledge of foreign cultures and ability to relate to NGOs." He also believed that leaders deploying to low-intensity conflict needed special education and that their units should train "on constabulary tasks, such as setting up roadblocks and checkpoints, patrolling in urban areas and aiding civilian refugees."[170]

In fact, even the chief of staff of the Army, General Reimer, admitted that Army units were not trained or prepared for low-intensity conflicts. In a postretirement interview he said, "These new missions required a lot of train-up to make sure that the people we were putting into these situations were prepared." He added that, because the Army wasn't organized for low-intensity conflict, "we also had to put together headquarters and staffs that didn't exist to command and control these task forces." Reimer finally acknowledged that many of the most needed capabilities in low-intensity conflict resided in the reserve component, where they were hardest to deploy.[171]

By the second half of the 1990s, many low-intensity conflict observers were beginning to ascribe cultural reasons to the Army's refusal to adapt to such conflict. In his 1996 *Military Review* article, John B. Hunt wrote that "the institutional Army is not eager to embrace" low-intensity conflict. "It is easier for the Army to continue in familiar ways, no matter what missions it is actually performing and whether or not current operating methods bring success," he said.[172] In another *Military Review* article Lawrence Yates wrote, "There are officers who are convinced that, for one reason or another, military history is largely irrelevant to their current concerns and operations, whether conventional or unorthodox." He noted, "In situations where there is no hostile force, there may be a tendency on the part of conventionally oriented officers and policy makers to create one." Finally, he explained, "The 'warrior' mindset so essential for combat operations can be the source of anger, confusion, frustration and failure when applied unmodified to OOTW [operations other than war]."[173]

Some low-intensity conflict observers also noted lessons about specific capabilities that the Army needed to successfully engage in such conflicts. Sean Maloney—reviewing in a 1996 *Military Review* article the Canadian peacekeeping doctrine to emerge from that country's involvement in Bosnia-Herzegovina, wrote, "National agendas may impinge on PK [peacekeeping] forces' effectiveness," While he did acknowledge that the UN "makes cosmetic attempts to screen out countries with ulterior motives," he also warned, "This cannot be totally successful because, for political reasons, the UN is multilateral and must consider all countries volunteering to participate."[174]

John Hillen of the Heritage Foundation was even more dismissive of the UN in an article for *Parameters* in summer 1996. He noted with satisfaction that the UN secretary-general had acknowledged the previous year that the UN was incapable of providing command and control to peacekeeping operations on the scale of Somalia or Bosnia-Herzegovina. Hillen called this "an explicit recognition of the failure of the UN's complex military operations and ambitious mandates of the peacekeeping missions in the former Yugoslavia and Somalia over the previous three years." He added, "The UN system is inherently dysfunctional in regard to mobilizing and controlling complex military operations in dangerous environments." Hillen appeared to endorse the approach of U.S. forces in Haiti and Bosnia-Herzegovina, where the United States demanded a lead role as the price for its participation in the low-intensity conflict.[175]

Hillen's thinking seemed more muddled when he tried to diagnose the reasons for the UN's dysfunction, blaming its ineffectiveness on "its political nature."[176] He seemed to claim that the UN had failed at military operations because it had attempted to engage in the political dimension of low-intensity conflicts. This passage seemed to perpetuate the myth that military forces had no role engaging in politics. But it was also, frankly, confusing, since many Army transformers considered operations like Haiti a success precisely *because* the Army had successfully dumped the political dimension of the low-intensity conflict on the UN.

Hillen's other explanation for the failure of UN-led low-intensity conflict missions was more plausible: "In dangerous . . . missions, the propensity for national contingents to 'phone home' rose accordingly." He added, "Competition from national chains of command was replete . . . with units in Cambodia, the former Yugoslavia, and Somalia at times flatly refusing to obey the UN chain of command."[177] The problem of national caveats and national chains of command outside the chain of command within the low-intensity conflicts itself was, indeed, a severe difficulty in multinational operations. Yet this problem was not unique to UN operations, as the United States would soon find out with Russia in Bosnia-Herzegovina and Kosovo.

As has been mentioned repeatedly throughout this book, Army transformers argued year after year that low-intensity conflict was merely a lesser included task of high-intensity conflict. As chief of staff, General Reimer provided one expression of this idea in a 1997 *Military Review* article describing the Army After Next. He believed that the Army units of 2020 needed to be "versatile," able to "perform multiple, disparate functions" and "flexible" to "deftly transition between the use of lethal and nonlethal force, as the situation dictates."[178]

In the very next issue of *Military Review*, Army colonel Steven Schook of the Joint Staff J7 attacked this idea of "versatility," though he challenged its use in the 1993 FM 100-5, *Operations*, rather than its use by the chief of staff of the Army. FM 100-5 defined versatility as the "ability of tactical units to adapt to different missions and tasks, some of which may not be on unit mission-essential task lists." The manual used "versatility" as a justification to ignore preparation for low-intensity conflicts. Schook concluded that "versatility . . . is really no more than a hiding place for a less-than-total commitment to OOTW." He wrote that the Army could "no longer delay the tough decisions"; it needed to dedicate some portion of its forces to low-intensity conflict.[179]

Throughout the 1990s other Army transformers argued that committing the Army to low-intensity conflicts hurt their readiness for high-intensity conflict. General Reimer, after leaving office, complained that in forming task forces for low-intensity conflicts the Army "had to pull people from other organizations to make a headquarters in Bosnia." This broke units and placed a strain on leaders, especially at the junior officer and NCO levels.[180]

Ralph Peters rejected this thinking, however: "If these peacekeeping and peacemaking, policing and observing missions [hurt] our readiness, it is because we have done nothing practical to fight the cancer." He added, "We fight against the missions, instead of facing the shortcomings inherent in a force that has contracted, but has not changed with the times." Peters accepted that the Army "continues to need heavy forces," but argued, "Their current configuration is fit for a museum, not for our likely missions." He concluded, "We prepare for our ideal missions, while the real missions must be improvised at great expense to readiness, unit integrity, and the quality of life of our service members."[181]

For many years Army transformers had argued that the Army should not only avoid preparing for low-intensity conflicts but also that it should not even be committed to these types of interventions. Ralph Peters insisted that low-intensity conflict *was* the Army's job—but a job that it was refusing to prepare to do:

> Our military is determined to be unprepared for missions it does not want, as if the lack of preparedness might prevent our going. We are like children who refuse to get dressed for school. . . .
>
> When the President is out of options and key interest groups or foreign leaders are clamoring for American action, we are going to go to school.

Peters added, "Military readiness is essential—but the military must be ready for reality, not for its fantasy war."[182]

Surprisingly, General Reimer agreed with Peters, but not for reasons Peters might have liked. In an interview after leaving office, Reimer unsurprisingly agreed with those who claimed that the Army's primary mission was to win the nation's wars. But he differed from those who argued that the Army should stay trained and ready for war. "And that meant high intensity conventional combat," Reimer explained, "and that's what we need to train against and that's all we should do." Instead he believed that the Army had to acquiesce to participating in low-intensity conflicts operations in order "to be relevant to the needs of the nation." His reasoning was simple: "From my perspective I was convinced that we needed to be relevant or accept the fact that we were going to be fairly small." Speaking specifically to the mission in Bosnia-Herzegovina, he added, "If you don't do those types of operations then somebody is going to come along and say, 'Why are we keeping this major force if we never use it?'"[183] (This is, in fact, almost exactly the same question Madeleine Albright had asked Gen. Colin Powell a few years earlier). For General Reimer, participating in low-intensity conflicts—while undesirable and outside the domain of legitimate uses for the Army—was at least a means to justify retaining a large standing Army.

By the 1990s, a new strain of thought was beginning to emerge among low-intensity conflict observers—the idea that the Army might need to bifurcate in order to create specialized forces for high- and low-intensity conflict, respectively. Steve Metz and his colleagues from SSI argued, "If tasks other than warfighting become more strategically important, the relationship between the Army's warfighting component and its peace operations/conflict resolution/grey area threat component may need radical change, perhaps to the point of separating the two into distinct organizations."[184]

Colonel Schook from the Joint Staff J7 likewise suggested creating forces specialized for high- and low-intensity conflict, organized within "'Holistic' Divisions." First he advocated "adding a light brigade to a heavy division." More radically, he suggested that "the heavy units continue to train for the next MTW [major theater war], while the light units focus on low-intensity conflict and OOTW." Within the division, he suggested, "An additional assistant division commander (ADC) for OOTW would be needed to conduct OOTW operations and direct the necessary training much the way the maneuver ADC does now." He also suggested changing training at the combat training centers, noting that their rotations "must include allies, strategic lift (both air and sea), staff representation from the unified commands and our sister services—either physical participation or as part of a simulation—as well as interagency and nongovernment organizations."[185]

Ultimately Ralph Peters dismissed the bifurcation of the Army into "a two-tier military establishment: ready, fully developed elite forces to fight our wars, and a secondary, cheaper, constabulary military to do the jobs the 'fighters' don't want to do." But the sarcasm with which he presented the idea made it a straw man, one easily knocked down by the "the impossibility of recruiting international garbage collectors . . . inevitable jealousies, and the damage that consequent reductions in the number of combat units would do to our forces." This observation, of course, ignored the empirical evidence; Army units that were deployed to low-intensity conflicts experienced higher rather than lower rates of reenlistment. He also dismissed forces specific to low-intensity conflict on other grounds, writing that "unprepared to conduct sustained combat operations, [they] would not only prove ineffective, but unwanted internationally." Elsewhere in the article he compared them unfavorably to UNPROFOR.[186] A more serious consideration of the proposal—with significantly less snark—may have yielded a more balanced analysis.

Low-intensity conflict observers continued to note that these types of conflicts happened in and among populations and that this facet of these conflicts had profound impacts on its nature. One such impact was the requirement for a nuanced application of force. In his 1996 *Military Review* article, Sean Maloney expressed the Canadian view on the use of force in low-intensity conflicts: force was "a last resort." But he also connected this to the political dimensions of the conflict, writing, "Long-term solutions to conflicts in PK cannot be imposed by force."[187]

Colonel Swannack likewise acknowledged that peace enforcement "requires soldiers to be more circumspect and discerning of their immediate environment to avoid injuring noncombatants and escalating the level of violence." He noted that this was, for many, "an unfamiliar psychological mind-set." To orient soldiers, units had to adjust their training program "to address peace operation-peculiar skills, reinforce disciplined ROE [rules of engagement] application and build soldiers' confidence in their ability to operate under diverse circumstances."[188]

Low-intensity conflict observers continued to emphasize the political dimension of such conflict. Though it was a bit belated, in his 1996 *Military Review* article Hunt finally complained in print about the 1993 edition of FM 100-5. Hunt—a longtime advocate of more focus on low-intensity conflict and, at the time, a member of the Concepts and Doctrine Directorate at Fort Leavenworth—wrote of FM 100-5, "The most serious fault is that the LIC [low-intensity conflict] imperatives are deleted, including 'political dominance,' the main imperative from which all others are derived." He explained that the political dimension was preeminent in low-intensity conflict because an adversary "does not merely hide his forces among the people; because of his political programs, his forces *are* the people."[189]

Maloney likewise highlighted the political dimension of such conflicts. Describing the interplay between politics and violence, he wrote, "Factions will use a 'fight-talk, talk-fight' technique to maneuver for geographic space and time, harass each other or garner media attention for their causes." This was reminiscent of the "fighting while negotiating" tactics of which the North Vietnamese were frequently accused during the Vietnam War. Maloney noted that the emerging Canadian doctrine on low-intensity conflict insisted that a military force must engage in negotiations: "The primary method of developing effective solutions to a conflict is negotiation."[190]

The Canadian peacekeeping doctrine was not all helpful. Like the U.S. Army, the Canadian Army had enshrined the idea of "mission creep," implying either that a peacekeeping mission could not legitimately be expanded after it began or that one could limit the political dimension of a conflict by artificially constraining the mission of the military force. Maloney posited a corollary: "Resist national aims if they do not coincide with the PK mission's objectives."[191] This likewise sounded precariously close to artificially limiting the mission to avoid the messy political dimension of the low-intensity conflict.

Lawrence Yates demonstrated a clearer understanding of what some called mission creep when he wrote, "Mission creep is currently portrayed as a negative phenomenon that can be eliminated through thorough planning and careful analysis." But Yates disagreed, writing, "Generally speaking, mission creep accrues more to the logic of a dynamic situation in which the success of the original mission depends on picking up additional missions." He concluded, "Rather than announce in advance that there will be no mission creep during a given operation, it is best simply to be prepared for it and make the necessary adjustments when it occurs."[192]

In addition to their flawed thinking on mission creep, the Canadians also struggled to move past an understanding of the political dimension of low-intensity conflict to the highest level of understanding: that an army needed to take sides in the conflict to bring it to a conclusion. One of their principles, according to Maloney, was impartiality. He wrote that Canadian doctrine insisted that impartiality "must be demonstrated by PK forces at all times and in all situations. . . . No belligerent point of view can be favored. To do so puts the mission in jeopardy. Peacekeepers must be patient and fair at all times."[193]

Asymmetry and Urban Operations, 1998–1999

As the Army continued to fight the low-intensity conflict in Bosnia-Herzegovina and the Army After Next program got underway, two points of convergence between Army transformers and low-intensity conflict observers began to come into focus: asymmetry and urban operations. While both camps understood these as problems for the future of the Army, each side saw both the problems and the solutions in very different ways.

Within the American defense establishment in 1996, the zenith of thought on the RMA was represented by the book *Shock and Awe: Achieving Rapid Dominance* by retired military officers Harlan Ullman and James Wade. Pointing to the Gulf War air campaign and envisioning a straight-line progression of military-technological advancement, they envisioned a day when the U.S. military could paralyze opponents with surgical strikes at enemy centers of gravity. But even Ullman and Wade acknowledged that some types of enemies would be impervious to this technology; in "situations such as guerilla war . . . most means of employing force to obtain Shock and Awe may simply prove inapplicable." They also acknowledged that as the U.S. military became ever-more capable in the realm of high-intensity conflict, it might "induce an adversary to move to a strategy that attempted to circumvent all this fighting power through other clever or agile means."[194] While Ullman and Wade never connected these two concepts—the idea that low-tech, low-intensity conflict capabilities defeated high-tech, high-intensity conflict means and that an adversary might seek ways to negate the Army's high-tech dominance of high-intensity conflict—they would eventually combine to form the core of asymmetry as it was conceived by low-intensity conflict observers.

In 1996, SSI researchers picked up on this theme in the institute's annual prediction of the future strategic environment. Steven Metz and his colleagues warned that the Army's "unquestioned superiority at mobile armored warfare will decline in strategic significance" as "challengers . . . seek low-tech, asymmetric responses to counterbalance the American advantage."[195] Quoting Brian Nichiporuk and Carl H. Builder, Metz and colleagues warned, "If the Army fixes itself too firmly on fighting and winning the nation's conventional wars as a way to husband its scarce resources, it may find that its market . . . is narrowing." Elsewhere Metz himself noted, "*Desert Storm* is not a prototype for all future wars."[196] Instead the future promised a low-intensity conflict conception of asymmetry—adversaries sidestepping the Army's unquestioned dominance of high-intensity conflict through low-tech, low-intensity conflict means.

As early as 1997 the QDR report began to describe terrorism against the homeland of the United States as an "asymmetric" threat, warning, "U.S. dominance in the conventional military arena may encourage adversaries to use such asymmetric means to attack our forces and interests overseas and Americans at home." But in other places where the report used the term *asymmetry* it was not employed to describe low-intensity conflict capabilities. Instead, the QDR report warned, "an aggressor may seek to avoid direct military confrontation with the United States, using instead means such as terrorism, NBC threats, information warfare, or environmental sabotage to achieve its goals." If these "strategic" means failed, the QDR report posited that an adversary's "asymmetric means" might be directed to "delay or deny U.S. access to critical facilities; disrupt our command, control, communications, and intelligence networks . . . or inflict higher than expected U.S. casualties in an attempt to weaken our national resolve."[197] These ideas—envisioning technological means by which an adversary might asymmetrically challenge the Army—formed the foundation of Army transformers' conception of asymmetry up until the beginning of the war on terror.

In a post-QDR article for *Military Review*, General Reimer broadly defined asymmetry: "Future adversaries will not try to match their forces directly against ours where the United States has overwhelming superiority. . . . Instead, they will exploit perceived political and operational weaknesses, thereby trying to negate US high-tech systems' advantages." Reimer repeated the QDR's list of technological "asymmetric challenges" but added "special operations or clandestine forces" as an additional asymmetric means that adversaries might choose.[198]

Not surprisingly, the conventional Army transformation view of asymmetry dominated the Army After Next experiments in their initial stages. As the commandant of the U.S. Army War College, Maj. Gen. Robert Scales largely drove this bent toward considering only technological asymmetry. In a PowerPoint briefing about the Army After Next, Scales listed "Geopolitical Trends" that would shape the future environment in which the Army After Next would operate. First, he assured listeners that "conflict [would continue] to center around States or State-Like Actors." But he also insisted that "Anti-Access Strategies" would proliferate in the future. This was simply a different expression of the QDR report's warning that adversaries would seek to "delay or deny U.S. access to critical facilities" or General Reimer's warning that enemies would seek asymmetric ways to counter U.S. capabilities, including "attempts to deny regional access."[199]

Some Army transformers did understand that low-intensity conflict was, in and of itself, a form of asymmetry that challenged the Army's unmatched dominance of high-intensity conflict. In a 1998 Army After Next project paper, TRADOC researchers warned that "a resourceful and dedicated foe may force an extended campaign. In such situations, quantity continues to have a quality all its own." Of course, quantity was the one thing that the Army After Next—which massed effects rather than forces—did not have. The paper also noted, "Even with very advanced capabilities, it will remain difficult to rapidly defeat resolute, well-prepared adversaries that possess large territories, enjoy sizeable populations, and fight in imaginative or unconventional ways." To illustrate their point, the researchers warned, "For every Panama and Desert Storm, there is a Korea or Vietnam" (though, of course, Panama was not nearly as neat in its post-high-intensity conflict phase, as this passage implies). They also cautioned that the problem might be unavoidable: "Once conflict termination is achieved, sustained presence by land power will provide the stability and security required for long-term resolution through the political process."[200]

But Army transformers generally focused on technological means that an adversary might use to defeat Army capabilities. In this regard, Army After Next researchers engaged in a great deal of mirror imaging—envisioning an adversary that was very similar to the U.S. Army in approach and technology. Hence, the asymmetries about which they most frequently warned were technological in nature. They warned that adversaries would attack U.S.

satellites to blind the Army After Next, use "cyberwar" to attack its networks, or try to achieve "soft kills" using means such as electromagnetic pulses to knock out U.S. electronics and communications equipment.[201]

TRADOC researchers described the "probable" emergence of "major competitors" who would "acquire significant capabilities to challenge the U.S. regionally and, to a selective degree, globally." The researchers predicted that these major competitors would "develop creative asymmetric strategies and employ niche capabilities aimed at avoiding U.S. strengths and capitalizing on U.S. vulnerabilities." Yet because "conflict [would] take place primarily within the nation-state structure" these challenges to U.S. capabilities would still be of a high-intensity conflict nature. Moreover, TRADOC warned of the proliferation of a number of decidedly symmetric technologies, including "precision fires with increased lethality at extended ranges," computers and surveillance technology, and the "exploitation of space platforms and the global information infrastructure."[202] Army After Next researchers were clearly not worried about future adversaries transforming into guerilla forces to avoid the Army's dominance of high-intensity conflict. Instead they worried about the emergence of a peer competitor with high-tech capabilities of its own that would challenge the Army After Next in a high-intensity conflict.

Army Vision 2010 saw the threat of high-tech asymmetric challenges not just from peer competitors but also from lesser competitors. It warned that even unstable, failing regimes, "while less capable militarily than wealthy democracies, have access to the most advanced military technology." The vision concluded, "This phenomenon creates a new danger in the future, i.e., conflict with a nation having a very sophisticated and asymmetric capability."[203]

The SSI's Robert Bunker penned a March 1998 monograph on the ways in which enemies might asymmetrically defeat the U.S. Army in future conflicts. Much of his thought was obscured by a cloud of technojargon thicker than any ever generated by an Army transformer describing the Army After Next. But buried within his generally indecipherable writings were many eerily prescient predictions that would be realized in the war on terror. He wrote that a future enemy might have "a heavily internetted command structure—its relationships . . . more weblike than hierarchical. . . . It does not field an army which can be decisively defeated in open battle, and its leadership is stealth-masked and transnational." He added, "This competitor is not a nation-state."[204]

In another prescient passage, Bunker warned that this form of enemy could create among foreign populaces in and around an area of operations "the political perception" that it was "a non-nation-state insurgency." He went on to describe a number of asymmetric means that low-tech, low-intensity conflict opponents might use to confound the Army After Next, describing international networking through the civilian internet that is very similar to the techniques used by al-Qaeda and the Islamic State in Iraq and Syria. Bunker provided the blueprints for a "$19.95 Military Robot" that was, in essence, a victim-activated improvised explosive device that, years later, would be immediately recognizable to any soldier from the wars in Afghanistan or Iraq.[205]

Bunker made it clear that this asymmetry would be a direct response to the Army's unmatched prowess in high-intensity conflict, writing that "the decisive American military victory in the Gulf War in 1991 and its ensuing Revolution in Military Affairs (RMA) has not been lost on" adversaries.[206]

It bears repeating that Bunker's observations were buried in a lengthy and largely inscrutable diatribe about "five-dimensional (cyber) warfighting," "stealthing," "cyber-shielding," and "cyber-maneuver." But Bunker *did* predict that the attack from this hypothetical, networked "non-nation-state-insurgency" would begin "with its initial terrorist campaign against [the U.S.] homeland."[207]

Instead of heeding these warnings from transformation's critics, Army transformers focused on developing a solution to the problem of "asymmetry" as they conceived it: as a technological challenge to the Army's technological overmatch in high-intensity conflict. An SSI study on asymmetry acknowledged that urban operations, insurgencies, and guerilla operations were asymmetric challenges to U.S. forces, but the study clouded the issue by also including *Joint Vision 2010*'s conception of asymmetry—based on the technologies of the RMA—that dominated thinking in the Army After Next project.[208]

But one form of low-tech, low-intensity conflict asymmetry to emerge from the Army After Next project was impossible for Army transformers to ignore.

The debate over urban operations began when the Army After Next project stumbled across the problem during wargames at the U.S. Army War College. The winter 1997 Army After Next wargame had tried to sidestep the sticky issue of urban operations by "assum[ing] that operations in urban terrain was not part of the AAN force's mission." Experimenters quickly found, however, that the Army After Next would not be able to avoid this mission. The report on the winter 1998 wargame described the problem: "In response to the threat of U.S. Battle Forces, the enemy may simply eschew maneuver and go to ground, an approach consistently taken by adversary players in AAN wargames." Every time the "red team" (enemy) was faced with techno-logically superior U.S. Army forces of 2025, it would "dive into cities." Army After Next researchers warned, "Urban operations could become as frequent and routine in the twenty-first century as operations in open terrain have been in the twentieth."[209]

For those who hoped they could avoid the problem by avoiding cities, the researchers added, "Given the apparent propensity for adversaries to dive into cities when AAN forces take the field, it is unlikely that AAN forces will avoid urban warfare in the future." The enemy chose this course "for both operational and political ends." The operational ends were to negate the Army After Next's "advantages in speed and mobility" and "diminish the effect of a U.S. information advantage because forces are more difficult to locate, target, and assess."[210] The political ends were to embroil the local population in the conflict. Because the enemy pulled the populace of the city into the conflict, the urban area would become an unavoidable political objective of a future war.

This was asymmetry rearing its head in a way that *Joint Vision 2010* had not anticipated: the enemy forcing the Army to fight a low-intensity conflict. The wargame report noted, "Urban operations will require a much higher degree of integration with local societies than has been the U.S. experience heretofore." Moving the conflict into an urban area would unavoidably make it a low-intensity conflict. The report concluded by lamenting that "inves-tigations into possible technological solutions provide no easy answers, thus far."[211]

As the Army After Next project stumbled upon the urban operations prob-lem, some low-intensity conflict observers began to reexamine the Russian experience in Chechnya for applicable lessons. Retired Army lieutenant col-onel Tim Thomas of the Foreign Military Studies Office at Fort Leavenworth believed the First Chechen War had raised serious questions about military operations in cities. Some of the lessons he enumerated were strictly tactical; in a 1999 article in *Parameters* he highlighted both the physical and psy-chological effects of both snipers and the vertical dimension—underground, ground-level, and above-ground—on soldiers in urban combat. He also noted the difficulties of using artillery in cities ("Indiscriminate bombing and shelling turned the local population against the Russians") and guerilla tactics ("Chechens deployed in the vicinity of a school or hospital, fired a few rounds, and quickly left. The Russians would respond by shelling the school or hospital, but usually after the Chechens had gone"). Elsewhere in the article he was more explicit: "The Chechen force was not a typical army but rather a composite force of armed home guards (guerrillas) and a few regular forces."[212]

Thomas's observations also touched on the low-intensity conflict nature of fighting in Grozny. This fight occurred in and among the populace of the city; people presented obstacles to detection of or fire on the guerillas. Thomas wrote that Chechen fighters "could blend in with the local population to their advantage." He also noted, "This not only continued to make it difficult to distinguish combatants from civilians, but it also helped the Chechens get the local population on their side." This effect was exacerbated by the fact that "Russian forces entered a city, destroyed property and buildings, and killed or wounded civilians while searching for their armed opponent." The Russian Army effort was made more difficult by the fact that the Russians lacked civil affairs units. Thomas noted that the Russian Army also desperately needed better nonlethal weapons to permit a more nuanced application of force when fighting in cities. He concluded, "The locals included some Russian citizens who were inhabitants of Grozny (and who found it incomprehensible that their own leaders had such disregard for the lives of civilians)."[213]

But Thomas believed that the core issue of urban operations was that the people who occupied the urban environment were politically important. This was just another form of asymmetry, using the physical and political complexity of urban areas to negate the technological advantages of the RMA-enabled Army focused on high-intensity conflict. Lest anyone misunderstand his point, he added, "As the United States learned in Somalia, it is not always the best-equipped force that wins." For those who might believe that the Army could somehow avoid urban combat in the future, he also added, "No army wants to engage in urban combat, but increasing urbanization and the danger of strikes from high-precision weapons may well force the fight into the city, where the defender has all the advantages."[214]

Ultimately, many of Thomas's best lessons were lost because he dismissed the Russian Army as inferior to the U.S. Army: "The Russian armed forces that attacked Grozny, while well-equipped, were not the same professional force that opposed the West during the Cold War." Citing Russian minister of defense Pavel Grachev, he wrote, "The combat capabilities of the armed forces were low, the level of mobilization readiness was poor, and the operational planning capability was inadequate. Soldiers were poorly trained. Their sui-cide rates as well as the overall number of crimes in the force were up."[215] Around the same time, Maj. Gen. Robert Scales, the commandant of the U.S. Army War College, also began to wrestle with the problem of urban opera-tions. Scales offered some hope to those who might want to wish away the problem of urban operations: "Urban warfare doesn't happen all that often. Both sides realize the destructive effects that street fighting may cause." He posited, "A casual glance at the last 500 years of major war history shows that as more of the world blankets itself in urban sprawl, the incidents of actual street fighting have declined."[216] Of course, this argument ignores the fact that *all* fighting had declined, at least since World War II. But of those wars that *had* occurred—especially those between technologically advanced and technologically disadvantaged adversaries—most featured at least some urban combat.

Major General Scales's next assertion was even more dubious. He assured readers, "Only a desperate enemy, defending at great disadvantage, willing to sacrifice [initiative], his cities, and a large portion of his military force, has taken to defending cities." Of course, this was no comfort at all. This was exactly the condition that the Army was trying to impose on future adver-saries; the purpose of the Army After Next was to so overmatch hypothetical future adversaries as to put all potential competitors "at great disadvantage." Scales was finally forced to admit as much, writing, "Military leaders who believe that future warfare will not encompass this unpleasant environment are deluding themselves."[217]

In explaining the challenge of urban operations, Scales noted, "The urban environment precludes mobility operations and largely negates the effects of weapons, while minimizing engagement ranges. The proximity of buildings plays havoc with communications, further adding to command and control difficulties. . . . The array of threats from multiple dimensions has a debili-tating effect on soldiers." He added that hiding in urban environments "may be the preferred approach of future opponents." An enemy would choose to "burrow his force in the urban terrain" because "an urban assault largely neutralizes American high-tech speed and mobility advantages." Cities also presented the Army After Next with "the added risk of excessive casualties and prolonged campaign timelines," both of which favored the enemy. Scales concluded, "An enemy occupies cities to slow us down and avoid our strengths," and "while he surrenders the tactical initiative, the close terrain offers protection from firepower and surveillance and allows further time to prepare a defense."[218]

Scales repeatedly conflated "extended urban conditions" and "large urban area[s]" with "complex terrain."[219] This conception of urban operations—as operations in complex terrain that created disadvantages for the highly mobile, highly networked Army After Next—was the preferred conception among Army transformers. Complex terrain could be overcome by technology, the transformers' preferred solution to military problems.

But this conflation of urban terrain with complex terrain obscured the most important facet of urban operations: the political importance of the city's populace. Some enemies—like the "red team" in Army After Next wargames—might choose to hide in cities as a direct reaction to America's uncontested, high-tech dominance in high-intensity conflict. But many more enemies—both insurgents and conventional armies defending their homeland—would operate in cities because of their need to influence or control the population. The true asymmetry of urban operations was not that urban areas were complex terrain, but that the Army was unprepared to engage in the political dimension of low-intensity conflicts.

Major General Scales did seem to vaguely understand this. He acknowledged that cities presented a challenge to the Army because they contained "millions of people that house [the enemy's] political, cultural, and financial centers of gravity." He noted, "Representing geo-strategic centers of gravity, these urban areas will contain all the vital functions of government, commerce, communication, and transportation activity."[220] But while he acknowledged that cities were important because of their political value to a conflict, he failed to account for this fact when positing solutions.

The Transformers' Take

To the extent that Army After Next researchers acknowledged the presence of civilians inside cities, they considered them only as impediments to target identification or unrestricted indirect fires. Thus, their options for solving "the urban problem" ignored the political importance of the populace of cities:

- First, U.S. forces could exploit superior mobility to preempt or deny enemy occupation of population centers: We get there first.
- Second, if neither the enemy force nor the city itself is of any particular value, it can be bypassed.
- Third, the U.S. could choose to contain, but not destroy, the enemy within the city
- Fourth, the reduction by stand-off strike operations, although the inevitable collateral damage may be unacceptable.

- Finally, U.S./allied forces could seize the city, a decision that may have a high cost in time, property, and lives.[221]

For these transformation enthusiasts, cities were complex terrain and people were impediments to indirect fires and target identification. The bottom line was that cities were to be avoided if at all possible.

Likewise, Scales's "best solution" to the problem of urban terrain was "to preempt the enemy from using complex terrain in the first place" by beating him to the city—through a variety of diplomatic, strategic mobility, or operational maneuver means—and seizing it before the enemy could enter. Barring this, Scales's solution was, essentially, to lay siege to the city: to sit outside it, "[let] the city collapse on itself," and wait for the enemy to quit. Of course, this missed the most important facet of the asymmetry of urban operations: controlling these "millions of people" and the "political, cultural, and financial centers of gravity" they represented were essential to the political ends that would prompt any U.S. military intervention in the first place.[222]

Army After Next experimenters Lt. Col. Robert Hahn and Bonnie Jezior tried to tackle the problem of urban operations in the pages of *Parameters*. They dismissed those who suggested avoiding cities altogether, writing that this solution would prevent the United States from "bring[ing] a military campaign to rapid conclusion and allow the enemy just the type of refuge he was seeking when he chose to enter the city." But they also explicitly rejected Scales's solution of laying siege to the city, explaining that eventually "political necessity, humanitarian concerns, and military requirements will force us to engage and rapidly defeat sizable enemy combat forces that have taken up positions within a large urban area."[223]

Because they believed the Army After Next would not be able to avoid urban warfare, Hahn and Jezior instead prescribed a dizzying array of high-tech salves—from jet packs to robots—to solve the urban operations problem. They promised, "High-technology weapon systems will fundamentally alter the course of urban warfare in the future." Their final recommendation was the "the 2025 Urban Warfighter System," which they described as "a revolutionary new man-machine fighting system with self-contained C4ISR [command, control, communications, computers, intelligence, surveillance and reconnaissance], lethality, mobility, survivability, and sustainability far exceeding those of the current and near-term systems." One could not help but picture Robert A. Heinlein's *Starship Troopers* as Hahn and Jezior described "a body suit with integral C4ISR, engagement, and active surviv-ability systems."[224]

In addition to their persistence in seeing a city as complex terrain rather than the home of a populace that was the political objective of the war, Army transformers also stumbled over another cognitive obstacle: they failed to conceive of an enemy fighting in any other way than as an organized military force. Hahn and Jezior noted that while some of those studying urban operations "presuppose that these forces will be fighting paramilitary elements operating in their own backyard," they believed "the most likely mid-intensity scenario that US forces will confront may be an enemy inva-sion and occupation of a city within a country whose population is generally favorably disposed to the United States."[225] Scales likewise demonstrated this misconception in his own writings on urban operations, describing the tactical situation that would give rise to an urban operation: "After a light-ning campaign lasting only days, the mobile formations of our future foe are decisively beaten in open warfare. To avoid total defeat, the enemy rushes his remaining forces into his capital city, a city of sprawling dimensions with millions of people that house his political, cultural, and financial centers of gravity." Scales believed this put an adversary at a disadvantage: "He loses the initiative. Time is now solely on the side of the intervening coalition. Without the capacity to maneuver, the enemy cannot escape." But Scales still expected this force to fight conventionally: "He arrays his forces throughout the capital to avoid creating lucrative targets for American precision weap-ons. He impresses the local citizenry into national service and appeals to the world to watch the impending slaughter of non-combatants."[226]

Of course, the first half of this hypothetical narrative is almost exactly what happened during the invasion of Iraq in 2003: the Iraqi Army was "beaten in open warfare." But instead of the Iraqi Army retreating in an orderly fashion, it disintegrated, its individual soldiers returning to their homes in the cities. When they reemerged as an adversary to the U.S. occupation, it was not in "arrays [of] forces" but as an insurgency. And this insurgency did not "impresses the local citizenry into national service"; it recruited them as willing fighters and suicide bombers against the "infidel" invaders. And time was decidedly *not* on the "side of the intervening coalition."[227]

Army After Next researchers remained blind to this possible outcome for many reasons. For one, they didn't see it because they had designed the Army After Next wargames to test their future force against a mirror image of itself. They also missed this possibility because their simulations did not replicate civilian populations or their influence on warfare. But, most significant, they missed it because the Army had, for decades, consistently refused to institutionalize the lessons of the very real low-intensity conflicts in which it had engaged or acknowledge the political dimension of each one of them.

Lester Grau and Jacob Kipp of the FMSO would not let Army transformers wish away the problem of urban operations. They predicted that the problem was not going to go away: "Its frequency and scale are likely to increase as emerging threats such as urban guerrillas, terrorists and underdog armies seek cover in the cities." They added, "Urban combat is increasingly likely, since high-precision weapons threaten operational and tactical maneuver in open terrain." In fact, they mused, the high-tech, high-intensity conflict capabilities that the Army had worked so hard to develop since the end of the Vietnam War might actually have turned modern warfare "on its head." They explained, "Maneuver by forces may now be possible *only in the cities* as long as high-precision systems dominate the open countryside." Because "precision strikes . . . cannot occupy and hold a city," U.S. forces would have no choice but to engage in urban operations.[228]

What Grau and Kipp espoused went directly to the heart of the "asymmetry" produced by urban operations: enemies would choose to fight in cities because they could "mobilize the city's resources and population to their purposes." Worse, they warned, an urban operation "at the lower end of the urban combat spectrum is more probable than at the upper end." "Thus," they concluded, "planners should consider how to fight criminal gangs, armed insurgents and urban guerrillas."[229]

Fighting in urban environments, Grau and Kipp insisted, would be in and among the populace; "civil affairs and psychological operations . . . [would] assume paramount importance." And, because the city was full of people, this warfare would "have greater political, economic, sociological and commercial consequences . . . than in the countryside." For this reason, they warned, the Army would need to understand "each city's complex social system reflecting social, ethnic, religious diversity and contradictions."[230] They were explaining that urban operations would, by their very nature, be low-intensity conflicts.

For Grau and Kipp, the population of the city was the political objective of a war. In light of this central fact, they insisted, both the Russian approach in Grozny—to destroy the city—and the approach suggested by Major General Scales—"don't go there"—suffered from "an utter disconnect between the political objective . . . and the military means." Likewise—and in direct response to Hahn and Jezior from the Army After Next program—Grau and Kipp warned that there was no "'silver bullet' technology" that could solve the problem. "However central terrain may be to the solution of tactical problems," Grau and Kipp explained, "a city's complex set of systems and high population densities poses the most daunting of problems in urban combat."[231]

Tim Thomas, talking about the problem of urban operations in 1999, noted that the Army lacked the ability to train its forces in a realistic, modern urban environment. He noted that, while Grozny was about one hundred square miles wide and had 490,000 residents in 1994, the urban combat facility at the JRTC was less than one-tenth of a square kilometer in size.[232]

Grau and Kipp advocated creating "an urban training center, similar to the combat training centers . . . to teach urban tactics, techniques and procedures." But this training center should not just teach the Army to conduct combat in complex terrain. "Such a training center would need to incorporate training models that include social, cultural, ethnic and political dynamics as well as urban terrain features."[233]

The Army After Next, 1998–1999

As researchers conducted the experimental regimen of the Army After Next program, the other revelation to emerge was that the new approach would require a much lighter force than the Army of the Big Five weapon systems that prevailed in the 1990s. These "very mobile light forces" would "destroy enemy ground forces" through the use of "sophisticated information systems that enable [them] to engage enemy forces with precision and at long range." But this force would be unable to fulfill the traditional "key role for Army forces"—namely, "the holding of terrain" (and presumably controlling populations). For this task it would have to rely on the "heavier forces" of the Army, which these experiments envisioned would still "comprise 70 percent of the 2020 force."[234]

Army transformers had consistently insisted that the Army After Next and its predecessor, Force XXI, were "full-spectrum" forces, capable of operating across the "spectrum of crises." But, as General Reimer had told an Association of the United States Army audience back in 1996 when announcing the Army After Next project, it "takes soldiers . . . to separate warring parties, to reassure fearful civilians, to restore public order, to keep criminals from taking advantage of the vacuum in civil order, to deliver humanitarian assistance, to prevent and win the nation's wars."[235] Instead, the force that researchers were envisioning, at least in these early wargames, would have been hard-pressed to fight a low-intensity conflict in which boots on the ground, in and among the population, was essential to success.

The idea that the contemporary heavy Army forces might remain in the nation's inventory for some time after the realization of fantastic future forces such as the Army After Next was an enduring concept that would continue well into the war on terror. Perhaps the first person to call this a "legacy force" was Major General Scales, who envisioned that the improvements of Force XXI would be grafted onto this legacy force and provide "improved situational awareness" and "strategic mobility" to these "legacy systems." Transformers envisioned their continued presence and utility up to or beyond 2015, when they would reach an envisioned "wear-out" date. But this idea of a remnant, heavy, legacy force militated against the idea of better deployability for the Army After Next—what Scales called "strategic speed" in a briefing on the topic. He described "power projection" as occurring from around the globe, across all domains, to arrive "First with the Most" to achieve "psychological domination" over adversaries.[236]

Yet Army transformers were perplexed by the fact that this speed would have to be balanced with the demand to deploy heavy legacy forces to hold ground and control populations. For this reason, RAND Corporation researchers warned that the Army After Next might have to moderate the pace of its operations—in essence, prolonging the war—to allow these legacy forces to arrive in the fight.[237]

The revelations of the first two winter reports of the Army After Next project emphasized the ideas of knowledge and speed. Reimer described knowledge as "the ability to answer three questions: Where am I? Where are my buddies? Where is the enemy?" Speed, on the other hand, was "tactical agility as well as rapid strategic responsiveness—getting there 'first with the most.'"[238]

The key to achieving knowledge and speed in the Army After Next—and the Force XXI that preceded it and continued development—was digitization. Through AWEs, the Army had progressed significantly in developing the varied computer- and network-enabled platforms of the Army Battle Command System, a system of systems that created, analyzed, and distributed information across an Army unit during a battle. As General Reimer's term as chief of staff of the Army drew to a close, the Army was on pace to field the first digitized division in 2000 and the first digitized corps in 2004.[239]

Reimer offered several parting shots on the Army After Next and why its continued development was important to the Army. Sustaining the myth that this hyperlethal, hypertechnological force was also useful for low-intensity conflict, Reimer began by meekly asserting that the Army After Next was needed to deal with "ethnic rivalries, national and religious tensions, international crime, terrorism, [and] drug trafficking." But he did not even complete the sentence before he was warning of "the rise of one or more major military competitors capable of challenging the U.S. regionally [that] will probably increase the likelihood of conflict over the next several decades." In this vision of the future, Reimer wrote, "Advances in precision weaponry and the proliferation of weapons of mass destruction will make the future battlefield a much more lethal place." Justifying the demand for expensive, high-tech weapons for his Army After Next, he concluded, "Tomorrow's tactical engagement areas will likely extend as far as today's operational and strategic distances."[240] This was hyperbolic, military technojargon for the contention that future peer competitors would be able to strike U.S. military forces on the battlefield with precision from virtually anywhere on earth.

At its core, the Army After Next program was an exercise to prepare the Army to fight a mirror image of itself. The future peer competitor that the program's wargames envisioned was capable of subjecting the Army After Next to "piecemeal destruction from a fire storm of precision munitions" if it attempted large-scale maneuver. The Army After Next would face a battlefield in which there would be "few or no sanctuaries," but this would not be due to guerilla forces or insurgents. Rather, "Forces deployed to any region in the theater will be vulnerable to a blend of conventional and unconventional attack by air, missiles, information operations (IO), special operations forces (SOF), and, potentially, space-based weapons." Future adversaries would be capable of inflicting "mass and shock" on the U.S. Army of the future by "massing and integrating the effects, in a short period of time, of variable range fires from a wide variety of air-, ground-, and sea-based platforms."[241]

To counter this profoundly conventional enemy, TRADOC researchers envisioned that the Army After Next would have to defeat "the enemy's precision system (particularly the information-based components of that system)." This fictional adversary was computer- and network-enabled just as the Army sought to become. To maintain "dominance across the air-, land-, sea-, space-, and cyber-domains," the Army would use superior "knowledge and speed" to dismantle the enemy. As TRADOC noted,

Distributed operations will be decentralized in execution, but carried out in accordance with a centralized, fully integrated joint plan, which is both orchestrated and supported by a pervasive and resilient C4I [command, control, communications, computers, and intelligence] network.

The tempo of operations will increase sharply, as tactical objectives are achieved in remarkably short bursts of time. Tactical successes, piled up nearly simultaneously across the entire battlespace, could then lead . . . to rapid operational-level disintegration as the enemy's . . . ability to control his own forces evaporates before he can respond.[242]

This was the realization of the RMA-enabled force that had been imagined ever since Air Force colonel John Boyd and his OODA (observe–orient–decide–act) loop: the dream of paralyzing an enemy by overwhelming his command and control capacity, and defeating the enemy by directly attacking his center of gravity.[243]

Conspicuously absent from this discussion of the future, given the behavior that was emerging from the Army After Next project's wargames, was the possibility that the enemy might hide in urban centers and fight from within a civilian populace, negating the Army After Next's "knowledge and speed."

Conclusion

By 1999 the institutional Army had almost completely isolated itself from the lessons of its low-intensity conflict operations in Bosnia-Herzegovina. Gen. Dennis Reimer had closed A-AFCLIC, and the CGSC had abolished its core course on low-intensity conflict. A new FM 100-5, *Operations*, was on the way that would further subordinate "stability and support operations" to the high-intensity conflict focus of the Army. And the Army After Next program—which would justify a whole new generation of weapon systems—was safely focused on shaping the Army for a hypothetical great power war against an imaginary future peer competitor. The only remaining issues to resolve were the specific shape of the Army After Next and the nagging issue of deployability—the significant gap in deployment timelines between the highly deployable Army After Next and the harder-to-deploy legacy forces that would remain in the inventory and be necessary for postconflict activities.

But try as they might, Army transformers could not make low-intensity conflict go away. Precisely because they had failed to institutionalize the lessons of low-intensity conflict, Army forces could not figure out how to bring Operation Joint Forge to a close and extricate the Army from Bosnia-Herzegovina. And the very experiments that General Reimer had hoped would solidify the Army's unmatched dominance of high-intensity conflict for the next three decades was instead revealing—through the problems of asymmetry and urban operations—troubling signs that the Army's ever-increasing dominance of high-intensity conflict might be increasingly irrelevant to a future ever-more dominated by low-intensity conflicts.

Meanwhile, toward the end of General Reimer's tenure as chief of staff of the Army, low-intensity conflict observers seemed once more ascendant. A growing chorus of observers was beginning to suggest that the Army needed to be segregated into two forces—optimized for high- and low-intensity conflict, respectively. Moreover, transformation's flagship project, the Army After Next project, was providing them with unexpected proof—in the form of insights into asymmetry and urban operations—that the Army needed to pay more attention to preparation for low-intensity conflict.

Ironically, it was another low-intensity conflict—the war in Kosovo—that would short-circuit the debate over such conflicts and launch an intense debate about the deployability—and, in fact, the relevance—of the U.S. Army. But the result of this debate would not be a refocus on low-intensity conflict; instead it would trigger the most dramatic equipment fielding and organizational change in the Army since the fielding of the Big Five in the late 1980s and early 1990s. Moreover, this dramatic turn of events would produce a road map that would give form to the nebulous Army After Next and chart the Army's course into 2020 and beyond.

As it turned out, this road map would arrive just in time to be made completely irrelevant by the war on terror.

NOTES

1. General Officer Management Office, *General Dennis Joe Reimer*; Proctor, *Containment and Credibility*, Kindle, 131.

2. Headquarters, 9th Infantry Division Artillery, "Operational Report for the Period Ending 15 July 1969," memorandum for record detailing the after-action review of 9th Infantry Division Artillery Operations in Vietnam, Republic of Vietnam, 9th Infantry Division Artillery, July 15, 1969, 2, 21.

3. General Officer Management Office, *General Dennis Joe Reimer*; Brown, *Kevlar Legions*, Kindle, 139.

4. Dennis J. Reimer, interview with Lewis Sorley, 2000, transcript, Senior Officer Oral History Project, U.S. Army Military History Institute, U.S. Army Heritage and Education Center, 206.

5. Brown, *Kevlar Legions*, Kindle, 139.

6. Dennis Reimer, "Where We've Been—Where We're Headed: Maintaining a Solid Framework While Building for the Future," *Army*, October 1995, reprinted in Reimer, *Soldiers Are Our Credentials*, 3; Brown, *Kevlar Legions*, Kindle, 151.

7. Chapman et al., *Prepare the Army for War*, 31–32.

8. Reimer, "Where We've Been—Where We're Headed," 21–26.

9. Joint Staff, *Expanding* Joint Vision 2010, I; Brown, *Kevlar Legions*, Kindle, 80–97.

10. Joint Staff, *Joint Vision 2010*, 1, 4.

11. Joint Staff, *Joint Vision 2010*, 2, 17, 25–27.

12. Joint Staff, *Joint Vision 2010*, 7, 8, 13, 18.

13. Joint Staff, *Joint Vision 2010*, 10, 13.

14. Bacevich, "Learning from Aidid," 32, quoted in Shimko, *The Iraq Wars*, Kindle, locations 2946–49; Record, "Ready for What and Modernized against Whom?," ix–x; Joint Staff, *Joint Vision 2010*, 10.

15. Joint Staff, *Joint Vision 2010*, 25–27.

16. Reimer interview, 203–7, 327–29, 398–99.

17. Reimer interview, 318–20.

18. Reimer interview, 200–202.

19. Reimer interview, 330–34; Brown, *Kevlar Legions*, Kindle, 5.

20. Brown, *Kevlar Legions*, Kindle, 140.

21. Brown, *Kevlar Legions*, Kindle, 127–28, 140, 144, 147.

22. Hunt, "OOTW: A Concept in Flux," 3–10; Joint Chiefs of Staff, *Joint Doctrine for Military Operations Other Than War*, JP 3-07, II-1, III-10.

23. Headquarters, U.S. Department of the Army, *Operations*, FM 100-5 (1993), 13-0–13-8; Joint Chiefs of Staff, JP 3-07, I-7.

24. Metz, "A Flame Kept Burning," 31–41.

25. Joint Chiefs of Staff, JP 3-07, I-7, emphasis in the original.

26. U.S. Army Command and General Staff College, *United States Army Command and General Staff College Catalog, Academic Year 1995–1996*, CGSC Circular No. 351-1, July 1995, Command and General Staff College Papers, Ike Skelton Combined Arms Research Library, 3–13; U.S. Army Command and General Staff College, *United States Army Command and General Staff College Catalog, Academic Year 1993–1994*, CGSC Circular No. 351-1, July 1993, Command and General Staff College Papers, Ike Skelton Combined Arms Research Library, 46, 49, 52–53, 63–64, 67, 70–71; U.S. Command and General Staff College, *United States Army Command and General Staff College Catalog, Academic Year 1991–1992*, CGSC Circular No. 351-1, July 1991, Command and General Staff College Papers, Ike Skelton Combined Arms Research Library, 42–43, 45, 49–51, 59–60, 62, 64–65, 68.

27. Joint Chiefs of Staff, JP 3-07, IV-13–IV-14, IV-1, emphasis in the original.

28. Metz, "A Flame Kept Burning," 31–41.

29. Tilford, *The Revolution in Military Affairs*, 6–7, 13–14.

30. Joint Chiefs of Staff, JP 3-07, IV-4.

31. Joint Chiefs of Staff, JP 3-07, I-2–I-3.

32. Joint Chiefs of Staff, JP 3-07, I-1.

33. Joint Chiefs of Staff, JP 3-07, IIII-9, III-10; Fitzgerald, *Learning to Forget*, Kindle, 79.

34. Joint Chiefs of Staff, JP 3-07, III-10.

35. Metz, "A Flame Kept Burning," 31–41.

36. Baumann, Gawrych, and Kretchik, *Armed Peacekeepers in Bosnia*, 2–3; Millett and Maslowski, *For the Common Defense*, Kindle, locations 12310–21.

37. Greenberg, *Language and Identity in the Balkans*, 1–17; Baumann, Gawrych, and Kretchik, *Armed Peacekeepers in Bosnia*, 3, 4.

38. Baumann, Gawrych, and Kretchik, *Armed Peacekeepers in Bosnia*, 1, 4; Crane, "Peace Dividends and Benevolent Interventions."

39. Baumann, Gawrych, and Kretchik, *Armed Peacekeepers in Bosnia*, 27, 40.

40. Baumann, Gawrych, and Kretchik, *Armed Peacekeepers in Bosnia*, 27, 40; Millett and Maslowski, *For the Common Defense*, Kindle, locations 12349–75.

41. Baumann, Gawrych, and Kretchik, *Armed Peacekeepers in Bosnia*, 25–26, 27–28, 42, 50; Brown, *Kevlar Legions*, Kindle, 154.

42. Millett and Maslowski, *For the Common Defense*, Kindle, locations 12349–75; Baumann, Gawrych, and Kretchik, *Armed Peacekeepers in Bosnia*, 29–30.

43. U.S. Army Europe, *Military Operations: The U.S. Army in Bosnia and Herzegovina*, AE PAM 525-100, 16; Baumann, Gawrych, and Kretchik, *Armed Peacekeepers in Bosnia*, 58–93.

44. Millett and Maslowski, *For the Common Defense*, Kindle, locations 12403–14; Baumann, Gawrych, and Kretchik, *Armed Peacekeepers in Bosnia*, 30–31.

45. Baumann, Gawrych, and Kretchik, *Armed Peacekeepers in Bosnia*, 37; Baumann, Gawrych, and Kretchik, *Armed Peacekeepers in Bosnia*, 94.

46. U.S. Army Europe, AE PAM 525-100, 12–13, 20–21; Baumann, Gawrych, and Kretchik, *Armed Peacekeepers in Bosnia*, 94.

47. Baumann, Gawrych, and Kretchik, *Armed Peacekeepers in Bosnia*, 95–96, 112.

48. Baumann, Gawrych, and Kretchik, *Armed Peacekeepers in Bosnia*, 121, 123; U.S. Army Europe, AE PAM 525-100, 21–22; Brown, *Kevlar Legions*, Kindle, 155, 156; Millett and Maslowski, *For the Common Defense*, Kindle, locations 12403–14.

49. U.S. Army Europe, AE PAM 525-100, 16–17; Baumann, Gawrych, and Kretchik, *Armed Peacekeepers in Bosnia*, 71, 105; Cherrie, "Task Force Eagle," 63–72; Fitzgerald, *Learning to Forget*, Kindle, 105; Phillips, *Operation Joint Guardian*, 14–17.

50. Manwaring, "Peace and Stability Lessons from Bosnia," 28–38.

51. U.S. Army Europe, AE PAM 525-100, 17–18, 19–20, 20–21; Baumann, Gawrych, and Kretchik, *Armed Peacekeepers in Bosnia*, 81–83.

52. Crane, "Peace Dividends and Benevolent Interventions"; Cherrie, "Task Force Eagle," 63–72.

53. Crane, "Peace Dividends and Benevolent Interventions"; U.S. Army Europe, AE PAM 525-100, 20–21.

54. Baumann, Gawrych, and Kretchik, *Armed Peacekeepers in Bosnia*, iii, 198; Reimer interview, 326.

55. U.S. Army Europe, AE PAM 525-100, 12–13; Brown, *Kevlar Legions*, Kindle, 159; Cherrie, "Task Force Eagle," 63–72.

56. Cherrie, "Task Force Eagle," 63–72; Olsen and Davis, *Training U.S. Army Officers for Peace Operations*.

57. Baumann, Gawrych, and Kretchik, *Armed Peacekeepers in Bosnia*, 114.

58. Cherrie, "Task Force Eagle," 63–72.

59. Crane, "Peace Dividends and Benevolent Interventions"; Baumann, Gawrych, and Kretchik, *Armed Peacekeepers in Bosnia*, 97–98.

60. Crane, "Peace Dividends and Benevolent Interventions"; Baumann, Gawrych, and Kretchik, *Armed Peacekeepers in Bosnia*, 223–24.

61. Baumann, Gawrych, and Kretchik, *Armed Peacekeepers in Bosnia*, 103.

62. U.S. Army Europe, AE PAM 525-100, 14–15, 20–21; Baumann, Gawrych, and Kretchik, *Armed Peacekeepers in Bosnia*, 79.

63. Manwaring, "Peace and Stability Lessons from Bosnia," 28–38.

64. Dennis Reimer, "Letter to Army General Officers, January 22, 1996, Observations from the Middle East, Bosnia, and Germany" in Reimer, *Soldiers Are Our Credentials*, 23.

65. Brown, *Kevlar Legions*, Kindle, 159–60.

66. Reimer interview, 350–55.

67. Cherrie, "Task Force Eagle," 63–72.

68. Cherrie, "Task Force Eagle," 63–72.

69. Manwaring, "Peace and Stability Lessons from Bosnia," 28–38.

70. Baumann, Gawrych, and Kretchik, *Armed Peacekeepers in Bosnia*, 126–27.

71. Baumann, Gawrych, and Kretchik, *Armed Peacekeepers in Bosnia*, 37; Crane, "Peace Dividends and Benevolent Interventions"; Walter Piatt, cited in Baumann, Gawrych, and Kretchik, *Armed Peacekeepers in Bosnia*, 196–97.

72. Reimer, "Letter to Army General Officers, January 22, 1996."

73. Cherrie, "Task Force Eagle," 63–72; Baumann, Gawrych, and Kretchik, *Armed Peacekeepers in Bosnia*, 100.

74. Baumann, Gawrych, and Kretchik, *Armed Peacekeepers in Bosnia*, 130, 95–96; Cherrie, "Task Force Eagle," 63–72; Crane, "Peace Dividends and Benevolent Interventions."

75. Manwaring, "Peace and Stability Lessons from Bosnia," 28–38.

76. Viney, *United States Cavalry Peacekeepers in Bosnia*, 223, quoted in Crane, "Peace Dividends and Benevolent Interventions."

77. Crane, "Peace Dividends and Benevolent Interventions"; Baumann, Gawrych, and Kretchik, *Armed Peacekeepers in Bosnia*, 102.

78. Baumann, Gawrych, and Kretchik, *Armed Peacekeepers in Bosnia*, 194, 196–97; Manwaring, "Peace and Stability Lessons from Bosnia," 28–38.

79. Crane, "Peace Dividends and Benevolent Interventions"; Baumann, Gawrych, and Kretchik, *Armed Peacekeepers in Bosnia*, 102, 175.

80. Baumann, Gawrych, and Kretchik, *Armed Peacekeepers in Bosnia*, 175, 176–77.

81. Baumann, Gawrych, and Kretchik, *Armed Peacekeepers in Bosnia*, 102.

82. Baumann, Gawrych, and Kretchik, *Armed Peacekeepers in Bosnia*, 102.

83. Cherrie, "Task Force Eagle," 63–72.

84. Brown, *Kevlar Legions*, Kindle, 154–55.

85. U.S. Army Europe, AE PAM 525-100, 16.

86. Gawrych, "Show of Force," 121; Baumann, Gawrych, and Kretchik, *Armed Peacekeepers in Bosnia*, 99.

87. Baumann, Gawrych, and Kretchik, *Armed Peacekeepers in Bosnia*, 175–76.

88. Gawrych, "Show of Force," 126; Thomas T. Smith, "Foreword," in Baumann, Gawrych, and Kretchik, *Armed Peacekeepers in Bosnia*, i–ii.

89. Baumann, Gawrych, and Kretchik, *Armed Peacekeepers in Bosnia*, 100–114; Nordland, "A Monster on the Loose."

90. Baumann, Gawrych, and Kretchik, *Armed Peacekeepers in Bosnia*, 158, 124; Crane, "Peace Dividends and Benevolent Interventions."

91. Manwaring, "Peace and Stability Lessons from Bosnia," 28–38, emphasis in the original.

92. Manwaring, "Peace and Stability Lessons from Bosnia," 28–38.

93. Reimer interview, 244–47.

94. Crane, "Peace Dividends and Benevolent Interventions"; Baumann, Gawrych, and Kretchik, *Armed Peacekeepers in Bosnia*, 217–18, 123.

95. Baumann, Gawrych, and Kretchik, *Armed Peacekeepers in Bosnia*, 52–53.

96. Reimer interview, 244–47.

97. Baumann, Gawrych, and Kretchik, *Armed Peacekeepers in Bosnia*, 133; John M. Shalikashvili, quoted in Priest, *The Mission*, 46; Baumann, Gawrych, and Kretchik, *Armed Peacekeepers in Bosnia*, 186.

98. Baumann, Gawrych, and Kretchik, *Armed Peacekeepers in Bosnia*, 133.

99. Cherrie, "Task Force Eagle," 63–72.

100. Baumann, Gawrych, and Kretchik, *Armed Peacekeepers in Bosnia*, 200.

101. Chapman et al., *Prepare the Army for War*, 52.

102. Chapman et al., *Prepare the Army for War*, 51, 52; Headquarters, U.S. Department of the Army, FM 100-5 (1993), 13-0-13-8.

103. Headquarters, U.S. Department of the Army, Department of the Army, *Intelligence and Electronic Warfare Support to Low-Intensity Conflict Operations*, FM 34-7, 1-1-1-5; Headquarters, U.S. Departments of the Army and the Air Force, *Military Operations in Low Intensity Conflict*, FM 100-20, 1–5; Joint Chiefs of Staff, JP 3-07, I-2.

104. Fitzgerald, *Learning to Forget*, Kindle, 98; Air Force News Service, "The Center for Low Intensity Conflict Closes after 10 Years," Federation of American Scientists, n.d. [June 1996], accessed April 21, 2018, https://fas.org/irp/news/1996/n19960626_960615.html; Hasskamp, *Operations Other Than War*, 17.

105. Headquarters, U.S. Department of the Army, *Operations*, FM 3-0 (2001).

106. Gordon Sullivan, "Foreword," in U.S. Army Training and Doctrine Command, *Force XXI Operations*, PAM 525-5, n.p.; Reimer, "Challenge and Change," 108–16, emphasis in the original.

107. Chapman et al., *Prepare the Army for War*, 52; Reimer, "Challenge and Change," 108–16.

108. Chapman et al., *Prepare the Army for War*, 52–53; "Major General Robert H. Scales," Strategic Studies Institute, n.d., accessed April 21, 2018, https://ssi.armywarcollege.edu/pubs/people.cfm?authorID=104.

109. U.S. Department of the Army, *Army Vision 2010*, 1.

110. U.S. Department of the Army, *Army Vision 2010*, 11–17; bracketed text in the original.

111. U.S. Department of the Army, *Army Vision 2010*, 1–3.

112. Brown, *Kevlar Legions*, Kindle, 163–64.

113. Aspin, *Report of the Bottom-Up Review*, 19; Brown, *Kevlar Legions*, Kindle, 165.

114. U.S. Department of Defense, *Report of the Quadrennial Defense Review*, emphasis in the original.

115. U.S. Department of Defense, *Report of the Quadrennial Defense Review*.

116. U.S. Department of Defense, *Report of the Quadrennial Defense Review*.

117. U.S. Department of Defense, *Report of the Quadrennial Defense Review*.

118. U.S. Department of Defense, *Report of the Quadrennial Defense Review*.

119. U.S. Department of Defense, *Report of the Quadrennial Defense Review*.

120. U.S. Department of Defense, *Report of the Quadrennial Defense Review*.

121. U.S. Department of Defense, *Report of the Quadrennial Defense Review*, emphasis in the original.

122. U.S. Department of Defense, *Report of the Quadrennial Defense Review*.

123. Reimer, "Challenge and Change," 108–16; Reimer interview, 324–28.

124. U.S. Department of Defense, *Report of the Quadrennial Defense Review*; Reimer, "Challenge and Change," 108–16.

125. Reimer, "Challenge and Change," 108–16.

126. Reimer, "Challenge and Change," 108–16, emphasis added.

127. Reimer, "Challenge and Change," 108–16, emphasis added.

128. Reimer, "Challenge and Change," 108–16.

129. Reimer, "Challenge and Change," 108–16; Brown, *Kevlar Legions*, Kindle, 144–45.

130. Reimer, "Challenge and Change," 108–16.

131. Macgregor, *Breaking the Phalanx*, 4.

132. Brown, *Kevlar Legions*, Kindle, 290.

133. U.S. Army Europe, AE PAM 525-100, 21–22.

134. Baumann, Gawrych, and Kretchik, *Armed Peacekeepers in Bosnia*, 199, iii; U.S. Army Europe, AE PAM 525-100, 204; Donald P. Wright and Timothy R. Reese, *On Point II: Transition to the New Campaign*, 58; U.S. Army Europe, AE PAM 525-100, 27; Baumann, Gawrych, and Kretchik, *Armed Peacekeepers in Bosnia*, 124.

135. James Dubik, interview with Robert M. Mages, transcript, 2008, Senior Leader Debriefing Program, U.S. Army Military History Institute, U.S. Army Heritage and Education Center, iv; Cucolo, "Grunt Diplomacy," 110–26; Freakley et al., "Training for Peace Support Operations," 17–24.

136. Baumann, Gawrych, and Kretchik, *Armed Peacekeepers in Bosnia*, 188, 199, 150–51.

137. Baumann, Gawrych, and Kretchik, *Armed Peacekeepers in Bosnia*, 151.

138. Baumann, Gawrych, and Kretchik, *Armed Peacekeepers in Bosnia*, 150–51, 178.

139. Baumann, Gawrych, and Kretchik, *Armed Peacekeepers in Bosnia*, 150, 158, 160; Agence France-Presse, "Croats Riot over Return of Serbs to a Village in Western Bosnia," *New York Times*, April 25, 1998, https://www.nytimes.com/1998/04/25/world/croats-riot-over -return-of-serbs-to-a-village-in-western-bosnia.html.

140. Crane, "Peace Dividends and Benevolent Interventions"; Maja Zuvela, "EU Peacekeepers Ready to Intervene in Bosnia in Case of New Strife," Reuters, March 28, 2017, https://www.reuters.com/article/us-bosnia-eufor/eu-peacekeepers-ready-to-intervene-in -bosnia-in-case-of-new-strife-idUSKBN16Z1YJ.

141. Cucolo, "Grunt Diplomacy," 110–26, emphasis in the original.

142. Freakley et al., "Training for Peace Support Operations," 17–24.

143. Freakley et al., "Training for Peace Support Operations," 17–24.

144. Freakley et al., "Training for Peace Support Operations," 17–24.

145. Baumann, Gawrych, and Kretchik, *Armed Peacekeepers in Bosnia*, 157, 167.

146. Baumann, Gawrych, and Kretchik, *Armed Peacekeepers in Bosnia*, 152, 158, 161.

147. Baumann, Gawrych, and Kretchik, *Armed Peacekeepers in Bosnia*, 167.

148. Baumann, Gawrych, and Kretchik, *Armed Peacekeepers in Bosnia*, 136–37.

149. Cucolo, "Grunt Diplomacy," 110–26.

150. Kretchik, "Force Protection Disparities," 73–76.

151. Cucolo, "Grunt Diplomacy," 110–26.

152. Cucolo, "Grunt Diplomacy," 110–26.

153. Cucolo, "Grunt Diplomacy," 110–26.

154. Reimer interview, 256–60.

155. James Dubik, "Subject: Thought Paper—Similarities: Haiti and Bosnia," unpublished paper for Lt. Gen. Thomas Burnette, Bosnia-Herzegovina, March 12, 1999, Folder 5, Unpublished Thought Papers, Box 29, Task Force Eagle and Multinational Division North Operation Joint Forge, B-H 7 Jan 1999–circa 1999, James M. Dubik Papers, U.S. Army Heritage and Education Center.

156. Dubik, "Subject: Thought Paper–Similarities," emphasis in the original.

157. Dubik, "Subject: Thought Paper–Similarities."

158. Dubik, "Subject: Thought Paper–Similarities."

159. Dubik, "Subject: Thought Paper–Similarities."

160. Dubik, "Subject: Thought Paper–Similarities."

161. Cucolo, "Grunt Diplomacy," 110–26.

162. Cucolo, "Grunt Diplomacy," 110–26.

163. U.S. Army Command and General Staff College, *States Army Command and General Staff College Catalog, Academic Year 1995–1996*, CGSC Circular No. 351-1, July 1995, Command and General Staff College Papers, Ike Skelton Combined Arms Research Library,

3-13; U.S. Army Command and General Staff College, *United States Army Command and General Staff College Catalog, Academic Year 1997–1998*, CGSC Circular No. 351-1, July 1997, Command and General Staff College Papers, Ike Skelton Combined Arms Research Library, 3-10, 3-12; U.S. Army Command and General Staff College, *United States Army Command and General Staff College Catalog, Academic Year 1998–1999*, CGSC Circular No. 351-1, August 31, 1998, Command and General Staff College Papers, Ike Skelton Combined Arms Research Library, 3-11, 3-13.

164. Metz et al., *The Future of American Landpower*, v–viii, 10–13, emphasis in the original; Metz, "Which Army After Next?," 15–26.

165. U.S. Department of the Army, *Army Vision 2010*, 6–11; Shimko, *The Iraq Wars*, Kindle, locations 2967–71; U.S. Department of Defense, *Report of the Quadrennial Defense Review*; U.S. Army Training and Doctrine Command, *Knowledge and Speed*, iv–v.

166. Hunt, "OOTW: A Concept in Flux," 3–10.

167. Reimer, "The Army After Next," *Strategic Review*, Spring 1999, reprinted in Reimer, *Soldiers Are Our Credentials*, 270.

168. Peters, "Heavy Peace," 71–79.

169. Swannack and Gray, "Peace Enforcement Operations," 3–10.

170. Swannack and Gray, "Peace Enforcement Operations," 3–10.

171. Reimer interview, 318, 319, 331.

172. Hunt, "OOTW: A Concept in Flux," 3–10.

173. Yates, "Military Stability and Support Operations," 51–61.

174. Maloney, "Insights into Canadian Peacekeeping Doctrine," 12–23.

175. Hillen, "Peace(keeping) in Our Time," 17–34.

176. Hillen, "Peace(keeping) in Our Time," 17–34.

177. Hillen, "Peace(keeping) in Our Time," 17–34.

178. Reimer, "Challenge and Change," 108–16.

179. Schook, "Paying the Price for Versatility," 19–25.

180. Reimer interview, 330.

181. Peters, "Heavy Peace," 71–79.

182. Peters, "Heavy Peace," 71–79.

183. Reimer interview, 327.

184. Metz et al., *The Future of American Landpower*, 10–13.

185. Schook, "Paying the Price for Versatility," 19–25.

186. Peters, "Heavy Peace," 71–79.

187. Maloney, "Insights into Canadian Peacekeeping Doctrine," 12–23.

188. Swannack and Gray, "Peace Enforcement Operations," 3–10.

189. Hunt, "OOTW: A Concept in Flux," 3–10, emphasis in the original.

190. Maloney, "Insights into Canadian Peacekeeping Doctrine," 12–23; Brigham, *Guerrilla Diplomacy*, 94.

191. Maloney, "Insights into Canadian Peacekeeping Doctrine," 12–23.

192. Yates, "Military Stability and Support Operations," 51–61.

193. Maloney, "Insights into Canadian Peacekeeping Doctrine," 12–23.

194. Shimko, *The Iraq Wars*, Kindle, locations 3109–15; Ullman and Wade, *Shock and Awe*: 26, 14, quoted in Shimko, *The Iraq Wars*, location 3129.

195. Metz et al., *The Future of American Landpower*, 10–13.

196. Metz et al., *The Future of American Landpower*, v–viii, 10–13; Nichiporuk and Builder, *Information Technologies and the Future of Land Warfare*, 83.

197. U.S. Department of Defense, *Report of the Quadrennial Defense Review*.

198. Reimer, "Challenge and Change," 108–16.

199. Robert H. Scales, "America's Army: Preparing for Tomorrow's Security Challenges," PowerPoint briefing, December 1, 1998, Strategic Studies Institute, U.S. Army War College, Carlisle Barracks, PA; U.S. Department of Defense, *Report of the Quadrennial Defense Review*; Reimer, "Challenge and Change," 108–16.

200. U.S. Army Training and Doctrine Command, *Knowledge and Speed*, 1–3, 6–7.

201. Perry and Millot, *Issues from the 1997 Army After Next Winter Wargame*, ix–xiii.

202. U.S. Army Training and Doctrine Command, *Knowledge and Speed*, iv–v.

203. U.S. Department of the Army, *Army Vision 2010*, 6–8.

204. Bunker, "Five-Dimensional (Cyber) Warfighting," 30.

205. Bunker, "Five-Dimensional (Cyber) Warfighting," 17, 19, 30.

206. Bunker, "Five-Dimensional (Cyber) Warfighting," 28.

207. Bunker, "Five-Dimensional (Cyber) Warfighting," 3, 28.

208. See Metz and Johnson, *Symmetry and U.S. Military Strategy*.

209. Perry and Millot, *Issues from the 1997 Army After Next Winter Wargame*, ix–xiii; U.S. Army Training and Doctrine Command, *Knowledge and Speed*, 19.

210. U.S. Army Training and Doctrine Command, *Knowledge and Speed*, 19.

211. U.S. Army Training and Doctrine Command, *Knowledge and Speed*, 19.

212. Thomas, "The Battle of Grozny," 87–102.

213. Thomas, "The Battle of Grozny," 87–102.

214. Thomas, "The Battle of Grozny," 87–102.

215. Thomas, "The Battle of Grozny," 87–102.

216. Scales, *Future Warfare*, 173–85.

217. Scales, *Future Warfare*, 173–85.

218. Scales, *Future Warfare*, 173–85.

219. Scales, "America's Army."

220. Scales, *Future Warfare*, 173–85.

221. U.S. Army Training and Doctrine Command, *Knowledge and Speed*, 19.

222. Scales, *Future Warfare*, 173–85.

223. Hahn and Jezior, "Urban Warfare and the Urban Warfighter of 2025," 74–86.

224. Hahn and Jezior, "Urban Warfare and the Urban Warfighter of 2025," 74–86.

225. Hahn and Jezior, "Urban Warfare and the Urban Warfighter of 2025," 74–86.

226. Scales, *Future Warfare*, 173–85.

227. Scales, *Future Warfare*, 173–85.

228. Grau and Kipp, "Urban Combat," 9–17, emphasis added.

229. Grau and Kipp, "Urban Combat," 9–17.

230. Grau and Kipp, "Urban Combat," 9–17.

231. Grau and Kipp, "Urban Combat," 9–17.

232. Thomas, "The Battle of Grozny," 87–102.

233. Grau and Kipp, "Urban Combat," 9–17.

234. Perry and Millot, *Issues from the 1997 Army After Next Winter Wargame*, ix–xiii.

235. Reimer, "Challenge and Change," 108–16; Brown, *Kevlar Legions*, Kindle, 168–69.

236. Scales, "America's Army."

237. Perry and Millot, *Issues from the 1997 Army After Next Winter Wargame*, ix–xiii.

238. Reimer, "The Army After Next," 270.

239. Reimer, "The Army After Next," 270.

240. Reimer, "The Army After Next," 270.

241. U.S. Army Training and Doctrine Command, *Knowledge and Speed*, iv–v, 8–10.

242. U.S. Army Training and Doctrine Command, *Knowledge and Speed*, 8–10.

243. John R. Boyd, "Organic Design for Command and Control," slide presentation, Defense and the National Interest, n.d., accessed September 21, 2019, https://www.ausairpower.net /JRB/organic_design.pdf.

4 KOSOVO, THE WAR ON TERROR, AND THE OBJECTIVE AND INTERIM FORCES

GEN. ERIC SHINSEKI'S TERM as chief of staff of the U.S. Army was marked by three dramatic strategic shocks and consumed by his desperate efforts to steer the Army through them.

Just before General Shinseki assumed office, the Army embarked on a military intervention in Kosovo to stop Serbia's ethnic cleansing of Kosovar Albanians. The intervention revealed deep issues that still plagued the Army's ability to deploy itself into an austere theater of operations. In response, Shinseki would embark on the creation of a new type of Army brigade combat team with a new combat vehicle as its foundation. He would also envision a new path for the next two decades of Army transformation.

The next strategic shock to confront Shinseki was the terrorist attacks on the United States on September 11, 2001, and the U.S. response. The U.S. Army would be challenged to deploy forces to Afghanistan and Iraq. The invasion of Afghanistan was a very different war from any Army transformers had ever envisioned in the 1990s. The invasion of Iraq seemed much more familiar, but a new administration and a new secretary of defense challenged the Army to do much more with much less. Like every other senior leader in the Army, General Shinseki was a product of the Army's neglect of low-intensity conflict since the end of the Gulf War. He thus lacked the vocabulary to articulate the requirements for the postinvasion conflict that would inevitably follow in Iraq. He failed to successfully steer the Army through this second strategic shock.

This failure created a third strategic shock: the simultaneous catastrophes of the Army's first year in Iraq and the unraveling of the war in Afghanistan. General Shinseki's successors continue to struggle with how to steer the Army through this crisis to this day.

Shinseki's term as chief of staff of the Army should have been a renaissance of rebalancing the Army's focus between high- and low-intensity conflict. Both he and his two most senior commanders— Gen. John Hendrix, commander of U.S. Army Forces Command, Gen. John Abrams, commander of U.S. Army Training and Doctrine Command (TRADOC)—had direct experience with the low-intensity conflicts in the Balkan States. Yet the culture of the Army and the momentum of transformation—both of which drove the Army toward ever-greater hyperpreparedness for a high-intensity conflict against a peer competitor—conspired to keep the Army from institutionalizing the growing mountain of lessons from the low-intensity conflicts of the 1990s.

Kosovo, Part One: Operation Joint Guardian

Kosovo, a province of Serbia, was the historic gateway for Yugoslavia's trade with Italy and the world. The vast majority of the population of Kosovo, as much as 90 percent, was ethnic Albanian, Muslim, and culturally distinct from Serbians. This ethnic and religious group was painfully aware of the history of the Serbian oppression of Kosovo and had support from neighboring Muslim Albania, which was sympathetic to its plight.[1]

The Serbian government, headed by President Slobodan Milosevic, continued to proclaim itself the legitimate government of all of Yugoslavia. In 1989, as Yugoslavia began to disintegrate, the Milosevic government in Belgrade significantly curtailed the traditional autonomy of Kosovo, creating severe political tension. Kosovo declared independence in 1990, but only Albania recognized the declaration and the Army of the Federal Republic of Yugoslavia (FRY) remained in the rebellious province. The war in Kosovo between Kosovar Albanians and the Serbian government in Belgrade began in earnest in 1996 with the formation of the Kosovo Liberation Army (KLA). The Kosovar Albanian separatist movement instigated a spectrum of resistance activities, from rallies and demonstrations to hit-and-run guerilla attacks, all encouraged by neighboring Albania; the KLA was able to use Albania as a sanctuary beyond the reach of the FRY throughout the conflict.[2]

Eager to exert his authority, Milosevic launched a counteroffensive against the separatists. FRY troops began a campaign of arrests and reprisals. The war featured ethnic cleansing and massacres by Serbian militias and FRY troops against Kosovar Albanian civilians. To escape the slaughter, Kosovar Albanians flooded the borders of Albania and Macedonia, destabilizing the region and creating a humanitarian crisis. By August 1998, nearly 100,000 Kosovar Albanians were rendered refugees or internally displaced people. The United Nations (UN) responded with UN Security Council Resolution 1199, calling for an end to the carnage and a negotiated settlement. China abstained in the vote. Behind the scenes, Russia tried but failed to block the motion.[3]

In March 1999, peace talks broke down in Rambouillet, France, over the issue of the deployment of North Atlantic Treaty Organization (NATO) peacekeepers to Kosovo. The Serbians responded with a spasm of arson, rape, and murder in Kosovo to complete the ethnic cleansing before NATO could intervene. Thousands of Kosovar Albanians died, and many tens of thousands more became refugees or internally displaced people.

As NATO contemplated intervention, Gen. Wesley Clark, Supreme Allied Commander Europe, advised that airpower would not be sufficient to force Milosevic to relent; he wanted the commitment of ground troops to compel the Serbians to capitulate. Secretary of Defense William Cohen and the Joint Chiefs of Staff balked, and the U.S. Senate—by a 58 to 41 margin—approved only an air campaign. President Bill Clinton pledged that U.S. ground troops would not be committed to the war, though that assertion would soften as the campaign wore on. When NATO finally intervened, 800,000 Kosovar Albanians were displaced from their homes, with the FRY Army and the KLA fighting in and among them.[4]

NATO began a sustained bombing campaign aimed at ending Milosevic's campaign against Kosovar Albanians. When the campaign began, nearly everyone believed it would last well less than a week.[5] Instead the air campaign lasted seventy-eight days (March 24–June 10, 1999); NATO launched thirty-eight thousand sorties, 60 percent of them American. As the campaign wore on, it became clear that bombing would not be sufficient. The Serbs had adopted precisely the tactics that Army After Next researchers had predicted; the Serbian Army was hiding in urban centers among the civilian population to avoid the coalition's precision targeting capabilities, using decoys to complicate targeting, and employing an array of other "asymmetric" measures designed to defeat the high-tech precision of the allies.[6]

Serbian militias and the FRY Army also responded to the NATO aerial onslaught by accelerating the pace of their rampage across Kosovo. The clock was ticking for the Kosovar Albanians; Operation Allied Force had to produce results fast or there might be no population left to save.[7]

The British pressed the Americans to open a ground war in Kosovo to expel the Serbians. President Clinton, facing pressure from the U.S. House of Representatives, resisted the British demands. Gen. Dennis Reimer, chief of staff of the Army, also cautioned against the use of ground troops, explaining that this would expand the war beyond the issue of Kosovo to "going after the Balkans and stabilizing Serbia."[8]

Instead, to counter Serbia's asymmetric tactics, Clinton authorized the deployment of AH-64 Apache attack helicopters, along with associated logistics and force protection support. Army helicopters would presumably hunt down and make more effective attacks against Serbian armor. On April 4, 1999, a Pentagon spokesman predicted that the task force would be deployed within eight days.[9]

The deployment soon devolved into a debacle. It was originally planned for the modest Camp Able Sentry in Macedonia, but at the last moment had to be redirected to Albania because Macedonia refused permission for use of its soil for offensive operations against Serbia. Facilities in and around the Albanian airfield were insufficient for the massive logistic requirements of the aviation unit or for the landing of large cargo planes. Two Army aviators were killed in Albania and their helicopters destroyed in a training accident while preparing for the operation. It would take three weeks just to deploy the attack helicopters and another two weeks before they were ready for employment.[10]

Ultimately, the attack helicopters never took part in the conflict. By the time the aviation unit was in place and ready to operate, the war had ended (with the capitulation of Milosevic in midsummer 1999). After weeks of NATO airstrikes, the North Atlantic Council was seriously considering a ground invasion. On May 27, 1999, Russian prime minister Viktor Chernomyrdin arrived in Belgrade and pressured Milosevic to agree to NATO's terms—particularly the withdrawal of FRY troops from Kosovo and the acceptance of a NATO peacekeeping force. Russia was Serbia's only ally; Slobodan Milosevic capitulated rather than lose its support.[11]

The FRY and Serbian militias had ravaged Kosovo; they were able to compel the vast majority of Kosovar Albanians to move out of their homes and into precarious mountain hideouts. The FRY and armed Serbian civilians murdered as many as ten thousand Kosovar Albanians, burned as many as five hundred villages, and displaced 1.5 million Kosovar Albanians (nearly 90 percent of the population) from their homes.[12]

The war ended with the establishment of a U.S.-led peacekeeping force—the Kosovo Force (KFOR)—and Kosovo achieving de facto independence as an autonomous region. NATO committed thirty thousand troops. Task Force Hawk, joined by mechanized ground forces from the 2nd Brigade, 1st Infantry Division, became the eight-thousand-strong Task Force Falcon and deployed into Kosovo to execute stability operations as part of Operation Joint Guardian. The deployment quickly ground to a crawl on streets clogged with refugees, and critics in the United States howled even louder than they had for the stalled deployment of Task Force Hawk.[13]

Members of KFOR were surprised by the level of violence they found in Kosovo, which far exceeded the level in Bosnia-Herzegovina, even at the outset of Operation Joint Endeavor. The primary work for KFOR was, ironically, protecting Kosovar Serbians from angry Kosovar Albanians returning to their homes. In the first week of the occupation, at least twenty-seven Kosovar Serbs were abducted by the KLA and never seen again. Across the U.S. sector, American forces darted from hot spot to hot spot, responding to reports of gunfire. In some cases they arrived only to find dead Kosovar Serbs or burnt homes. In other cases they arrived in the middle of firefights or were met with gunfire themselves.[14]

The deployability woes that plagued the U.S. Army happened in full display of many of the senior Army leaders who would lead the Army into the twenty-first century, including Lt. Gen. John Hendrix, the commander of the V Corps and the future commander of U.S. Army Forces Command (FORSCOM).[15]

The debate over the lessons of Kosovo in its immediate aftermath circled around two issues. The first was whether airpower alone had won the war, and the second was the deployability issues that had plagued both Task Force Hawk and Task Force Falcon. Ultimately, both questions spoke to the relevance of the U.S. Army in the post–Cold War strategic environment; if the Army was slow and cost-inefficient in deploying to conflicts and airpower could win the war on its own, then why did the United States need an Army? Secretary of Defense William Cohen ordered a comprehensive review of the Army's readiness. Talk began to circulate about cutting spending on the Army's modernization program for its current fleet of AH-64 helicopters.[16]

The question of whether airpower had won the war was rather easily dispatched. As Daniel L. Byman and Matthew C. Waxman wrote, "NATO threats and bombing did not halt the ethnic terror for seventy-eight days, more than enough time for Serbia to displace almost a million Kosovar ethnic Albanians and kill thousands in Kosovo." General Klaus Naumann, the German chairman of the NATO Military Council, was even blunter: "They want us to stop the individual murderer going with his knife from village to village and carving up some Kosovars; that you cannot do from the air."[17] The debate would continue, but, while Stephen Hosmer would (in a publication prepared for the U.S. Air Force) claim, "Milosevic and other senior officials

Figure 4.1. Operation Joint Guardian. *Source:* U.S. Army Center of Military History, *Operation Joint Guardian: The U.S. Army in Kosovo,* 20-1.

have consistently asserted that the primary reason [they capitulated] was to avoid the destructive bombing that a failure to yield would have inevitable unleashed," it was the potential of being abandoned by the Russians that ultimately convinced Milosevic to comply with NATO demands.[18] And the Russian demands were, in turn, a response to the threat of the introduction of NATO ground forces.

The deployment problems encountered by Task Force Hawk, on the other hand, created a crisis of relevancy for the Army that overshadowed all other lessons from the conflict. There were certainly factors beyond the Army's control that complicated the deployment. While the Pentagon had pledged eight days to deploy the force, the Army had never agreed to nor felt

298

obligated to meet that timeline. Interservice rivalry almost certainly delayed the deployment. The Air Force repeatedly delayed the movement of Army equipment either due to weather conditions or because it had supposedly been directed to give higher priority to humanitarian aid shipments than to Task Force Hawk. Given that the Air Force had already expressed misgivings about the Army even participating in the air war, it is at least equally likely that it slowed down the deployment of Army forces to give itself time to win the war without the Army's help.[19]

But the Army also bore some of the blame; fears over security in Albania—similar to the overconcern for force protection that had handicapped the Army's mission in Bosnia-Herzegovina—compelled the Army to expand the mission from two thousand to five thousand service members and add artillery, armor, and other ground combat assets.[20]

Yet beyond all of these factors, Army forces—more than five thousand troops, twelve M1A1 Abrams main battle tanks, forty-two Bradley fighting vehicles, twenty-four AH-64 Apache helicopters, thirty-seven other helicopters, and dozens of smaller vehicles and pieces of equipment—were simply too heavy to move quickly. The logistics effort required to move such large forces and sustain them in an austere environment was mind-boggling. It would ultimately take 475 sorties of heavy, C-17 cargo aircraft from the Air Force to transport Task Force Hawk to Albania.[21]

Critics used the episode to argue that the Army was too heavy and too slow, rapidly becoming irrelevant to modern warfare. Vice president of the Council on Foreign Relations and former assistant secretary of defense Lawrence Korb told interviewers that Task Force Hawk was "a metaphor for how heavy the army is." He added, "If you can't get it to where you want to, it's no good."[22]

Recriminations abounded, made worse when an after-action review from the aviation task force, intended for the chief of staff of the Army, was leaked to the press. The consensus from reporters and analysts was that the Army was too heavy and too slow, in danger of becoming irrelevant to twenty-first-century warfare. One critic later wrote, "To many reformers both inside and outside the US Army, the problems the service encountered while trying to deploy a regiment of AH-64 *Apache* attack helicopters and support units have come to symboli[z]e its inability to respond rapidly to crises across the globe."[23]

Deployment of Task Force Falcon into Kosovo for Operation Joint Guardian gave critics even more ammunition to attack the Army's "weight." The elements of the 1st Infantry Division designated for Task Force Falcon were moved to Camp Able Sentry in Macedonia before their ground convoy into Kosovo. The lone two-lane road into Kosovo was already choked with refugees when Task Force Falcon began its movement with M1A1 Abrams tanks. The movement ground to a halt amid hours-long traffic jams stretching as far as five miles each. The deployment was also delayed by border guards stopping contractor vehicles at border crossings for days at a time. It took two days to get the first two hundred soldiers into Camps Bondsteel and Monteith in Kosovo.[24]

Inside the Army, the response to the episode was alarm that the Army still had not solved its deployability issues. Planners noted that had the seventy-ton M1A1 Abrams tanks been forced to deploy to Kosovo directly from Albania without first moving to Macedonia, four heavy engineer battalions would have required four months to build sufficient bridges to get the tanks into the country.[25]

While the issues surrounding the relevancy of the Army—raised by the perceived effectiveness of bombing and the difficulties deploying Task Forces Hawk and Falcon—overshadowed nearly all other lessons of the conflict, some astute observers did note other lessons from the war in Kosovo.

The asymmetric methods that the FRY Army and Serbian militias used to avoid NATO airpower—hiding in cities among civilian populations to avoid the precision strike capabilities of the U.S. military—sparked some debate over the similar observations that Army After Next researchers had noted in wargames. The Serbs used camouflage, decoys, and close proximity to Kosovar Albanian civilians inside the cities to avoid attack. FRY air defense systems—especially heat-seeking surface-to-air missiles—forced NATO aircraft to fly at higher altitudes, making Serbian forces even more difficult to detect. Strict NATO rules of engagement (ROE) magnified the effectiveness of these simple techniques.[26]

This asymmetry was not lost on low-intensity conflict observers. In a Spring 2000 article in the journal of the U.S. Army War College, *Parameters*, Tim Thomas of the Foreign Military Studies Office at Fort Leavenworth, Kansas, warned that "Kosovo . . . exposed problems" with the transformation ideal of "Information Superiority, the cornerstone of Force XXI." He added, "There is much to learn from Kosovo about the current myth of information superiority, particularly that simple human innovations can severely degrade digital dominance, and that human interpretation of data is a science worth reinvigorating."[27]

Because of the issues deploying U.S. forces to Kosovo, many of the lessons about the low-intensity conflict's political dimension were lost. General

Reimer did at least recognize the political difficulties in Kosovo. After leaving office, he said in an interview, "I don't see any real way of extracting from [the conflict in Kosovo] unless we're willing to let the cycle of revenge continue again. If we don't want the cycle of revenge to continue, then we're going to be there a long time. Those are the problems associated with Kosovo, and the problems associated with so many of the deployments we did in this post–Cold War world."[28] As with Bosnia-Herzegovina, Reimer saw the Army almost as a victim of the low-intensity conflict in Kosovo, as if it had no agency in creating the political solutions that might allow it to leave Kosovo.

General Eric Shinseki

The furor surrounding the deployment of U.S. Army forces to Albania and Kosovo had an especially significant impact on Army transformation since, in the midst of this furor, Gen. Eric Shinseki became the thirty-fourth chief of staff of the Army on June 22, 1999.[29]

Shinseki, a Japanese American from Hawaii, received a commission as a field artillery officer in December 1965 and was sent to Vietnam, where he was wounded twice, losing part of his foot while serving as a forward observer for an infantry company. After a prolonged convalescence and recovery in Hawaii, Shinseki transferred to the armor branch and served as an armor officer until he became a general officer.[30]

General Shinseki had a front row seat to transformation. He had held two general officer field commands: as the 1st Cavalry Division commander and later as the commander of U.S. Army Europe (USAREUR) and the NATO Stabilization Force (SFOR) in Bosnia-Herzegovina. Otherwise, since 1993, General Shinseki had been a creature of the Pentagon. Before becoming chief of staff of the Army, he served as director of training for the Office of the Deputy Chief of Staff for Operations (DCSOPS), then assistant DCSOPS, then DCSOPS, and finally as vice chief of staff of the Army.[31]

Before assuming his role as chief of staff, Shinseki directed a remarkably extensive survey of the Army's junior and senior leadership on the health of the Army. A team called the Special Staff Study Group—led by Brig. Gen. John S. Brown—conducted sensing sessions with general officers, junior officers, noncommissioned officers, and soldiers. It also surveyed Army families. To gauge opinions outside the Army, this team of more than a dozen officers and command sergeants major spread out beyond the Army to interview members of the secretary of the Army's staff, members of the U.S. Congress, congressional staffers, academics, and Army critics. This was, in short, the most exhaustive review of the state of the Army since the end of the Cold War.[32]

Just after assuming office, and based on this review, General Shinseki would complain that the Army was "smaller and busier, with a reduced budget and a tremendous increase in mission requirements." This analysis was borne out by the numbers. By 2000 the U.S. military would comprise 1.36 million men and women. The Army had borne the brunt of manpower reductions across the services over the course of the 1990s; as Shinseki assumed office, the active component of the Army had been cut by one-third, to 480,000.[33]

Defense budgets were, however, beginning to rise as Shinseki assumed office. The defense budget had rebounded from a low of $267.2 billion to $318 billion as the internet revolution began to manifest in a new industry, a booming economy, and federal budget surpluses for the first time in recent memory. Yet while the overall defense budget was growing, the Army was gradually losing its share of it to the other services; the Army's percentage of the total defense budget shrank from 27 percent to less than 25 percent during the same period. Since the end of the Cold War, the Army's budget had shrunk by well over a third—from $102 billion to $64 billion between 1989 and 1999.[34]

Meanwhile, the pace of operations was picking up. The ability to deal with two nearly simultaneous "major regional conflicts" developed in the Bottom-Up Review (BUR) at the beginning of the decade remained the hypothetical sizing structure for the Army, but senior Army leaders and pundits increasingly doubted that goal was achievable. And with the Army decisively engaged in Bosnia-Herzegovina and Kosovo, U.S. European Command expressed serious doubts that it could disengage from either mission should a high-intensity conflict erupt that required a draw of forces from Europe.[35]

And the clock was ticking; the next Quadrennial Defense Review (QDR) would begin in months. General Shinseki told subordinates that they could only take one year—until June 2000—to establish a change of direction for the Army and convince Army leaders of the need for said change. That would leave three years of Shinseki's term to try to institutionalize any recommended changes. Shinseki called this the "1 to 4 Rule."[36]

From the beginning, Shinseki had a very clear understanding of where he wanted to take the Army and its transformation. His focus was squarely on addressing the crisis of relevance that had emerged following the Army's disastrous deployment to Kosovo. During his tenure he would create a whole new organization, the Interim Force, to fill what he perceived as a deployability gap in the Army's capabilities.[37]

Shinseki today claims that his conviction that the Interim Force needed was born of the six-month buildup of U.S. Army forces required to launch Operation Desert Storm against Iraq in 1990 rather than the troubled deployment of Task Force Hawk to Kosovo. When a PBS *Frontline* interview asked Shinseki in 2000 if "the transformation that is underway is a very short-term answer to the deficiencies of Task Force Hawk," the general would only allow, "No effort to transform ever ignores any of those experiences where we lacked capability that we would have liked to have."[38]

Lt. Gen. (Ret.) James Dubik, on the other hand, is more direct, today admitting that the impetus for the Interim Force was the debacle surrounding the deployment of Task Force Hawk to support Operation Allied Force in Kosovo. In an interview after leaving office, General Reimer acknowledged that his successor "needed to do something a little bit more dramatic" than the normal trajectory in which transformation was headed. He added, "With the publicity we took on Task Force Hawk [the Interim Force] got him over the mobility hump." It is hard to see how the disastrous deployment of Task Force Hawk, culminating mere days before General Shinseki became chief of staff of the Army, could not have been dominant in driving his decisions concerning transformation.[39]

Shinseki's target for unveiling his plan was the October 1999 annual meeting of the Association of the United States Army (AUSA). In the four months preceding the event, a working group chaired by the vice chief of staff of the Army, Gen. Jack Keane, fleshed out the plans for the Interim Force while General Shinseki enlisted the support of Secretary of the Army Louis Caldera, the secretary of defense, and the chairman of the Joint Chiefs of Staff.[40]

The Objective and Interim Forces, Part One

The opening salvo of General Shinseki's "strategic communications" campaign was his statement of intent, meant to communicate to the Army, the American national security establishment, Congress, and the American people his vision for the future of the Army. First on Shinseki's list of "objectives" was addressing the deployability woes that had come to light during the Army's participation in the war in Kosovo. He wrote that a key goal of the Army should be "increasing strategic responsiveness," adding, "Heavy forces must be more strategically deployable and more agile with a smaller logistical footprint, and light forces must be more lethal, survivable, and tactically mobile." The vice chief of staff of the Army, Gen. Jack Keane, echoed those sentiments a few weeks later.[41]

Shinseki finally revealed his vision for the future of the Army at the AUSA's annual Eisenhower Luncheon on October 12, 1999. He began by repeating the mantra that had driven all of his predecessors: "The fundamental business of the Army is to fight and win our Nation's wars. Warfighting remains job #1." And lest anyone argue that current low-intensity conflict responsibilities should receive equal billing, he added, "There is no greater peacetime priority than preparing . . . for the next war."[42]

General Shinseki then described the key problem that he believed transformation needed to address. He believed that the "heavy divisions remain unequalled in their ability to gain and hold ground" but worried that these forces "are challenged to get to . . . contingencies where we have not laid the deployment groundwork," adding that "it takes significant effort and cost to sustain them." On the other hand, he said that "light forces . . . can strike lightning fast but lack staying power, lethality, and tactical mobility once inserted." He also worried that the Army's "logistical footprints for deployed forces" were "unacceptably large [and] driven sometimes by unrealistic replenishment demands but also by a complex inventory of multiple types of equipment."[43]

To address these problems General Shinseki envisioned "a strategically responsive force that is dominant across the full spectrum of operations." This force would combine all of the benefits of the heavy and light forces while avoiding their weaknesses. He described the force in terms of its characteristics. To accomplish "log support reductions" this force would use "common platform / common chassis / standard caliber designs by which to reduce our stockpile of repair parts." In other words, to reduce the logistical requirements for different kinds of repair parts and fuel, the force would use one common vehicle platform for all of its systems—direct fire, indirect fire, logistics, and troop-carrying systems. This new common combat vehicle would be "smaller, lighter, more lethal, yet more reliable, fuel efficient, and more survivable." This envisioned future vehicle would come to be referred to as the future combat system (FCS). The hypothetical vehicle would be equipped "with internetted C4ISR [command, control, communications, computers, intelligence, surveillance and reconnaissance]" to harness the power of the revolution in military affairs (RMA).[44]

In a shocking suggestion from a former armor officer, Shinseki asked, "Can we, in time, go to an all wheel vehicle fleet where even the follow-on to today's armored vehicles can come in at 50%–70% less tonnage?" He

responded, "I think the answer is yes." Because he was a former armor officer, General Shinseki's suggestion of replacing tracked vehicles with wheeled
vehicles would be immune to charges that it was an attempt to destroy the
armor branch; this was a proposal that perhaps only Shinseki could successfully make.[45]

Shinseki's vision went beyond General Reimer's Army After Next;
Shinseki's "Objective Force" would be vastly more deployable. He explained
that this new force would be able to "put a combat capable brigade anywhere
in the world in 96 hours . . . a division on the ground in 120 hours, and
five divisions in 30 days." This requirement was a direct response to the
asymmetric tactic of "delay[ing] or deny[ing] U.S. access to critical facilities"
that the 1997 QDR had predicted might complicate future wars. Shinseki
explained, "We intend to get to trouble spots faster than our adversaries can
complicate the crisis."[46]

Shinseki described the Objective Force as being the future of the Army.
He explained, "We intend to transform the Army, all components, into a
standard design," and made it clearer that he was talking about transforming
the entire Army: "When technology permits, we will erase the distinctions,
which exist today, between heavy and light forces and review our requirements for specialty units."[47]

In a conference at the U.S. Army War College a few weeks later, Shinseki
elaborated on the technologies he envisioned for the Objective Force. To
increase deployability, he noted, "We will look for future systems which can
be strategically deployed by C-17, but also be able to fit in a C-130 . . . for
tactical intra-theater lift." He also described other technologies that he wanted to integrate into this force. He explained that the Objective Force needed
"technologies that will provide survivability through low observable ballistic
protection, long-range acquisition, deep targeting, early attack, [and] first-
round-kill at smaller caliber." The Objective Force was an unequivocally
high-intensity conflict force. Shinseki continued, "If deterrence fails, we will
be postured to prosecute war with an intensity that wins at least cost to us
and our allies and sends clear messages for all future crises."[48]

The Objective Force could not be realized immediately because the technology to achieve the envisioned requirements for the FCS did not yet exist.
What really made the FCS difficult to realize was the balance between its
survivability and its weight. In a *Military Review* article in October 2000,
Maj. Gen James Dubik explained that the FCS needed to be "lethal and
survivable, but lighter and deployable," adding that the vehicle had to be "as
sustainable, lethal and survivable as the Abrams and the Bradley are right

now." Yet, Dubik wrote, "Such a vehicle should weigh 20 to 25 tons and fit into C-130 aircraft so it can get anywhere" and concluded, "That vehicle is not available now, because there is no technology to do it. We are looking for that answer by 2003."[49]

Congress, the defense industry, and the U.S. Army had heard such promises before. The Army had been talking about Force XXI and, later, the Army After Next throughout the 1990s. What was truly revolutionary about Shinseki's vision was that it was going to begin immediately. General Shinseki had lived through a decade of experimentation and wargames about the future of the Army. He would later say in the *Frontline* interview, "I am not sure what we gain out of experimentation. We had our experiment in the desert ten years ago [during the Gulf War], and we didn't like it."[50] Shinseki was ready to stop experimenting and act.

Thus, while Shinseki inherited transformation from his predecessors, he would do more than any of them to *actually transform* the Army—change its manning, equipment, organization, and training. As he explained at the 1999 AUSA Eisenhower Luncheon, "We will begin immediately to turn the entire Army into a full spectrum force which is strategically responsive and dominant at every point on the spectrum of operations." He proposed that the Army use "today's 'off-the-shelf' equipment to stimulate the development of doctrine, organizational design, and leader training even as we begin a search for the new technologies that will deliver the material needed for the objective force." General Shinseki would begin to field a new, medium, Interim Force within one year.[51]

The Interim Force would fulfill two roles. First, it would rapidly field units equipped with lighter, more deployable vehicles that could fill the gap between light forces, which were rapidly deployable but not survivable or self-sustaining beyond a few days, and heavy forces, which were lethal, mobile, and survivable but took months to get to a theater of operations. Second, it would fill the gap in time between the deployment of light and heavy forces—from ninety-six hours to 180 days—by providing a force with the deployability of light forces and the lethality and staying power of heavy forces.[52]

Yet the Interim Force was also frequently described as the "vanguard of the Objective Force." It would provide a platform to test and validate the technology and tactics of the Objective Force and its FCS.[53]

The Interim Force was not just a vision but also a course of action. General Shinseki explained, "As quickly as we can, we will acquire vehicle prototypes, in order to stand up the first units at Fort Lewis, Washington." He added, "It is our intent to have an initial set of prototype vehicles beginning to arrive at Fort Lewis this fiscal year."[54]

A few days later, at a conference at the U.S. Army War College, Shinseki made it clear that this Interim Force was not just an experimental force in the vein of Gen. Gordon Sullivan's experimental force (EXFOR) at Fort Hood, Texas—the target of Force XXI experimentation. The Interim Force would be a deployable force that would be put into operation in real-world contingencies. He also made clear that this force was not just aspirational. The first two Brigades would be stood up—with their new vehicles—within a year. Eventually, Shinseki hoped, at least six Interim Brigade Combat Teams (IBCTs) would be built.[55]

To quickly communicate this vision, General Shinseki's staff produced a slick PowerPoint visual. This one-slide show almost immediately began to be called the Trident slide, and it quickly became ubiquitous across the Army.[56]

While the urgency was novel, many of the concepts that fueled the design of the Objective and Interim Forces were not new. Rather, they were a fulfillment of Army XXI (the end state of the Force XXI transformation process)

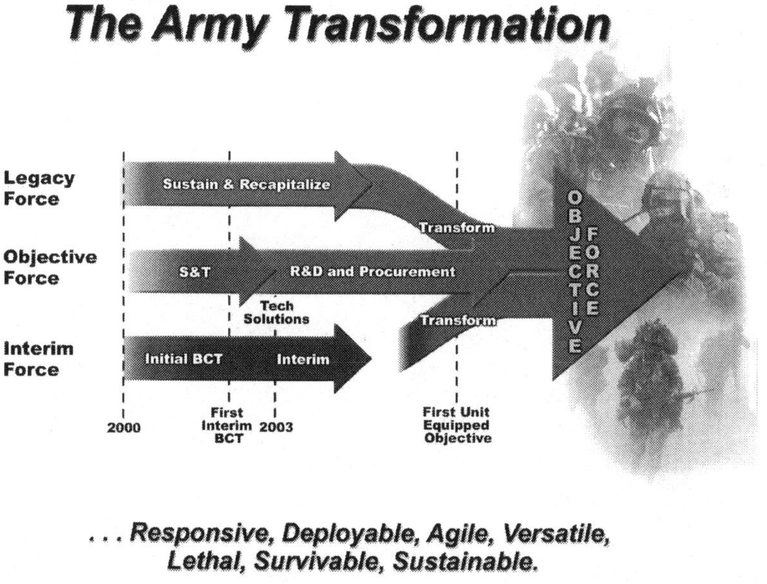

Figure 4.2. The Objective and Interim Forces. *Source:* John Sloan Brown. *Kevlar Legions: The Transformation of the U.S. Army, 1989–2005*, 195. *Note:* BCT = Brigade Combat Team; R&D = research and development; S&T = science and technology

and the culmination of the advanced warfighting experiments. The IBCT organization and operational (O&O) concept would say explicitly that the IBCT was "specifically designed as an Army XXI organization." In a *Military Review* article in October 2000, Col. Michael Mehaffey explained that the Interim Force's "unique, evolving, commander- and execution-centric C2 [command and control] environment builds on lessons learned during Force XXI experimentation." Interim Brigades would be "equipped with appropriate Army Battle Command System[s]"[57] that had emerged through a decade of prototyping, experimentation, and fielding with the EXFOR for Force XXI.

Another concept that pervaded Shinseki's vision for the future, more deployable Army was that the increased situational awareness and precision strike capability provided by the technology of the RMA would compensate for less protective armor to make the force lighter and more deployable. As the O&O concept for the Interim Brigades explained, "Although the [Interim] Brigade Combat Team must have the capability for information dominance, it will not always enjoy combat platform overmatch." It also noted that "situational understanding is the fundamental force enabler across all Brigade Combat Team battlefield operating systems and the foundation for risk mitigation with respect to Brigade Combat Team vulnerabilities."[58]

The networked, interconnected Interim Brigades would be able to see the enemy and themselves so well that they could avoid attack and mass their own firepower at a time and place of their own choosing—a time and place of maximum vulnerability for the enemy and minimum vulnerability for this future Army force. This was a realization of the ideas of knowledge and speed developed in the Army After Next wargames.[59]

The adoption of a new combat system—let alone an entirely new type of unit—normally takes a decade or more. In that light, the creation of the Interim Brigades was a masterpiece of strategic leadership in a massive bureaucracy worthy of a study all its own. At the strategic level, General Shinseki enlisted the support of all of the other services behind the initiative to ensure its success. To convince the U.S. Navy to sign on, Shinseki met with U.S. Pacific Command (PACOM) commander Admiral Dennis C. Blair and agreed to build the first four IBCTs in the PACOM area of responsibility. To enlist the support of the U.S. Air Force, Shinseki promised to require that the Interim Combat Vehicle be air-transportable in a C-130 aircraft. And, presumably, to enlist the support of the U.S. Marine Corps, the vehicle eventually selected as the Interim Combat Vehicle shared a common chassis with the Marines' Light Armored Vehicle (LAV) III, decreasing the cost of both vehicles and spare parts for the Marines. By enlisting the support of the other services up front, before beginning his public campaign, Shinseki was able to move to solicitation of proposals—a process that normally took years—in only seven months.[60]

Shinseki's efforts inside the Army were no less impressive. On November 9, 1999, he appointed Maj. Gen. James Dubik to oversee the actual manning, organization, and training of the IBCTs, charging him to "create irreversible momentum in one year." Designated as the deputy commanding general for transformation of TRADOC, Dubik was sent to Fort Lewis to construct the first two brigades. The post was the perfect place to start; it had one armor brigade (the 3rd Brigade, 2nd Infantry Division) and one light infantry brigade (the 1st Brigade, 25th Infantry Division). Here the expertise of heavy and light brigades could be fused into the new medium brigades. As Dubik explained in the *Military Review*, "The Army leadership chose Fort Lewis partly because it is home to both a heavy brigade and a light brigade. Both are going to be transformed into interim brigades." He added, "We are merging cultures and the strengths of these forces into new operational capabilities."[61]

Dubik's immediate supervisor was the FORSCOM commander, General John W. Hendrix, while his second-line supervisor was the TRADOC commander, General John N. Abrams—investing both of the Army's highest commanders in the outcome of the effort. The effort was also overseen in monthly video teleconferences and on-site visits by a "board of directors" consisting of the chief of staff of the Army, the FORSCOM and TRADOC commanders, and all of the Army's other commanders of major commands.[62]

Dubik established a "brigade coordination cell" to develop a vision and doctrine for the new IBCTs and a training plan to realize that vision. In February 2000, because a platform for the IBCTs had not yet (officially) been selected—let alone purchased and fielded—the cell borrowed thirty-two LAV III light wheeled vehicles from the Canadian Armed Forces that it could use to train the IBCT.[63]

Congress approved funding for the first two IBCTs in spring 2000, and Dubik immediately set about transforming the first Interim Brigade. Before its reorganization, the 3rd Brigade Combat Team (BCT), 2nd Infantry Division was a typical heavy BCT, equipped with M1A1 Abrams main battle tanks and M2 Bradley fighting vehicles. To create "irreversible momentum" Dubik began by divesting the armor brigade of all of its equipment, rendering it combat ineffective. Until it was equipped with Interim Combat Vehicles and manned and trained, it would be rated at C5—the lowest combat readiness rating—which would generate pressure from Congress and the U.S. Department of Defense to get it equipped.[64]

The 3rd BCT, 2nd Infantry Division was the first unit to transform into an IBCT, but it would not be the last. It would be followed by the 1st BCT, 25th Infantry Division, also at Fort Lewis. After work was complete at Fort Lewis, the Army would move on to the conversion of other units: the 172nd Infantry Brigade (Separate) in Alaska and the 2nd Armored Cavalry Regiment at Fort Polk, Louisiana. Others would follow: the 2nd BCT, 25th Infantry Division at Schofield Barracks, Hawaii, and the 56th Brigade of the 28th Infantry Division in the Pennsylvania Army National Guard.[65]

A key to the IBCT design was its mobility and flexibility. The IBCT was to be "high mobility at all three levels of operations. Strategically . . . at the operational level . . . [and] 100-percent tactical mobility." It would be an infantry-centric force, increasing the number of dismounted infantrymen from the 243 normally present in a heavy brigade to the 972 in the IBCT organizational structure.[66]

A *Military Review* article reprinted in a CALL newsletter spelled out other capabilities that would characterize the Interim Brigades. The IBCT would use its enhanced "intelligence, reconnaissance and surveillance (ISR) capabilities and digitized battle command systems to develop and disseminate a common operational picture throughout the force" and "a robust array of direct and indirect fire systems . . . to shape the battlespace and achieve decision in the close fight." To compensate for being "a force equipped with medium-weight armored and thin-skinned vehicles," it would use an array of technological and tactical methods to avoid engagements until it could achieve overmatch. And to reduce its logistical footprint, it would make "use of common platforms" across the force.[67]

But above all of the capabilities resident in the IBCT, its main focus was on deployability. In his *Military Review* article in 2000, Dubik noted that "there is a gap between the heavy and light forces that we need to fill; something that can get somewhere fast, that has more combat punch than a light force." He continued, "The IBCTs are being designed, manned and equipped to fill the gap." They would do so by being "capable of deploying anywhere in the world in 96 hours to immediately begin operations across the full spectrum of possible contingencies."[68]

Still, Dubik made it clear that the Interim Brigades were only a stopgap measure: "The IBCTs . . . will provide a complementary capability to our current light and mechanized forces, serving as a bridging force until science and technology allow the U.S. Army to achieve objective force capabilities."[69]

The O&O concept enumerated two other features of the IBCT that facilitated deployability. First, like the Objective Force, the Interim Brigades would be outfitted with a single, common combat chassis—the Interim Combat Vehicle. The O&O concept explained, "Demand reduction, and sustainment efficiency further drive the Brigade Combat Team to exploit *commonality of vehicular platforms* for combat." Second, the O&O concept explained that because of the networked capabilities of the IBCT, "functions that can be accomplished out of theater or through *reach-back* to higher echelons with the joint contingency force will not be incorporated in the Brigade Combat Team's organic force structure."[70] Both of these features would reduce the number of people, computers and vehicles, and the amount of repair parts required to deploy with the brigade.

At its core, the IBCT—and transformation, for that matter—was about preparing the Army to do the next Gulf War–like high-intensity conflict better. Dubik warned, "The next [Saddam] Hussein is not going to wait six months to attack. Whoever the next thug is has already learned that. So, speed is essential for us now; that is why the objective force has to have a vehicle as fearsome as the Abrams, but as deployable as the HMMWV [high-mobility, multipurpose wheeled vehicle]."[71] General Shinseki made a very similar argument in his interview with *Frontline* in 2000. When asked, "What would you say then was the biggest lesson that enemies watching took away from the Gulf War?" He answered, "That when they commit to battle, they should not take a six-month pause." He added, "They should follow up their early victories with sustained momentum, because the pause is what gave us the opportunity to structure the outcome of that war."[72]

Like the Objective Force, the Interim Force was unequivocally designed and optimized for high-intensity conflict. According to Col. Michael Mehaffey, "The IBCT has a pronounced offensive orientation." He added, "Its key operational capabilities are deliberately designed to enhance its offensive power, with clear benefits for deterrence, conflict prevention, containment or conflict resolution."[73] Secretary of the Army Louis Caldera and General Shinseki penned a joint article for the *Military Review* in which they also made this fact clear. They wrote that their goal was to maintain "the world's finest land force for the next crisis, the next war and an uncertain future."[74]

The IBCT was designed to operate in "small scale contingencies" like the Gulf War rather than "major theater wars" (great power wars against peer competitors) or "stability and support operations"—that is, low-intensity conflicts. While the O&O concept for the IBCT repeatedly referred to the IBCT as a "full spectrum, combat force," it also acknowledged that the

IBCT was "designed and optimized primarily for employment in small scale contingency operations." It went on to explain that the IBCT would need "additional enabling capabilities, for operations outside the scope of SSCs [small-scale contingencies], such as stability and support operations . . . and major theater war." The O&O concept added that the IBCT could not succeed in "stability and support operations" without significant "augmentations."[75]

The O&O concept explained that "SASO [stability and support operations] contingencies will often require additional expertise in the areas of information operations, psychological operations, civil affairs, public affairs, and legal affairs." The brigade would also need augmentation from engineers for "meeting requirements with respect [to] construction, facilities repair and management, infrastructure improvements (roads, bridges, etc.), sanitation, water supply, provision of shelter, and real estate management."[76]

The O&O concept revealed its authors' bias toward shielding "combat" soldiers such as infantry from messy low-intensity conflict tasks. In describing why the IBCT would need military police, it listed the tasks for which military police were needed: "Significant numbers of refugees and detainees; security and force protection; maneuver and mobility support; security for nongovernmental and private volunteer organizations; black market and other criminal activities; establish[ing] linkages with local or multi-national law enforcement agencies for the restoration of order in the absence of local police; and route and physical security."[77] Essentially, the Interim Brigade would need augmentation to do anything other than purely combat tasks within a low-intensity conflict. In this regard, the IBCT was no better at prosecuting such conflicts than the Army's so-called legacy force BCTs.

Moreover, the O&O concept explained that in low-intensity conflicts, the IBCT was intended only to serve "as an initial entry force and/or as a guarantor to provide security for stability forces."[78] It was silent on what mythical other force would arrive to do the hard work of engaging the political dimension of low-intensity conflicts so that the IBCT's leaders didn't have to.

Almost no one questioned the Army's continued diminution of low-intensity conflict in favor of a focus on high-intensity conflict. Instead Army transformers and Army critics argued about whether the Army's changes would be sufficient to prepare it for a coming great power war. Lt. Col. (Ret.) Andy Krepinevich argued that the Army was not going far enough to change, writing, "It sounds like the Army is talking the talk, but it is taking baby steps on the walk."[79]

In fact, the experiments that fixed the design of the IBCTs did not even test the formation against low-intensity conflicts. Instead, the BCT was tested against "a high end SSCO [small-scale contingency operation]" and "within an MTW [major theater war]." This was simply one more expression of the Army transformers' misconception that low-intensity conflict tasks were lesser included subtasks of high-intensity conflict competency. The "overriding need" that drove the design of the IBCTs was "to *achieve balance between capabilities for strategic responsiveness and requirements for battlespace dominance*," not achieving a balance between low- and high-intensity conflict capacity.[80]

To the extent that the U.S. Army's 1990s low-intensity conflict experiences had any impact on the design of the IBCTs at all, it was in that the M1A1 Abrams tanks and M2 Bradley fighting vehicles were too heavy for the roads in places like the Balkans. In the *Frontline* interview, General Shinseki said of operations in Bosnia-Herzegovina, "We discovered that most of our heavy equipment, in a country that was wrestling to reestablish itself economically, tore their roads up so badly that commerce could not get through." He added, "We put most of that equipment on a ready status inside our installations in Bosnia, and really went to patrolling with much lighter wheeled vehicles. Our Humvees are fine for driving the roads, but when you go to a hot situation you would revert back to those heavier pieces of equipment." Shinseki explained that the Army needed lighter vehicles with the same protection as the Army's present heavy armor. An AUSA pamphlet on transformation made a similar observation about Kosovo, adding that the U.S. Army's armored vehicles took a long time to deploy to the country.[81]

Yet while the IBCTs had lighter vehicles and were easier to deploy, they were not intended as low-intensity conflict forces. Even when asked directly—and with the benefit of hindsight—if the Interim Brigades were intended to address shortfalls in executing operations other than war, General Shinseki still insists that the Interim Brigades were intended for "conventional" operations. He adds that his quest to outfit the Army in lighter vehicles with a common chassis was about increasing the Army's deployability, not its ability to operate in and among the populace in urban environments.[82]

The same AUSA pamphlet on transformation was adamant that the IBCT should not be misconstrued as a low-intensity conflict force. In response to the "misconceptions & myths" that "the interim brigade combat teams are just peacekeeping forces," the pamphlet answered, "Wrong again. They will have major hell-bent-for-leather, go-to-war missions in addition to being able to handle operations other than war. They are combat formations first and foremost and will have a substantial amount of firepower. There are a number of roles for them on the conventional battlefield."[83] The IBCTs were not about rebalancing the Army to better execute low-intensity conflict; they were about getting to high-intensity conflicts faster.

By late 1999, when General Shinseki announced his vision for the Objective and Interim Forces, the problems of asymmetry and urban operation—points of convergence in the otherwise divergent debate between Army transformers and low-intensity conflict observers—had become central to transformation. Thus, the O&O concept for the IBCT explicitly addressed both topics. It repeatedly claimed that the IBCT was "designed and optimized primarily for employment . . . in complex and urban terrain, confronting low-end and mid-range threats that may employ both conventional and asymmetric capabilities."[84] For an assertion so frequently and forcefully posited by transformers, it was surprising how many conflicting views there were on exactly what these terms meant. Indeed, there was very little agreement across the Army—and especially between Army transformers and low-intensity conflict observers—as to what asymmetry and urban environments actually were.

By 1999 the Joint Staff, at least, seemed to have settled on a definition for asymmetry, noting, "Asymmetric approaches are attempts to circumvent or undermine US strengths while exploiting US weaknesses using methods that differ significantly from the United States' expected method of operations. [Asymmetric approaches] generally seek a major psychological impact, such as shock or confusion that affects an opponent's initiative, freedom of action, or will. Asymmetric methods require an appreciation of an opponent's vulnerabilities. Asymmetric approaches often employ innovative, nontraditional tactics, weapons, or technologies, and can be applied at all levels of warfare—strategic, operational, and tactical—and across the spectrum of military operations."[85]

In the summer of 2000, the Joint Staff published *Joint Vision 2020*, which claimed that it "buil[t] upon and extend[ed] the conceptual template established by *Joint Vision 2010* to guide the continuing transformation of the Armed Forces." The vision maintained *Joint Vision 2010*'s focus on the principles of "Dominant Maneuver," "Precision Engagement," "Full-Dimension Protection," and "Focused Logistics," now enabled by "Information Superiority and Technological Innovations."[86]

There were really only two significant innovations in *Joint Vision 2020*. First, it added three contexts for operations in 2020, "Peacetime Engagement," "Deterrence and Conflict Prevention," and "Fight and Win."

These corresponded roughly to peace, low-intensity conflict, and high-intensity conflict, respectively. Second, it purported to respond to "the appeal of asymmetric approaches" among adversaries and the supposed increasing "focus on the development of niche capabilities." *Joint Vision 2020* claimed, "The potential of such asymmetric approaches is perhaps the most serious danger the United States faces in the immediate future." But it saw asymmetry in strictly technological terms, including "long-range ballistic missiles and other direct threats to U.S. citizens and territory."[87]

In January 2001, Steven Metz and Douglas Johnson were commissioned by the Strategic Studies Institute (SSI) at the U.S. Army War College to consolidate an Army understanding of asymmetry. Their definition varied somewhat from that of the Joint Staff: "Asymmetry is acting, organizing, and thinking *differently* than opponents in order to maximize one's own advantages, exploit an opponent's weaknesses, attain the initiative, or gain greater freedom of action." Metz and Johnson also listed different categories and levels in which asymmetry could occur.[88] A key innovation of this definition was its removal of U.S. forces as the victim of asymmetry—an acknowledgment that the U.S. military, too, could have asymmetric advantages.

In fact, perhaps their greatest contribution to the subject of asymmetry was to hint that it was the U.S. Army's own hyperpreparedness for a high-intensity conflict against a mythical peer competitor that was driving adversaries to seek asymmetric solutions. Citing the National Defense Panel, they warned that America's adversaries had watched and understood the ramifications of the Gulf War. No enemy would ever face the United States in that way again.[89]

Yet Metz and Johnson's effort to generalize asymmetry instead reduced asymmetry to any successful tactic or operational approach. Some of the techniques they identified were "anti-access or counter-deployment techniques using missiles, mines, terrorism, and other weapons"; "Maoist People's War, blitzkrieg, and Massive Retaliation"; the North Vietnamese "painting themselves as a victim and gaining the 'moral high ground'"; guerrilla wars; and "protracted wars."[90] Rather than revealing the true nature of asymmetry, Metz and Johnson succeeded in revealing that asymmetry was a buzzword without any tangible meaning beyond the emotional response that it invoked—fear of the unknown.

Nonetheless, a technological conception of asymmetry dominated Army transformation thinking on the topic. The IBCT O&O concept warned of a variety of asymmetric threats that might emerge among adversaries, including new ballistic missile technology and advanced air defense capabilities or even precision guided weapons "such as laser guided mortar rounds." It cautioned that future adversaries would have comparable unmanned aerial vehicle and night vision capabilities. TRADOC planners also predicted that "low cost GPS jammers would be employed to disrupt precision munitions targeting, sensor to shooter links, and navigation."[91]

The O&O concept even warned that future enemies would be able to compete with the United States in space, using "satellite reconnaissance, communications, and navigation" available due to the "commercialization of space" that would make "these capabilities available to all . . . [and] allow even low-tech forces to enter the world of information age capabilities." Particularly interesting was that the TRADOC planners identified space as an "asymmetrical advantage" that the United States rather than its adversaries then enjoyed.[92]

The TRADOC planners were particularly worried about what they called "technological surprise." They feared that future adversaries might employ unanticipated or unexpected technologies "through hybridization or the purchase of off-the-shelf systems." One example the O&O concept provided was that "a hand-held GPS can be purchased at Wal-Mart for $49.00." Such tactics might allow an adversary to "achieve parity and even fleeting superiority in portions of an AO [area of operations]."[93]

Describing the threat that the Interim Brigades would be designed to combat, the TRADOC planners also identified some low-tech means that adversaries might use to combat the brigades, though they did not always identify these means as asymmetry. They wrote, "The threat recognizes the strengths and weaknesses of their adversary and adjusts their tactics accordingly." Planners repeatedly warned that the enemy would use "ambush techniques" in order to avoid "decisive combat with superior forces." The intent of these attacks would be to "inflict mass casualties" to wear down the U.S. "national will" and "reduce the U.S. will to continue the fight." They also expected adversaries to exploit the ROE that the United States and its allies might impose upon themselves, writing, "Opponents will attempt to operate outside the limits of our ROE and influence the establishment of more restrictive ROE through information operations that attack national will and coalition sensitivities."[94]

To the extent that IBCT planners thought about low-tech, low-intensity conflict means at all, they did so as unconventional components of an otherwise conventional adversary, used for the purpose of slowing or blocking the deployment of the Army. Denying the Army access to theaters of war continued to be a point of particular concern for Army transformers considering asymmetry. In describing the most likely adversaries in both major theater wars and small-scale contingencies, the O&O concept for the IBCT predicted that at most this enemy would "present the U.S. [with] a non-linear, simultaneous battlefield" by using "coordinated operations between police, paramilitary, special purpose, guerrilla forces, mercenaries, terrorists, and conventional forces, and WMD [weapons of mass destruction]" to attack airfields and seaports where the U.S. Army was attempting to enter the theater of war. The planners noted that "SLOCs (sea lines of communication) will be interdicted by naval mines. Through astute use of diplomacy (use of threats, mutual interest, cultural homogeneity, etc.), the threat will deny the U.S. bases in U.S. friendly countries, or delay U.S. entry (through protracted negotiation, hostage-taking, or appearing to modify policy to seemingly conform to U.S. demands)." This was a subsumption of low-tech, low-intensity conflict asymmetry into the Army transformer's comfortable conception of denial of access as a form of asymmetry. Army transformers could not conceive of an adversary using low-tech, low-intensity conflict means as the *only* means to challenge the United States in a future war, refusing to present the Army with any conventional force at all to fight. Instead low-tech means were seen solely as a supporting effort that had to be overcome before the Army could reach and destroy the "conventional forces" that the enemy would inevitably present.[95]

TRADOC planners wrote in the O&O concept that "a terrorist threat to U.S. civilians or soldiers not directly connected to the intervention *could* be used [as] a device to change the fundamental nature of the conflict." They did call this threat "asymmetrical in nature," but they could not conceive of it being carried out against large, vulnerable civilian targets. Instead, they mainly saw the threat as one to "continental U.S. or overseas staging areas" in order to "disrupt power-projection capabilities by attacking installations, information systems, or transportation nodes."[96] In other words, the planners saw terrorism as a tool to assist a conventional enemy in protecting its country from invasion rather than as an enemy attack in and of itself.

In their SSI study on asymmetry, Metz and Johnson likewise noted that while the Army was much more comfortable with technological asymmetry, "Less attention has been given to other, equally feasible asymmetric challenges, especially protracted warfare, political constriction, and organizational asymmetry based on the emergence of networked enemies." Among these were "*normative asymmetries*" such as "terrorism, ethnic cleansing, human shields, and the like" which exploited the United States' higher moral threshold for military activities. Yet Metz and Johnson also echoed the Interim Force planners' conception of low-intensity conflict as a form of asymmetry—as augmentation to the operations of a conventional adversary. They warned of "guerrilla operations in an enemy's rear area as an adjunct to conventional operations."[97]

While Interim Force developers could at least conceive of an asymmetric, low-tech, low-intensity conflict adjunct to a high-intensity conflict enemy, Objective Force planners were laser focused on a high-tech conception of asymmetry. Among the "effective asymmetric doctrine and capabilities" that enemy forces possessed in Objective Force wargames at the U.S. Army War College were "a 'web-like' concept of operations that sought to mitigate the vulnerability of its decisive points and centers of gravity"; "precision fires, relying heavily on cruise and theater ballistic missile strikes"; and "offensive air defense strikes against U.S. forces." This imagined enemy would also employ "sophisticated decoys, spoofing, and other creative methods to counter American ISR [intelligence, surveillance, and reconnaissance], and precision engagement systems."[98] This was the technological conception of asymmetry that had dominated Army transformation thinking on the subject since the concept of asymmetry first crystallized in *Joint Vision 2010*, years earlier.

Notably, Interim Force planners did acknowledge that an "Asymmetric Adversary" would seek to prolong a conflict to "compel" the United States "to cease resistance and agree to seek conflict resolution through a negotiated settlement." The planners explained that, for such an enemy, "decisive operations extend further in time (beyond a cease-fire, for example) to include the post-conflict stability operations."[99] This was a surprisingly low-tech, low-intensity conflict conception of asymmetry not normally recognized by Army transformers.

Yet the planners offered no solution to this low-intensity conflict form of asymmetry. Instead the Objective Force would continue to be optimized through technology, doctrine, and training for high-intensity conflict. Future soldiers, proficient in high-tech, high-intensity conflict, would just have to deal with these low-tech, low-intensity conflict challenges as best they could.

Low-intensity conflict observers had a different conception of asymmetry. Increasingly they began to see the U.S. Army itself as having a suite of asymmetric capabilities in the realm of high-intensity conflict that would cause adversaries to avoid high-intensity conflict and instead seek low-tech, low-intensity conflict means to sidestep this asymmetry and create an asymmetry of their own.

During the Iraq War, Lt. Col. (Ret.) Conrad Crane of the SSI would go on to be one of General David Petraeus's "insurgents," coauthoring Field Manual (FM) 3-24, *Counterinsurgency*.[100] But throughout much of Gen. Eric Shinseki's tenure as chief of staff of the Army, Crane, writing for SSI, was one of the more influential low-intensity conflict observers. When Crane weighed in on Kosovo, he never used the word *asymmetry*, but still wrote of the unintended consequences of U.S. Army dominance of high-intensity conflict: "One of the paradoxes of conventional deterrence is that the stronger U.S. military forces are, the less likely we are to have to use them." He added, "No aggressor since the Vietnam War has risked attacking where deployed American ground forces blocked the way."[101]

For low-intensity conflict observers, the response of potential adversaries to this Army asymmetry in its dominance of high-intensity conflict would be to resort to low-tech, low-intensity conflict means. Max Manwaring, in an SSI publication released just before 9/11, warned, "The better a power such as the United States becomes at the operational level of conventional war, the more a potential opponent turns to asymmetric solutions." He added, "There appears to be no recognition of the fact that the lessons of the Vietnam War, the Persian Gulf War, and any of the hundreds of smaller conflicts that have taken place over the past several years are not being lost on the new political actors emerging into the contemporary multipolar global security arena." Manwaring repeatedly used the conflation "asymmetric guerrilla threats" and warned of the dangers of neglecting them.[102]

In a monograph on asymmetry published by the SSI, Melissa Applegate of the Defense Intelligence Agency likewise warned that future enemies would use low-tech, low-intensity conflict measures as a counter to the Army's high-tech, high-intensity conflict dominance. She explained, "The strengths inherent in [*Joint Vision*] *2020* operational and enabling concepts are countered by a relatively rigid reliance on—and anticipation of—familiar forms of conflict" and the presumption that the enemy would not present the U.S. Army with these challenges. Applegate noted that "disorganized, civilianized, and primitive warfare, conducted by individuals and nonstate actors" could have "potentially crippling effects" on the future U.S. military envisioned in *Joint Vision 2020*. Her solution: "Asymmetry cannot be treated as a 'lesser included case.'" Instead, she posited, "The United States must decide whether conflicts that fall short of [the high-intensity conflict] threshold are legitimate missions worthy of the same levels of effort . . . that we have placed for years on conventional warfighting." But while the solution was simple, adopting it was difficult. Applegate believed that the senior leadership of the Army was "not ready to take the chance that we will not be ready for war."[103]

Applegate's final warning would seem eerily prescient less than a year later. She noted that *Joint Vision 2020* "acknowledges the emerging threat asymmetry brings to the environment but does not acknowledge the possibility that asymmetry *may be all there is*."[104] In other words, future enemies might present the United States with no conventional force at all to fight.

Another point of convergence that continued to mature between Army transformers and low-intensity conflict observers was the challenge that urban operations presented to transformation—now embodied by the Objective and Interim Forces. When asked about urban operations at a conference at the U.S. Army War College in November 1999, General Shinseki's response betrayed that he shared the Army transformers' predilection for seeing urban terrain simply as complex terrain that impeded movement and observation. He explained that urban operations complicated "situation awareness" and frustrated the Army's "intelligence systems," which were "very well focused on large units, large platform formations." For Shinseki, the problem of conducting urban operations was one that could be solved by technology.[105]

The O&O concept for the IBCT tried to address those who warned about the challenge posed by urban operations, though the document did little more than acknowledge that the environment posed problems. Promisingly, Interim Brigade planners did note that urban terrain contained civilian population and infrastructure, which made them "centers of gravity and therefore required areas of operation." Planners understood that because "45% of the world population currently resides in urban settings and it is projected that in the next ten years that this percentage will increase to 60%" it would "become increasingly more difficult to avoid operations in an urban environment."[106]

The writers of the O&O concept even acknowledged that the presence of civilians might "add the aspect of humanitarian crisis conditions requiring population management and/or support and control." These crises, in turn, would draw in nongovernmental organizations (NGOs) and international organizations that might or might not support U.S. goals. The closest that the Interim Brigade planners came to acknowledging that urban operations might involve low-intensity conflict was to say that, because of the people living in the urban area, the IBCT might be required to execute "stability and support and sustainment tasks, to include population and area control, support to humanitarian operations and peace enforcement" at the same time that it was conducting "combat operations."[107]

Yet while its authors repeatedly insisted that the IBCT was designed to dominate in "urban and complex terrain," the O&O concept did not give any explanation of how the IBCT would reduce this complexity. In fact, to the extent that the population was addressed at all, it was as an obstacle to fire and movement. The civilian populace complicated target detection and engagement. The document posited that the presence of civilians would allow the enemy to mix "with civilian population as a method of concealing a light infantry force." In explaining how enemies might use that environment asymmetrically, the Interim Brigade planners noted that a future adversary "will focus on urban areas (or complex terrain) to negate technological overmatches in intelligence and weapon systems and as a means of creating strongholds where he can achieve sanctuary from our effects. Such settings degrade weapons system standoff, are troop and supply-intensive, and add complexity to the application of firepower to avoid collateral damage and non-combatant injuries."[108] Apparently, then, people in cities had no significance beyond their use by the enemy as human shields.

The conflation of the terms *urban* and *complex terrain* is telling (and in fact, the two terms appear only in conjunction with each other throughout the entire O&O concept). Probably looking to the asymmetric example of the Serbs in Operation Allied Force, planners warned that the future augured "increased use of urban areas as sanctuary for conventional capability."[109]

For TRADOC planners, urban terrain could be overcome by mobility and technology. The O&O concept explained some of the capabilities the IBCT would need to overcome "complex and urban terrain," writing, "The Brigade Combat Team must possess a robust array of direct and indirect fire systems adequate to . . . achieve decision in the close fight inherent within complex and urban terrain. . . . All elements of the team must be internetted to supporting fires to achieve 'point and shoot' level of responsiveness. Requirements to minimize collateral damage and noncombatant casualties require accuracy to the level of 'the third window on the right of the second floor.' "[110]

In fact, to Army transformers—who saw urban areas solely as complex terrain—the IBCT itself was the solution to the problem of urban operations. Colonel Mehaffey's October 2000 *Military Review* article declared that the IBCT was "optimized primarily for employment in smaller scale contingencies (SSC) in complex and urban terrain." Mehaffey claimed that the IBCT was more than suited to urban operations because of its "high dismounted infantry strengths for close combat in urban and complex terrain."[111] The article didn't connect the large number of infantry soldiers to their ability to interact with the populace. Instead, it appears that Mehaffey believed that additional infantry made the brigade more capable in urban terrain merely because foot soldiers could maneuver through structures and narrow paths inside cities to hunt down enemies.

It was easy for Army transformers to wish away the problem of urban operations because wargames testing the Objective Force ended before the messy business of stabilization after high-intensity conflict began. The report on one iteration of a 2001 exercise at the U.S. Army War College noted that, at the end of the exercise, "Although deprived of freedom of maneuver and external support and sustainment, locally dangerous forces remained scattered throughout the area of operations, most in fortified urban enclaves." The report continued, "While incapable of major offensive operations, these residual forces posed a persistent hazard to Blue's lines of communication," and the researchers also acknowledged that "their elimination or suppression would be a necessary precondition to conflict termination."[112] Yet this most difficult and perhaps lengthy phase of the operation was not even tested in the wargame. One reason the researchers felt justified in ignoring this low-intensity conflict phase of the operation was almost certainly because they did not understand how difficult it can be to successfully engage the political dimension of such a conflict; they didn't understand that wars are not truly won until the urban areas are stabilized and a political settlement has been reached.

While publicly Army transformers might have dismissed the problem of urban operations, behind the scenes the problem was sufficiently disconcerting to prompt TRADOC commander Gen. John Abrams to commission a study, the results of which decidedly did not offer Army transformers any solace. The report submitted by Roger Spiller, a researcher with the U.S. Army Combined Arms Center (CAC) at Fort Leavenworth, echoed earlier arguments by Les Grau and Jacob Kipp that the essential property of a city was its nature as a human environment. Spiller wrote, "An urban environment is not defined by its structures or systems but by the people who compose it."[113]

But Spiller added yet another layer to the urban operations problem by using historical examples to show that an urban center becomes an even messier, more complex military problem as it begins to collapse under the stresses of war. "Cities are, after all, built to function in peace," he wrote. "The machinery of the essential and the commonplace—civil order, power, distribution of food and water, transport, medical care, communications—grinds toward an eventual halt. Then, the city *in extremis* becomes a different entity altogether."[114]

In their treatise on asymmetry, the SSI's Steven Metz and Douglas Johnson had identified urban operations as a form of asymmetry: "Urban operations . . . counterbalance a military force with superior mobility and long-range fires." Spiller quipped that Army transformers had taken to calling anything they did not understand "asymmetry," adding, "That asymmetric warfare would be associated with urban warfare is significant."[115]

Spiller warned that the Army would not be able to avoid cities in the future: "It is strategic or operational inferiority that drives an enemy to resort to such desperate measures [as hiding in cities] in the first place." Yet, he added, "Cities have always been important because we have made them so." Spiller concluded his study by urging the Army to stop all transformation until it could come to grips with the problem of urban operations.[116]

Ralph Peters likewise noted in an article for *Parameters* that the crux of the asymmetry presented by urban areas was that they were full of people: "Tasked with urban operations, soldiers think of buildings. . . . This focus on 'terrain' leads to the assumption that military operations would be more challenging in a Munich than in a Mogadishu." Of course, Peters pointed out, Munich submitted with barely a fight, while Mogadishu dealt the U.S. Army a humiliating defeat. The difference, he explained, "lay not in the level of physical development, but in the human architecture." He added that it was "the people, armed and dangerous, watching for exploitable opportunities, or begging to be protected, who will determine the success or failure of" an urban operation.[117]

Peters's categorization of different types of cities was of limited use, but his appreciation of the complexity of these human environments captured the true challenge of military operations in cities. "Cities are far more complex organisms than any text can suggest," he wrote. "Suffice to say that the greatest illustration of the human ability to self-organize shows in the daily functioning of cities." Peters warned, "The increasing size and number of cities pose practical challenges for urban operations. Even in the smoothest operation, cities consume troops; in combat, they devour armies." He concluded, "The center of gravity in urban operations is never a presidential palace or a television studio or a bridge or a barracks. It is always human."[118]

As had been the case throughout the 1990s, both Army transformers and low-intensity conflict observers continued to acknowledge that the future held many more low-intensity conflicts. Mehaffey's article on the IBCT in the *Military Review* acknowledged that "the high frequency of joint contingency operations in the 1990s" was "expected to continue and perhaps rise during the 21st century."[119] Major General Dubik, in his *Military Review* article, asked, "What is the future Kosovo? What is the future Bosnia? What is the future Somalia? We do not know what they are, but we know they are going to be out there."[120] Even General Shinseki acknowledged that there were likely many more low-intensity conflicts in store for the U.S. Army. As he told *Frontline*, "We've seen a bit of what [the] next century is going to look like, and the kinds of deployments we've had in the last ten years." He added, "Yes, it is Desert Storm. But it's also Somalia and Haiti and Panama, Bosnia, Kosovo, and East Timor."[121]

Not surprisingly, though, Shinseki's solution to the problem was transformation. In October 2000, at his second AUSA Eisenhower Luncheon speech, he admitted that because of the low-intensity conflicts of the 1990s, "The agility and versatility inherent in our Cold War formations have been stretched beyond design limits."[122] Elsewhere Shinseki explained that his new force would "dominate across the full spectrum of operations by providing them the versatility and the agility to transition rapidly from one point on that spectrum to another with least loss of momentum." Here, "versatility" and "full spectrum" were both used as euphemisms for the ignoring of low-intensity conflicts. As Caldera and Shinseki noted in their *Military Review* article, such conflicts could not be allowed to interfere with "fighting and winning our Nation's wars, our nonnegotiable contract with the American people."[123]

At the core of Shinseki's logic was also the belief that within the constellation of low-intensity conflict tasks the most dangerous and difficult—and thus the most important—among them were the combat tasks. In the *Frontline* interview, he explained, "The conditions change very quickly—as they can in places like Kosovo—in 20 minutes. You find yourself having to go very quickly intellectually and physically from what was a peacekeeping mission into war fighting. . . . How have you prepared your youngsters . . . intellectually . . . to be able to very quickly prevail in that more intense higher mission requirement?"[124] But this logic did not comport with the Army's actual experiences throughout the 1990s. While combat was indeed the most dangerous facet of a low-intensity conflict in terms of bodily harm to soldiers, unless your only goal was avoiding casualties—which many senior Army leaders perceived to be the case—combat was not the most dangerous element in terms of success in such a conflict. In Somalia, Haiti, Bosnia-Herzegovina, and Kosovo (among many other places), it was the operation in and among the populace and the wrestling with the political dimension of low-intensity conflict, rather than combat tasks, that proved most difficult. Being untrained and ill equipped for these more difficult tasks in such a conflict was infinitely more dangerous to the United States' strategic objectives in a conflict. And, ultimately, by failing to end the conflict, Army units prolonged it, creating more danger to more soldiers as they were continually rotated through another forever war.

Behind the scenes there was some debate within the Army leadership on whether to alter the training of the force to better deal with low-intensity conflicts. A memorandum for the chief of staff of the Army from the CAC proposed three options: "Train for War" (with mission readiness exercises, or MREs, when low-intensity conflict deployments were imminent); "Train for MOOTW [military operations other than war]" (with MREs when "war" was imminent); or "Let the Commander decide the METL [mission essential task list] focus." After a cursory analysis, the CAC recommended letting commanders decide the METL because of three drawbacks to training for low-intensity conflict: "First, it reduces the deterrent value of Army forces because units will be less ready for offense and defense actions. Second, it reduces responsiveness in case Army forces are alerted for missions that require offensive and defensive actions. . . . Third, Army forces not trained for offense and defense cannot be effective in stability operations because unit effectiveness in stability operations often depends on the ability to threaten and to employ the use of overwhelming deadly force."[125] The underlying assumption of this analysis, of course, was that all of the Army's forces should be trained in the same way rather than bifurcating forces—that is, designating Army units to specialize in either low- or high-intensity conflict. But on a deeper level, this analysis took for granted that low-intensity conflict was a lesser included subset of high-intensity conflict, for which it was easier to reorganize, reequip, and retrain on short notice.

There were a number of admissions buried in this memorandum that senior Army leaders had directly contradicted in public statements. The CAC authors admitted that Army units were unprepared to participate in low-intensity conflicts in the initial phases of operations—making them less useful to joint force commanders. They also noted that for combat units "the difference between executing tasks in war and in MOOTW can be significant," adding, "While many of the tasks are similar, conditions can be dramatically different." The memorandum admitted, "If Army forces must always conduct MREs before deployment then Army forces will, at best, have marginal usefulness to the joint force commander for initial phases of crisis response operations and for any short term operation" and noted that "MREs require time to prepare the exercise and to complete the training—time that joint force commanders often cannot afford."[126] All of these admissions contradicted the repeated assertion on the part of senior Army leaders that low-intensity conflict tasks were a lesser included subtask of high-intensity conflict proficiency.

In fact, Maj. Gen. James Dubik would make this exact claim in a briefing to students at the U.S. Army School of Advanced Military Studies at Fort Leavenworth. One of his briefing slides would argue that the Army should tailor its forces to the specifications of the BUR, to fight "conventional, theater war" in "two theaters." He did acknowledge that many of the innovations envisioned for the IBCT and the Objective Force would have negligible impact on low-intensity conflict: "Technological superiority DOES NOT offset numerical inferiority in these arenas as much as it does in the realm of conventional combat." Yet, in Dubik's view, the Army would have to muddle along with the forces "justified in terms of the 'high-intensity' warfight." He believed this solution was adequate because "the size of the 'warfight' force is large enough and diverse enough to handle lesser included, low-intensity contingencies."[127]

Thus, General Shinseki insisted that Army units train only on high-intensity conflict tasks unless they were one of the small percentage of units designated for an impending deployment to support a low-intensity conflict. In the October 2001 AUSA *Army Green Book*, in an article that had been written before 9/11, Shinseki noted that units would only "train on warfighting METL tasks unless ordered to change to stability operations tasks or support operations tasks by the Corps Commander."[128]

This official diminution of low-intensity conflict continued to be reflected in the Army's formal education. In the 1999–2000 academic year, the U.S. Army Command and General Staff College (CGSC) at Fort Leavenworth still did not have a block of instruction dedicated to low-intensity conflict in its curriculum. The core curriculum still retained one practical planning exercise in a low-intensity conflict scenario, but the lessons learned from such conflicts in the 1990s were otherwise absent from field grade officer education at the institution. This practice continued into the 2000–2001 academic year.[129]

Low-intensity conflict observers, like Army transformers, believed that the future held more such conflicts. In a PBS *Frontline* interview Ralph Peters said, "I'm as confident as one can be betting that we are going to see a lot more Sierra Leones, East Timors, Kosovos, Bosnias, Kashmirs, Chechnyas, and Colombias." He added, "Inevitably for national interests, or because the genocide is so horrific we cannot stay out, we're going to get involved in some." Notably, Peters seemed to think that IBCTs were being built for this low-intensity conflict future. He concluded, "Our nation . . . needs these medium-weight forces."[130]

General Shinseki had expressly rejected the idea that the Interim Brigades were peacekeeping forces. When asked to respond to this idea in the *Frontline* interview, Shinseki commented, "They're intended to have a war fighting responsibility." He did allow that these forces might "perform peacekeeping missions as well," but he believed that it was their combat capabilities and their strategic mobility—not any greater capacity to operate in and among the populace or to engage the political dimension of low-intensity conflicts—that would be valuable in this environment.[131]

Yet even though he denied that the Interim Brigades were built for peacekeeping, Shinseki held out the possibility of such missions as a justification for building more IBCTs. Raising the specter of operations in Bosnia-Herzegovina, which were still ongoing, he explained, "If you're going to have a long-term rotation policy as we do now—six months per unit—you'll need more than just three interim brigades. You'll need something on the order of five, as a minimum."[132]

Venerable Vietnam-era defense analyst Jeffrey Record contended, "The post–Cold War world is a world predominantly of *low-technology* wars *within* states, not high-tech combat between states." He noted, "Clausewitzian great-power clashes have been superseded by smaller, politically messy wars, many of them fought by irregular forces within failed states," adding that "policing such states has become more time- and force-consuming than preparing to refight the Korean and Gulf wars."[133]

Army major David W. Shin cited Samuel Huntington and Robert Kaplan in warning of a "coming anarchy" caused by "scarcity of resources, cultural and racial conflicts and geographic destiny." He predicted that the technology of the RMA, which the Army was chasing as quickly as it could, would be of limited if any use in the age of low-intensity conflicts that most likely awaited the Army.[134]

In a conference at the U.S. Army War College in late 2000 sponsored by the SSI and the Georgetown University Center for Peace and Security Studies entitled "Alternative Military Strategies for the United States," a series of national defense luminaries informed the Army that since it undeniably faced a future rife with low-intensity conflict, it ignored it at its peril. Retired ambassador Robert Oakley, who had served as President George

H. W. Bush's envoy to Somalia, believed that the United States would face many more such conflicts in the foreseeable future. He wrote that the Army needed to reorganize into more modular formations and increase its ability to conduct civil-military planning and operations. He added that the Powell Doctrine's insistence on clear-cut political objectives was an anachronism in the modern age of low-intensity conflict warfare. Michael O'Hanlon, a senior fellow at the Brookings Institution, contended that the strategic location of the fault lines along which low-intensity conflicts were likely to emerge would also make it impossible for the United States to avoid becoming entangled in them. Carl Conetta of the Commonwealth Institute in Cambridge, Massachusetts, believed that the Army had both overestimated the likelihood of two major regional conflicts and the forces required to deal with them. He was convinced that the Army was in danger of becoming irrelevant in a future dominated by low-intensity conflicts and he, too, advocated smaller, more flexible tactical formations.[135] Former assistant director of the Congressional Budget Office, Cindy Williams, was blunter, questioning both the force-sizing construct of two major regional conflicts and the suitability of the Army that it necessitated to the operations in which the Army was actually engaged and likely to remain engaged in the foreseeable future.[136]

In a 2001 monograph about the low-intensity conflicts of the 1990s, the SSI's Conrad Crane noted, "Since mid-1993, American military forces have engaged in 170 separate SSCs, ranging from humanitarian assistance to peacekeeping, averaging between 20 and 30 a month." He added, "The Army had an important role in the vast majority of the recorded contingencies, especially in peacekeeping or show-of-force missions and cases of domestic support or humanitarian assistance, and usually bore the brunt of major operational requirements." Crane also cautioned that there would be many more low-intensity conflicts in the future. For Crane, the consequences were clear: "*The Army must be trained and structured to execute some degree of nation-building during the stabilization phase of SSCs.*"[137]

In the introduction to a collection of essays on the future of the BUR's force-sizing construct of two major regional conflicts, the SSI's Steven Metz wrote, "Cross-border invasion by a rogue state with a Soviet-style military is much less likely today than a decade ago." Instead, he noted, "Humanitarian intervention and protracted peacekeeping are increasingly important tasks for the militaries of many nations, including the United States." Metz also cautioned that it would be harder to ignore these conflicts: "Globalization and the information revolution have linked the world more closely than at any time in history, thus making it more difficult to ignore aggression or instability in less vital regions."[138]

Many of the contributors to this SSI publication were equally convinced that the future held not two major regional conflicts but a flood of new low-intensity conflicts. Richard Krugler, a professor at the National Defense University explained that it was fantasy for the Army to believe it could be "kept [on] the shelf, standing guard against two regional wars that are unlikely to erupt, and being unavailable to deal with important events and missions that actually are occurring." He added, "Military forces are created in order to be useful when they are needed, not treated as a precious asset that can be applied only in a few extreme cases that seldom, if ever, occur."[139]

In this same collection, Robert David Steele wrote, "Severe low-intensity conflicts (defined as conflicts with over 1000 casualties per year), have leveled off." "However," he added, "lesser low-intensity conflicts are increasing steadily in number each year." He even provided a chart that showed that the number of high-intensity conflicts in the world had remained steady, while "internal political & ethnic violence [was going] through the roof." He advocated making the requirements for low-intensity conflict coequal with high-intensity conflicts in sizing the force.[140]

In his SSI publication, venerable low-intensity conflict observer Max Manwaring noted that "over half the countries in the international community are faced with one variation or another of asymmetric small (i.e., guerrilla) wars." He added that such conflicts had become "the most pervasive and likely type of conflict" and concluded, "It is almost certain that, sooner or later, the United States will become involved in some of these small (i.e., guerrilla) wars."[141]

For Manwaring, the implication was clear: "There is a high probability that the President and Congress of the United States will continue to require military participation in small internal guerrilla wars well into the future." He added, "In this security environment, military . . . forces have little choice but to rethink security as it applies to guerrilla menaces that many governments have tended to wish away."[142]

Writing on *Joint Vision 2020*, Melissa Applegate acknowledged that high-intensity conflicts might still occur. Yet she warned, "Within the [*Joint Vision*] *2020* timeframe, the objectives, motives, and intent of others are much more personal and less cataclysmic in nature; the scale and scope of conflict more localized." This was not, she added, "the type of conflict envisioned by [*Joint Vision*] *2020*."[143]

Applegate suggested that the Army give greater focus to low-intensity conflict. She noted that the Army's high-intensity conflict dominance had "not proven to be an effective conflict prevention mechanism among lesser powers and sub-state warring factions," adding, "The existence of a powerful U.S. military [cannot] deter nonstate actors from pursuing their objectives." "Indeed," she noted ironically, "such a force, globally deployed, offers lucrative targets of opportunity to those dedicated to demonstrating American weakness (Khobar Towers and the U.S.S. *Cole*, for example) or circumventing military operations through asymmetric approaches."[144]

Deployability

Fueled by the ongoing research and development of the Objective and Interim Forces among Army transformers, deployability continued to be a topic of discussion among low-intensity conflict observers as well. Army brigadier general Virgil L. Packett II, assistant division commander for support of the 101st Airborne Division, penned an article with Capt. Timothy M. Gilhool in early 2000 for the *Military Review* in which he recalled the division's deployability challenges during its deployment to Central America to provide relief in the aftermath of Hurricane Mitch. Writing about the deployment of almost five thousand service members to the region, he noted, "The limited number and quality of ports and airfields presented significant challenges for JTF [Joint Task Force] *Aquila*."[145] The Army had experienced very similar issues during its deployment to Somalia.

Col. George Shull, chief of staff of the Missouri Army National Guard (ARNG), raised another issue with Army deployability: the continuing challenge with deploying reserve component soldiers. He wrote, "The 15 ARNG enhanced brigades are not planned for deployment until 90 days into any future conflict." Other reserve component units, he added, "have been shelved as a strategic hedge with no real relevance." Referencing the BUR force-sizing construct of two major regional conflicts, he noted that "it would have been very difficult for the Army to have withdrawn the division from Bosnia for redeployment to either Saudi Arabia or Korea." Shull advocated "earlier deployment of the 15 ARNG enhanced brigades to meet the US Army MOOTW . . . tempo without jeopardizing a response to two nearly simultaneous MTWs."[146] But this discussion missed the much more critical lesson that the Army had learned from its 1990s low-intensity conflicts: that the true problem with mobilizing reserve component soldiers was not combat forces but the many capabilities like civil affairs, psychological operations (PSYOP), engineers, and logistics—all of which resided primarily or wholly in the reserve component. This is a clear indication that at least this lesson had not been institutionalized across the total Army.

At the beginning of the new century, several low-intensity conflict observers began to reexamine the conflicts of the 1990s for lessons that could be applied to the Objective and Interim Forces. Chief among these observers was Conrad Crane of the SSI, who completed his detailed treatise on these conflicts in early 2001. Writing about the rotation of forces into and out of Somalia, Crane noted, "There was a poor transition from one force to another, and a lack of appreciation for the increasing security problems and capabilities of the armed threats in the country." He added, "One problem with short rotations is the loss of institutional knowledge that results." Rotations had also led to "the failure to properly coordinate humanitarian, military, and diplomatic requirements."[147]

The Army Unprepared

Low-intensity conflict observers continued to note—with growing exasperation—that the U.S. Army was not prepared to execute such conflicts. Army major David Shin warned that the Army's hyperpreparedness for high-intensity conflict actually made it less prepared to execute low-intensity conflicts. In a late 1999 article in the *Military Review*, he wrote that the idea that superior knowledge would allow the Army to overmatch any adversary was not universally applicable to every enemy in every kind of conflict. He asked, "This approach may be valid with an enemy such as the Iraqi army of *Desert Storm*, but what if we are involved in another low-intensity conflict like Vietnam or Afghanistan?" Taking up the argument of many observers who had warned of a low-tech, low-intensity conflict form of asymmetry, he added, "We will still face great difficulty with irregular warfare because the very nature of such conflicts tends to negate the technological advantages of our future force." Shin concluded, "[The] Army is becoming a narrow-spectrum force designed primarily to deal with an enemy similar to the one we faced in the Iraqi desert."[148]

Like many other observers before him, Shin recommended revisions to the Army's education system. He wrote that soldiers need to "understand different cultures and religion[s]." He added that soldiers also needed more education in "regional history, culture and the religions of the world," and concluded that the Army needed to "institutionalize" this education.[149]

The Defense Intelligence Agency's Melissa Applegate gave similar warnings: "Steadfast pursuit of new and improved conventional capabilities—essentially getting better and better at what we are already the best in the world—may leave the United States without the proper tools and techniques to fight the emerging threat." She added, "Worse, following a predetermined course based on past successes may create a false sense of security and blind the United States to problems just over the horizon."[150]

Defense analyst Jeffrey Record argued that the Army was irrelevant to the current security environment, noting, "The Army remains structured primarily to fight a big conventional war in Europe. . . . The active-duty Army remains organized around ten logistically and bureaucratically ponderous divisions, six of them heavy divisions trained and equipped for combat against like armies employing modern tanks and armored fighting vehicles. . . . Heavy forces cannot be deployed quickly from the United States, are terrain-constrained, have limited utility in small wars against irregular adversaries, and are ill-suited for either peacekeeping or peace-enforcement operations." Record argued that the logic for the massive, mechanized U.S. Army disappeared with the collapse of the Soviet Union. Record wrote that war with Russia was extremely unlikely, adding, "Nor can significant, if indeed any, US Army force requirements be squeezed out of the postulation of China's emergence as America's next global strategic rival over the next 20 to 50 years." He noted, "China-as-the-new-Soviet-Union also ignores China's critical dependence on the global capitalist economy." Of the prospect of "another Korean or Gulf war requiring a major Army commitment," Record argued that they were both unlikely and that a much smaller conventional U.S. would be more than able to win them.[151] These were direct assaults on Army transformers and their myth of a great power war against an imagined future peer competitor.

Record argued that the United States' unmatched dominance of high-intensity conflict meant it would never have to fight such a war: "Demonstrated US conventional military supremacy . . . has driven our adversaries into the search for effective supra- and sub-conventional alternative weapons and strategies." He added, "The American lead in the technologies of the so-called 'Revolution in Military Affairs' is incontestable because no other state can afford to compete . . . [but] mastery of the RMA is mastery of a war that likely will never be fought." Record stopped short of claiming "that the United States has fought the last of its big conventional wars," but he did insist that "there is no potentially hostile national army in the world that comes anywhere close to matching both the human and technological quality of the US Army." He added, "Full note must be taken of the fact that we are in the midst of fundamental structural change in the international political system which is producing less and less interstate war and more and more conflicts within states."[152]

Lest Army transformers dismiss his arguments as discounting General Shinseki's new Objective and Interim Forces, Record took direct aim at this new strategy for transformation, writing that, even as the Army talked about lighter formations, it "plans no fundamental reorganization, such as discarding divisions in favor of smaller, more strategically mobile combat groups." More important, he added, "Nor does the Army plan any fundamental restructuring, such as . . . creating units specifically trained and equipped for peace operations."[153] For Record, fundamental change could only be achieved when the Army acknowledged the necessity of proficiency in low-intensity conflicts and began to adapt itself to better execute them.

Conrad Crane wrote in his treatise on the conflicts of the 1990s about the consequences of the Army's unpreparedness for the low-intensity conflict phase of the war in Panama, "Due to a focus on conducting a decisive operation and not the complete campaign, the aftermath of this SSC did not go as smoothly." He continued, "Though guidance from SOUTHCOM [U.S. Southern Command] on post-hostility missions was clear, tactically oriented planners at the 18th Airborne Corps in charge of JTF South gave it short shrift." As a result, he wrote, chaos reigned once the Panamanian Defense Force disintegrated and military police were both unprepared and numbered too few to restore law and order. Crane concluded, "Senior commanders admitted afterwards that they had done poorly in planning for the stabilization phase and hoped the Army would remedy that situation in the future."[154] All of these facts would have been a revelation to many Army officers; news about the aftermath of the war in Panama had largely been lost in the din of the Gulf War.

Regarding Somalia, Crane made other observations about the Army's unpreparedness for low-intensity conflict. He argued that U.S. forces "were not structured or resourced to accomplish all their required missions, and this culminated in the debacle in Mogadishu in October 1993." He added, "This SSC illustrates the importance for peacekeepers also to be capable of warfighting, and that task forces configured primarily for humanitarian missions might not be able simultaneously to conduct effective peace enforcement.[155]

Addressing the challenges and shortfalls in Bosnia-Herzegovina, Crane noted that the Army was short on military police, engineers, civil affairs personnel, military intelligence soldiers, and linguists. He also asserted that the focus of military intelligence on the "enemy" was completely unsuited to the complex, sectarian political environment and the primarily human intelligence means of collecting information in the country.[156]

Crane also made a more general observation about the Army's capacity to execute low-intensity conflicts: "American military forces have not done . . . well with post-hostility planning and execution." He attributed this situation to "shortfalls in Army attitudes, resourcing, and force structure" while participating in low-intensity conflicts, and attributed this weakness to "an operational focus concerned primarily with the conflict phase of SSCs." He blamed "the disappointing experience in Vietnam" for causing the Army to focus "almost exclusively on winning major wars" and believing that "nation-building missions are to be avoided." His study concluded, "The U.S. Army . . . must be trained and structured to execute some degree of nation-building during the stabilization phase of SSCs." For those who thought that this responsibility could be dumped on other U.S. government agencies such as the State Department, he added, "inadequacies of civilian organizations . . . insure that the Army will not be able to avoid such missions in a future."[157]

While Crane praised General Shinseki's Interim Force as a first step in reorganizing to better address low-intensity conflicts (a contention Shinseki himself had at least partially discounted), Crane advocated more changes, including "employ[ing] combat engineers with armored personnel carriers for peacekeeping, relying on their secondary mission as infantry while retaining their building skills." Alternatively, he suggested using military police officers equipped with armored security vehicles (ASVs).[158] Yet these suggestions would have required less than a complete commitment by the Army to low-intensity conflict. The Army's experience in the many such conflicts of the 1990s had already shown that there were far too few military policemen or engineers (or civil affairs or PSYOP officers) to do all of the required low-intensity conflict tasks alone; infantrymen would have to condescend to performing these tasks as well if the Army hoped to ever bring such a conflict to a successful conclusion.

Crane also advised, "Army schools at all levels will have to prepare soldiers better" to meet the challenge of low-intensity conflict, and "units will have to adjust METLs accordingly." He recommended moving more logistics and combat support units (such as engineers, military police, civil affairs, and PSYOP soldiers) from the reserve component back to the active component of the Army. Finally, Crane believed that the Army needed to "increase the ability of units at all levels to train for, plan, and execute stabilization phase tasks" and "ensure adequate focus is placed on the planning and execution of stabilization phase tasks at the Command and General Staff College and the Army War College."[159]

In the same vein, Max Manwaring wrote, "Given today's realities, failure to prepare adequately for small war (i.e., guerrilla) contingencies is unconscionable." He advocated a number of educational reforms "to modify Cold War mind-sets and to develop the leader judgment that is needed to deal effectively with ambiguous, complex, politically dominated, multidimensional, multi-organizational, multinational, and multicultural internal war situations." After suggesting a long list of specific points on which Army education needed to begin to focus to better prepare leaders for low-intensity conflicts, Manwaring asserted that "the study of conventional war has always been considered to be essential preparation for leaders involved in war. The study of 'unconventional' asymmetric war is no less essential."[160]

At the beginning of the war on terror, Army lieutenant colonel John Nagl wrote the influential book, *Learning to Eat Soup with a Knife*. During the Iraq War, Nagl would go on to be one of General David Petraeus's "insurgents," and another coauthor of FM 3-24. But in 2000 Maj. John Nagl was still an instructor at the U.S. Military Academy at West Point, where he penned a *Military Review* article with a West Point cadet, Elizabeth Young, that was considerably less diplomatic than Jeffrey Record's article on the misdirected energy of Army transformation.[161]

In the article Nagl and Young attacked Army transformation directly: "While the US Army prepares to fight and win two nearly simultaneous major theater wars, it will frequently be called upon to provide the military forces necessary to implement our nation's multifaceted response to CHEs [complex humanitarian emergencies].... Even though peace operations and preventing deadly conflict are becoming increasingly common missions, the Army currently treats each CHE as an exception." They noted that "the Army ... engages in little routine preparation for such events."[162]

Nagl and Young admitted that the Army conducted "ad hoc" training for units when they were designated to deploy to a low-intensity conflict. They also acknowledged that the Joint Readiness Training Center (JRTC) at Fort Polk, Louisiana, was training units in "mission rehearsal exercise[s]" prior to their deployment to Bosnia-Herzegovina. But, they cautioned, "While training prior to scheduled deployment on peacekeeping operations is certainly both sensible and appropriate, it is insufficient."[163]

It was arguably Nagl and Young who first advocated a complete restructuring of the Army— "a two-tier military establishment complete with a constabulary force" and "specialized units whose primary mission is to respond to CHEs." Barring this restructuring, they advocated the complete reengineering of the Army's training regimen. If some units were not going to be specially designated for low-intensity conflict operations, they contended, then every unit would have to be trained in such conflict. Nagl and Young asserted that rotations at the Joint Multinational Readiness Center, the JRTC, and the National Training Center (NTC) should include a low-intensity conflict scenario "both leading up to and building down from a typical mid-intensity conflict . . . scenario." They argued that these rotations "should integrate . . . multinational, NGOs, PVOs [private volunteer organizations], UN participants and relevant US agencies."[164]

Col. Mark Vinson of USAREUR likewise advocated a bifurcated Army, "with unit-specific MTW or smaller-scale contingency mission focus and organization." He claimed this would solve the problem of readiness degradation that occurred when units that were trained for high-intensity conflicts were committed to low-intensity ones. Vinson advocated using the reserve components to build "rotation forces for extended contingencies"—providing all but the initial entry portion of a low-intensity conflict force—while the vast majority of the active component continued to prepare for a great power war against an imaginary peer competitor.[165] This solution ignored, of course, a key lesson from all of the Army's 1990s low-intensity conflict experiences: many of the most essential capabilities for such conflicts already resided in the reserve components and were slow and difficult to mobilize when needed. But on a deeper level Vinson's solution was really just another form of diminution of low-intensity conflict as a less important mission—relegating such conflict to the "weekend warriors" so that the "professional soldiers" could continue to train for "real" wars.

Col. George D. Shull, chief of staff of the Missouri ARNG, made a more serious proposal than Vinson for using reserve component soldiers to restructure the Army to better address low-intensity conflicts. He recommended reorganizing two National Guard divisions to form special purpose divisions, organized with military police and engineer brigades optimized for rear area tasks during high-intensity conflict and postconflict stability and low-intensity conflict tasks after or in the absence of a high-intensity conflict.[166] While less condescending than Vinson's suggestion, Shull's proposal likewise ignored the repeated problems that the Army had experienced with the slow deployment of critical combat support capabilities such as military police, engineers, civil affairs, and PSYOP soldiers, which resided primarily in the reserve component.

Other observers also believed that the time had come to consider building separate U.S. military forces for low- and high-intensity conflicts. Former commander in chief of U.S. Central Command (CENTCOM), retired Marine Corps general Anthony Zinni, believed that the U.S. military might need to bifurcate, with one force focused on preparedness for two major regional conflicts and the other ready to respond to low-intensity conflicts. U.S. Army War College professor David Jablonsky suggested that the time had come for "the transformation effort to consider such experimental concepts as a two-force army" to address the divergent challenges of low- and high-intensity conflict. Robert David Steele believed the U.S. military should actually be trisected, with a "force-on-force" contingent fighting large high-intensity conflicts; a "small wars" force operating under U.S. Special Operations Command, but with three divisions of marines to help with the more violent low-intensity conflicts; and a "constabulary" force—weighted with military policy and civil affairs soldiers—conducting the less violent low-intensity conflicts. In this vision, other U.S. government agencies would be subsumed into the U.S. military's low-intensity conflict strategy and swoop in to rescue U.S. forces from the messy political dimension of the conflict.[167] Steele's vision ignored the repeated lessons of the Army's 1990s conflicts—lessons through which other agencies of the U.S. government had shown themselves wholly inadequate to meet this challenge.

The SSI's Conrad Crane rejected the bifurcation of the Army on several grounds. First, Crane argued, "The overall conventional deterrent value of today's relatively small Army will be significantly reduced if some units are perceived as having more limited capability for offensive and defensive operations, unless these constabulary units are an addition to the existing force structure." He also argued that the Army could not build enough units specific to low-intensity conflict to address the growing number of such conflicts around the world. Crane did look positively on the prospect of using reserve component forces to form low-intensity conflict forces, though he again worried about the scale of the demand.[168]

While the arguments for and against bifurcating the Army were generally considered and insightful, they all shared the presumption that the BUR requirement "to respond to two nearly simultaneous MTWs" was a valid and necessary requirement.[169] None of these observers questioned the assumption that the Army had to be ready to fight two great power wars—presumably against not one but two imaginary peer competitors—at the same time.

Max Manwaring was asking a much more fundamental question, one that even most low-intensity conflict observers were not. Of Army transformers' quest for ever-greater capacity to fight high-intensity conflict, he wrote, "It would appear that the civilian and military leaders . . . are still expecting some equivalent of Soviet combined arms armies to come crashing into West Germany, or Iraqi armies to again maneuver in the open desert." Manwaring concluded with a series of very fundamental questions about the Army's proposed future Objective and Interim Forces: "Why? For what purposes? Against whom? And, how will these forces be used to achieve a political end?"[170] No answer ever came from Army transformers to these very good questions.

To the extent that Army transformers did address low-intensity conflict in the O&O concept for the Interim Brigade, it was only to explain that the brigade was not designed to participate in these conflicts without additional augmentation. One reason Army transformers gave for dismissing prepa-ration for low-intensity conflict in favor of hyperpreparedness for a great power war went to the heart of the misconception that Army senior leaders held about the future. As the O&O concept for the IBCT cautioned, "The United States must remain vigilant and attuned to the activities of advancing competitors or potential competitors over the next two decades. History has proven time and again that some actors can quickly develop the military might to challenge world powers. Today's asymmetric threat capabilities improve the capability of even regional adversaries to effectively oppose our interests and military forces." The writers added that while "conflicts with non-state actors such as international terrorist groups and drug cartels will increase, state-on-state conflict will persist and will remain the most haz-ardous mode of war." The Army had been caught unprepared for two world wars at the cost of grievous loss of life. In the future, this line of thinking held, the Army had to ignore the distractions of the present and continue to prepare for "the most hazardous mode of war."[171]

This was the idea that drove not just the creation of the Objective and Interim Forces but the entirety of Army transformation since the end of the Cold War. Upon assuming office as chief of staff of the Army, General Shinseki sent copies of *America's First Battles, 1775–1965* to members of both the House of Representatives and the Senate. This 1986 book, edited by Lt. Col. Charles E. Heller and Brig. Gen. William A Stofft, chronicled how America's lack of preparation in peacetime had repeatedly cost the Army dearly in war. The editors' conclusion: "For the military leader confronting the potential of war in a prewar environment, constant effort is required to reduce the gap between training and battle."[172] When asked in the *Frontline* interview why he had sent the book, General Shinseki said he had wanted to communicate to the congresspeople and senators that "the first battles of all of the wars we have fought have seen tremendous price and human loss because of our lack of preparedness for that war." Echoing the warnings of Gen. Gordon Sullivan a half-decade earlier, he added, "It's about not repeating the Task Force Smith experience."[173]

General Shinseki made this same point when announcing the Objective and Interim Forces at the AUSA Eisenhower Luncheon in 1999. He insisted that a great power war against a peer competitor was imminent. To prove this point he compared himself, having assumed office at the end of the twentieth century, to Maj. Gen. Nelson Miles and Elihu Root, commanding general of the Army and secretary of war, respectively, who had held office at the end of the nineteenth century. When these men led the Army it had "ended a war of near-global proportions [the Spanish-American War] just the year before. They also had led an Army . . . stretched thin by its post-conflict peacekeeping and peace enforcement responsibilities" in Cuba, Puerto Rico, and the Philippines. Yet, General Shinseki added, "The First World War . . . was then just 15 years away." He continued by asking whether these men, at the end of World War I, "could . . . have foreseen a second global war that would end with the birth of atomic weaponry."[174]

Shinseki's implication was clear; the end of one war (the Cold War) simply augured the coming of the next war. The Army could not, therefore, be distracted by the current peacekeeping responsibilities (in the Balkans); it needed to continue to prepare for that next great power war.

The belief that it was simply a matter of time before the next great power war was something that pervaded the U.S. Army's culture—inside and outside the service. The cover of the AUSA pamphlet on transformation explained that the Army had "To Be Ready for a Battle That Could Break Out Today." Inside the pamphlet posited, "The Army must maintain the capabilities to fight the nation's wars." The present was "a time during which experts believe the United States will not face a major adversary, which gives America a little breathing room to make shifts in its defenses." The pamphlet did acknowledge that the proliferation of low-intensity conflicts meant the present was "peacetime, but that does not necessarily mean it will be peaceful." It concluded, "During a period in which the United States faces no peer enemy, the Army must take advantage of science and technology breakthroughs to create the next generation of equipment while developing training and doctrine advancements to go with it."[175]

This logic pervaded the thinking of the Army's senior leadership. Maj. Gen. James Dubik put it more succinctly in his 2000 article in the *Military Review*: "If history has taught us anything, it is that somewhere, at some time, the United States will confront a regional, and eventually, a near-peer competitor, so we must prepare for that inevitability now." If a great power war occurred, Dubik noted, "It is not like World War II when the United States had the opportunity to adjust its tactics after it saw what the Nazi blitzkrieg did in Belgium and France. We will have to come as we are, so we have to get [transformation] right enough to use."[176]

Like the rest of the Army's senior leadership, General Shinseki seems to truly have believed that a great power war was imminent. In his *Army Green Book* article he wrote, "There will most certainly be another war in our future." He added, "If history is any indicator, it will happen sometime in the early decades of this century."[177] When Shinseki and other Army transformers talked about the Army of 2020, they envisioned it arriving at the precipice of another world war. For the senior leaders of the Army, transformation was a race against time.

From this perspective, the 1990s were not a preview of things to come but the closing of a window of opportunity to get ready for the imminent next great power war against an as yet unidentified future peer competitor. As Shinseki explained in the *Frontline* interview, "Our country is at peace, and we lead the world economically. . . . There is a pause in world affairs where we can advantage ourselves to make some changes with minimal risk." The alternative was unacceptable: "We will go through transformation at some later date when the risk is much higher. And if our history of first battles is any suggestion, that may come on the eve of the next war. And that would be unfortunate."[178]

From this perspective, it isn't surprising that Army transformers ignored those saying that a great power war was unlikely. The O&O concept authors wrote, "Current assessments indicate that while [major theater war] remains on the least likely end of the spectrum, its potential for occurring with little warning is of concern." They added, "A Major Theater of War represents the most serious conventional military scenario the Army would have to face. With the vital interests of the nation at stake, it is the one scenario that must result in victory."[179] If one assumes—as Army transformers did—that the arrival of another great power war was a certitude, then the news that it was presently unlikely was good news indeed; the Army still had time to prepare.

But it had to prepare. These core tenets of Army transformation—that a great power war *was* coming, that it was the most dangerous form of war for the nation, and that it might finally arrive without warning, leaving the Army with no time to prepare—meant that the Army had no choice but to chase ever-greater capacity to fight high-intensity conflicts. These had become metaprinciples—beliefs so deeply held that they had become underlying, unquestioned assumptions.

So fully had this tenet permeated the culture of the Army that even low-intensity conflict observers felt compelled to apologetically acknowledge that high-intensity conflict was, of course, the most important mission. For instance, Manwaring began his broadside on the Army's unpreparedness for low-intensity conflict with the refrain, "Prudent armies must prepare for high risk low-probability conventional war." In the foreword to Conrad Crane's treatise on the conflicts of the 1990s, SSI director Douglas Lovelace—in the middle of contending that "the services will have to find a way to accomplish these most-likely [low-intensity conflict] missions"—acknowledged that the Army must retain "full ability to win those major wars that remain the most dangerous threat to national security." In her withering assault on *Joint Vision 2020*, Applegate conceded that it was premature to say that "global or regional warfare with global implications will not occur." She followed with the refrain, "We cannot ignore the potential for a conventional competitor to emerge on the horizon with the intent and willingness to fight the American Way of War, although at present this scenario appears unlikely." In the midst of an article-long assault on U.S. unpreparedness for low-intensity conflict, Nagl and Young paused to write, "Clearly, training and preparation for peace operations should not detract from a unit's primary mission of training to fight and win in combat."[180] The pervasiveness of this core cultural metaprinciple—that the Army must prepare for the coming great power war—was so powerful that it blinded even the most diehard low-intensity conflict observers to the possibility that the two-major-regional-conflict sizing construct might be an anachronism that needed to be abandoned.

Given the power of these underlying assumptions about the inevitability of another great power war, it is not surprising that Army transformers refused to budge on their contention that the Army should not prepare for—or even be committed to—low-intensity conflicts lest that impair or retard its degree of readiness for high-intensity conflict. Col. Mark E. Vinson decried the "negative readiness effects on the Army of frequent deployments to smaller-scale contingencies." Contradicting the contention of some Army transformers, Vinson claimed that participation in low-intensity conflict training and execution yielded "little or no relevant combat training." He cited an Army inspector general report from early 1999 as saying "entry-level performance at the combat training centers 'continues to decline,' in part because units have fewer opportunities to train at home station." Vinson concluded, "As long as the Army continues to deploy its first-to-fight MTW forces to smaller contingencies, it still faces a significant, long-term readiness challenge."[181]

Others made similar arguments. Alan Goldman, a senior intelligence analyst at the Forces Directorate of the National Ground Intelligence Center, speaking at a conference at the U.S. Army War College on the force-sizing construct of two major regional conflicts attacked the constant commitment of Army forces to low-intensity conflict, arguing that the Army should remain hyperfocused on preparation for a high-intensity conflict. Goldman enumerated strategy goals that ignored all low-intensity conflict threats to stability and dismissed as shortsighted those who doubted the emergence of a future peer competitor to the United States. He explained that a coalition of adversary nations could amalgamate the equivalent of a peer threat capability. Echoing the deepest fears of Gen. Eric Shinseki and his cadre of transformers, Goldman warned that this peer competitor could emerge by surprise, requiring that the Army remain ever-prepared for this dubious possibility.[182]

At the same conference, Timothy D. Hoyt, a visiting assistant professor in the National Security Studies Program at Georgetown University, agreed with Goldman. He saw China and Russia as potential future peer competitors against which the best deterrent was an Army that remained dominant in high-intensity conflict. He explained that the highly technological bent of the Army, combined with the shrinking industrial base in the United States, meant that the nation could not rapidly arm, mobilize, and train for war. Citing the disintegration of the British military in the interwar period, Hoyt was convinced that the Army had to maintain and continue to grow its hyperpreparedness for high-intensity conflict.[183]

And it was at this same conference where Michèle A. Flournoy, a distinguished research professor from the Institute for National Strategic Studies at the National Defense University, likewise voiced a profoundly conventional vision of the future. She warned that the rash of low-intensity conflicts had hurt preparation for the coming great power war in two ways. First, she argued, the call for more capability to execute such conflicts bred complacency in preparing to face a peer competitor. Second, she complained that the higher operational tempo created by constant deployments to such conflicts had sapped energy from testing and training new capabilities for hyperpreparedness to wage for high-intensity conflict. In essence, like so many of her fellow transformers, she argued that low-intensity conflicts were a distraction from preparation for the real threat.[184]

In an article for an SSI essay collection, School of Advanced Military Studies founder Brig. Gen. (Ret.) Huba Wass de Czege and the SSI's Antulio J. Echevarria II pointed out to Army transformers that they were never going to be able to totally avoid low-intensity conflicts. They acknowledged that participation in and preparation for the conflicts "indisputably detracts from some aspects of warfighting preparations." Yet they noted that participation in "peacetime engagement" and low-intensity conflicts also "helps build a basis for cooperative action with regional friends and neighbors that in the long run can reduce the expense of crisis response to the American taxpayer and the burden on American forces."[185] In other words, participation in low-intensity conflicts might create a strategic environment that would actually make future high-intensity conflicts easier to prosecute.

Wass de Czege and Echevarria also argued that "peacetime engagement builds strategically valuable military skills and capabilities that can enable and assist the U.S. military's capacity to wage war." Nagl and Young made a similar point, noting that "90 percent of the training for peacekeeping is also training for general combat capability."[186] Packett and Gilhool, in their *Military Review* article on the U.S. military response to Hurricane Mitch, similarly argued—with the hindsight of practical experience—that the skills his force acquired in training and execution were also applicable to high-intensity conflict: "Units that can move, operate and communicate well in a stressful peacetime deployment are well on their way to being ready for war."[187] This was, essentially, turning one of the Army transformers' primary arguments against them; if low-intensity conflict tasks were a subset of high-intensity conflict tasks, then training for and prosecuting these conflicts was also training for and gaining experience useful for high-intensity conflicts.

Nagl and Young also argued that the refusal to prepare for low-intensity conflicts made them even more costly: "As we prepare for the missions we would like to fight, the real missions we are currently conducting, responses to CHEs, are improvised at great expense to our readiness, unit integrity and quality of life of our service members."[188]

As the Army tried to determine the required capabilities for the Interim Brigades, the transformers did acknowledge a few specific capabilities that the Army lacked to engage in low-intensity conflicts. In the O&O concept for the IBCT, TRADOC acknowledged, "Adversaries understand the weakness of coalitions and alliances and will seek to force the creation of alliances with nations who are more sympathetic to their cause." The planners also expected that "interoperability issues will be exploited tactically and operationally." Additionally, the O&O concept reflected some lessons that the Army had learned from its 1990s low-intensity conflict experiences: "Mistrust, classification, language, [and] lack of a common operational framework are all characteristics that have been present in every U.S. operation and will become more prevalent in the future."[189]

Yet the writers of the O&O concept had learned some negative lessons from Haiti and Bosnia—particularly that the Army should eschew operating within a multinational force, instead demanding a separate, all-U.S. chain of command in exchange for its participation. The planners wrote, "The SSCO environment will often require [the IBCT] to maintain direct links with multinational forces . . . operating in the theater." The planners noted, however, that "the Brigade Combat Team is expected to always operate under ARFOR [a U.S. Army force] command."[190]

A number of low-intensity conflict observers likewise noted that the Army still faced challenges in participating in multinational operations, even in a separate chain of command. In his September 2001 SSI publication, Manwaring wrote that "unity of effort" was the key to effective coalitions. He added, "The many problems of the U.N. operation in the Congo (UNOC), the U.N. operation in Somalia (UNOSOM II), the U.N. operations in the former Yugoslavia, and the NATO operations in Bosnia and Kosovo stem from a lack of unity of effort among the various players." Melissa Applegate noted that as the Army became ever-more networked and computerized, it would be more and more difficult for the United States' multinational partners to "to expend the resources necessary to become full partners under the doctrine of overwhelming force as we have defined it in [*Joint Vision*] *2020*."[191]

Operations in and among the Populace

Low-intensity conflict observers likewise noted many capabilities that the U.S. Army still lacked for executing operations in and among the populace. Packett and Gilhool noted that civil affairs "quickly became the focal point for coordination between the JTF and numerous HN [host nation] government and nongovernment organizations, as well as several international relief and private volunteer organizations." They added, "The successes flowed from CA's [the civil affairs team's] linguistic skills and the enduring relationships from previous deployments to the region."[192] These were skills and relationships that the rest of his troops presumably lacked.

Nagl and Young were more direct in addressing the Army's lack of preparedness to engage in coordination with the other civil-military operations required of low-intensity conflict, acknowledging that dealing with NGOs could be "frustrating and confusing for both the military and its civilian counterparts." They believed, however, that "military forces . . . cannot themselves build peace." Success in such conflict, they contended, "requires increased coordination with NGOs, PVOs and other US government agencies." In addition to their ability to help with reconstruction, the authors noted, "Many of these agencies will already have established a close rapport with the belligerents and local nationals in the area." They recommended better training at combat training centers on interoperating with NGOs and more effort during low-intensity conflicts "to begin working together."[193]

The TRADOC planners who wrote the IBCT O&O concept seemed to have absorbed some of the lessons from the Army's 1990s low-intensity conflict experiences about civil-military operations, an essential facet of operating in and among the populace. The O&O concept authors noted, "Some NGOs or PVOs are favorable to U.S. efforts and are willing to provide assistance regarding the culture, languages, and peculiarities of the local population for military forces." But they also warned, "Alternatively, an NGO or PVO may have the same ideology as a potential adversary and may adversely affect U.S. mission accomplishment. A third alternative is that NGOs or PVOs make mistakes based upon inexperience." The planners also acknowledged that, "in many situations, the Brigade Combat Team will benefit from exploiting the knowledge and capabilities residing within these organizations." They added, "Consequently, [the IBCT] must be organized to interact effectively with these significant actors in the battlespace." These TRADOC planners did not, however, see any role for the IBCT in actually assisting these NGOs with resources or reconstruction support. Instead, NGOs were a security concern: "U.S. forces may . . . be required to divert resources from their assigned missions toward rescues or providing security."[194]

The Army transformers' conception of media and information, on the other hand, seemed not to have been influenced at all by the Army's low-intensity conflict experiences in the 1990s. The TRADOC planners saw media only in regard to its ability to influence the American public, writing, "From the U. S. perspective, media allows our opponents the best means of attacking the national will of our population." Similarly, the planners saw information operations only in relation to how it influenced the IBCT's ability to gain and maintain dominant situational awareness. The main information threat from future adversaries would be that "hacking, viruses, physical destruction and nonnuclear EMP [electromagnetic pulse] will cause U.S. commanders to lose the use of some or all of their automated systems for a period of time."[195] Absent from this analysis was any concern for the ability of the U.S. Army or adversaries to use media or information to influence the populace inside an area of operations.

The O&O concept did recount a vignette about PSYOP from Operation Restore Hope in Somalia: "U.S. information operations planners were initially hampered by a lack of understanding of . . . culturally-specific factors applicable in Somalia. Initial operation leaflets were prepared using type styles, language, and illustration styles alien to the average Somali, and thus, except for their novelty were not as effective." Yet aside from acknowledging this example, the authors described no additional capabilities or training that the IBCT would possess over and above legacy forces to allow it to avoid this negative example. Instead, the focus of psychological impacts was entirely on adversaries.[196] This was a conception of psychological impacts on the battlefield that would have been immediately familiar to any cold warrior, but it ignored the Army's low-intensity conflict experiences throughout the 1990s.

Of course, low-intensity conflict observers did note the importance of media and PSYOP products to success in these conflicts. As Manwaring wrote, "Willing support to the state on the part of a majority of the populace, *motivated by legal, democratic, and honest informational actions* on the part of the government are directly related to the synergism and effectiveness of a counterguerrilla war."[197]

Metz and Johnson highlighted the way in which the Army's tactical operations could impact the perceptions of the populace inside an area of operations during a low-intensity conflict: "Military strategists and commanders must think in terms of *psychological* precision . . . shaping a military operation to attain the desired attitudes, beliefs, and perceptions on the part of both the enemy and other observers, whether noncombatants in the area of operations or global audiences."[198]

The Political Dimension

Low-intensity conflict observers continued to note the inherently political nature of these conflicts. As Manwaring explained, guerrilla wars "are the organized application of violent or nonmilitary coercion or threatened coercion intended to resist, oppose, change, or overthrow an existing government, and to bring about political change." Recalling the Vietnam War, he noted, "The guerrillas operated in relatively small units with political, psychological, and military objectives—in that order." He added, "'Armed propaganda' was conducted not to 'win,' but to further discredit the South

346

Vietnamese government and the Americans." He concluded that "an adequate response must be essentially a strategic political-economic-social-psychological-security effort. The most refined tactical doctrine, operational expertise, and logistical backup that are carried out by the optimum military or police structure in pursuit of a policy that ignores the strategic whole—to include the populace—will be irrelevant." Manwaring believed that the political acumen required to be successful in low-intensity conflict required "leadership development"—to build the required "leader judgement"—that had heretofore not been present in Army education and training.[199]

Crane's 2001 reexamination of the Army's 1990s low-intensity conflict interventions highlighted the connection between the political dimension of these conflicts and ultimate conflict resolution: "The Army makes perhaps its greatest contribution towards accomplishing national policy objectives in the stabilization phase of SSCs after most air and naval forces have gone home." Crane explicitly stated that only the Army could accomplish political tasks: "The Army will . . . have to facilitate the organization of a regime to replace the one it helped destroy." He added that, in the absence of a political solution, "when ground troops leave, as in Haiti or Somalia, the situation soon reverts to the conditions that sparked the crisis in the first place." Crane concluded that the Army was required in low-intensity conflicts both because of "the deterrent effect of their offensive and defensive abilities" against a resumption of conflict and later as "requirements for rebuilding infrastructure and restoring normal life increase." Yet, despite the repeated requirement for these capabilities, Crane noted, "Generally the Army has planned poorly for the stabilization phase after crisis resolution, and is not properly resourced or structured to handle the growing number of such overlapping commitments accruing to it."[200]

Nagl and Young fell into a cognitive pitfall while examining the political dimension of the U.S. commitment to Bosnia-Herzegovina. While they acknowledged that the Army had failed to engage the political dimension of the conflict, they seemed convinced that it could not possibly engage this dimension without the help of civilian U.S. government representatives: "US policy implementation in Bosnia lacks a mechanism to ensure effective integration of the civilian and military peacebuilding programs at the tactical, operational or strategic level." As a result, they noted, "the military conditions for success of the Dayton Peace Accord were largely met, but the situation on the ground was never transformed into a condition from which the military could withdraw." Nagl and Young blamed the Army's inability to leave Bosnia-Herzegovina on its inability to integrate with other U.S.

government agencies, as if the Army had no ability, absent these civilians, to forge a political solution to end the conflict.[201]

Crane disagreed, insisting that the U.S. government's representatives in Bosnia were totally incapable of meeting this demand: "By late 1997 it became apparent to stabilization forces (SFOR) that a large disparity existed between the ability of military forces to achieve their initially assigned tasks of the General Framework Agreement for Peace (GFAP) and that of their less-capable civilian counterparts to meet their own implementation requirements." He added, "SFOR realized it could not disengage with such a large 'GFAP Gap' remaining, and expanded its mission," and also noted that "U.S. military leaders on the scene recognized they were moving into the area of nation-building, but saw no alternative if SFOR was ever going to be able to withdraw or significantly reduce its commitment without risking the peace."[202] For Crane, the problem in Bosnia-Herzegovina was that Army leaders did not go far enough soon enough to engage the political dimension of the conflict themselves.

While Crane acknowledged that there could be a "handoff to civilian and indigenous agencies" of low-intensity conflicts over "a very long time," he still believed that this handoff "rarely reduces or eliminates all military requirements." He cited the conclusions of the Post Conflict Strategic Requirements Workshop at the Army War College's Center for Strategic Leadership in contending, "The lack of quick response capability of civilian agencies and problems coordinating them . . . insure that the military [will] bear the brunt of all essential tasks in rebuilding and reorganizing a failed or war-torn state for a long time." He concluded that "the interagency process still has far to go" to be able to assume significant responsibility in low-intensity conflicts.[203]

Major General Dubik seemed to understand the essentially political nature of low-intensity conflicts better than his fellow senior Army leaders. Specifically, he grasped that the Army was essential to forging a political settlement because it could exert force. "When the United States sends the Army," he wrote, "we go there to force somebody to do something. Our success depends on the certain ability to impose our will. Combat capability is why we are in the mission, whether peacekeeping or peace enforcement. People must be afraid not to obey us." He added, "Terms like peacekeeping and peacemaking are deceptive euphemisms; what we do is all about force." He concluded that, "in places like Kosovo," low-intensity conflicts began in the first place because of the "absence of legitimate force to maintain peace and order." This represented a profound breakthrough for an Army transformer—especially one who, only a few years earlier, in the midst of the war in Bosnia-Herzegovina, had written that some "National Judicial Force" should be created to swoop into low-intensity conflicts and relieve the Army of the responsibility to engage the political dimension.[204] Unfortunately, Dubik was the only senior Army leader to reach this conclusion—or at least to express it so publicly.

And, unfortunately, the ability and willingness to wield force or the threat of force in a low-intensity conflict was necessary but not sufficient alone to bring such conflicts to a successful conclusion. The Interim Brigades that Dubik was building at Fort Lewis lacked any of the other capabilities—such as civil affairs, military police, engineers, and PSYOP—required to operate in and among the people. And the IBCTs had no more capacity than the legacy force to engage the political dimension of low-intensity conflicts. In their October 2000 *Military Review* article, Caldera and Shinseki promised that, in the IBCTs, they would "develop leaders at all levels and in all components who can prosecute war decisively and who can negotiate and leverage effectively in those missions requiring engagement skills." Yet, in fact, on the ground at Fort Lewis no specialized training in these skills was happening. The IBCTs were conducting live fire exercises and demonstrating the deployability of new platforms—preparing only for high-intensity conflict.[205]

This was not an oversight; the TRADOC planners that designed the IBCT did not envision these brigades engaging the political dimension of low-intensity conflicts. In the O&O concept, the planners wrote that the IBCT's role in "stability and support operations" was to be "a 'guarantor combat force.'" Their job was not to engage in the political dimensions of the conflict but "to permit peacekeeping and stability forces (and cooperating interagency, international, private, and non-governmental organizations) to carry out their missions in a secure environment." Mehaffey's October 2000 *Military Review* article repeated this dubious claim. The O&O concept and this article were both silent on what "peacekeeping and stability forces" would miraculously appear to relieve the IBCT of the responsibility to forge a political settlement to the conflict.[206]

The political dimension of low-intensity conflicts continued to confound Army transformers (with the exception of Dubik), and they saw no role for the U.S. Army in engaging it. As General Shinseki noted in his *Frontline* interview, "We have been in Kosovo now a year. We're coming up on five years in Bosnia. The Sinai Desert is 18 years, and Korea is 40." Yet instead of musing on how the Army might go about ending any of these conflicts by helping to forge a political settlement, he treated them like an immutable condition of the modern strategic environment. In a sense they were simply

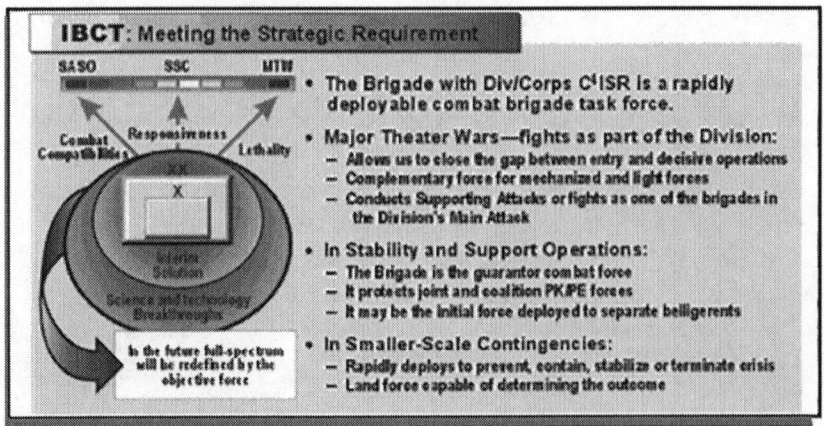

Figure 4.3. The Interim Brigade Combat Team in each conflict environment. *Source:* Michael Mehaffey, ed., *Army Transformation Taking Shape . . . Interim Brigade Combat Team*, CALL newsletter no. 01-18.

taxes on the Army that "strip away inventory and capability." Absent here was any discussion about how the Army might, through its own actions, engage the political dimension of low-intensity conflicts and end them by forging a political settlement. Instead, Shinseki made an implicit plea for more soldiers—people to garrison these and future forever wars, as if they were immune to conclusion.[207]

One common tack that Army transformers took in avoiding the political dimension of low-intensity conflicts was to advocate artificially limiting their mission. Closely related to this tendency was their condemnation of so-called mission creep. Conrad Crane dismissed the idea of mission creep as a fallacy; writing about the necessity to engage the political dimension of such conflicts, he noted, "Being prepared to conduct such operations will avoid a sense of 'mission creep' when they inevitably have to be performed." Speaking specifically to Operation Restore Hope in Somalia—perhaps the genesis of the term *mission creep*—Crane turned the transformers argument upside down when he posited, "Resistance to nation-building contributed to 'mission creep' as those tasks were forced upon unprepared American units or fell to them by default."[208] Because the U.S. Army actively avoided engaging the political dimension of the conflict, it was surprised and unprepared when forced to do so.

Similarly, Max Manwaring rejected the idea that international and NGO partners created mission creep, insisting that new tasks and missions would

350

naturally emerge as "the roles and missions of the various civilian and military elements evolve deliberately" and "as the situation changes to accommodate progress toward the achievement of a mutually agreed political vision."[209]

Army critic Ralph Peters came tantalizingly close to realizing the ultimate level of understanding in low-intensity conflict—the realization that the U.S. Army needed to take sides in a conflict. Recounting recent peacekeeping operations, he noted that warring parties often engaged in one last surge of murder and ethnic cleansing as the deployment of international peacekeeping forces became imminent (as had happened in Kosovo). The reason for this phenomenon, Peters argued, was that "a declared or perceived partisanship on the part of peacemaking or peacekeeping forces prior to deployment creates a window of slaughter, during which the threatened group accelerates ethnic cleansing operations."[210] This was at least an acknowledgment that the parties to a low-intensity conflict *perceived* the U.S. Army and coalition forces as having taken sides. Yet Peters's analysis unfortunately stopped short of urging that the Army indeed needed to back a winner in a conflict in order to bring it to a conclusion.

The Administration of George W. Bush

In the contentious 2000 presidential campaign, Texas governor George W. Bush rejected the Clinton-era foreign policy of nation building. Bush complained that President Clinton's foreign policy had been dangerously ambitious and wasteful of the United States' finite military resources. He promised "a more humble foreign policy" were he elected. In a *Foreign Affairs* article, Bush's national security adviser, Condoleezza Rice, wrote that the U.S. Army was "not a civilian police force. It is not a political referee. And it is most certainly not designed to build a civilian society." She added, echoing the Powell Doctrine principle of clear, achievable objectives, "Military force is best used to support clear political goals," and also quipped, "We don't need to have the 82nd Airborne escorting kids to kindergarten." Former Secretary of Defense and Bush supporter Donald Rumsfeld called low-intensity conflicts "nonmilitary functions," arguing, "To the extent we can have as few people in uniform doing nonmilitary functions, I think we better serve ourselves, our country and our personnel."[211] Army transformers—who generally believed the Army should not be employed in such conflicts—surely must have welcomed these promises of a more restrained foreign policy.

As one would expect from his campaign rhetoric, President George W. Bush came into the White House promising an end to nation building—the engagement of the U.S. Army in low-intensity conflicts. Bush promised not to commit U.S. forces to "missions without end" or "to stop ethnic cleansing and genocide" in areas of dubious U.S. national interest, as he claimed his predecessor had. He added, "Nor do I think we ought to try to be the peacekeepers all around the world." He essentially endorsed the Powell Doctrine that had prevailed during his father's presidency: "When America uses force in the world, the cause must be just, the goal must be clear, and the victory must be overwhelming."[212]

Rice was similarly skeptical of the wisdom of committing U.S. forces to low-intensity conflicts, noting, "It takes courage to set priorities." She argued, "Using the American armed forces as the world's '911' will degrade capabilities, bog soldiers down in peacekeeping roles, and fuel concern among the great powers that the United States has decided to enforce notions of 'limited sovereignty' worldwide in the name of humanitarianism."[213]

While the new Bush administration might have eschewed nation building, many low-intensity conflict observers clearly understood that the Army wouldn't be able to avoid such conflicts in the future. As Crane warned, "While military leaders and security advisers for the incoming Bush administration have often expressed resistance to employing the U.S. Army in nation-building, recent history demonstrates it will occur anyway."[214]

Jablonsky was equally skeptical that the Bush administration would be able to avoid low-intensity conflicts, asserting, "With the click of a TV remote control or a computer mouse, the American public can be face-to-face with the realities of the post–Cold War world." He added, "The current President may choose his use of force at the lower end of the operations spectrum more consistently and effectively than his predecessor. What he cannot do is choose not to choose."[215]

Despite these warnings, in anticipation of an end to the worldwide commitment of the Army to low-intensity conflicts, General Shinseki's staff began the process to close the Peacekeeping Institute at the U.S. Army War College. A Realignment Task Force designed to harvest manpower from institutional Army facilities to man the operational Army apparently concluded that that even the meager ten-man institute was an overcommitment of resources to the topic of low-intensity conflict. The institute was saved only months before it closed its doors after post-9/11 wars prompted forty members of Congress to write a letter to the Defense Department objecting to the closure. Ironically, it was Secretary of Defense Donald Rumsfeld—who entered office decrying the wasteful commitment of the Army to low-intensity conflicts—who finally saved the newly renamed Peacekeeping and Stability Operations Institute in mid-2003.[216]

When new President Bush introduced Donald Rumsfeld as his secretary of defense, he explained that he had given Rumsfeld "a broad mandate to challenge the status quo." Bush endeared himself to Army transformers by professing his own belief in the promise of the RMA.[217] Bush echoed the wildest hopes of Army transformers over the last decade, proclaiming a coming "revolution in the technology of war, [with] powers increasingly defined not by size, but by mobility and swiftness." He added that in this future, "safety is gained in stealth and forces projected on the long arc of precision-guided weapons." To achieve this vision, Bush intended "to move beyond marginal improvements to harness new technologies that will support a new strategy."[218] But Army transformers' hopes would be dashed as they got to know the instrument of President Bush's envisioned transformation of the U.S. military.

Donald Rumsfeld's first twenty-year career in Washington began in 1957. His first position in the capital culminated in his appointment as secretary of defense during the administration of President Gerald R. Ford. After his term ended, Rumsfeld returned to his native Chicago to serve as the chief executive officer of a pharmaceuticals company for twenty-five years. When he returned to Washington at the age of sixty-nine to once more serve as the secretary of defense, Rumsfeld arrived with some very definite opinions about how the Pentagon should be run and the path that transformation of the American national defense establishment should take. And that path was a radical—even revolutionary—reshaping of the U.S. military to harness a vision of the RMA that exceeded the wildest imaginings of Army transformers.[219]

Secretary Rumsfeld would pursue a lighter, faster, smaller, and more lethal U.S. military with a zeal that had not been seen in the halls of the Pentagon since at least the era of Robert McNamara. He didn't just want to acquire new technologies; he wanted to totally reshape the structure of the military. No traditions, practices, or ideas were sacred. As General Reimer had, Rumsfeld sought "leap-ahead" technologies that would allow him to "skip a generation" of transformation. Rumsfeld and his acolytes were particularly hostile to the massing of ground forces that characterized contemporary Army operations.[220]

From the beginning it was clear that Rumsfeld's relationship with the services and the Joint Chiefs of Staff would be adversarial. Presumably because he believed that an Air Force officer would be more sympathetic to his ambitions for innovation in airpower and space platforms—and diminution of ground forces—Rumsfeld bypassed the chief of naval operations and instead chose Air Force general Richard B. Myers to succeed Army general Hugh Shelton as chairman of the Joint Chiefs of Staff. The new secretary also created two new offices—for "force transformation" and "special plans"—to bypass the services as he charted the course for his own vision of transformation. At a gathering at the Pentagon after taking office he warned attendees of a threat "plac[ing] the lives men and women in uniform at risk." Attendees were shocked when he revealed the identity of that threat: "it's the Pentagon bureaucracy."[221]

A half year after taking office, Rumsfeld had so alienated the service chiefs, the Joint Staff, and even the Office of the Secretary of Defense that he seemed destined for an early exit. But dramatic events in late summer 2001, only one day after he declared war on "the Pentagon bureaucracy," would postpone his dismissal.[222]

Kosovo, Part Two: From Mid-2000 to 9/11

While the humanitarian disaster that prompted Operation Allied Force had ended, a political solution to the tensions between Kosovar Albanians and Serbians had proved harder to find. The war criminals on both sides of the conflict had been rounded up and tried. Former Yugoslav President Slobodan Milosevic died in prison in 2006. The parliament in Kosovo again declared the country's independence in 2008. But the low-intensity conflict continues in Kosovo to this day; KFOR and the UN Interim Administration Mission in Kosovo were supposed to only last twelve months, but as of this writing, both are still in force in Kosovo.[223]

Many lessons might have been learned from this conflict. But the dramatic events of 9/11 would completely overshadow events in Kosovo, robbing the Army of valuable insights into the nature of low-intensity conflict that might otherwise have emerged from this war.

In mid-2001, a full two years after the beginning of the NATO intervention in Kosovo, a handful of articles written about lessons learned from Operation Joint Guardian in Kosovo did begin to get published.

On the eve of the war on terror, in September 2001, the *Military Review* published an entire issue on peace operations. Included in the issue was a case study on the low-intensity conflict in Kosovo written by Lt. Col. Joseph Anderson. (Anderson would go on to achieve the rank of lieutenant general, and as of this writing serves as the Army's deputy chief of staff for operations and training.) In his article Anderson listed a number of organizational shortfalls in Task Force Falcon's ability to conduct low-intensity conflict operations in Kosovo. He noted that there were insufficient military police in Kosovo to maintain "law and order" in the country and those military police that the Army did have arrived late. But Anderson's solution betrayed an assumption that performing certain low-intensity tasks was beneath the dignity of infantrymen. He explained that "regular combat troops" should only "neutralize ethnic violence." The necessary task of "preventing domestic crime" should not begin, Anderson wrote, until the arrival of "sufficient MPs [military police] in the AO to perform this task." He reached this conclusion because "MPs are better trained in traffic control, arrest, detention and investigation than combat soldiers."[224]

Lieutenant Colonel Anderson also betrayed a predilection to dump the low-intensity conflict in Kosovo on civilian agencies as soon as possible. He felt that the Army should stop enforcing law and order as soon as "an international police force arrives." Only this force should, according to Anderson, be expected to train and integrate "a local police force."[225] This thinking made it clear that the Army had failed to institutionalize the lessons of previous conflicts; in Panama and Haiti—and to lesser degrees in other conflicts—the Army had already discovered that international police were insufficient alone to train local police; they needed the active assistance of U.S. Army personnel to be successful.

Anderson did address capabilities the Army lacked that were required to operate in and among the population, noting that "military ties to NGOs in PKO are poorly structured." While he did acknowledge the role of civil affairs in maintaining these ties, he was silent on why this wasn't occurring in Kosovo or why other troops couldn't perform these tasks.[226]

Anderson did speak to information operations but seemed to revert to the Cold War–era conception of PSYOP—that this capability should be directed against an enemy force rather than directed toward influence civilian populaces. He wrote, "In accordance with joint doctrine, information operations (IO) are used in Kosovo to degrade Albanian and Serbian abilities to respond to KFOR operations to restore peace." As an example, Anderson explained,

"A message supporting the theme used in Kosovo was that military leaders who violate provisions of the [cease-fire agreement] will be prosecuted under international law."[227]

On a positive note, Anderson did appreciate the practice of embedding media in units. "Military commanders host reporters daily and give them open access to unit activities," he wrote, noting that embedding "allows reporters to live and travel with commanders and units throughout the AO, and to see and feel the emotions and difficulties that [a unit] faces each day. The embedded media program is a meaningful and effective way to decipher and exploit information under favorable conditions. Open access tells the real story."[228] While there is no direct evidence that these practices were implemented as a result of lessons learned in Haiti and Bosnia-Herzegovina, their presence here at least hints that Army public affairs professionals were better at sharing lessons with each other than was the Army at large.

While the title of Anderson's article was "Military Operational Measures of Effectiveness for Peacekeeping Operations," measures of effectiveness were, in and of themselves, a cognitive obstacle to moving from an under-standing of operating in and among the population to a full understanding of the inherently qualitative political dimension of low-intensity conflicts. The misconceptions embedded in the concept of measures of effectiveness are perfectly illustrated in Anderson's thesis statement: "Events during *Joint Guardian* indicate that military peacekeeping operations (PKO) require iden-tifying specific operational measures of effectiveness (MOEs) to determine when conditions are established for transferring control to legitimate civilian authorities or other political organizations." This single sentence is replete with wrongheaded thinking. Besides the fallacious concept of measures of effectiveness—the idea that one can quantify "winning" in the inherently qualitative political dimension of a low-intensity conflict—the sentence also betrayed the misconception that the Army could avoid the political di-mension by dumping it on "legitimate civilian authorities" such as the U.S. Department of State or the UN. It also betrayed the misconception—born of the Nixon Doctrine and the Reagan-era model for counterinsurgencies—that the Army could avoid the political dimension by handing it off to "other political organizations" like the host nation government. One need look no further than this sentence to see that the Army had failed to institutionalize the lessons of its 1990s low-intensity conflict experiences.[229]

Unfortunately, these were some of the last lessons to be widely dissem-inated from the conflict in Kosovo. Events in the Balkans would soon be completely overshadowed by the war on terror.

Field Manual 3-0, *Operations*

TRADOC commander Gen. William Hartzog began the writing of the new FM 100-5 in 1996, with an intended publication date of 1998. The manual would ultimately not be published until June 2001, under the supervision of a new TRADOC commander, Gen. John N. Abrams. Moreover, the numbering system for Army field manuals was realigned to bring them in line with joint doctrine, so the new manual was published as FM 3-0, *Operations*.[230]

General Hartzog had envisioned that this new manual would be profoundly conventional, erasing the remaining distinctions between low- and high-intensity conflict in Army doctrine. The final product was more balanced. With the arrival of General Shinseki and his TRADOC commander, General Abrams, the manual—as Thomas White and General Shinseki put it in an Army posture statement to Congress in August 2001—more forcefully emphasized "the Army's ability to apply decisive force through network-centric capabilities." It also presaged and facilitated the development of the Objective Force. But much of the low-intensity conflict material was still retained from the earlier 1993 edition of FM 100-5. In fact, even some of the material from Joint Publication 3-07, *Joint Doctrine for Military Operations Other Than War*, written by the now defunct Army–Air Force Center for Low Intensity Conflict (A-AFCLIC), was added. Especially important was the return of the political dimension of low-intensity conflict to Army doctrine.[231] On balance, the 2001 edition of FM 3-0 essentially restored the low-intensity doctrine of the 1990 edition of FM 100-20, *Military Operations in Low Intensity Conflict*, which had also been written by A-AFCLIC. In other words, despite eleven years of continuous operations in low-intensity conflicts—after a brief regression with the publication of the 1993 edition of FM 100-5—Army doctrine on such conflicts had basically not advanced at all since 1990.

The 2001 edition of FM 3-0 represented the Army's last iteration of its vision for transformation before the beginning of the war on terror. But it also represented the Army's last opportunity to capture the lessons of its 1990s low-intensity conflict experiences. The results were mixed. One of the more important contributions of the 2001 edition was to segregate low-intensity conflict into two categories of operations: "stability" operations—such as peacekeeping and support to counterinsurgency—and "support" operations—such as disaster relief and humanitarian assistance. The manual defined "stability" operations as the use of the Army to "promote and protect US national interests by influencing the threat, political, and information dimensions of the operational environment." Among the tasks in stability operations were "developmental, cooperative activities during peacetime and coercive actions in response to crisis." The manual added, "The military activities that support stability operations are diverse, continuous, and often long-term." The return of a political dimension to low-intensity conflict in an Army manual—after its eight-year hiatus after the publication of the 1993 edition of FM 100-5—was a significant triumph for observers who had complained of its absence ever since it was removed.[232]

"Support operations" could happen inside or outside the United States and were defined as the "use [of] Army forces to assist civil authorities, foreign or domestic, as they prepare for or respond to crises and relieve suffering." For the authors of FM 3-0, the key to support operations was to hand over the operation to civilians as soon as possible. While FM 3-0 acknowledged, "In extreme or exceptional cases, Army forces may provide relief or assistance directly to those in need," it also explained that "the purpose of support operations is to meet the immediate needs of designated groups for a limited time, until civil authorities can do so without Army assistance."[233]

These two categories of operation were listed alongside offense and defense, but they were not coequal. As the manual made clear, "Army forces cannot train for every possible mission; they usually train for war and prepare for specific missions as time and circumstances permit."[234]

Notably, FM 3-0 did not remove the problematic "operations other than war" that first appeared in the 1993 edition of FM 100-5. Instead it segregated operations by whether they did or did not involve the use of force. "Military operations other than war" encompassed operations with a "general US goal" of "deterring war and resolving conflict" or "promoting peace," but stability and support operations could occur in either "war" or "military operations other than war."[235]

The authors of the 2001 edition of FM 3-0 echoed low-intensity conflict observers when they argued, "Even during major theater wars, Army forces conduct stability operations. These occur during combat operations and throughout the post-conflict period. The US strategy of promoting regional stability by encouraging security and prosperity means Army forces will be engaged in stability operations for the foreseeable future."[236]

Yet the Army's "focus on warfighting" remained firmly codified in FM 3-0 through an unusually prescriptive restriction on the scope of Army training. Reflecting the August 1999 discussion between new chief of staff of the Army Eric Shinseki and the CAC, FM 3-0 directed, "Major Army command (MACOM), ASCC [Army service component command], continental

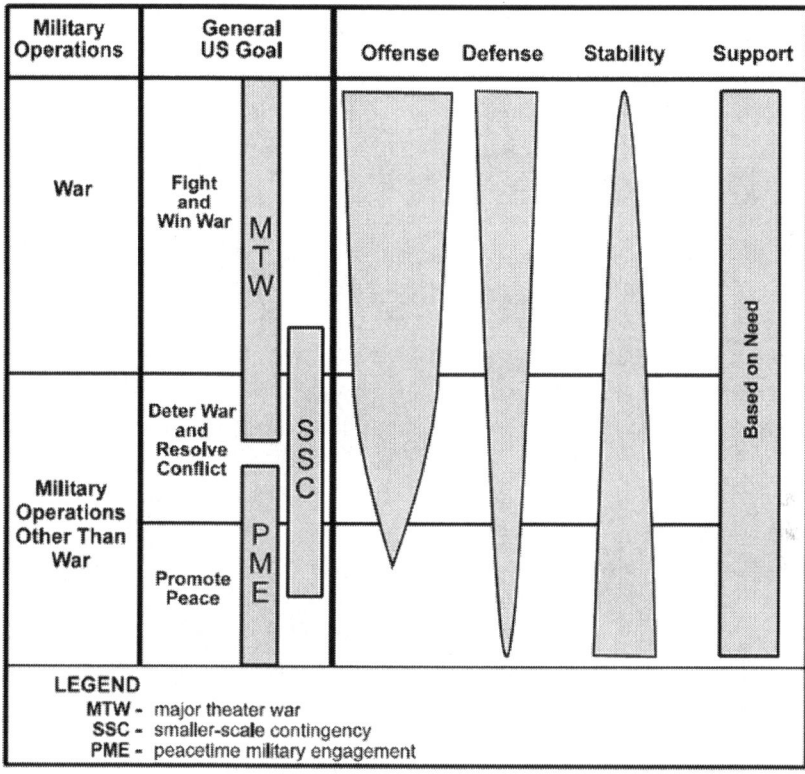

Figure 4.4. The Range of Army Operations. *Source:* Headquarters, U.S. Department of the Army, *Operations*, FM 3-0 (2001).

US Army, and corps commanders determine the battle focus, resources, and METL that maintain the required readiness posture for anticipated operations in war or MOOTW." It added, "Commanders at lower levels conduct battle focused training unless otherwise directed."[237] Incredibly, this abbreviation-laden directive effectively reserved the authority to permit Army units to train for low-intensity conflict to lieutenant generals and above. No lower commanders were authorized to decide independently to train their units in preparation for such conflicts without permission from a three-star general.

FM 3-0 rejected those who argued that the Army was not properly manned, equipped, or trained for low-intensity conflict. Instead, the manual contended, "Army forces are trained, equipped, and organized to control land, populations, and situations for extended periods," adding, "Although

Army forces focus on warfighting, their history and current commitments include many stability operations." The manual did admit that "the Army is not specifically organized, trained, or equipped for support operations. Army forces are designed and organized for warfighting." But the authors of the manual also believed that the Army's capabilities made it ideal for use in support tasks. It was "a disciplined force" with "a functional chain of command, reliable communications, and well-trained and well-equipped organizations." Army units could "operate and sustain themselves in austere environments" and "rapidly move large forces to the affected location." Other capabilities such as "engineer, military police, medical, transportation, aviation, and civil affairs assets [were] especially valuable for support operations." FM 3-0 was silent on the lack of training readiness for these missions—which the CAC had privately acknowledged to the chief of staff of the Army nearly two years earlier.[238] But this admission that the Army was well suited to at least some low-intensity conflict tasks was still striking for an official Army publication.

The manual reflected some positive lessons that the Army had learned about operating in and among the populace in the low-intensity conflicts of the 1990s. Writing on stability operations, FM 3-0 warned of the "potential for unintended consequences of individual and small unit actions" and noted that "a single act of indiscipline or rash application of force can undo months and years of disciplined effort." "Likewise," the manual explained, "actions that are destructive to the natural or cultural environment may introduce negative perceptions that must be overcome." The solution was to "apply force selectively and discriminately" using "ROE to guide tactical application of combat power." FM 3-0 also suggested that "nonlethal capabilities can provide additional tools to augment, but not replace, the traditional means of deadly force. Nonlethal means expand the number of options for confronting situations where deadly force is not warranted."[239] This was at least one clear example of lessons from Somalia, Haiti, and the latter years of Army operations in Bosnia-Herzegovina being carried into Army doctrine.

FM 3-0 also weighed in on civil-military operations, noting, "Coordination makes unity of effort and effective integration work in environments where unity of command is not possible." Taking a lesson from recent low-intensity conflicts, the manual noted that "coordinating centers such as civil-military operations centers (CMOCs) accomplish this task. CMOCs include representatives from as many agencies as required."[240] This was another example of lessons from the Army's low-intensity conflicts being carried over into Army doctrine.

The manual also acknowledged the political dimension of low-intensity conflict. In its discussion of counterinsurgency, it explained that the goal was to "protect the people from insurgent violence and separate them from insurgent control." FM 3-0 posited that "the fundamental cause of insurgent activities is widespread dissatisfaction with standing ethnic, religious, political, social, or economic conditions by some sizable portion of the population." Both insurgencies and counterinsurgencies, the manual concluded, were inherently political in nature.[241]

Elsewhere, however, the manual enshrined cognitive obstacles to a complete understanding of the political dimension of low-intensity conflict. Echoing the Nixon Doctrine and the Reagan-era model for counterinsurgency, FM 3-0 insisted that "Army forces help host governments" rather than engaging in the political dimension themselves. The manual explained that it was the job of the "host government . . . to address the problems that underlie the insurgency." And it preserved the idea of "foreign internal defense" (FID) as the proper use of U.S. Army forces in a counterinsurgency, noting that "FID is participation by civilian and military agencies of one government in programs taken by another government to free and protect its society from subversion, lawlessness, and insurgency. . . . Army forces participating in FID normally advise and assist host nation forces conducting operations." The manual added, "Normally, using US forces in combat operations is a temporary measure. . . . Poorly executed, direct involvement by the US military can damage the legitimacy and credibility of the host government and host nation security forces."[242]

In fact, the flurry of caveats about placing the responsibility for low-intensity conflict on the host nation were so pervasive that they largely obscured the manual's acknowledgment of the political dimension of such conflicts. It cautioned, "Commanders must not allow stability issue solutions to become a US responsibility. Within their capabilities, the host nation must take the lead, in both developmental and security activities." The Vietnam War was highlighted as a negative example wherein U.S. forces tried to do too much: "The majority of South Vietnamese people came to rely on US forces for their protection, eroding their confidence in their own government to provide for their security."[243] The manual was silent on what to do if there was no host nation government on which to dump the politics of low-intensity conflict—as was the case in Somalia and would be the case in Afghanistan and Iraq.

FM 3-0 advocated dumping the messy political dimension of low-intensity conflicts not only on the host nation but also on special operations forces

(SOF). Reflecting the relegation of such conflicts to SOF that was ostensibly dictated by the Nunn-Cohen Amendment to the 1987 National Defense Authorization Act, the manual noted that "FID is also a specified and significant mission for selected Army special operations forces (ARSOF)." In a highlighted text box, the manual specifically praised the Reagan-era U.S. counterinsurgency effort in El Salvador, where "the advisors, primarily ARSOF . . . served with El Salvadoran units to support small unit training and logistics."[244]

Unfortunately—in this last significant manual before the start of the war on terror—the Army failed to achieve a complete understanding of low-intensity conflict. FM 3-0 did return the political dimension of low-intensity conflict to Army doctrine, but with a mountain of caveats that diluted its impact on the Army.

The 2001 Quadrennial Defense Review

To lead his charge to reshape the military, Secretary of Defense Rumsfeld turned to a fellow precision-guided munitions enthusiast, Andrew Marshall, the director of the Office of Net Assessment. Rumsfeld's first target for re-shaping the U.S. military was the 2001 QDR. To make sure the QDR report reflected his own—rather than the services'—vision for transformation, Rumsfeld put Andrew Marshall in charge of rewriting it before it went to Congress.[245]

The entire review had already been completed under his predecessor, William Cohen. But the Clinton administration had not yet submitted the report to Congress when Clinton left office. This opened the door for the new secretary of defense to throw out everything and start over, as he believed was his mandate from President Bush. Rumsfeld's QDR abandoned all previous analysis and closeted discussions away from service participation whenever possible. In a process that was already engineered to pit the interests of Congress, the services, the Joint Staff, the defense industry, and the administration against one another, these added layers of secrecy and tension created an atmosphere of suspicion and recrimination that guaranteed that the initial results generated by Rumsfeld and his compatriots would be rejected by all other parties.[246]

For the Army, the results of Rumsfeld's view were particularly abhorrent; they suggested deep cuts in force structure in a service that already believed it was carrying the majority of the low-intensity conflict burden around the world with too few soldiers. The secretary's new QDR proposed cutting at least two divisions and one corps from the Army inventory. The AUSA immediately mobilized and attacked the findings in the halls of Congress, encouraging members of the House Armed Services Committee and dozens of other congresspeople to sign a letter protesting Rumsfeld's proposed cuts.[247]

Ultimately, the argument between the Army and Secretary Rumsfeld over the 2001 QDR was never settled. The attacks on 9/11 short-circuited the debate, and Cohen's draft QDR report—which endorsed a slightly modified version of the BUR's force-sizing construct of two major regional conflicts— was submitted to Congress. Only the introduction, written after the attacks of 9/11—was changed from Cohen's original version.[248]

The Objective and Interim Forces, Part Two

One of the hallmarks of General Shinseki's tenure overseeing the Army's transformation was the incredible consistency of his vision throughout his term. Through experimentation—via the Louisiana Maneuvers, the Advanced Warfighting Experiments, and the Army After Next project— General Sullivan's and General Reimer's visions for transformation had slowly evolved over the course of their terms. General Shinseki arrived in office knowing the change he wanted to make and never wavered from that vision.

Even two years after taking office, in their Army posture statement to Congress, White and Shinseki's language in describing the objective force had changed very little from when Shinseki had first proposed it two years earlier. In noting how the Army would blend the best features of the legacy force, White and Shinseki explained,

Our heavy forces are the most formidable in the world . . . but they are severely challenged to deploy to all the places where they might be needed. Conversely, our magnificent light forces are agile and deployable. They are particularly well suited for low-intensity operations but lack sufficient lethality and survivability. There is, at present, no rapidly deployable force with the staying power to provide our national leadership a complete range of strategic options.[249]

In his *Army Green Book* article, General Shinseki repeated the six buzzwords from the "Trident slide" produced two years earlier: "The Objective Force will be more responsive, more deployable, more agile, more versatile, more lethal, more survivable, and more sustainable than today's force." The dubious claim that "light forces" were "particularly well suited for low-intensity operations" aside, in two years of Army transformation toward

363

ever-greater capacity to fight high-intensity conflicts—and in the face of the first two years of low-intensity conflicts in Kosovo and two more years in Bosnia—all that had changed was the font and background colors on General Shinseki's ubiquitous "Trident slide."[250]

While the Army awaited the arrival of equipment for the Interim Brigades being built at Fort Lewis, the program of defining the shape and size of the Army of 2020 continued. Maj. Gen. Robert Scales, still serving as the commandant of the U.S. Army War College, oversaw a new series of wargames—dubbed Vigilant Warrior—to define the shape of the Objective Force. Notably, these wargames were the first to test the Army of 2020 against low-intensity conflicts. In addition to high-intensity conflicts in Azerbaijan, Iraq, and Korea, the Army of 2020 was also tested against Albanian criminal armies, Colombian narco-terrorists, an Indonesian insurgency, and attacks by domestic terrorists. What these tests revealed was that the envisioned Army of 2020 would still require that more than one-half of its forces remain legacy force units. The Army found that it would have to continue to rely on these forces to manage low-intensity conflicts; experimenters found that the Objective Force was even less capable than the current Army to deal with them.[251] The Objective Force of 2020 was even more hyperspecialized for high-intensity conflict against a mythical future peer competitor than the legacy army of 2001.

Even before these experiments, the Army had acknowledged that it would need the legacy force well into the new century. In his October 2000 *Military Review* article, Maj. Gen. James Dubik explained that the Army had to "retain some of the current forces. . . . While we transform as a hedge against potential trouble." He reminded readers, "North Korea has not gone away; Southwest Asia has not gone away; the requirements for these forces around the world have not gone away." Dubik concluded, "We cannot erode this capability; we need to keep that warfighting capability, the forced-entry capability."[252]

As an AUSA pamphlet on transformation from 2001 explained, "The Army envisions an active-Army 'counterattack corps' (III Corps) consisting of three heavy divisions and one armored cavalry regiment to be the fully modernized legacy force, thus providing the heavy force capable of decisive victory in a major theater war through the period anticipated for Army Transformation." The remainder of the legacy force would slowly transform to IBCTs, while the eight initial IBCTs would be the first transformed to Objective Force units. The pamphlet promised that "around . . . 2031" the entire Army would finally transform to the Objective Force.[253] But it was silent on what force would continue to do the low-intensity conflict tasks for which the Objective Force had already proven ineffective.

The Army would also have to continue to rely heavily on the reserve component because of the civilian skills these soldier brought to low-intensity conflicts. Moreover, transformation would do nothing to cure the Army's critical shortage of linguists, civil affairs, PSYOP, military police, and explosive ordinance disposal soldiers that the Army had identified both in every low-intensity conflict of the 1990s and in the Objective Force wargames.[254]

In fact, the Objective Force itself was never tested against a low-tech, low-intensity conflict adversary. Instead lower-end threats were relegated to the legacy force; the Objective Force in this wargame fought essentially a mirror image of itself. The exercise report described the enemy that the Objective Force faced in the exercise as "a free-thinking and technically competent adversary possessing powerful ground forces, able to threaten and conduct out-of-region attacks including the use of weapons of mass effects. It was also capable of conducting complex and innovative offensive and defensive operations employing a variety of conventional and asymmetric means." This enemy had "a modernized military and [was] a major regional competitor"; it possessed "sufficient *mass and technology to inflict highly visible and embarrassing losses on an unprepared foe!*"[255]

Low-intensity conflict observers Steven Metz and Douglas Johnson of the SSI were quick to point out that the Army had chosen an opponent with which it was completely comfortable: "Today the enemy in most Service and DoD experiments or wargames remains a traditional mechanized, state military which has invaded a neighboring state." They added, "Rather than seeking to confirm or endorse existing transformation and modernization programs, joint wargames should [be] a robust test of them." This was the principle of *falsification*—the idea that when testing a hypothesis one should try to prove it wrong rather than prove it right, thus avoiding confirmation bias. Metz and Johnson explained, "A blue team 'defeat' in a wargame does not invalidate a transformation or modernization program, but simply provides a means of adjustment and refinement."[256]

Senior leaders of the Army didn't heed such advice because, ultimately, they weren't trying to build a force suitable for low-intensity conflict; they were trying to prepare the Army for a great power war against a peer competitor. To be prepared, they would have to harness the promise of the RMA that had fueled Army transformation throughout the 1990s. Army transformers sought a force that could "See First . . . Understand First . . . Act

First . . . Finish Decisively," one that could exploit a "layered constellation of advanced sensors" and use the networked technology of the RMA to "rapidly process and distribute knowledge" across the force.[257]

The construction of the first two Interim Brigades proceeded at a brisk pace throughout late 2000 and early 2001. Congress authorized $600 million for acquisition of a new Interim Armored Vehicle, and the selection in November 2000 of the LAV III as the platform of choice was not without controversy. Many current and former officers from the Army's armor branch instead favored the M113A3 armored personnel carrier—the latest generation of the tracked vehicle that had been replaced by the M2 Bradley fighting vehicle during the fielding of the Big Five weapon systems in the 1980s and 1990s. These advocates were able to reopen discussions on the vehicle and get a belated test of it, but still eventually lost to the LAV III. Frankly, much of this controversy was probably manufactured; General Shinseki acknowledged that he had wanted to use the LAV chassis for the Interim Armored Vehicle in order to enlist the support of the Marine Corps for the IBCTs, and the Interim Brigades at Fort Lewis had been practicing with borrowed LAV IIIs for months.[258]

While platoons and companies trained on how to use their new vehicles and integrate them with dismounted infantry maneuvers, the battalion and brigade headquarters conducted simulated command post exercises culminating in September 2001 in a brigade-level warfighter exercise facilitated by the Army's Battle Command Training Program from Fort Leavenworth.[259]

September 11 and the War on Terror

When the terrorist attacks on the United States came on September 11, 2001, the stage was set for a slow-motion military disaster. Worse, it was a disaster in which all of the scariest predictions of low-intensity conflict observers over the preceding decade came true. The first, most obvious prediction that came true on 9/11 was the terrorist attack on the American homeland itself. Both low-intensity conflict observers and Army transformers had predicted such an attack, but Army transformers had predicted that a terrorist attack would be an asymmetric adjunct to hostilities from a conventional enemy. Instead the attacks unfolded in a manner predicted by low-intensity conflict observer Melissa Applegate when she wrote, only weeks before 9/11, "Asymmetry *may be all there is*."[260] There was no country to invade or conventional army to fight or government to depose that could stop this enemy from attacking the United States.

Yet as the saying goes, to a man with a hammer, everything looks like a nail. So the U.S. military did invade a country, defeat an army (such as it was), and depose a government. In a lightning campaign, U.S. SOF and airpower, partnered with the patchwork forces of the Northern Alliance, swept the Taliban out of power and chased both the Taliban and al-Qaeda out of Afghanistan and into the mountains of western Pakistan. The apparent "cheap" win in Afghanistan through SOF and airpower further validated Rumsfeld and his transformers in their conviction that technology could supplant numbers.[261]

The Objective and Interim Forces, Part Three

General Shinseki and the Army's senior leaders refused to wrestle with the attacks of 9/11 or how they called into question the Army's misguided transformation. And, incredibly, even though the Army had become very much engaged in a new, massive low-intensity conflict in Afghanistan—prompted by the greatest strategic shock to the United States since the attacks on Pearl Harbor sixty years earlier—the Army *still* marched forward with transformation and its quest to prepare for a great power war against a mythical future peer competitor.

Instead, incredibly, in their testimony before the Senate Armed Services Committee in February 2002—after the invasion and occupation of Afghanistan had begun—White and Shinseki argued, "The attacks of 11 September 2001 were more than just the first salvo in a new war; they validated the direction of our Vision and the need to accelerate Army Transformation." In a tortured explanation of how the Objective and Interim Forces, profoundly focused on high-intensity conflict, might be in any way relevant to the war in Afghanistan and other looming low-intensity conflicts, they claimed that "See First" technology would enable leaders to "gain a greater situational awareness of themselves, their opponents, and the battle space." They could then "Understand First, to assess and decide on solutions to the tactical and operational problems at hand faster than our opponents—to gain decision superiority over our opponents." This future Army could then "Act First, to seize and retain the initiative, moving out of contact with the enemy to attack his sources of strength or key vulnerabilities at a time and place of our choosing." White and Shinseki added that future forces would be able to switch "seamlessly from stability operations to combat operations and back again," concluding, "When we attack, we destroy the enemy and Finish Decisively. Absent from this soliloquy was any acknowledgment that a low-tech, low-intensity conflict opponent presented an asymmetry to the high-tech Objective and Interim Forces and their focus on high-intensity conflict. This transformed Army would not be able to understand first, act first, or finish decisively if it couldn't see the opponent, because the enemy was indistinguishable from and hiding in and among the populace—exactly the kind of asymmetry that Army After Next researchers had uncovered with the problem of urban operations more than three years earlier.[262]

Even several months after the initial invasion of Afghanistan, as it became clear that conventional forces would have to be deployed and remain in the country for some time to make permanent the gains from the first days of the war, Shinseki and his transformers continued to tout a future Objective Force optimized for a high-intensity conflict against a mythical future peer competitor. In the same testimony to Congress in February 2002, White and Shinseki asserted, "The Army will possess the lethality and speed of the heavy force, the rapid deployment mentality and toughness of our light forces, and the unmatched precision and close combat capabilities of our special operations forces."[263] In the afterglow of the SOF-enabled routing of the Taliban, General Shinseki had added SOF to his list of capabilities that the Objective Force would supersede. But in every other way, this was the same description of the Objective Force that Shinseki had given in 1999, two years before the war on terror had begun.

In March 2002, only a month later, Shinseki took another tack, trying to argue that the Army needed to look beyond the war on terror in order to prepare for the imminent great power war still awaiting the United States on the other side of the very real low-intensity conflict occurring in Afghanistan. As White and Shinseki testified before Congress, "While we fight and win the global war on terrorism, The Army must prepare itself to handle demanding missions in the future strategic environment."[264] The Army had just begun a war that, as of this writing, has lasted nearly two decades and continues with no end in sight. Yet the chief of staff of the Army still considered it a distraction from the need to prepare for a great power war.

The AUSA pamphlet on transformation published just before September 11, 2001, featured an epigraph from English military historian B. H. Liddell Hart intended to condemn those who did not subscribe to General Shinseki's vision for transformation. Ironically, the quote perfectly described the psychosis that would, after 9/11, beset General Shinseki and the other senior leaders of the Army: "The only thing harder than getting a new idea into the military mind is to get an old one out."[265]

The quest to realize an Interim Force hyperspecialized for high-intensity conflict likewise continued in the face of the decidedly low-intensity conflict growing in Afghanistan.

Construction of the first two Interim Brigades at Fort Lewis was largely complete by early 2002. Training at the platoon, company, and battalion levels culminated in mid-2002, when Joint Forces Command sponsored Millennium Challenge 2002, a live and virtual exercise at multiple locations across the United States. The Army demonstrated the deployability of its Interim Combat Vehicle, now called the Stryker, deploying it by different means up to and including Air Force C-130s. Rather than citing the folly of testing high-intensity conflict capabilities in the midst of the growing low-intensity conflict in Afghanistan, Army transformers clucked about the fact that the Stryker's appliqué armor and spare infantrymen had to travel in a separate C-130 from the Stryker.[266]

As the Interim Force completed its validation exercises, it began to transform from a prototype, experimental organization into an operational force in the Army's inventory. A big step in this institutionalization was the publication of doctrine for the IBCTs. The Stryker company manual was published in January 2003. The brigade-level manual was published in March. Finally, the battalion manual was published in April 2003.[267]

Operation Iraqi Freedom

Two months after the 9/11 attacks, and in the midst of reports of anthrax attacks across the country, 74 percent of Americans supported invading Iraq with ground troops. By October 2002, however, support had cooled markedly; only 51 percent of Americans supported an invasion. Still, Congress—wary of the political fallout for those who had failed to endorse the Gulf War over a decade earlier—enthusiastically authorized military action, even if the UN failed to endorse an invasion.[268]

To the extent that there was a public debate about the war in Iraq before it began, it swirled around the potential cost of an invasion. There were fears that Saddam Hussein might use chemical weapons or that the fighting to take Baghdad might devolve into a door-by-door, street-by-street bloodbath.[269]

Yet there were other voices before the start of the war that were warning that the war would devolve into a low-intensity conflict. On the eve of war with Iraq, SSI researcher Conrad Crane was commissioned to write a monograph on the pitfalls of postconflict stabilization. His work, *Avoiding Vietnam: The U.S. Army's Response to Defeat in Southeast Asia*, was an indictment of the Army's neglect of low-intensity conflict doctrine since the end of the Vietnam War and a prescient warning that another such conflict awaited the Army on the far side of its impending invasion of Iraq. Crane began by reminding the Army leadership of a fact that was not yet widely reported in the press: the dazzling, high-tech victory in Afghanistan was giving way to a low-intensity conflict. He wrote, "Global missions in Operation ENDURING FREEDOM are evoking increasing comparisons with past experience in Southeast Asia." He added, "The American Army can no longer run away from Vietnam. For it has found us in Afghanistan."[270]

Crane cited Andrew Bacevich in a frank declaration that the Army's un-challenged dominance of high-intensity conflict would drive adversaries to accelerate "the blurring or elimination of the boundaries between war and peace, soldiers and civilians, and military and political spheres." Crane cau-tioned that the enemy in Iraq was unlikely to fight the U.S. Army in the way that it wished to be fought: "One of the oft-repeated current justifications for the utility of ground forces in a modern campaign is that their presence will force the enemy to mass, thus providing a better target for long-range precision strikes." He noted, however, that "North Vietnamese did not con-centrate in response to large American ground formations, and neither has Al-Qaeda." Echoing some of the language of Army transformers concerns about urban operations, Crane added, "Smart enemies will force pursuers to find them and dig them out. Ground forces are necessary for close combat, not as decoys."[271]

Crane believed that the Army had failed to institutionalize the lessons of the low-intensity conflicts of the 1990s and was headed for disaster in Iraq. He asserted that "to truly be a Full Spectrum force" the Army needed to immediately institute "reforms in training, doctrine, and force structure" and needed to accept that it would have to engage in "missions including counterinsurgency and some degree of nation-building."[272]

One of Crane's most important contributions—even if it did not ultimately impact planning for the Iraq War or its early execution—was to remind the Army of the inherently political dimension of low-intensity conflict. Quoting a 1973 monograph by Army colonel Francis Kelly, Crane wrote, "In Vietnam military decisions were viewed in terms of the political consequences they might have, a situation to which the average military professional was unac-customed." He concluded, "The Army is still uncomfortable with the highly charged political atmosphere of a Bosnia or Haiti mission, but that is the future we must face."[273]

In January 2003 Conrad Crane and W. Andrew Terrill got yet another opportunity to warn the U.S. Army of the coming disaster. Their brief monograph, *Reconstructing Iraq: Challenges and Missions for Military Forces in a Post-Conflict Scenario*, applied the lessons of the Army's 1990s low-intensity conflict operations (especially in Haiti) to a hypothetical—but very likely—postinvasion low-intensity conflict in Iraq.[274]

The monograph was intended as a planning tool, a list of considerations as the Army planned for the postconflict stabilization of Iraq. It was by no means perfect; its first recommendation was to get out of the business of engaging the political dimension of low-intensity conflict as "expeditiously as possible."[275]

On balance, however, the monograph had a number of prescient cautions for U.S. Army officers hoping to forge a political settlement in Iraq. First, Crane and Terrill warned, "Tensions among Iraqi religious, ethnic, and tribal communities are expected to complicate both the occupation and efforts to build a viable postwar government." For those hoping to build a liberal democracy in Iraq, the two researchers noted, "Power-sharing among groups is a new and untested concept in Iraq that could well be subverted by elite political instincts to struggle for power before rivals achieve opportunities to consolidate their own gains." For those who expected to be greeted as liberators, they cautioned, "Even the most benevolent occupation will confront increasing Arab nationalist and religious concerns as time passes." Finally, the timeline that Crane and Terrill constructed to sequence the myriad of tasks required to stabilize Iraq anticipated an occupation of four to five years or more.[276]

Crane and Terrill also believed that the Army would not be able to avoid direct involvement in civil-military operations in Iraq. They explained that conditions in Iraq were so dire even before the invasion that the scale of the humanitarian crisis would be beyond the capability of other agencies of the U.S. government or NGOs to handle. After listing a dizzying array of civil-military tasks for the Army to accomplish in postinvasion Iraq, Crane and Terrill added, "Civil affairs, engineer, military police, and transportation units will be in high demand."[277]

What would have been shocking in this document to most Army officers, having been insulated from the lessons of 1990s low-intensity conflicts, were the huge number of tasks to be accomplished in the political dimension of the impending low-intensity conflict in postinvasion Iraq. As Crane and Terrill explained, "First priority . . . will be establishing viable local governments, relying as much as possible on existing institutions." They added, "Civil affairs units will be needed throughout the country to assist this process." For those who might look at the list of political tasks that Crane and Terrill had outlined and conclude that they were not the Army's job, the two noted, "While recent experiences in the Balkans and Afghanistan appear to indicate that civilian agencies are now better prepared to take over transition responsibilities from military forces, this should not be assumed for Iraq." They concluded, "The U.S. Army has been organized and trained primarily to fight and win the nation's major wars. Nonetheless, the Service must prepare for victory in peace as well."[278]

Little did Crane and Terrill know that in January 2003, when their monograph was published, they were virtually the only people actually planning for the low-intensity conflict certain to follow the invasion of Iraq.[279]

Gen. Tommy Franks, the commander of CENTCOM, entered the Army as an enlisted soldier before going to Officer Candidate School and being commissioned as a field artillery officer. His tour in Vietnam was decidedly high-intensity in nature; he earned three Purple Hearts while serving as a forward observer for an infantry company, aerial observer, and assistant S-3 for an artillery battalion. The remainder of his career—including service as an assistant division commander of the 1st Cavalry Division during Operation Desert Storm and commander of the 2nd Infantry Division in Korea—was equally dominated by high-intensity conflict experiences.[280] And because the Army had stalwartly refused to institutionalize the lessons of its 1990s low-intensity conflict experiences, General Franks was just as oblivious as the overwhelming majority of Army officers to the challenges the Army had faced in these operations.

Thus, when the CENTCOM staff convened to plan the invasion of Iraq, Franks was intensely involved in the details of the high-intensity conflict phase of the operation, a concerned bystander for the debate over troop numbers, and completely uninterested in planning for postconflict stability. Franks largely refereed the debate between Lt. Gen. David McKiernan, the commander of the Third Army and Coalition Force Land Component Command, and Secretary Rumsfeld and his cronies over troop numbers.[281]

Lieutenant General McKiernan had served in the Allied Rapid Reaction Corps headquarters in Sarajevo, Bosnia-Herzegovina, in 1996 and in USAREUR headquarters in Germany from 1998 to 1999 while the U.S. Army was deploying ground forces to and establishing security in Kosovo. These experiences probably armed him with a better appreciation of the manpower-intensive demands of low-intensity conflicts. Yet McKiernan— steeped in the same Army culture as all of his peers—could not articulate the inevitability or the demands of a postinvasion low-intensity conflict.

Instead he couched his pleas for more troops in terms of the Powell Doctrine principle of overwhelming force; his objections were based on not having sufficient troops to maintain the tempo of his attack all the way to Baghdad rather than on the future demand for troops to stabilize Iraq.[282]

While Army senior leaders and Rumsfeld and his team argued over how many hours of aerial bombardment should precede the ground invasion and how many troops should cross the berm into Iraq, no one was thinking about, let alone planning for, Phase IV—military jargon for the postinvasion low-intensity conflict phase that would make permanent the coalition's gains in Iraq. In August 2002 General Franks told his commanders that the State Department would figure out the plan for postinvasion Iraq. Franks expected the other agencies of the U.S. government to "pay attention to the day *after*" so that he could focus on the "day *of*."[283]

In October 2002, Donald Rumsfeld "won" the bureaucratic battle to freeze the Department of State out of responsibility for planning or executing the postinvasion phase in Iraq. Planning effectively ended until the following year.[284]

The debate within the American national security establishment over the invasion of Iraq should have been about the low-intensity conflict that would follow the initial invasion. Up until the eve of the invasion, low-intensity conflict observers warned that a postinvasion Iraq would descend into ethnic and sectarian anarchy. But the Army had failed to institutionalize the lessons of the low-intensity conflicts of the 1990s, and thus the debate over the invasion of Iraq instead circled around the RMA. How much had the U.S. Army transformed? How much could modern technology—information collection and sharing and precision targeting—replace massed ground forces in modern warfare?

To the extent that the Army was engaged in a conversation about a low-intensity conflict in Iraq at all, the debate centered on the number of troops required to stabilize the country in a postinvasion phase. General Shinseki and his transformers tried to argue that the Army needed more troops to garrison the country.[285] Yet, having ignored low-intensity conflict for decades, these senior Army leaders lacked the vocabulary to explain why these forces were needed, what they would do, or how long they would be needed.

On the other side of the debate, Secretary of Defense Rumsfeld and his disciples of the RMA made convincing arguments that the U.S. military could do more with less. Armed with the evidence of the supposed "cheap" win in Afghanistan—using only SOF and airpower—and the earlier, dazzling victory of the Gulf War, they made compelling arguments using the same transformation vocabulary perfected by Army transformers over the preceding decade.[286]

The debate came to a head when General Shinseki testified before Congress on February 25, 2003, only days before the war. Senator Carl Levin asked Shinseki for his opinion as to the "magnitude of the Army's force requirement for an occupation of Iraq following a successful completion of the war." Shinseki tried to deflect the question, explaining that he preferred to rely on the opinion of the combatant commander—General Tommy Franks, commander of CENTCOM. When Senator Levin insisted that General Shinseki give a number, the chief of staff of the Army replied, "I would say that what's been mobilized to this point, something on the order of several hundred thousand soldiers are probably a figure that would be required. We're talking about post hostilities, control over a piece of geography that's fairly significant with the kinds of ethnic tensions that could lead to other problems. And so it takes a significant ground force presence to maintain a safe and secure environment, to ensure people are fed, that water is distributed—all the normal responsibilities that go along with administering a situation like this." Senator Levin asked what impact such a large commitment might have on the Army and its other missions. Shinseki answered, "If it were an extended requirement for presence of U.S. only Army forces, it would have significant long-term effect. Therefore, I think the kind of assistance from friends and allies would be helpful."[287]

General Shinseki has been rightly praised for his candor in this exchange, but some important caveats about this testimony must be highlighted. First, this was a hearing on the Army budget for 2004 and the transformation program—which were continuing in spite of the war on terror. Second, General Shinseki did not offer this testimony freely or go to Congress with the intent of speaking truth to power; in fact, the testimony had to be coaxed out of him by Senator Levin. Third, Shinseki's testimony failed to mention that while "several hundred thousand soldiers" might have been preparing for combat, nowhere near that number were headed to the Persian Gulf.[288] This single exchange between General Shinseki and Senator Levin during hours of congressional testimony on other topics was less an alarm shouted from the mountain tops than an objection to be noted for the historical record in case everything went awry.

More important, once General Shinseki was cornered into answering the question, he could not articulate why so many troops were needed beyond hazy allusions to humanitarian assistance and vague worries that Iraq was a big country with multiple ethnic groups. He also could not articulate how long it would take; he simply raised the possibility that it might be an "extended requirement."[289]

One can only wonder how his testimony might have differed if the U.S. Army had institutionalized the lessons of its 1990s low-intensity conflicts. Shinseki might have pointed out that the Army's previous experiences had showed that the scope of building a new government in Iraq would be beyond the capability of the Department of State and that the "assistance from friends and allies" that he wished aloud for had not proven "helpful" at all in the past.[290] He might have reminded Senator Levin that it had taken a year to reestablish a government in Panama that had already held power before Manuel Noriega and had the support of the populace. He might have noted that the UN was still trying to stabilize the government in Haiti that the Army had established nearly a decade earlier. He might have also noted that the Army was still trying to stabilize the Balkans after nearly a half decade of low-intensity conflict. And he might have mentioned that the Army found itself incapable of building a government from scratch in Somalia—precisely the feat it would have to accomplish in Iraq.

One final caveat on this testimony is also in order. Shinseki did not give it until February 25, only weeks before the U.S. invasion of Iraq. One might rightly ask what other actions he—or any other senior Army leader—had taken to affect the size of the occupation force in Iraq before this testimony. As of this writing, the historical record on this question is still incomplete, but it doesn't appear that any objections or concerns were raised by the chief of staff of the Army to the president, as was Shinseki's constitutional duty if he had such reservations. The U.S. Army Center for Military History—charged by General Shinseki to determine the number of troops required for the postinvasion occupation of Iraq—arrived at a number of around 350,000.[291] But there is no evidence that this number was ever shared with the secretary of defense or the president. In an email exchange with a reporter from *Newsweek* from 2006, Shinseki was asked to respond to charges that he had failed to protest to the president about the lack of forces to stabilize Iraq. He responded, "Probably that's fair. Not my style." According to CNN's Jamie McIntyre—citing other senior military leaders who were in the room for key briefings before the invasion of Iraq—General Shinseki never objected to the troop numbers, even when asked directly by the chairman of the Joint Chiefs of Staff, Gen. Richard Myers, if he had any concerns about the plans for the invasion before they were taken to the president.[292]

In *Cobra II: The Inside Story of the Invasion and Occupation of Iraq*, General Bernard Trainor and Michael R. Gordon offer this account of what was perhaps General Shinseki's best opportunity to provide his military advice on the matter to the president:

> [President George W.] Bush . . . called the Joint Chiefs to a January 30 meeting at the White House to hear what they had to say about the war plan. Myers, Rumsfeld's appointee as JCS [Joint Chiefs of Staff] chairman, was happy with the plan. Vern Clarke, the chief of naval operations, and John Jumper, the Air Force chief of staff, praised it. Shinseki, the Army chief, gave a more qualified assessment. Shinseki said that he would have liked to have more forces in place before kicking off the attack, and Turkey's agreement to open a northern front. He also cautioned that the logistics would be crucial and it would be important to keep reinforcements flowing. . . . With these caveats, Shinseki advised that the plan was executable and that [Gen. Tommy] Franks had the situation in hand.[293]

In this account, General Shinseki never even raised the possibility of a low-intensity conflict that might follow the invasion; he wanted more troops for the invasion, not the occupation. In that argument, he was doomed to be at a decided disadvantage to Secretary Rumsfeld. After all, as an Army transformer, Shinseki had for years been making all of the same RMA-enabled arguments about the power of technology to replace mass that Rumsfeld was making in favor of smaller numbers of ground troops for the invasion.

Vice President Dick Cheney, Secretary of Defense Rumsfeld, and Deputy Secretary of Defense Paul Wolfowitz believed that after they had invaded Iraq and ousted the dictator Saddam, Americans would be greeted "as liberators," that Iraq's oil revenues would pay for reconstruction (leaving the United States off the hook financially for rebuilding postwar Iraq), and that France or the Arab states would pitch right in and do much to effect the postwar stabilization of Iraq.[294] Because Shinseki lacked the vocabulary to articulate the demands of the low-intensity conflict or why such a conflict would inevitably follow the initial invasion, he instead sounded like one of the dinosaurs of which Rumsfeld constantly warned, too conservative to understand or realize the promise of the RMA.

Ultimately, Shinseki decisively lost this argument. Deputy Secretary of Defense Wolfowitz responded to General Shinseki's testimony by saying it was "hard to conceive that it would take more forces to provide stability in post-Saddam Iraq than it would take to conduct the war." He called Shinseki's estimate of a requirement for hundreds of thousands of troops "wildly off the mark."[295]

The refusal of the Army to institutionalize the lessons of the low-intensity conflicts of the 1990s had finally taken its toll. In addition to being tragically ill prepared for the character of warfare that it ultimately faced, the Army that invaded Iraq in March 2003 was woefully undersized. By May 2003 there were 150,000 American troops—the great majority contingent of a coalition of 173,000 troops from around the world—in Iraq. The deployment of the nearly 100,000 troops that were supposed to follow the initial invasion force was canceled by the secretary of defense after the invasion was complete. CENTCOM and Coalition Force Land Component Command planners complained about this move,[296] but there is no evidence that senior Army leaders objected directly to the secretary or the president.

According to McKiernan's lead planner, Col. Kevin Benson, the plan for Phase IV was not actually completed until two or three days before the invasion of Iraq began, by which time McKiernan and his subordinate commanders were "engaged with the details of impending D-Day" preparations and not giving much thought to what would happen after the completion of the invasion. President Bush didn't even receive a brief on the plans for postwar Iraq until March 10–12, 2003, a little over a week before the invasion.[297]

Thus it is not surprising that no one knew the plan when the invasion ended and Phase IV began. The Army that invaded Iraq in March 2003 was tragically unprepared for the character of warfare that it would face. As the Army advanced, it was challenged by the Fedayeen Saddam guerilla force, which struck from within the populace and urban centers against vulnerable rear areas—just as Army After Next project researchers had predicted over a half decade earlier.

And while the depleted Iraqi Army rapidly melted before the advance of the vastly superior U.S. Army, it did not disappear. Instead—again, just as Army After Next project researchers had predicted—it hid in the urban centers, among the populace, evading America's high-tech surveillance and precision strike capabilities. Once Saddam's regime was toppled, the Iraqi Army reemerged not as a conventional military threat but as an insurgency that severely challenged U.S. efforts to establish a new Iraqi government. Other insurgent groups and militias also emerged, including Shiite Arab militias, Sunni Arab Iraqi Islamists, and foreign terrorist groups.[298]

When Coalition Provisional Authority chief Paul Bremer disbanded the 300,000-man Iraqi Army on May 23, 2003, it did upend CENTCOM's postwar plan, which had hinged on Iraqi troops to assist with stabilization.[299] But Bremer could be forgiven for not knowing this; most of the U.S. Army in Iraq didn't know the postwar plan either.

The End of Transformation

When Shinseki testified before Congress, his replacement as chief of staff of the Army had already been named—a year in advance. The conflict between General Shinseki and Secretary Rumsfeld likewise stunted the careers of many of Shinseki's protégés. The general could take some solace in having created "irreversible momentum" for his effort to build an Interim Force.[300] Yet much of the transformation toward hyperpreparedness for a future great power war that he had shepherded during his tenure would be decimated by the very present low-intensity conflicts that the U.S. Army faced—and still faces today—in Afghanistan and Iraq.

While the Interim Force was indeed fielded, it fell short in two significant regards. Major General Dubik wrote in October 2001 that the Interim Brigades provided the Army with a "twofer": the brigades would provide immediate capabilities for the Army that could be employed in the current operating environment and would provide an interim organization for testing the tactics and organization of an eventual Objective Force.[301]

It failed in its first purpose: the Interim Brigades proved no more capable than had their legacy counterparts to contend with the war in Iraq. On May 28, 2003, the first Interim Brigade (increasingly being referred to as a Stryker Brigade for the name given to the brigade's wheeled vehicles), completed two months of combat training center rotations at the NTC at Fort Irwin and the JRTC at Fort Polk and was declared ready for combat. The vice chief of staff of the Army, Gen. Jack Keane, who was serving as acting chief of staff, announced in July 2003 that the Stryker Brigade would deploy to Iraq to fill the Army's desperate demand for troops in the manpower-intensive counterinsurgency developing there. When it arrived in Iraq,[302] the troops of the Stryker Brigade where no better prepared to fight a low-intensity conflict than the rest of the Army.

But the Interim Brigade also failed on a deeper level insofar as the Objective Force never materialized. As it became clear that the wars in Afghanistan and Iraq were not going to end anytime soon, the Army no longer had the time, resources, or intellectual energy to chase imaginary future peer competitors. In July 2003 the Objective Force Task Force convened to write an *Army in*

2020 White Paper. The tone and language seemed quaintly anachronistic in light of the present and apparent future reality of the Army. The white paper was published on November 1, 2003. The Task Force limped along for another four months before being unceremoniously closed on March 1, 2004. In a testament to the irreversible momentum—at least of defense spending—development of the FCS, the central platform of the Objective Force, was not finally canceled until 2009.[303]

As Gen. Peter Schoomaker succeeded Shinseki as chief of staff of the Army, he did not have the luxury that his predecessors had enjoyed of imagining future wars. Whereas Generals Sullivan, Reimer, and Shinseki had each come into office touting the promise of the RMA and warning of an imminent great power war against a peer competitor, General Schoomaker was face-to-face with two very real, very present wars, both of them decidedly low-intensity in nature. He was too busy generating forces for the fight against insurgents in Iraq while trying not to lose ground to the Taliban in Afghanistan to worry about enemies that had yet to—and might never—materialize.[304]

The Army was finally faced with low-intensity conflicts too big to ignore, and it became clear that the two major regional conflicts that had actually awaited the Army in the new millennium were counterinsurgencies.

Conclusion

The Army transformers' core metaprinciples—that a great power war was inevitable and imminent, that it was the least likely but most dangerous threat to the nation, and that it would come without warning, leaving the Army no time to prepare—caused the senior leaders of the Army to repeatedly and forcefully reject preparation for low-intensity conflict.

Because these were metaprinciples—seldom spoken underlying assumptions—they were almost never questioned by Army transformers or low-intensity conflict observers. Many deep problems with this logic stand out, even without the benefit of hindsight. Why, exactly, was a great power war imminent? There had not been a great power war anywhere in the world since the end of World War II and the advent of the nuclear age. In fact, the global trend was toward fewer and smaller wars, with all of the great powers relying on proxy wars and peacetime competition rather than direct warfare against one another.

And even if a great power war were imminent, was a conventional, high-intensity conflict with a great power really the least likely but most dangerous threat to the country? Doesn't that better describe a global nuclear war? If the United States were going to insist on hyperpreparedness for the least likely and most dangerous threat, why not pour all of its resources into the Strategic Defense Initiative and other programs designed to prepare the nation to survive *that* threat?

And, finally, the assumption that a peer competitor would appear out of nowhere, leaving the Army no time to prepare, was the most dubious proposition of all. To accept this assumption, one would first have to ignore the fact that the same internet revolution that was fueling the RMA was also making the world so interconnected and transparent that it was increasingly difficult to keep anything a secret, from anyone in the world, and especially the secret construction of a massive military capability equal to the capacity of the greatest military force in the history of the world. And even if China, Russia, or some other nation could suddenly and unexpectedly build such a capacity, why couldn't the United States, the most economically and militarily powerful country on Earth, quickly mobilize just a few more resources to top that new peer competitor capability, even at the last minute?

Army transformers sometimes argued—on the infrequent occasion when this final assumption was questioned at all—that modern warfare and weapons were so complicated and so technically challenging that soldiers required long training and preparation time to be able to understand them. But the Iraq War would reveal this assertion to be patently untrue; the Army had 480,000 active personnel the day before the surprise attacks of September 11, 2001. By the beginning of the Iraq War in early 2003, it had grown to 499,000 and, by 2008, at the height of the Iraq "surge," to nearly 544,000.[305]

Shackled by these faulty assumptions, the Army never solved the problems of low-intensity conflict, particularly the political dimensions that its devotees described in terms of the challenges of urban operations and the asymmetry. Army transformers couldn't claim that they were oblivious to the challenges of low-intensity conflict. Major General Dubik, General Hendrix, General Abrams, and General Shinseki had all served in the Balkans only a few years before they began work on the IBCTs. They were simply so deeply invested in U.S. Army culture—the underlying assumption of which was that preparing for and fighting great power wars was its sole purpose—that they were impervious to conflicting data.

And there was a *mountain* of conflicting evidence. Plenty of observers pointed out that the Army was not manned, equipped, or trained for low-intensity conflicts. Throughout the 1990s and early 2000s observers pointed out that the Army was ill prepared to operate in and among populaces. Many low-intensity conflict observers tried to convince Army leaders that there was a political dimension to such conflict and that it was the Army's job to engage it. And a few observers even acknowledged that the Army needed to stop worrying about neutrality and back a winner if it wanted to end the interminable low-intensity conflicts in which it was trapped.

But Army transformers refused to see this evidence—a phenomenon that organizational culture theorist Edgar Schein calls "resistance to change" and historian Thomas Kuhn calls the "suppress[ion of] fundamental novelties because they are necessarily subversive of . . . basic commitments."[306] The Army—the actual organization itself—defended itself against change. And the Army's senior leaders were the chief defenders of the faith—faith that their present rash of low-intensity conflicts would pass and the Army could get back to the kind of fighting that it preferred.

Even after 9/11 and the war on terror made it clear that the two major regional conflicts that the Army would face at the beginning of the twenty-first century—the stabilization of Afghanistan and the postinvasion stabilization of Iraq—would be low- rather than high-intensity conflicts, Army transformers refused to question its foundational metaprinciples.

To justify continuing to ignore low-intensity conflict, even in the face of a looming massive conflict of that type in Iraq, Army transformers listened to the dubious promises of Bush administration officials that the Iraqis would welcome the U.S. Army as liberators and that some other force would arrive to rescue the Army from the responsibility to rebuild Iraq. To the extent that senior Army leaders were fooled by these promises, it was because they wanted to be fooled. Steeped in their culture of hyperpreparedness for an imminent great power war, they saw *all* low-intensity conflicts, even conflicts that could consume the entire Army, as distractions.

As of this writing, America has spent eighteen years paying the price for its Army's initial unpreparedness for the low-intensity conflicts it has faced in Afghanistan and Iraq. And there is no sign that that bill will be paid in full anytime soon.

NOTES

1. Shimko, *The Iraq Wars*, Kindle, locations 3216–21; Millett and Maslowski, *For the Common Defense*, Kindle, locations 12324–28.

2. Millett and Maslowski, *For the Common Defense*, Kindle, locations 12403–14, Shimko, *The Iraq Wars*, Kindle, locations 3216–21; "A Kosovo Chronology," *Frontline*, PBS, n.d., accessed April 12, 2017, http://www.pbs.org/wgbh/pages/frontline/shows/kosovo/etc/cron.html; Perritt, *Kosovo Liberation Army*, 13–24; Chris Hedges, "Conflict in the Balkans: In Albania; Serbs Using Land Mines in Effort to Seal Kosovo-Albania Border," *New York Times*, June 12,

1998, http://www.nytimes.com/1998/06/12/world/conflict-balkans-albania-serbs-using-land -mines-effort-seal-kosovo-albania.html.

3. Perritt, *Kosovo Liberation Army*, 61–87; "A Kosovo Chronology"; Millett and Maslowski, *For the Common Defense*, Kindle, locations 12413–19; Shimko, *The Iraq Wars*, Kindle, locations 3223–29; Harsch, *The Power of Dependence*, 57–102.

4. Phillips, *Operation Joint Guardian*, 7–14.

5. Millett and Maslowski, *For the Common Defense*, Kindle, locations 12419–41; Shimko, *The Iraq Wars*, Kindle, locations 3241–44, 3253–59.

6. Gowan, "The War and Its Aftermath"; Larson et al., *Interoperability of U.S. and NATO Allied Air Forces*, 95; Millett and Maslowski, *For the Common Defense*, Kindle, locations 12417–441.

7. Crane, "Peace Dividends and Benevolent Interventions."

8. Millett and Maslowski, *For the Common Defense*, Kindle, locations 12419–41, Dennis J. Reimer, interview by Lewis Sorley, 2000, transcript, Senior Officer Oral History Project, U.S. Army Military History Institute, U.S. Army Heritage and Education Center, 381–83.

9. Dana Priest, "Army's Apache Helicopter Rendered Impotent in Kosovo," *Washington Post*, December 29, 1999, http://www.washingtonpost.com/wp-srv/WPcap/1999-12/29/014r -122999-idx.html.

10. Phillips, *Operation Joint Guardian*, 14–17; Shimko, *The Iraq Wars*, Kindle, locations 3440–48; "A Kosovo Chronology."

11. "A Kosovo Chronology"; Shimko, *The Iraq Wars*, Kindle, locations 3286–92.

12. "Ethnic Cleansing in Kosovo: An Accounting," U.S. Department of State, n.d., accessed April 15, 2017, https://www.state.gov/www/global/human_rights/kosovoii/homepage.html; U.S. Department of State, *Ethnic Cleansing in Kosovo: An Accounting*, Washington, DC: U.S. Department of State, December 1999, 7.

13. "A Kosovo Chronology"; Crane, "Peace Dividends and Benevolent Interventions"; Phillips, *Operation Joint Guardian*, 14–30.

14. Phillips, *Operation Joint Guardian*, 22–24.

15. Phillips, *Operation Joint Guardian*, 14–30.

16. Armed Forces Newswire Service, "Kosovo Fallout: Army Readiness under Intense Scrutiny," news release, July 2, 1999.

17. Byman and Waxman, "Kosovo and the Great Air Power Debate," 16, quoted in Shimko, *The Iraq Wars*, Kindle, locations 3358–72.

18. Hosmer, *The Conflict over Kosovo*, 92; Shimko, *The Iraq Wars*, Kindle, locations 3379–94.

19. Brown, *Kevlar Legions*, Kindle, 169–70; Reimer interview, 373–80.

20. Brown, *Kevlar Legions*, Kindle, 169–70.

21. Shimko, *The Iraq Wars*, Kindle, locations 3440–48; Brown, *Kevlar Legions*, Kindle, 169–70.

22. "Interview: Lawrence Korb," *Frontline*, PBS, n.d. [2000], accessed April 30, 2018, https:// www.pbs.org/wgbh/pages/frontline/shows/future/interviews/korb.html.

23. Hunter Keeter, "Cody: Apache Units Need Equipment, Training," *Defense Daily*, July 2, 1999; Armed Forces Newswire Service, "Kosovo Fallout"; "Army Says Apaches Were Ready for Action in Kosovo," *Helicopter News*, July 16, 1999; Shimko, *The Iraq Wars*, Kindle, locations 3448–62; "The NATO Capability Gap."

24. Phillips, *Operation Joint Guardian*, 14–17, 20–25.

25. Shimko, *The Iraq Wars*, Kindle, locations 3448–62.

26. Brown, *Kevlar Legions*, Kindle, 155; Shimko, *The Iraq Wars*, Kindle, locations 3266–69; Millett and Maslowski, *For the Common Defense*, Kindle, locations 12419–41.

27. Thomas, "Kosovo and the Current Myth of Information Superiority," 13–29.

28. Reimer interview, 380–85.

29. Shinseki, "The Army—Intent of the Chief of Staff."

30. James Dao and Thom Shanker, "No Longer a Soldier, Shinseki Has a New Mission," *New York Times*, November 10, 2009, https://www.nytimes.com/2009/11/11/us/politics/11vets.html; General Officer Management Office, *General Eric K. Shinseki*; Philip Rucker, "Obama Picks Shinseki to Lead Veterans Affairs," *Washington Post*, December 7, 2008, http://www.washingtonpost.com/wp-dyn/content/article/2008/12/07/AR2008120701487.html?noredirect=on; Millett and Maslowski, *For the Common Defense*, Kindle, locations 12667–71.

31. General Officer Management Office, *General Eric K. Shinseki*.

32. Brown, *Kevlar Legions*, Kindle, 191–92.

33. High, "Talking with . . . the New Chief of Staff"; Millett and Maslowski, *For the Common Defense*, Kindle, locations 12228–32; Bradley Graham, "Army Chief May Seek Hike in Troop Levels; Gen. Shinseki Awaits Study Findings," *Washington Post*, June 24, 1999.

34. Millett and Maslowski, *For the Common Defense*, Kindle, locations 12228–32; Vinson, "Structuring the Army for Full-Spectrum Readiness," 19–32.

35. Aspin, *Report of the Bottom-Up Review*; Millett and Maslowski, *For the Common Defense*, Kindle, locations 12228–32; Crane, "Peace Dividends and Benevolent Interventions"; Crane, *Landpower and Crises*, 30.

36. Objective Force Task Force, *Objective Force Task Force*, 2.

37. Eric K. Shinseki, interview with the author, Falls Church, VA, October 21, 2016; Brown, *Kevlar Legions*, Kindle, 195.

38. Shinseki, interview with the author; "Interview: General Eric K. Shinseki," *Frontline*, PBS, n.d. [October 2000], accessed October 15, 2016, http://www.pbs.org/wgbh/pages/frontline/shows/future/interviews/shinseki.html.

39. James M. Dubik, interview with the author, Arlington, VA, November 17, 2016; Reimer interview, 398–400; Dennis Steele, *The Army Magazine Hooah Guide to Army Transformation: A 30-Minute Course on the Army's 30-Year Overhaul* (Arlington, VA: Association of the United States Army, 2001), Combined Arms Center Historical Archive, U.S. Army Combined Arms Center.

40. Brown, *Kevlar Legions*, Kindle, 195.

41. Brown, *Kevlar Legions*, Kindle, 194; Shinseki, "The Army—Intent of the Chief of Staff"; "Keane: Army Must Change to Stay Relevant," *Defense Daily*, September 3, 1999.

42. Brown, *Kevlar Legions*, Kindle, 195; Eric K. Shinseki, "Address to the Eisenhower Luncheon, 45th Annual Meeting of the Association of the United States Army (as Prepared for Presentation)," transcript, October 12, 1999.

43. Shinseki, "Address to the Eisenhower Luncheon."

44. Shinseki, "Address to the Eisenhower Luncheon"; Feickert, *The Army's Future Combat System*, 2–7.

45. Shinseki, "Address to the Eisenhower Luncheon"; Shinseki, interview with the author.

46. Eric K. Shinseki, comments in "1999 Fletcher Conference, Panel 4: Serving the Nation in the 21st Century, with General Eric K. Shinseki, General James L. Jones, Admiral Donald L. Pilling, General Lester L. Lyles," November 3, 1999, transcript, U.S. Army War College, Carlisle Barracks, PA; Shinseki, "Address to the Eisenhower Luncheon"; U.S. Department of Defense, *Report of the Quadrennial Defense Review*.

47. Shinseki, "Address to the Eisenhower Luncheon."

48. Shinseki, "1999 Fletcher Conference."

49. Dubik, "IBCT at Fort Lewis," 17–23.

50. "Interview: General Eric K. Shinseki."

51. Shinseki, "Address to the Eisenhower Luncheon"; Brown, *Kevlar Legions*, Kindle, 149.

52. Objective Force Task Force, *Objective Force Task Force*, 2; Shinseki, "Address to the Eisenhower Luncheon."

53. Shinseki, "The Army Vision," 33; Objective Force Task Force, *Objective Force Task Force*, 2.

54. Shinseki, "Address to the Eisenhower Luncheon."

55. Shinseki, "1999 Fletcher Conference"; Brown, *Kevlar Legions*, Kindle, 197.

56. Brown, *Kevlar Legions*, Kindle, 195.

57. Brown, *Kevlar Legions*, Kindle, 149; U.S. Army Training and Doctrine Command, "The Brigade Combat Team, Organizational and Operational Concept," January 6, 2000, Box 2, Official Correspondence Email Traffic Received from 20 to 5 January 2000, Folder 7, Official Correspondence—Email Traffic Received in January 2000 (Part 19 of 20), James M. Dubik Papers, U.S. Army Heritage and Education Center, 26–31; Mehaffey, "Vanguard of the Objective Force," 6–16.

58. U.S. Army Training and Doctrine Command, "The Brigade Combat Team," 14–21.

59. U.S. Army Training and Doctrine Command, "The Brigade Combat Team," 14–21; Reimer, "The Army After Next," reprinted in Reimer, *Soldiers Are Our Credentials*, 270.

60. Shinseki, interview with the author; Kennedy, "Army Approaches Decision on Interim Combat Vehicle"; Brown, *Kevlar Legions*, Kindle, 202; Shinseki, "The Army Vision," 33.

61. U.S. Army Training and Doctrine Command, "The Brigade Combat Team," 1–13; Brown, *Kevlar Legions*, Kindle, 202; Dubik, interview with the author; Dubik, "IBCT at Fort Lewis," 17–23.

62. Dubik, interview with the author.

63. Brown, *Kevlar Legions*, Kindle, 204.

64. Shinseki, "The Army Vision," 33; Brown, *Kevlar Legions*, Kindle, 205; Dubik, interview with the author.

65. Brown, *Kevlar Legions*, Kindle, 202; Shinseki, "The Army Vision," 33.

66. Mehaffey, "Vanguard of the Objective Force," 6–16; Brown, *Kevlar Legions*, Kindle, 205.

67. Mehaffey, "Vanguard of the Objective Force," 6–16.

68. Dubik, "IBCT at Fort Lewis," 17–23.

69. Dubik, "IBCT at Fort Lewis," 17–23.

70. U.S. Army Training and Doctrine Command, "The Brigade Combat Team," 1–13, emphasis in the original.

71. Dubik, "IBCT at Fort Lewis," 17–23.

72. "Interview: General Eric K. Shinseki."

73. Mehaffey, "Vanguard of the Objective Force," 6–16.

74. Caldera and Shinseki, "Army Vision," 3–5.

75. U.S. Army Training and Doctrine Command, "The Brigade Combat Team," 1–13.

76. U.S. Army Training and Doctrine Command, "The Brigade Combat Team," 63–66.

77. U.S. Army Training and Doctrine Command, "The Brigade Combat Team," 63–66.

78. U.S. Army Training and Doctrine Command, "The Brigade Combat Team," 1–13.

79. Andrew Krepinevich, quoted in Steven Lee Myers, "Army Is Restructuring with Brigades for Rapid Response" *New York Times*, October 13, 1999.

80. U.S. Army Training and Doctrine Command, "The Brigade Combat Team," 1–13, emphasis in the original.

81. "Interview: General Eric K. Shinseki"; Steele, *The Army Magazine Hooah Guide*, n.p.

82. Shinseki, interview with the author.

83. Steele, *The Army Magazine Hooah Guide*, n.p.

84. U.S. Army Training and Doctrine Command, "The Brigade Combat Team," 1–13.

85. Joint Staff, "Joint Strategy Review," 1999, quoted in Metz and Johnson, *Symmetry and U.S. Military Strategy*, 5, brackets in the original.

86. Joint Staff, "Joint Vision 2020," 57–76.

87. Joint Staff, "Joint Vision 2020," 57–76.

88. Metz and Johnson, *Symmetry and U.S. Military Strategy*, 5, emphasis in the original.

89. Metz and Johnson, *Symmetry and U.S. Military Strategy*, 3.

90. Metz and Johnson, *Symmetry and U.S. Military Strategy*, 9, 10.

91. U.S. Army Training and Doctrine Command, "Chapter 2: Assessment of Operational Environment," November 2, 1999, Box 2, Official Correspondence Email Traffic Received from 20 to 5 January 2000, Folder 6, Official Correspondence—Email Traffic Received in January 2000 (part 18 of 20), James M. Dubik Papers, U.S. Army Heritage and Education Center, 1–14.

92. U.S. Army Training and Doctrine Command, "Chapter 2: Assessment," 1–14.

93. U.S. Army Training and Doctrine Command, "Chapter 2: Assessment," 22–24.

94. U.S. Army Training and Doctrine Command, "Chapter 2: Assessment," 1–14.

95. U.S. Army Training and Doctrine Command, "Chapter 2: Assessment," 1–14.

96. U.S. Army Training and Doctrine Command, "Chapter 2: Assessment," 1–14, emphasis in the original.

97. Metz and Johnson, *Symmetry and U.S. Military Strategy*, 9, emphasis in the original.

98. U.S. Army War College, *Army Transformation Wargame 2001*.

99. U.S. Army Training and Doctrine Command, "The Brigade Combat Team," 26–31.

100. Crane, *Cassandra in Oz*, Kindle, 43–60.

101. Crane, *Landpower and Crises*, 5–9.

102. Manwaring, *Internal Wars*, vii-34.

103. Applegate, *Preparing for Asymmetry*, 1–25.

104. Applegate, *Preparing for Asymmetry*, 1–25, emphasis in the original.

105. Shinseki, "1999 Fletcher Conference."

106. U.S. Army Training and Doctrine Command, "Chapter 2: Assessment," 1–14.

107. U.S. Army Training and Doctrine Command, "Chapter 2: Assessment," 1–4; U.S. Army Training and Doctrine Command, "The Brigade Combat Team," 26–31.

108. U.S. Army Training and Doctrine Command, "The Brigade Combat Team," 1–13; U.S. Army Training and Doctrine Command, "Chapter 2: Assessment," 1–14.

109. U.S. Army Training and Doctrine Command, "The Brigade Combat Team," 1–13; U.S. Army Training and Doctrine Command, "Chapter 2," 1–14.

110. U.S. Army Training and Doctrine Command, "The Brigade Combat Team," 14–21.

111. Mehaffey, "Vanguard of the Objective Force," 6–16.

112. U.S. Army War College, *Army Transformation Wargame 2001*.

113. Grau and Kipp, "Urban Combat," 9–17; Spiller, *Sharp Corners*, 5–6.

114. Spiller, *Sharp Corners*, 5–6.

115. Metz and Johnson, *Symmetry and U.S. Military Strategy*, 9. See also Spiller, *Sharp Corners*.

116. Spiller, *Sharp Corners*, 73.

117. Peters, "The Human Terrain of Urban Operations," 4–12.

118. Peters, "The Human Terrain of Urban Operations," 4–12.

119. Mehaffey, "Vanguard of the Objective Force," 6–16.

120. Dubik, "IBCT at Fort Lewis," 17–23.

121. "Interview: General Eric K. Shinseki."

122. Eric K. Shinseki, "Eisenhower Luncheon Speech, Annual Meeting 2000 Association of the United States Army," October 17, 2000, transcript.

123. Shinseki, "1999 Fletcher Conference"; Caldera and Shinseki, "Army Vision," 3–5.

124. "Interview: General Eric K. Shinseki."

125. U.S. Combined Arms Center, "Subject: Balancing Operations, Leadership, and Training Doctrine," doctrine position paper sent from Lt. Gen William Steele to Gen. Eric Shinseki, August 1999, Folder 3, Official Correspondence—ACoS, Incoming and Outgoing 23–25 August 1999, Box 8, Series I Correspondence, Official Army Chief of Staff 4 Aug–15 Oct 1999, Eric K. Shinseki Collection, U.S. Army Heritage and Education Center.

126. U.S. Combined Arms Center, "Subject: Balancing Operations."

127. James M. Dubik, Transformation Brief to the School of Advanced Military Studies, PowerPoint briefing, n.d. [1999–2000], Folder 9, Briefing regarding Force Structure and Change, Given to SAMS, Box 30, Deputy Commanding General for Transformation, TRADOC and 25th Infantry Division after Sep 1999–23 Sep 2001, James M. Dubik Papers, U.S. Army Heritage and Education Center, emphasis in the original.

128. Shinseki, "The Army Vision," 33.

129. U.S. Army Command and General Staff College, *United States Army Command and General Staff College Catalog, Academic Year 1999–2000*, CGSC Circular No. 351-1, n.d. [1999], Command and General Staff College Papers, Ike Skelton Combined Arms Research Library, 3-1, 3-13; U.S. Army Command and General Staff College, *United States Army Command and General Staff College Catalog, Academic Year 2000–2001*, CGSC Circular No. 351-1, n.d. [2000], Command and General Staff College Papers, Ike Skelton Combined Arms Research Library, 3-13.

130. "Interview: Ralph Peters," *Frontline*, PBS, n.d. [October 2000], accessed October 15, 2016, http://www.pbs.org/wgbh/pages/frontline/shows/future/interviews/peters.html.

131. "Interview: General Eric K. Shinseki."

132. "Interview: General Eric K. Shinseki."

133. Record, "Operation Allied Force," 15–23, emphasis in the original.

134. Shin, "Future War," 63–79.

135. Crane, *Alternative National Military Strategies*, iii, 1–15, 16–22.

136. Crane, *Alternative National Military Strategies*, 10–11.

137. Crane, *Landpower and Crises*, 1–4, 5–9, 30–37.

138. Aspin, *Report of the Bottom-Up Review*, 19; Metz, "Introduction," in Metz, ed., *Revising the Two MTW Force Shaping Paradigm*, 3–4.

139. Aspin, *Report of the Bottom-Up Review*, 19; Kugler, "Replacing the 2 MTW Standard, 48, 54–55, 59–60, 62–64.

140. Steele, "Threats, Strategy, and Force Structure," 139, 141–42, 145–50, 160–62.

141. Manwaring, *Internal Wars*, vii-34.

142. Manwaring, *Internal Wars*, vii-34.

143. Applegate, *Preparing for Asymmetry*, 1–25.

144. Applegate, *Preparing for Asymmetry*, 1–25.

145. Packett and Gilhool, "Diplomacy by Other Means," 81–84.

146. Shull, "Correcting the Force Structure Mismatch," 31–39.

147. Crane, *Landpower and Crises*, 16–25.

148. Shin, "Future War," 63–79.

149. Shin, "Future War," 63–79.

150. Applegate, *Preparing for Asymmetry*, 1–25.

151. Record, "Operation Allied Force," 15–23.

152. Record, "Operation Allied Force," 15–23.

153. Record, "Operation Allied Force," 15–23.

154. Crane, *Landpower and Crises*, 16–25.

155. Crane, *Landpower and Crises*, 16–25.

156. Crane, *Landpower and Crises*, 16–25.

157. Crane, *Landpower and Crises*, 1–4, 25–28.

158. Crane, *Landpower and Crises*, 35–37.

159. Crane, *Landpower and Crises*, 25–28, 35–37.

160. Manwaring, *Internal Wars*, vii-34.

161. Crane, *Cassandra in Oz*, Kindle, 43–60; Nagl and Young, "*Si Vis Pacem, Para Pacem*," 31–34.

162. Nagl and Young, "*Si Vis Pacem, Para Pacem*," 31–34.

163. Nagl and Young, "*Si Vis Pacem, Para Pacem*," 31–34.

164. Nagl and Young, "*Si Vis Pacem, Para Pacem*," 31–34.

165. Vinson, "Structuring the Army for Full-Spectrum Readiness," 19–32.

166. Shull, "Correcting the Force Structure Mismatch," 31–39.

167. Crane, *Alternative National Military Strategies*, 1–15; Aspin, *Report of the Bottom-Up Review*, 19; Jablonsky, "Army Transformation," 139, 141–42, 145–50, 160–62.

168. Crane, *Landpower and Crises*, 25–28.

169. Shull, "Correcting the Force Structure Mismatch," 31–39.

170. Manwaring, *Internal Wars*, vii-34.

171. U.S. Army Training and Doctrine Command, "Chapter 2: Assessment," 1–14.

172. "Interview: General Eric K. Shinseki"; Heller and Stofft, *America's First Battles*, xi.

173. "Interview: General Eric K. Shinseki."

174. Theo Farrell, Rynning, and Terriff, *Transforming Military Power since the Cold War*, 39–44; Shinseki, "Address to the Eisenhower Luncheon."

175. Steele, *The Army Magazine Hooah Guide*, n.p.

176. Dubik, "IBCT at Fort Lewis," 17–23.

177. Shinseki, "The Army Vision," 33.

178. "Interview: General Eric K. Shinseki."

179. U.S. Army Training and Doctrine Command, "Chapter 2: Assessment," 1–14.

180. Manwaring, *Internal Wars*, vii-34; Applegate, *Preparing for Asymmetry*, 1–25; Lovelace, "Foreword," iii–iv; Nagl and Young, "*Si Vis Pacem, Para Pacem*," 31–34.

181. Vinson, "Structuring the Army for Full-Spectrum Readiness," 19–32.

182. Crane, *Alternative National Military Strategies*, 1–15.

183. Crane, *Alternative National Military Strategies*, 1–15.

184. Crane, *Alternative National Military Strategies*, 1–15.

185. Wass de Czege and Echevarria, "A New Strategy and Military Logic for the 21st Century," 72–74, 77–78.

186. Wass de Czege and Echevarria, "A New Strategy and Military Logic for the 21st Century," 72–74, 77–78; Nagl and Young, "*Si Vis Pacem, Para Pacem*," 31–34.

187. Packett and Gilhool, "Diplomacy by Other Means," 81–84.

188. Nagl and Young, "*Si Vis Pacem, Para Pacem*," 31–34.

189. U.S. Army Training and Doctrine Command, "Chapter 2: Assessment," 1–14.

190. U.S. Army Training and Doctrine Command, "The Brigade Combat Team," 14–21.

191. Manwaring, *Internal Wars*, vii-34; Applegate, *Preparing for Asymmetry*, 1–25.

192. Packett and Gilhool, "Diplomacy by Other Means," 81–84.

193. Nagl and Young, "*Si Vis Pacem, Para Pacem*," 31–34.

194. U.S. Army Training and Doctrine Command, "The Brigade Combat Team," 14–21; U.S. Army Training and Doctrine Command, "Chapter 2: Assessment," 22–24.

195. U.S. Army Training and Doctrine Command, "Chapter 2: Assessment," 1–14, 26–28.

196. U.S. Army Training and Doctrine Command, "Chapter 2: Assessment," 1–14, 26–28.

197. Manwaring, *Internal Wars*, vii-34, emphasis added.

198. Metz and Johnson, *Symmetry and U.S. Military Strategy*, iii-23, emphasis in the original.

199. Manwaring, *Internal Wars*, vii-34.

200. Crane, *Landpower and Crises*, 5–9, 25–28.

201. Nagl and Young, "*Si Vis Pacem, Para Pacem*," 31–34.

202. Crane, *Landpower and Crises*, 16–25.

203. Crane, *Landpower and Crises*, 30–37.

204. Dubik, "IBCT at Fort Lewis," 17–23; James Dubik, "Subject: Thought Paper—Similarities: Haiti and Bosnia," unpublished paper for Lt. Gen. Thomas Burnette, Bosnia-Herzegovina, March 12, 1999, Folder 5, Unpublished Thought Papers, Box 29, Task Force Eagle and Multinational Division North Operation Joint Forge, B-H 7 Jan 1999–circa 1999, James M. Dubik Papers, U.S. Army Heritage and Education Center.

205. Caldera and Shinseki, "Army Vision," 3–5; Dubik, "IBCT at Fort Lewis," 17–23.

206. U.S. Army Training and Doctrine Command, "The Brigade Combat Team," 1–13; Mehaffey, "Vanguard of the Objective Force," 6–16.

207. "Interview: General Eric K. Shinseki."

208. Crane, *Landpower and Crises*, 16–25, 35–37.

209. Manwaring, *Internal Wars*, vii-34.

210. Peters, "The Human Terrain of Urban Operations," 4–12.

211. Condoleezza Rice and Donald Rumsfeld, quoted in Fitzgerald, *Learning to Forget*, Kindle, 121.

212. George W. Bush, quoted in Jablonsky, "Army Transformation," 54.

213. Condoleezza Rice, quoted in Jablonsky, "Army Transformation," 54.

214. Crane, *Landpower and Crises*, 35–37.

215. Jablonsky, "Army Transformation," 43–62.

216. Schultz, "Ten Years Each Week," 278–80.

217. George W. Bush, quoted in Jablonsky, "Army Transformation," 54.

218. Fitzgerald, *Learning to Forget*, Kindle, 114.

219. Millett and Maslowski, *For the Common Defense*, Kindle, locations 12644–53.

220. Shimko, *The Iraq Wars*, Kindle, locations 3840–47; Millett and Maslowski, *For the Common Defense*, Kindle, locations 12644–53.

221. Millett and Maslowski, *For the Common Defense*, Kindle, locations 12653–66; Shimko, *The Iraq Wars*, Kindle, locations 3851–54.

222. Shimko, *The Iraq Wars*, Kindle, locations 3860–64.

223. Crane, "Peace Dividends and Benevolent Interventions"; "NATO's Role in Kosovo," North Atlantic Treaty Organization, March 9, 2017, https://www.nato.int/cps/en/natolive/topics_48818.htm; "Mandate," United Nations Mission in Kosovo, n.d., accessed April 29, 2018, https://unmik.unmissions.org/mandate.

224. Anderson, "Military Operational Measures of Effectiveness for Peacekeeping Operations," 36–44; "New Academy Will Train NCOs."

225. Anderson, "Military Operational Measures of Effectiveness for Peacekeeping Operations," 36–44.

226. Anderson, "Military Operational Measures of Effectiveness for Peacekeeping Operations," 36–44.

227. Anderson, "Military Operational Measures of Effectiveness for Peacekeeping Operations," 36–44.

228. Anderson, "Military Operational Measures of Effectiveness for Peacekeeping Operations," 36–44.

229. Anderson, "Military Operational Measures of Effectiveness for Peacekeeping Operations," 36–44.

230. Chapman et al., *Prepare the Army for War*, 52; White and Shinseki, "A Statement on the Posture of the United States Army 2001," 3.

231. Chapman et al., *Prepare the Army for War*, 52; White and Shinseki, "A Statement on the Posture of the United States Army 2002," 6; Joint Chiefs of Staff, *Joint Doctrine for Military Operations Other Than War*, JP 3-07, I-2.

232. Headquarters, U.S. Department of the Army, *Operations*, FM 3-0 (2001), 9-1; Headquarters, U.S. Department of the Army, *Operations*, FM 100-5 (1993), 13-0–13-8.

233. Headquarters, U.S. Department of the Army, FM 3-0 (2001), 10-1–10-4.

234. Headquarters, U.S. Department of the Army, FM 3-0 (2001), 3-0.

235. Headquarters, U.S. Department of the Army, FM 3-0 (2001), 1-14–1-17.

236. Headquarters, U.S. Department of the Army, FM 3-0 (2001), 9-1.

237. U.S. Combined Arms Center, "Subject: Balancing Operations"; Headquarters, U.S. Department of the Army, FM 3-0 (2001), 9-1, 1-14–1-17.

238. Headquarters, U.S. Department of the Army, FM 3-0 (2001), 9-1, 10-1–10-4; U.S. Combined Arms Center, "Subject: Balancing Operations."

239. Headquarters, U.S. Department of the Army, FM 3-0 (2001), 9-1–9-17.

240. Headquarters, U.S. Department of the Army, FM 3-0 (2001), 9-1–9-17.

241. Headquarters, U.S. Department of the Army, FM 3-0 (2001), 9-1–9-17.

242. Headquarters, U.S. Department of the Army, FM 3-0 (2001), 9-1–9-17.

243. Headquarters, U.S. Department of the Army, FM 3-0 (2001), 9-1–9-17.

244. Headquarters, U.S. Department of the Army, FM 3-0 (2001), 9-1–9-17.

245. Brown, *Kevlar Legions*, Kindle, locations 5023–45.

246. Brown, *Kevlar Legions*, Kindle, locations 5023–45.

247. Brown, *Kevlar Legions*, Kindle, locations 5023–45; Objective Force Task Force, *Objective Force Task Force*, 6.

248. Brown, *Kevlar Legions*, Kindle, 225, 227; Fitzgerald, *Learning to Forget*, Kindle, 115.

249. White and Shinseki, "A Statement on the Posture of the United States Army 2001," 3.

250. Shinseki, "The Army Vision," 33; White and Shinseki, "A Statement on the Posture of the United States Army 2001," 9.

251. Brown, *Kevlar Legions*, Kindle, 198.

252. Dubik, "IBCT at Fort Lewis," 17–23.

253. Steele, *The Army Magazine Hooah Guide*, n.p.

254. Brown, *Kevlar Legions*, Kindle, 198.

255. U.S. Army War College, *Army Transformation Wargame 2001*, emphasis in the original.

256. Metz and Johnson, *Symmetry and U.S. Military Strategy*, 16.

257. U.S. Army War College, *Army Transformation Wargame 2001*.

258. White and Shinseki, "A Statement on the Posture of the United States Army 2001," 3; Brown, *Kevlar Legions*, Kindle, 203–4; Kennedy, "Army Approaches Decision on Interim Combat Vehicle."

259. Brown, *Kevlar Legions*, Kindle, 206.

260. U.S. Army Training and Doctrine Command, "Chapter 2: Assessment," 1–14; Applegate, *Preparing for Asymmetry*, 1–25, emphasis in the original.

261. Millett and Maslowski, *For the Common Defense*, Kindle, locations 12781–84.

262. White and Shinseki, "A Statement on the Posture of the United States Army 2002," introduction, n.p.

263. White and Shinseki, "A Statement on the Posture of the United States Army 2002," 8, 9.

264. Shinseki, "Statement by General Eric K. Shinseki, Chief of Staff, United States Army," 2.

265. B. H. Liddell Hart, quoted in Steele, *The Army Magazine Hooah Guide*, n.p.

266. White and Shinseki, "A Statement on the Posture of the United States Army 2002," 11; Brown, *Kevlar Legions*, Kindle, 206, 208.

267. Brown, *Kevlar Legions*, Kindle, 206.

268. Lydia Saad, "Top Ten Findings about Public Opinion and Iraq," Gallup, October 8, 2002, http://www.gallup.com/poll/6964/top-ten-findings-about-public-opinion-iraq.aspx; Michael Elliott, "Bush Isn't as Lonely as He Looks"; Authorization for Use of Military Force against Iraq Resolution of 2002, Pub. L. No. 107-243, October 16, 2002, https://www.govinfo.gov/content/pkg/PLAW-107publ243/pdf/PLAW-107publ243.pdf.

269. Reuters, "Saddam's Promises and Threats."

270. Crane, *Avoiding Vietnam*, v–vi.

271. Crane, *Avoiding Vietnam*, 1–19.

272. Crane, *Avoiding Vietnam*, v–vi.

273. Francis Kelly, "U.S. Army's Special Forces 1961–1971," manuscript, 1973, quoted in Crane, *Avoiding Vietnam*, 1–19.

274. Crane and Terrill, *Reconstructing Iraq*, 1–4.

275. Crane and Terrill, "*Reconstructing Iraq*," 1.

276. Crane and Terrill, *Reconstructing Iraq*, 2–3.

277. Crane and Terrill, *Reconstructing Iraq*, 4–9.

278. Crane and Terrill, *Reconstructing Iraq*, 4–9.

279. Benson, "OIF Phase IV," 61–68.

280. "Biography + Military Career: General Tommy R. Franks, US Army Retired," General Tommy Franks Leadership Institute and Museum, n.d., accessed May 13, 2018, http://www.tommyfranksmuseum.org/biography--military-career.html.

281. Gordon and Trainor, *Cobra II*, Kindle, locations 1400–1600.

282. "GEN David D. McKiernan," Military Hall of Honor, n.d., accessed May 13, 2018, http://www.militaryhallofhonor.com/honoree-record.php?id=282; Gordon and Trainor, *Cobra II*, Kindle, locations 1562–80.

283. Gordon and Trainor, *Cobra II*, Kindle, locations 2722–39; Tommy Franks, quoted in Fitzgerald, *Learning to Forget*, 126.

284. Gordon and Trainor, *Cobra II*, Kindle, locations 2785–807; Benson, "OIF Phase IV," 61–68.

285. Shimko, *The Iraq Wars*, Kindle, locations 4069–98.

286. Shimko, *The Iraq Wars*, Kindle, locations 4051–55; Crane, "Peace Dividends and Benevolent Interventions."

287. Eric Shinseki and Carl Levin, "2003: General Shinseki Tells Senate Several Hundred Thousands Troops Needed to Occupy Iraq," February 25, 2003, video of congressional testimony, YouTube, May 18, 2014, https://www.youtube.com/watch?v=HjpCfzY4SKo.

288. Shinseki and Levin, "2003."

289. Shinseki and Levin, "2003."

290. Shinseki and Levin, "2003."

291. Brown, *Kevlar Legions*, Kindle, 231.

292. Barry, "Anatomy of a Revolt"; Jamie McIntyre, "McIntyre: Myth of Shinseki Lingers," CNN, December 8, 2008, http://www.cnn.com/2008/POLITICS/12/08/obama.shinseki/index.html.

293. Gordon and Trainor, *Cobra II*, Kindle, locations 2023–29.

294. Calamur, "Oil Was Supposed to Rebuild Iraq"; Shimko, *The Iraq Wars*, Kindle, locations 4104–11.

295. Crane, "Peace Dividends and Benevolent Interventions"; Shimko, *The Iraq Wars*, Kindle, locations 4069–98.

296. O'Hanlon and Livingston, *Iraq Index*, 13; Fitzgerald, *Learning to Forget*, Kindle, 119.

297. Benson, "OIF Phase IV," 61–68; Gordon and Trainor, *Cobra II*, Kindle, locations 9766–89.

298. See, for example, Gordon and Trainor, *Cobra II*; Ricks, *Fiasco*; Filkins, *The Forever War*; Woodward, *State of Denial*; Baker and Hamilton, *Iraq Study Group Report*; and Gordon and Trainor, *The Endgame*.

299. Gordon and Trainor, *Cobra II*, Kindle, locations 9387–410.

300. Millett and Maslowski, *For the Common Defense*, Kindle, locations 12665–76; Brown, *Kevlar Legions*, Kindle, 237.

301. Dubik, *The Army's "Twofer."*

302. Brown, *Kevlar Legions*, Kindle, 206.

303. Objective Force Task Force, *Objective Force Task Force*, 23, 22; Greg Grant, "It's Official: FCS Cancelled," Military.com, June 23, 2009, http://www.dodbuzz.com/2009/06/23/its-official-fcs-cancelled/.

304. Brown, *Kevlar Legions*, Kindle, 248.

305. Crane, *Alternative National Military Strategies*, 1–15; Defense Manpower Data Center, cited in David Coleman, "U.S. Military Personnel 1954–2014," History in Pieces, n.d., accessed May 16, 2018, https://historyinpieces.com/research/us-military-personnel-1954-2014.

306. Schein, *Organizational Culture and Leadership*, 8; Kuhn, *The Structure of Scientific Revolutions*, 5.

5 "THE FIRST, THE SUPREME, THE MOST FAR-REACHING ACT OF JUDGMENT"

THE U.S. ARMY's persistent challenge throughout the 1990s and into the war on terror has been its inability to achieve a political solution and terminate conflicts. In Somalia, Haiti, Bosnia-Herzegovina, and Kosovo, the Army was unable to successfully engage the political dimension of the conflict to reach a settlement. In Afghanistan, Iraq, and now Syria, the United States is trapped in forever wars because, despite nearly eighteen years of low-intensity conflict, the Army has not been able to compel every faction to accept a political order that would end the fighting.

The reasons for this are legion, but at their core the difficulties boil down to two problems. First, the Army has consistently been reluctant to engage the political dimension, and has instead looked to others—other agencies of the U.S. government, the host nation, the United Nations (UN), or allies—to engage this dimension on its behalf. But even when the Army has reluctantly waded into the political dimension of low-intensity conflicts, it has refused to take sides and back a designated winner in the conflict, instead trying to forge a solution in which every party won. Together these behaviors are a recipe for forever war. As nineteenth-century Prussian military theorist Carl von Clausewitz wrote in his seminal work, *On War*, "The first, the supreme, the most far-reaching act of judgment that the statesman and commander have to make is to establish . . . the kind of war on which they are embarking; neither mistaking it for, nor trying to turn it into, something that is alien to its nature."[1]

But this is exactly the wrongheaded thinking in which the Army engaged throughout the 1990s when it imagined the wars of the future. It was repeatedly forced to fight low-intensity conflicts. And everyone—both Army transformers and low-intensity conflict observers alike—agreed that the Army was likely to continue to be asked to fight such conflicts for the foreseeable future. Yet the Army stubbornly refused to prepare to fight such conflicts. As irascible Army critic Ralph Peters wrote in 1999, "Our military is determined to be unprepared for missions it does not want, as if the lack of preparedness might prevent our going. We are like children who refuse to get dressed for school."[2] Instead the Army continued to prepare for an imagined future great power war as if hoping that preparation might make it come to pass. In the end, the Army had optimized itself to fight and win the nation's *battles* at the cost of fighting and winning the nation's *wars*.

The Army's fixation on a great power war as the prototype for warfare lay at the core of its reluctance to prepare for or prosecute low-intensity conflicts. But this fixation was based on a misperception of history. According to orthodox Army thinking in the second half of the twentieth century, the Army's role in a great power war was to end it as quickly as possible by destroying an enemy's ability to resist—vanquishing an enemy's military forces and its ability to rearm, crushing the will of its people to continue to resist, and forcing its government from power. In this idealized model there would be no postwar low-intensity conflict and no role for the Army in postwar stability. The U.S. Department of State, an intergovernmental organization such as the UN, or even allies could do the heavy lifting of postwar reconstruction. Not surprisingly, this was the Army's preferred model for warfare.

Of course, this idealized vision of a great power war, in which the Army would have no role in postwar stability and reconstruction, flies in the face of the Army's own history. Before the Cold War, the Army had a long tradition of military governance of conquered lands after great power wars, including the South after the Civil War and the Philippines after the Spanish-American War. Even in the biggest great power wars in human history—World War I and World War II—there was still a significant role for the U.S. Army in postwar stabilization that went well beyond providing security for civilian governmental and intergovernmental agencies.[3]

But even if the Army were completely freed of its responsibility for overseeing postwar stability in great power wars, there has not been a great power war since 1945. And what, realistically, are the chances that there will be one in the nuclear age? Since the end of World War II, the Army has either fought limited high-intensity conflicts against second-tier competitors that devolved into lengthy low-intensity conflicts or it has been asked to intervene in conflicts that were from their outset of a low-intensity nature. And the foreseeable future portends many more such conflicts for the Army.

In the absence of an acknowledgment of its responsibility to engage the political dimension of low-intensity conflicts, the Army's "measure of effectiveness" was not reaching a sustainable political solution that would permit it to withdraw. Instead, "winning" was defined as avoiding casualties and dumping the messy business of achieving a political settlement on someone else as quickly as possible. From this perspective, Somalia was a "loss" because of the relatively high casualties. Bosnia-Herzegovina and Kosovo were "draws" because there were few casualties but the Army found itself trapped in each conflict. And Haiti was a "win" because there were few casualties and the Army was relatively quickly able to dump the conflict on the UN and leave. Senior Army leaders did not understand that all four of these conflicts were "losses" because the Army failed to achieve the United States' political ends in each country.

Yet from the perspective of Senior Army leaders, their actions were imminently logical and, in fact, vital to the survival of the United States. If, as the senior leadership of the Army contended, the lessons of the twentieth-century high-intensity conflicts was that the Army had to be prepared because peer competitors could emerge with little warning, then it was critical that the Army ignore low-intensity conflicts and single-mindedly continue to prepare for an imminent great power war against a completely hypothetical future peer competitor.

Armed with this conviction, the Army defended its culture against contradictory thinking with an aggressiveness approaching zealotry. The first way that senior Army leaders protected the Army culture against change was by actively disputing evidence that might otherwise force the Army to change course away from hyperpreparedness for high-intensity conflict.

Reality—the loss of the United States' only true peer competitor and the unbroken series of low-intensity conflicts in which the Army was forced to engage—should have forced the Army to rethink its course. Low-intensity conflict observers and Army transformers both acknowledged that the future held many more such conflicts. How were senior Army leaders able to justify ignoring these realities?

The first, least avoidable reality with which the Army had to contend was the collapse of the Soviet Union. The threat of aggression by the communist world had been the focal point of the Army's training and preparation for a half century. A few Army leaders made the argument that Russia had not gone away and could still be a dangerous, near-peer competitor. But even the chief of staff of the Army, Gen. Carl Vuono, acknowledged that this was a remote possibility. It seemed as though the Army was about to have a serious discussion about how it might radically reshape itself for the future.[4]

Just in time, Army transformers were saved by the Gulf War. As if designed to confirm all of the Army's most deeply held biases, the Iraqi Army—a Soviet-style, heavy, mechanized force—happily obliged the U.S. Army by fighting it in exactly the way the U.S. Army had prepared to be fought for over fifty years. Even better, the chairman of the Joint Chiefs of Staff, Gen. Colin Powell—of the eponymous Powell Doctrine—ensured that President George H. W. Bush and his administration set clear, limited objectives and used overwhelming force to eject the Iraqis from Kuwait in the most dramatic fashion imaginable.[5] The American public and the national security establishment were treated to a six-month-long, televised justification for the Army to remain focused on high-intensity conflict—to the exclusion of all other priorities.

The subsequent decade of transformation toward ever-greater capacity to replicate—or even surpass—the Army's dramatic Gulf War victory was an exercise in dogged determination to ignore mounting evidence and rising voices recommending an alternative path.

First, there were the many low-intensity conflicts in which the Army was engaged throughout the 1990s. Somalia, Haiti, Bosnia-Herzegovina, and Kosovo were only the most visible of those conflicts. There were well over one hundred other crises and missions, from partnered military training, to hurricane and disaster relief, to humanitarian assistance on nearly every continent on Earth. Each of these low-intensity conflicts ably and repeatedly demonstrated that the Army was not properly manned, equipped, or trained to prosecute these conflicts.

Then there were the voices calling for change. Many warned that the Army's own asymmetric advantage—its hyperpreparedness for high-intensity conflict—would drive enemies to produce their own asymmetric responses that avoided U.S. capabilities.

While Army transformers and low-intensity conflict observers agreed that adversaries would seek asymmetric solutions to sidestep the Army's dominance in high-intensity conflict, they differed sharply on what those measures would be. Army transformers—focused on a future great power war against a peer competitor—predicted adversaries would acquire high-tech means of defeating the Army. Transformers worried about weapons and technologies that could deny the Army's access to theaters and key deployment facilities like airfields and seaports. They worried about mass casualty–producing weapons that could create negative reactions in the United States. And they worried about precision munitions and new air defense systems that could level the playing field by stripping away the Army's key capabilities. Of course, low-intensity conflict observers predicted completely different forms of asymmetry—those of a low-tech, low-intensity conflict adversary—that would be even more effective in sidestepping the Army's dominance in high-intensity conflict. Guerilla tactics, hiding in and among the populace, and using decoys and terrain to disperse and hide personnel and equipment were all cheap means to frustrate Army capabilities. The Serbians provided the U.S. military with a proof of principle of some of these asymmetric tactics in 1999 during Operation Allied Force.

As thought on asymmetry progressed, some observers began to warn that one potential avenue for adversaries to achieve asymmetry was through urban operations. The challenge of urban operations was perhaps the greatest cognitive threat to Army transformation to emerge throughout the 1990s. It was particularly dangerous because it was Army transformers themselves—through the Army After Next project wargames—who first identified the phenomenon; every time the Army After Next confronted an enemy, the enemy rushed into a city and hid among the populace.

Low-intensity conflict observers immediately recognized that the true source of the asymmetric advantage imparted to an adversary by urban operations was the people in the city. The Army was not properly manned, equipped, or trained to fight and operate in and among a civilian populace. Moreover, in a city, the Army would be unable to avoid the political dimensions of the conflict. In short, these observers immediately recognized that urban operations forced the Army to engage in its least preferred type of warfare: a low-intensity conflict.

Army transformers stubbornly refused to acknowledge the low-intensity conflict aspects of operating in an urban environment. Some tried to wish away the whole problem, arguing that the Army could avoid urban centers or that the civilian populace would reject the enemy military in their midst and expel them from the town themselves. These dubious arguments were easily dismissed; control of the cities was the political objective of a war and could not be ignored. Army transformers quickly moved from these arguments to the conception of urban areas as "complex terrain,"[6] difficult to move through and complicating the detection and attack of enemy targets. To the extent that civilian populaces in cities were considered at all, they were thought of as concealment within which enemies could hide or as human shields. This conflation of urban and complex terrain lent itself to technological solutions that would allow the Army to continue to ignore low-intensity conflict and remain focused on high-intensity conflict.

In the end, Army transformers never really came to grips with the problem of urban operations. Instead researchers simply ended wargames when all of the enemy outside the cities had been destroyed, declaring victory and cataloging their findings with a brief footnote about the necessary "mop-up" of enemy forces remaining in the urban centers.

This deliberate refusal by Army transformers to acknowledge the glaring arguments piling up against Army transformation went beyond a simple inability to see the issue from the perspective of low-intensity conflict observers. Army transformers—including successive chiefs of staff of the Army—were willfully ignoring evidence that contradicted the Army culture of hyperpreparedness for a great power war.

The Solution: Don't Institutionalize

Another way that senior Army leaders defended the Army culture against the challenge presented by low-intensity conflict observers was to pay lip service to the problem without actually fixing it (which would require institutionalizing low-intensity conflict as a mission coequal with high-intensity conflict).

The Army did produce low-intensity conflict doctrine. But as David Fitzgerald—citing Lt. Col. (Ret.) Andrew Krepinevich—has pointed out, "Field manuals can be used as a smoke screen to persuade political masters that a military is making desired changes without actually necessitating real change."[7] Moreover—with the exception of a brief regression in low-intensity conflict doctrine with the publication of the 1993 edition of Field Manual (FM) 100-5, *Operations*—the Army's doctrine regarding low-intensity conflicts remained virtually unchanged between the publication of the 1990 edition of FM 100-20, *Operations Other than War*, and the publication of the 2001 edition of FM 3-0, *Operations*. The lessons of the Army's 1990s low-intensity conflict experiences had almost no impact on Army doctrine.

Additionally, those 1990s experiences were barely reflected in the Army's education system. At the beginning of the decade, the Army taught only a few hours on low-intensity conflict at the Command and General Staff College (CGSC) at Fort Leavenworth, Kansas, and the U.S. Army War College (USAWC) at Carlisle Barracks, Pennsylvania. By the end of the decade, low-intensity conflict didn't even warrant its own course at the CGSC; it was relegated to a single practical planning exercise.

Low-intensity conflict was not institutionalized in unit training or essential tasks, either. The only units that had such conflict added to their list of assigned missions or received specialized training in such conflict were those units earmarked for actual deployment to a low-intensity conflict operation. And, as soon as they returned—or sometimes even before they returned—these units went back to training for high-intensity conflict missions, which remained their directed essential tasks even during their low-intensity conflict deployments. The Joint Readiness Training Center (JRTC) at Fort Polk, Louisiana, was purportedly created to train light infantry divisions to execute low-intensity conflict, but the actual training that occurred there was decidedly high-intensity in nature. The JRTC and, more frequently, the Joint Multinational Readiness Center (JMRC) at Hohenfels, Germany, did occasionally execute low-intensity conflict training rotations, but only for units deploying to these types of conflicts.

The Army likewise did not institutionalize low-intensity conflict in any of its organization. It did establish two offices to capture lessons from its low-intensity conflict experiences: the Army–Air Force Center for Low Intensity Conflict (A-AFCLIC) and the Peacekeeping Institute (PKI). A-AFCLIC was closed by Gen. Dennis Reimer in the mid-1990s and the PKI was almost closed by Gen Eric Shinseki; it only survived because of the beginning of the war on terror. The Army never created any operational units specifically earmarked to execute low-intensity conflicts. Some believed that the Interim Brigades were intended to specialize in such conflicts, but senior Army leaders—most notably the chief of staff of the Army, General Shinseki—along with internal and public writings about the Interim Brigades confirm that this organization was designed for high-intensity conflicts.

Finally, the Army never established incentives to encourage expertise or experience in low-intensity conflict. In fact, complaining too loudly about the Army's diminution of low-intensity conflict proficiency was a likely path to an early end for an Army officer's career; most notably, it had nearly derailed Major General Sullivan's.

And this is the final way in which the senior leadership of the Army protects the Army culture from contradictory ideas: by preventing innovators from entering the senior leadership.

Stephen Gerras and Leonard Wong conducted a statistical analysis of personality test results for students at USAWC—a select population of Army lieutenant colonels and colonels from which the vast majority of Army brigade commanders and general officers are selected—and compared them to the general population of the United States. Gerras and Wong described people who scored high in "openness" as having "a strong intellectual curiosity, creativity, and a comfortable relationship with novelty and variety." They explained that "people scoring high in openness . . . are more likely to hold unconventional beliefs," adding, "Officers with higher levels of openness would be expected to have more potential to change their minds, if needed." Conversely, they wrote, "People with low scores on openness tend to have more conventional, traditional interests, preferring familiarity over novelty."[8]

Gerras and Wong found that the Army officers who scored lower in "openness" were selected for advancement at higher rates than other officers. Summarizing the test data, they observed, "The most successful officers score lower in openness than the general U.S. population. . . . To make matters worse, though, those USAWC students selected for brigade command score even lower than the overall USAWC average."[9] From the empirical evidence, the Army appears to be protecting its culture from innovation by, consciously or unconsciously, selecting the least innovative and imaginative leaders for advancement into its senior leadership. Gerras and Wong were silent on the mechanism by which the Army advantages closed-minded leaders over other officers, but their findings clearly show that being less open to new ideas or contradictory evidence apparently makes senior officers more successful in the Army culture.

Thus it is not terribly surprising that senior Army leaders zealously defended the Army culture against any evidence that might have otherwise forced the Army to change the course of its transformation. It appears that they were selected for their position at least in part based on their propensity to actively reject novel or contradictory ideas.

Low-Intensity Conflict Observers Couldn't Articulate the Costs

Low-intensity conflict observers, faced with an Army culture that was actively defending itself against dissent, would have probably found the challenge of changing the direction of Army transformation insurmountable under any circumstances. But these observers also internally wrestled with some cognitive obstacles that made it more difficult for them to communicate the risk to the Army of failing to build a capacity to engage in low-intensity conflicts.

First, low-intensity conflict observers failed to conceive of a "big" low-intensity conflict. Throughout the 1990s Army doctrine segregated the "spectrum of conflict" or the "range of military operations" into great power war ("war" or "major theater war"), limited high-intensity conflicts ("small-scale contingencies"), and low-intensity conflicts ("operations other than war" or "stability and support operations"), implying through this ordering that these conflicts would be smaller than small-scale contingencies. Moreover, throughout the 1990s, low-intensity conflicts were all small in scale; Operation Uphold Democracy in Haiti had the largest U.S. contingent, with twenty-one thousand troops—both soldiers and Marines.[10] Low-intensity conflict observers could not conceive of such a conflict that was also a major theater war—a big war that consumed the entire Army and all of its resources, as was eventually the case in Afghanistan and Iraq.

Yet on a deeper level, low-intensity conflict observers struggled internally to free themselves from the cognitive shackles of the ubiquitous Army culture. These observers were reluctant to question the core tenets of the Army culture. While they did occasionally object to the Army's specific plans for transformation and questioned the specific arguments that Army transformers made for ignoring low-intensity conflicts, they refused to question the Army culture's chief metaprinciples: that a great power war was imminent and would come without warning.

This self-censorship is especially tragic because these ideas are so very vulnerable to scrutiny. Was a great power war inevitable and imminent? Even without the benefit of hindsight, the answer is clearly no. The consensus among historians and international relations experts outside the national security establishment was (and is) that nuclear weapons precluded a great power war. And, of course, this assertion was also contradicted by the growing number of low-intensity conflicts in which the Army was engaged around the world. Would a peer competitor emerge without warning, leaving the Army no time to prepare? No. This flew in the face of all of the scholarship on the post–Cold War world. Numerous influential observers from outside the national security establishment, witnessing the new strategic environment and the advent of the internet age, argued that the entry fee to the new global economy was opening up one's country to the world and playing by the rules of the international order.[11] In short, the means to become a peer competitor were unachievable in the secrecy and shadows required to build a peer military capacity in secret.

Without conceiving of a "big" low-intensity conflict, and without rejecting the increasingly dubious assumptions of the Army culture, low-intensity conflict observers could not effectively communicate the cost of refusing to develop a capacity to engage in such conflicts. Had they been able to overcome these two cognitive obstacles, they could have communicated that, while there was not going to be another great power war, it was only a matter of time before the Army was faced with a low-intensity conflict too big to ignore.

Operating under the cognitive shackles of the Army culture, they were instead relegated to equivocation. Of course, they were forced to say, the Army should continue to prepare for a great power war against a mythical peer competitor. If it had time, however, the Army should also try to get better at the small, low-intensity conflicts that it would continue to fight for the foreseeable future. This was an utterly ineffective argument to make against senior Army leaders who were actively and aggressively defending their culture against just such arguments. Moreover, their detractors were armed with both the authority of senior leadership and the excuse that they already had too few resources and too little time to prepare for an imminent third world war.

There Will Never Be Another Great Power War

But the fact remains: there will not be another great power war, at least as long as nuclear weapons remain the dominant feature of the strategic landscape.

That is not to say that the United States will never have a peer competitor. The economic and international political order that the United States and its allies established after World War II was designed to open markets, standardize and regulate monetary exchanges, and prevent the outbreak of another great power war. Theoretically, any country with the resources, determination, and internal stability to effectively compete in this economic and international political order could eventually match or surpass the United States in economic and military power.

Yet as long as nuclear weapons remain an ultimate threat against which there is no counter, there will not be a great power war. And, as such, exorbitant expenditures on "readiness" for a great power war are both wasteful and unnecessary.

And what, for that matter, does "readiness" for a great power war against China or Russia even *mean*? Throughout his tenure, the Army chief of staff, Gen. Mark A. Milley, repeatedly insisted that the Army "must always be ready to fight tonight." The United States put over sixteen million men in uniform for World War II, as much as 12 percent of the total population of the United States. That was one-sixth of all men in the United States. To match this level of "readiness" today, America would have to reinstate conscription and put over thirty-seven million Americans—men and women—in uniform.[12] Such a level of perpetual preparedness would quickly plunge the country into economic ruin. What's more, it is completely unnecessary.

Despite the exhortations of the senior Army leaders of the 1990s or today, another great power war against a peer competitor is neither imminent nor inevitable. In fact, the trend in the world is not toward another great power war but toward low-intensity conflicts, ever diminishing in scale and scope. World War II was the bloodiest war in the history of mankind. Around 3 percent of the world's population was killed in the Second World War, most of them from the Soviet Union and China. But that war ended with the advent of nuclear weapons and, ever since, wars have been shrinking in scale and violence. It is a phenomenon that historians and political scientists have taken to calling the Long Peace.[13]

The trend for the United States since the end of World War II has not been toward an imminent great power war, either. Since the end of the Vietnam War, the United States has spent a total of less than two months engaged in high-intensity conflicts—the invasions of Grenada and Panama, the liberation of Kuwait, and the invasion of Iraq. In contrast, the Army has cumulatively spent decades engaged in low-intensity conflicts in Central America, Africa, the Caribbean, the Balkans, and now in the Middle East and Central Asia.

On one level this reflects the observations of both low-intensity conflict observers and Army transformers throughout the 1990s: the U.S. Army's asymmetric dominance in high-intensity conflict—most dramatically demonstrated during the stunning victory of the Gulf War—has prompted U.S. adversaries to seek asymmetric means of their own to negate this dominance. And, as low-intensity conflict observers predicted, the asymmetry that adversaries chose was to eschew high-tech, high-intensity conflict means all together and instead confront the Army with low-tech, low-intensity conflict challenges. Moreover, as the Defense Intelligence Agency's Melissa Applegate predicted only days before 9/11, in the war on terror, asymmetry has turned out to *be all there is.*[14] The latest generation of adversaries the United States faces—al-Qaeda and the Islamic State in Iraq and Syria—are not even countries with armies; they are small networks of like-minded individuals, connected only by their ideology and a desire to oppose the West.

But on another level, the present strategic environment is what winning looks like. The United States has so established its hegemonic power—economically, diplomatically, culturally, and militarily—that no country can prevail against it in any of these domains. As a result, the only entities in the world that still directly challenge the United States militarily are the nonstate organizations that cannot be deterred by any instrument of power that a nation-state possesses.

Undue Deference

Despite these realities, the outgoing chief of staff of the Army, General Milley, still told the nation that Russia was "literally an existential threat" and plead with the U.S. Congress for the dollars to develop and build the Next

Generation Combat Vehicle to keep up with China and Russia. Moreover, despite the fact that the United States spent $611 dollars on defense in 2016—36 percent of the world's total defense spending, and more than the next eight countries combined—General Milley still complained that he received $7–$9 billion less than he needed for modernization.[15]

Pundits and politicians frequently lament the dwindling number of Americans who are serving or have served in uniform. Presently less than 1 percent of Americans, around 2.2 million, serve in uniform either on active duty or in the reserve components. A little over 7 percent of Americans have ever served in uniform. Others note that the military profession is increasingly becoming a family business, with children following their parents into service. These observers worry of a growing rift between the military and the society that it serves and object that the burden of wartime service has fallen on a too-small minority of Americans. Others worry that the separation between society and those who protect it makes Americans too cavalier about the use of force.[16]

These are legitimate concerns, but the greatest danger of the divide between the military and the society it protects is that civilians—and their elected representatives—give U.S. military leaders undue deference in military affairs, feeling unqualified to question the military's views and advice on the defense of their nation. This carte blanche given to senior Army leaders is both a historical anomaly and an infinitely more dangerous threat to the health and survival of the republic than the risk of unwise employment of military force.

Healthy skepticism of advice from senior military leaders has served the nation well throughout its history. One can look to just a few of the many examples in the Cold War. President Harry S. Truman prevented a nuclear war over Korea in the 1950s by rejecting the advice of—and firing—Gen. Douglas MacArthur.[17] Presidents John F. Kennedy and Lyndon Johnson arguably prevented America's war in Vietnam from turning into a great power war with China or the Soviet Union when they ignored the U.S. military's repeated calls to escalate the conflict by invading North Vietnam or bombing North Vietnamese ports full of Soviet ships. Today's civilian leaders give much greater deference to military leaders than their counterparts in earlier periods of America's history.

This excessive deference to senior Army leaders since the end of the Cold War has been disastrous for the country. After the collapse of the Soviet Union, the U.S. Army flouted repeated calls by the president and Congress to increase its capacity to prosecute low-intensity conflicts. Still, Congress showered money on the Army that the Army then used to hyperprepare for a third world war. Civilian leaders remained silent while Army leaders groused and dragged their feet on engaging in the low-intensity conflicts of the 1990s. The Army refused to take even basic steps to increase its competency to execute President Bill Clinton's strategy of global engagement. And as the Army prepared to fight the war in Iraq, Congress and President George W. Bush failed to exercise even the most basic oversight, while senior Army leaders demonstrated gross negligence in failing to prepare for—or even plan for—Phase IV, the postinvasion, low-intensity conflict in Iraq.

And now Congress and succeeding administrations are doing it again. Despite the fact that the Army is still engaged in the forever wars in Afghanistan and Iraq—and a low-intensity conflict in Syria—the civilian leadership of the United States remains silent, refusing to intervene while the Army resumes preparation for a supposed World War III.

Failure to act will have real strategic consequences. The United States' inability to successfully prosecute low-intensity conflicts is a slow-motion catastrophe that will eventually destroy the nation's influence in the world. Because the Army refuses to build a capacity to engage in low-intensity conflicts, this will be the only way in which adversaries will challenge the Army for the foreseeable future. The United States has absolute supremacy above all other countries in economic power, diplomatic heft, media reach, and high-intensity conflict capacity. But the Army is totally incompetent at fighting low-intensity conflicts. Increasingly, this is the only way that adversaries will challenge the United States.

Second-tier powers will attempt to operate below the threshold for a U.S. military response, trying not to present the United States with affronts that would constitute casus belli for a high-intensity, U.S. military intervention. Iran and Saudi Arabia have already begun to embroil the entire Muslim world in proxy conflicts. While the Army struggles to keep its head above water in the Sunni Arab regions of Iraq and Syria, Iran has nearly completely subverted the government of Iraq in Baghdad.

If, through miscalculation, a second-tier power accidentally triggers a U.S. military response, a U.S. adversary will not present the Army with a Soviet-style defense that the Army is hyperprepared to defeat. Instead, the adversary's government and military will melt away into the populace— probably in urban centers—and confront the Army with an insurgency that the Army is supremely ill suited to meet. This is exactly what the Sunni Arab Iraqis did in Iraq during the invasion.

The great powers have already begun to adapt their military strategies to exploit the Army's inability to fight in low-intensity conflicts. Russia has begun to spread its influence across eastern Europe, the Caucasus region, and Central Asia through low-intensity conflict proxy wars and subversion—what in military circles is being called gray zone or phase zero operations.[18] China has likewise begun subverting governments politically and economically across Central, East, and Southeast Asia and even into Africa. These tactics will expand and proliferate as the United States increasingly demonstrates its inability to combat these activities.

As low-intensity conflicts pile up, one on top of the other, and the Army repeatedly proves itself incapable of engaging in them, countries at risk of subversion will begin to realign themselves, becoming less responsive to the desires of the United States and trying to balance themselves between Chinese and Russian interests. The United States will slowly recede from the world stage, unable to defeat challengers to its will in the international arena or convince allies that it can protect their sovereignty from low-intensity conflict threats. The United States will be forced into a kind of neoisolationism not by its unwillingness to engage in the world but by its inability and ineffectiveness to influence events that are increasingly determined by low-intensity conflicts.

Permanent Change

This does not have to be the United States' future. To avoid it, however, the Army must change. And, if this book has shown anything, it is that the Army is not going to change on its own; it is still engaged in low-intensity conflicts in Afghanistan, Iraq, and now Syria, yet it is already throwing away the lessons of these conflicts and returning to hyperpreparedness for an imagined future great power war. The Army will only change if it is forced to do so by the nation's civilian leadership.

U.S. history has repeatedly demonstrated that permanent, comprehensive change to the U.S. military only comes from outside the U.S. Department of Defense—from the president or Congress. Many changes that the Army now takes for granted were instituted by executive orders from the president. The Army was racially integrated after World War II by President Truman. The draft was ended, and the all-volunteer force created by President Richard M. Nixon in the latter days of the Vietnam War.[19] President Clinton established the so-called Don't Ask, Don't Tell policy that allowed gay people to serve, if not openly, in the U.S. military. President Barack Obama made a number of changes that further diversified the U.S. military, finally opening nearly every job—including virtually all combat jobs—to women and allowing gay and transgender people to serve openly.

But changes to the military by presidential edict are more fragile than those imposed by Congress. In just one example, President Donald Trump has, as of this writing, already begun to roll back President Barack Obama's policy of allowing transgender people to serve openly.

The changes made to the U.S. military by Congress have proven more permanent and more comprehensive. In 1947, immediately following World War II, Congress established the Department of Defense and the Joint Chiefs of Staff and established the Air Force as its own separate service, no longer part of the Army. In 1950 Congress imposed the Uniform Code of Military Justice on the services, subjecting their power to discipline their service members to standardization and due processes. In 1986 the Goldwater-Nichols Defense Reorganization Act imposed "jointness" on the U.S. military, increasing the power of the geographic combatant commands at the expense of the services and forcing officers from each service to become more experienced in joint operations as a condition of promotion. The following year, the Nunn-Cohen Amendment to the 1987 National Defense Authorization Act established U.S. Special Operations Command (SOCOM) as both a global combatant command and as a service-like headquarters with responsibility over all special operations forces (SOF).[20] (The amendment also inadvertently gave the Army an excuse to abdicate its responsibility for low-intensity conflict to SOF.) Nearly all of these changes made to the U.S. military were made over the objections of senior military leaders. But they endured because they were codified in U.S. law.

How should the Army be reshaped to contend with its most likely future? What changes should Congress make?

To reshape the Army for the strategic environment it will actually face in the remainder of this century—rather than the fantasy of a future great power war—is to finally institutionalize the lessons of low-intensity conflicts of the 1990s and the war on terror. First, the Army must revise its doctrine to reflect an acknowledgment that low-intensity conflict is an essential Army mission, that the Army must itself engage the conflict's political dimensions, and that—in order to bring these conflicts to a successful conclusion—the Army must back a winner rather than chase the chimeras of "neutrality" or "legitimacy."

407

Doctrine

For decades, the Army's doctrine has declared, "The Army's primary focus is to fight and win the nation's wars."[21] The key admission that the Army must make—from its capstone doctrine to the foundational declarations of its chief of staff in every one of his speeches—is to acknowledge that fighting and winning a war includes both winning the initial, high-intensity conflict battles and securing a political settlement favorable to the United States' strategic goals through an equally well-resourced low-intensity conflict phase. It is time for the Army to admit that no military strategy for a war, no matter how well formulated or resourced, will allow the Army to avoid a low-intensity conflict phase.

The next fundamental acknowledgment that the Army must make is that, just like high-intensity conflict, low-intensity conflict *is war*—not "operations other than war." As Carl von Clausewitz wrote, "War is . . . an act of force to compel our enemy to do our will."[22] Just like high-intensity conflict, low-intensity conflict involves the use of force to achieve a political objective; only the magnitude of that force is different. The Army must overcome this cognitive obstacle in order to truly embrace low-intensity conflict as a coequal task to high-intensity conflict within the array of Army missions.

Within low-intensity conflict doctrine, there are some other important acknowledgments that the Army must make to bring these conflicts to a successful conclusion. First, it must admit that the political dimension of low-intensity conflict is *the Army's* responsibility. No one else is going to swoop in and prosecute the political dimension of low-intensity conflict on the Army's behalf. SOF is too small to effectively engage this dimension by itself. Civilian agencies such as the State Department cannot wield the violence required to effectively engage the political dimension of a low-intensity conflict. If the host nation could have effectively engaged the political dimension, the Army wouldn't be involved in the conflict in the first place. And the UN or U.S. allies will not pursue U.S. interests when seeking a political solution. The U.S. Army must engage this dimension itself in order to forge a political settlement because it is the only entity on the battlefield that is able to wield violence or the threat of violence to compel parties to comply with U.S. objectives in the conflict. As Chinese Communist Party leader Mao Zedong wrote, "Political power grows out of the barrel of a gun,"[23] at least in a low-intensity conflict environment.

The next critical change that the Army needs to make in its doctrine is to purge the idea of *neutrality*—often used synonymously or in conjunction with the idea of *legitimacy*—from its lexicon. The Army must back a winner in order to bring a low-intensity conflict to a successful conclusion. If the United States' civilian national leadership did not want one party to prevail over others in a conflict—if it did not care who won—it would not have sent the Army into harm's way in the first place. And legitimacy in a low-intensity conflict comes from the Army's ability to wield force or the threat of force in order to compel compliance, not how nice it is to each party in the conflict. It is necessary and appropriate for senior military leaders to demand that their civilian masters articulate which side they want to win. But failure to back a winner is the recipe for a forever war.

There are some smaller, but no less essential, acknowledgments that must be made in Army doctrine. First, the idea of measures of effectiveness and measures of performance must be stricken from Army doctrine. While harmless in and of themselves, in implementation these measures become a more comfortable alternative to the messy business of engaging the political dimension of low-intensity conflicts. Army units busy themselves with chasing better statistics on weapons collected, streets repaired, wells dug, and schools built but never address the inherently qualitative political problem that must be solved in order to actually bring the low-intensity conflict to a successful conclusion.

Next, the civilian populace—especially in urban areas—are not just obstacles to fire and movement or concealment for an adversary or hungry mouths to be fed; the populace in a low-intensity conflict is the political objective of the conflict. Their acceptance of the political settlement that the Army seeks to impose in the conflict is essential to ending that conflict. All of the Army's efforts in a low-intensity conflict must be gauged against whether they bring that goal closer or push it farther away; this is the only true "measure of effectiveness."

And, finally, the lessons of the Army's 1990s low-intensity conflict experiences (as well as its experiences in the war on terror) point to two other issues that must also be addressed in Army doctrine. First, in a low-intensity conflict, concern for force protection must be subordinate to the ultimate goal of achieving a political settlement and ending the conflict—which requires operating in and among the people and engaging the conflict's political dimension. In high-intensity conflicts, wars are not won on defense but by risking offensive action to seize the initiative and defeat the enemy. Likewise, in a low-intensity conflict, one must take prudent risks—exposing forces to possible attack by leaving fortified bases—in order to effectively operate in and among the populace and engage the political dimension of the conflict. This principle is already reflected in the Army's FM 3-24, *Counterinsurgency*. The manual's "paradoxes of counterinsurgency" include the insight, "Sometimes, the More You Protect Your Force, the Less Secure You May Be." In another paradox—"The More Successful the Counterinsurgency is, the Less Force Can Be Used and the More Risk Must Be Accepted"—the manual explains, "Soldiers and Marines may . . . have to accept more risk to maintain involvement with the people."[24] This idea should be reflected in the rest of Army doctrine.

But the U.S. Army has changed its doctrine before. Throughout the 1990s, the Army used doctrine as a way to pay lip service to low-intensity conflict without actually institutionalizing this mode of warfare in any other way. Changing doctrine is not nearly enough to change Army culture. The Army must also be forced to acknowledge that low-intensity conflict tasks are not simply lesser included tasks of high-intensity conflict capacity—the panacea of "versatility" and "full spectrum" dominance that Army transformers used as a euphemism for ignoring low-intensity conflict throughout the 1990s.[25]

To truly address this crisis of relevancy, the Army must be forced by congressional mandate to reshape itself. As numerous low-intensity conflict observers suggested throughout the 1990s, the Army should be bifurcated into high- and low-intensity conflict forces. Some percentage of the overall force should retain the Army's unmatched capacity to dominate high-intensity conflicts while the larger remainder of the Army is reorganized into a force optimized for low-intensity conflicts.

Organization and Tasks

A strategy for manning and equipping the U.S. Army for the future should acknowledge that both former secretary of defense Donald Rumsfeld *and* former chief of staff of the Army, Gen. Eric Shinseki, were correct on the eve of the invasion of Iraq. During the planning and preparation for the invasion of Iraq, Rumsfeld insisted that the Army could do more with less—that the revolution in military affairs (RMA) had produced an Army that was so fast and lethal that it could defeat the second-tier Iraqi Army with a very small force. Yet General Eric Shinseki famously told Congress that the stabilization of Iraq after the invasion would require hundreds of thousands of troops.[26] The course of the war that followed proved both men right. A future U.S. Army should be shaped by this insight.

The High-Intensity Conflict Army

The Army will continue to need a high-intensity conflict capability for the foreseeable future. Nuclear arms deter great powers from actions—such as an attack on U.S. territory, a NATO ally, or Taiwan—that would bring them into direct military conflict with the United States. But nuclear arms do not deter second-tier powers from actions counter to U.S. interests—such as North Vietnam's invasion of South Vietnam or Saddam Hussein's invasion of Kuwait. There will continue to be a need for a conventional capability to deter and, if deterrence fails, defeat second-tier powers. A smaller high-intensity conflict army—retaining the organization and missions of the present Army—should remain prepared for high-intensity conflict.

The design of this force should benefit from the insights of Army trans-formers who, throughout the 1990s, designed, tested, and refined the shape of the Army to deal with such threats. The Army will need light infantry forces that can deploy on a moment's notice to austere locations to execute "forcible entry" as part of a joint force comprising the Army, the Navy, the Air Force, and the Marine Corps. The mission of this initial entry force will be to seize key infrastructure such as seaports, airfields, government buildings, and broadcast media facilities and secure key individuals such as American citizens and government officials. The high-intensity conflict army will also need a highly deployable, lightly armored, medium-weight force such as the Stryker Brigades to follow these light forces, expand the lodgments created at seaports and airfields, and continue the momentum of the attack into the country. And the Army will continue to need heavy armored forces—the arm of decision—to outmatch any enemy and conclude the high-intensity conflict phase on U.S. terms. Many things can be done to speed the deployment of these forces, such as forward-basing some forces and establishing pre-positioned stocks in strategic locations around the world. But heavy forces will unavoidably take longer to deploy than light or medium forces.

Yet this high-intensity conflict army can also benefit from the insights of Secretary of Defense Donald Rumsfeld and his ilk: in a high-intensity conflict, an RMA-enabled Army can do more with less. As national security analyst Chris Schnaubelt wrote, "The initial phases of OIF [Operation Iraqi Freedom] appeared to validate the ability of U.S. (and coalition) forces to rapidly defeat a much larger military."[27] Daniel Benjamin and Steve Simon observed, "Rumsfeld proved his point. The force that deployed to Iraq was a fraction of what 'old think' commanders would have used, and it was suffi-cient to topple the Iraqi regime in no time."[28] Networked and highly mobile, benefiting from all of the trappings of the RMA, and fighting as part of the joint force, the high-intensity conflict army will not have to be that big; the key attributes of this force are strategic mobility, speed, and lethality rather than mass.

Of course, the authors of the Bottom-Up Review (BUR) were also correct in noting that the United States must maintain the capability to simultaneously fight two high-intensity conflicts. This is true for two reasons. First, as the BUR authors observed, the capacity to fight two high-intensity conflicts prevents a second adversary from seizing an opportunity to act while the Army is engaged with another foe. This capacity is also required, however, to prevent what Army colonel Walter Anderson called "self-deterrence." He explained, "If you have one arrow, when do you fire it?"[29] If the Army is only able to engage in one high-intensity conflict at a time, national leaders might be stricken by strategic paralysis, unable to use the high-intensity conflict army for fear of what other crisis might emerge while it is busy.

Much more study and experimentation is needed to flesh out the specific shape of the high-intensity conflict army. But the active component of the high-intensity conflict force could consist of as little as two corps headquarters and four Army divisions—each with one light infantry brigade combat team (BCT), one Stryker BCT, and one heavy, armored BCT. As long as the United States can continue to rely on the support of its allies and security partners, this should be sufficient to meet the demands of any two limited high-intensity conflicts that might emerge in the foreseeable future.

With the huge stockpile of equipment that will go unmanned when this force is right-sized, the Army could establish additional prepositioned stocks of brigade or division size near trouble spots around the globe. And with the savings in personnel, the Army can build its decisive, low-intensity conflict force.

The Low-Intensity Conflict Army

The majority of the Army—those forces not dedicated to the smaller high-intensity conflict army—should be transformed into a low-intensity conflict army, with the mission of operating in and among foreign populaces and applying nuanced force and threat of force to compel the parties to a low-intensity conflict to accept a political settlement. As of this writing, eighteen years of fighting in Afghanistan and Iraq have taught that these are not lesser included tasks of high-intensity conflict proficiency. If the Army does not arrive in the next theater of war with forces trained and ready to execute these tasks, it will condemn America to yet another forever war in the future.

In the run-up to the Iraq War, Thomas P. M. Barnett of the U.S. Naval War College wrote an article for *Esquire* magazine titled "The Pentagon's New Map," laying out a new geostrategy for the post–Cold War and post-9/11 world. In 2004 Barnett published a book by the same title that fleshed out his ideas. One of those ideas was the idea of bifurcating the Army into a "Leviathan force" (a high-intensity conflict army) and a "System Administrator," or "Sys Admin," force (a low-intensity conflict army).[30]

At the time, the chief criticism of this idea was that no one would ever want to serve in the Sys Admin force. This was a fair criticism; besides its regrettably wonky name, the Sys Admin force, as described in the book, was barely even a military force. Barnett wrote of the force, "Where the Leviathan projects power menacingly, the Sys Admin will export security nonthreateningly." The Sys Admin force would "necessarily stay on the defensive, guarding sites versus killing bad guys." Barnett added, "Unlike those in the Leviathan force, personnel in the Sys Admin force will alternate service in the ranks with periods of work outside in normal society."[31] Rather than convincing the Army to change, such descriptions gave Army transformers ammunition with which to attack and defeat the idea of specializing some segments of the Army for low-intensity conflict.

On a deeper level, conceptions such as Barnett's misunderstand the essential role of violence in a low-intensity conflict. Low-intensity conflicts differ from political processes within a peaceful country in that they *are war*. The essential requirement for an Army unit to compel parties to accept a political settlement in a low-intensity conflict is that unit's ability to wield force or the threat of force. Barnett's conception of a low-intensity conflict force misses this core fact.

The low-intensity conflict army *does* need to be able to wield force. With the lessons of the war on terror as a guide, one can conceive of a low-intensity conflict force that balances the ability to wield force with the ability to operate in and among the populace and engage the political dimension of a conflict. The low-intensity conflict army needs to be mobile, lightly armored, and ready to fight, much like the Army—equipped with the mine-resistant, ambush protected vehicle (MRAP)—from the latter years of the Iraq War. But its training must be balanced to wield force in a nuanced and restrained way to avoid collateral damage and civilian casualties that would make a political solution harder to achieve.

This low-intensity conflict force also needs to arrive in a theater of operations ready to perform the full array of tasks that this environment demands. The 1990s provides a clear blueprint of the capabilities that the low-intensity conflict army needs. The force must be able to interoperate as part of a multinational coalition force. The force must be able to integrate SOF operations into its own. The low-intensity conflict force must have the capacity to execute security force assistance—standing up, training, and operating jointly with host nation police and military forces.

The low-intensity conflict army must also be optimized to operate in and among the populace. This force needs the ability to influence foreign populations using psychological operations (PSYOP). It must also be able—through interaction with the local and international media—to effectively inform the American public, communicate to international audiences, and influence the population inside a theater of low-intensity conflict operations. This force must be able to execute civil-military operations, coordinating with and providing resources and support to nongovernmental organizations, the host nation (if present), other agencies of the U.S. government present in theater, and intergovernmental organizations like the UN. And the low-intensity conflict army should be equipped and trained to wield nonlethal force and apply lethal force in a nuanced manner within the rules of engagement.

A few organizational changes to the broader U.S. military are required to create this low-intensity conflict army. First, all of the Army's active PSYOP and civil affairs forces should be moved from SOCOM and placed back under the full control of the U.S. Department of the Army. Keeping these capabilities in SOCOM perpetuates the myth that low-intensity conflict is the responsibility of SOF rather than the Army. But, more important, these capabilities need to be integrated into the low-intensity conflict army so that they will be better integrated for training and during operations. Of course, this change reverses elements of the Nunn-Cohen Amendment and would particularly require congressional action.

Organizational changes to the Total Army will also be required. Those construction engineers, military police, PSYOP, civil affairs, public affairs, and logistics units that reside in the reserve component of the Army need to be moved to the active component. Throughout the 1990s, low-intensity conflict observers repeatedly noted that these capabilities—desperately needed at the very beginning of such a conflict—were slow to mobilize and deploy. Moving these capabilities back into the active component will relieve this problem. In exchange, more high-intensity conflict capabilities, such as heavy brigades and combat aviation brigades, can be moved into the reserve component.

As with the high-intensity conflict force, much more study and experimentation is needed to sketch out the specific shape of this force. But General Shinseki was right in his congressional testimony on the eve of the Iraq War: it does take significantly more troops to prosecute a manpower-intensive low-intensity conflict than it does to execute a relatively short, RMA-enabled high-intensity conflict. In the active component, the low-intensity conflict army needs at least two corps-level headquarters and six division headquarters, each with a light infantry BCT and three hybrid BCTs—containing a mix of light infantry, civil affairs, PSYOP, public affairs, military police, engineers, and logistics forces. The entire low-intensity conflict army should be MRAP mounted for mobility and armor protection.

Incentives

Institutionalizing low-intensity conflict in the U.S. Army also requires incentivizing individual proficiency in prosecuting such conflicts. While the Army should bifurcate itself into forces specific to high- and low-intensity conflict, it should not specialize its personnel. Throughout their careers, officers, noncommissioned officers, and soldiers should move back and forth between assignments in the high- and low-intensity conflict armies. (This, of course, excepts those serving in the most specialized occupational fields such as civil affairs and PSYOP, who would serve only in the low-intensity conflict army.)

Primarily, this cross-pollination is intended to share expertise between the two forces. High-intensity conflict army leaders need to understand the specific challenges that the low-intensity conflict army faces in executing its mission. Decisions that leaders in the high-intensity conflict force make during the prosecution of their phase of the war can have a profound, possibly insurmountable negative effect on the low-intensity conflict phase of the war. Army leaders in the high-intensity conflict army need to be cognizant of this as they plan and execute their operations. Likewise, leaders in the low-intensity conflict army need to understand high-intensity conflicts and how they will unfold as they are planning their postinvasion low-intensity operations. Additionally, there is frequently combat within a low-intensity conflict. A basic understanding of battalion- and brigade-level tactics for the synchronization of lethal and nonlethal effects into combined arms maneuver will occasionally be valuable to leaders in the low-intensity conflict army.

But there is another, more cultural reason for forcing Army personnel to move back and forth between these forces. As this study has clearly shown, the Army has a deep cultural aversion to low-intensity conflict. If Army leaders are not forced to move back and forth between the high- and low-intensity conflict armies, the members of the latter army will inevitably become second-class citizens to members of the former. Assignments to the low-intensity conflict army will be considered less desirable and reserved for less competitive, less capable Army leaders. The only way to ensure that talent is truly spread evenly between these two forces is for Congress to mandate that leaders get significant experience in both forces as a precondition for promotion, just as the Goldwater-Nichols Act mandated that Army officers have joint experience—serving alongside members of sister services—as a precondition for promotion to the rank of general officer.

This is the "stick" of incentivizing low-intensity conflict proficiency, but there is also a "carrot," enticing Army offers, noncommissioned officers, and soldiers to *want* to serve in the low-intensity conflict force. The Army only fights a high-intensity conflict about once every decade or two, but it is almost perpetually engaged in low-intensity conflicts. The low-intensity conflict army will be engaged in operations around the world much more frequently than will the high-intensity conflict army. If personnel are forced to move back and forth between the high- and low-intensity conflict forces, they will inevitably come to see their service in the high-intensity conflict army as "preparing for the unlikely" and their service in low-intensity conflict army as "serving in real-world operations." As Army leaders found in the 1990s low-intensity conflict deployments, the sense of "making a difference" in a real-world mission can have a positive effect on soldier morale and retention (reenlistment).

Congress can further foster this perception by expanding the definition of "wartime service," "imminent danger," and "combat zones" to include a broader range of low-intensity conflicts. The Army already incentivizes combat service, favoring Army leaders who have more combat service with faster promotion. Congress has also incentivized combat service by providing for "imminent danger pay" and waiving income taxes on salaries for service members serving in combat zones. Soldiers join the Army to train for and fight in wars. Calling low-intensity conflicts what they are—*wars*—will go a long way toward changing the way soldiers and leaders see participation in such conflicts.

Education

Congress must also force the Army to institutionalize low-intensity conflict in its education system. In its effort to refocus the Army on high-intensity conflict, senior leaders have once again purged low-intensity conflict from the Army's premier educational institutions—the CGSC and USAWC. At USAWC in the 2016–17 academic year, only a single day of an eight-month core curriculum was dedicated to low-intensity conflict—and this day was combined with a discussion of SOF, perpetuating the myth among Army colonels that low-intensity conflict is a SOF function. Moreover, multiple electives focused on low-intensity conflict were canceled due to insufficient enrollment.[32]

Education in subjects related to low-intensity conflict that were in vogue during the height of the Iraq War—such as anthropology, political science, public policy, business administration, and economics—have disappeared from the Army educational system. A radical solution is required.

The Army already has the infrastructure for delivering the core curricula for the CGSC and USAWC via "satellite course" (i.e., a shortened resident course) or "correspondence course" (i.e., distance learning). Presently these methods are used to teach a small minority of field grade officers falling in to one of two classes: those not successful enough in their Army careers to be selected for attendance at the full resident course or those highly successful officers selected for attendance at civilian postgraduate education, sister service colleges, or foreign military colleges in lieu of attending Army educational institutions.

The Army should instead educate *all* field grade officers through the satellite or distance learning courses. Those officers who would have been selected for the resident course, a civilian advanced degree program, a sister service college, or foreign military college—about the top half of Army majors and the top 10–15 percent of Army lieutenant colonels and colonels—should subsequently be sent for a master's degree at a civilian institution in a discipline related to low-intensity conflict or military governance. At present, only a tiny fraction of this number are afforded the opportunity for civilian education—nearly always in fields unrelated to low-intensity conflict.

In addition to the better education that officers will receive in fields related to low-intensity conflict, this will also help address the crisis of closed-mindedness that Gerras and Wong observed in their study *Changing Minds in the Army*. By exposing Army officers to civilian professors and students with different perspectives on topics related to low-intensity conflict, Army field grade officers will be forced to see the world in a different way—a behavior that the Army currently discourages through its promotion selection processes.[33]

Initially, this will require significant funding from Congress. Eventually— as the culture of the Army changes to embrace low-intensity conflict as a coequal Army mission—this education can return to Army institutions. Through education and focused effort, suitable curricula can eventually be developed for these low-intensity conflict subjects to be taught within the Army's education system. But, until then, the Army must rely on civilian universities to fill the void.

Training

To truly institutionalize low-intensity conflict proficiency in the Army, Congress must also compel the Army to reshape its training. Established as part of the so-called training revolution" in the 1980s, the Army's combat training centers (CTCs)—the National Training Center (NTC) at Fort Irwin, California; the JRTC at Fort Polk, Louisiana; the JMRC at Hohenfels, Germany; and the Mission Command Training Program (MCTP) at Fort Leavenworth, Kansas—are the centerpieces of Army training. They are the ultimate, objective, external test of the Army's readiness for war. As such, the missions they include in (and excluded from) their exercises and the standards by which they measure units—chiefly, Army doctrine—drive the training for every brigade, division, and corps in the Army.

Both before the war on terror and after the recent refocusing of the Army on high-intensity conflict, the CTCs almost exclusively trained Army units to fight high-intensity conflicts. But the CTCs also have experience in training units for low-intensity conflicts. During both the 1990s and the war on terror, those units earmarked for deployment to low-intensity conflicts were trained to fight such conflicts in so-called mission readiness exercises. During the height of the war on terror, when virtually *every* Army unit was earmarked for deployment to either Afghanistan or Iraq, the CTCs trained exclusively in low-intensity conflict.

Scenarios at the CTCs should be redesigned to train on *both* high- and low-intensity conflict in *every* rotation. At the "dirt" CTCs—NTC, JRTC, and JMRC—where BCTs are trained, rotations should be extended from two to four weeks. In the first two weeks, a BCT from the high-intensity conflict army would fight a simulated high-intensity conflict against a high-intensity conflict opposing force (OPFOR). At the end of these two weeks, a hybrid brigade from the low-intensity conflict army would arrive, relieve the high-intensity conflict force, and train in a two-week, simulated low-intensity conflict, complete with role-players replicating civilians, host nation government and military leaders, U.S. government officials, and nongovernmental organizations. The members of the OPFOR would take off their uniforms and act as guerillas, operating from within the populace to oppose the hybrid brigade's efforts. This expanded CTC rotation would train both the high- and low-intensity conflict armies. But it would also train

Army leaders in the difficult and all-important transition between high- and low-intensity conflict during a war.

A similar model should be applied for MCTP warfighter exercises. A corps and two division headquarters from the high-intensity conflict army would wage simulated high-intensity conflict operations against a high-intensity conflict-focused OPFOR. After ten days, a corps and two divisions from the low-intensity conflict army would relieve the high-intensity conflict force and conduct ten days of low-intensity conflict to achieve (or at least make progress toward) a political settlement to end the war.

Such a model would do more than just train these forces; it would imprint in the minds of a whole generation of Army leaders a new paradigm for war beyond the Army's present great power war conception that underpins the current Army culture. Army leaders would learn that war is not just a series of high-intensity conflict battles; it is also low-intensity conflict stabilization that solves the political problem that started the war in the first place.

Conclusion

History has shown that the Army won't change itself; only legislation will create the permanent changes required to reshape the Army's culture.

And the clock is ticking on these changes. What awaits the U.S. Army is not the imminent great power war for which senior leaders insist on preparing to the exclusion of all other priorities. What actually awaits the Army is an interminable series of low-intensity conflicts. If the Army does not radically reshape itself to address this future, it will prove itself increasingly irrelevant to the contemporary strategic environment. The cost will be a loss of the Army's share in the defense budget to the other services. But, more important, it will cost the Army in the loss of soldiers wounded or killed.

Make no mistake; the refusal by senior Army leaders to build a capacity to fight low-intensity conflicts *is* killing soldiers. The Army continues to lose soldiers in Afghanistan, Iraq, Syria, and even Kosovo because it cannot figure out how to bring these conflicts to a successful conclusion.[34] And, if nothing changes, the Army will lose more soldiers unnecessarily when the next low-intensity conflict becomes a forever war.

But for the nation, the cost will be even greater. This slow-burn crisis of relevancy will relegate the United States to the role of spectator on an international stage increasingly dominated by growing peer competitors that prosecute their foreign policies through low-tech, low-intensity conflict means that the U.S. Army stubbornly remains powerless to combat.

419

NOTES

1. Clausewitz, *On War*, 88, quoted in Summers, *On Strategy*, Kindle, location 1508.

2. Peters, "Heavy Peace," 71–79.

3. Schadlow, *War and the Art of Governance*, 1–13; Hudson, *Army Diplomacy*, 1–27.

4. Brown, *Kevlar Legions*, Kindle, 49, 65; Millett and Maslowski, *For the Common Defense*, Kindle, locations 4600–4602.

5. Crane, *Avoiding Vietnam*, 1–19.

6. U.S. Army Training and Doctrine Command, "The Brigade Combat Team, Organizational and Operational Concept," January 6, 2000, Box 2, Official Correspondence Email Traffic Received from 20 to 5 January 2000, Folder 7, Official Correspondence—Email Traffic Received in January 2000 (Part 19 of 20), James M. Dubik Papers, U.S. Army Heritage and Education Center.

7. Fitzgerald, *Learning to Forget*, Kindle, 15.

8. Gerras and Wong, *Changing Minds in the Army*, 8–9.

9. Gerras and Wong, *Changing Minds in the Army*, 8–9.

10. 10th Mountain Division, "10th Mountain Briefs," in *Operation Uphold Democracy: US Forces in Haiti*, CD-ROM, Norfolk, VA: U.S. Atlantic Command, September 1997.

11. Fukuyama, *The End of History and the Last Man*; Friedman, *The Lexus and the Olive Tree*.

12. Mark A. Milley, "General Mark A. Milley, AUSA Eisenhower Luncheon," October 4, 2016, transcript, http://wpswps.org/wp-content/uploads/2016/11/20161004_CSA_AUSA _Eisenhower_Transcripts.pdf; Kennedy, *The American People in World War II*, 285; U.S. Census Bureau, "1940 Census," database, n.d., accessed December 10, 2017, https://1940cen sus.archives.gov/index.asp; Millett and Maslowski, *For the Common Defense*, Kindle, locations 7618–19; U.S. Census Bureau, "American FactFinder," table for 2016, n.d., accessed December 10, 2017, https://factfinder.census.gov/faces/tableservices/jsf/pages/productview .xhtml?pid=PEP_2016_PEPANNRES&src=pt.

13. Priscilla Roberts, "Introduction," in Roberts, ed., *Voices of World War II*, xii. See also Gaddis, "The Long Peace"; Gaddis, "The Cold War, the Long Peace, and the Future"; and Holmes, "Preserving History and The Long Peace."

14. Applegate, *Preparing for Asymmetry*, 1–25, emphasis in the original.

15. Sarah Sicard, "Top Army General: These 4 Nations Are the Most Dangerous to US Security," Business Insider, June 20, 2016, http://www.businessinsider.com/top-army-general-these-4 -nations-are-the-most-dangerous-to-us-security-2016-6; C. Todd Lopez, "Army 'Confident in Current Capabilities' Chief of Staff Says," Army News Service, April 16, 2018, https://www.army .mil/article/203942/army_confident_in_current_capabilities_chief_of_staff_says; Emmanuel Ocbazghi, "The US Spent $611 Billion on Its Military in 2016—More Than the Next 8 Countries Combined," Business Insider, May 1, 2017, http://www.businessinsider.com/us-spent-611 -billion-on-military-2016-army-defense-missile-trump-money-arms-politics-2017-4; Natalie Johnson, "The US Army Is Reportedly $7 Billion to $9 Billion Short of the Money It Needs to Start Modernizing the Force," Business Insider, August 11, 2017, http://www.businessinsider .com/army-7-billion-to-9-billion-short-of-money-needed-to-modernize-2017-8.

16. Ricks, "The Widening Gap"; Mona Chalabi, "What Percentage of Americans Have Served in the Military?" FiveThirtyEight, March 19, 2015, https://fivethirtyeight.com/features /what-percentage-of-americans-have-served-in-the-military/; Thompson, "Here's Why the

U.S. Military Is a Family Business"; Ricks, "The Widening Gap"; Fallows, "The Tragedy of the American Military."

17. Millett and Maslowski, *For the Common Defense*, Kindle, locations 9203–11.

18. Chambers, *Countering Gray-Zone Hybrid Threats*; Kyle Johnston, "U.S. Special Operations Forces and the Interagency in Phase Zero," 76–104.

19. Brown, *Kevlar Legions*, Kindle, 19, 24.

20. Millett and Maslowski, *For the Common Defense*, Kindle, locations 9036–42, 11684–87.

21. Headquarters, U.S. Department of the Army, *Operations*, FM 100-5 (1993), 13-0–13-8.

22. Clausewitz, *On War*, Kindle, 75.

23. Mao Tse-Tung, "Problems of War and Strategy," November 6, 1938, in *Selected Military Writings of Mao Tse-Tung*.

24. Headquarters, U.S. Department of the Army, *Counterinsurgency*, FM 3-24, 7-1–7-2.

25. Eric K. Shinseki, comments in "1999 Fletcher Conference, Panel 4: Serving the Nation in the 21st Century, with General Eric K. Shinseki, General James L. Jones, Admiral Donald L. Pilling, General Lester L. Lyles," November 3, 1999, transcript, U.S. Army War College, Carlisle Barracks, PA.

26. Millett and Maslowski, *For the Common Defense*, Kindle, locations 12981–83.

27. Schnaubelt, "Wither the RMA," 95–107.

28. Benjamin and Simon, *The Next Attack*, 187.

29. Aspin, *Report of the Bottom-Up Review*, 19; Crane, *Alternative National Military Strategies*, 1–15.

30. Barnett, *The Pentagon's New Map*, Kindle, xix.

31. Barnett, *The Pentagon's New Map*, Kindle, 320–22.

32. The author was a resident student at USAWC in the 2016–17 academic year. These insights are the product of direct observation of the course of study and electives at USAWC from the author's perspective as a student.

33. Gerras and Wong, *Changing Minds in the Army*, 8–9.

34. Kyle Rempfer, "Pentagon Releases Identity of Soldier Who Died in Kosovo," *Army Times*, May 25, 2018, https://www.armytimes.com/news/2018/05/25/pentagon-releases-identity -of-soldier-who-died-in-kosovo/.

BIBLIOGRAPHY

ARCHIVES

American Presidency Project. University of California–Santa Barbara, Santa Barbara, CA. http://www.presidency.ucsb.edu/.

Carl E. Vuono Papers. U.S. Army Heritage and Education Center, Carlisle Barracks, PA.

Colonel Robert R. McCormick Research Center. First Division Museum at Cantigny Park, Wheaton, IL.

Combined Arms Center Historical Archive. U.S. Army Combined Arms Center, Fort Leavenworth, KS.

Command and General Staff College Curricular Collection. U.S. Army Heritage and Education Center, Carlisle Barracks, PA.

Command and General Staff College Papers. Ike Skelton Combined Arms Research Library, Fort Leavenworth, KS.

Dennis J. Reimer Papers. U.S. Army Heritage and Education Center, Carlisle Barracks, PA.

Eric K. Shinseki Collection. U.S. Army Heritage and Education Center, Carlisle Barracks, PA.

Gordon R. Sullivan Papers. U.S. Army Heritage and Education Center, Carlisle Barracks, PA.

James M. Dubik Papers. U.S. Army Heritage and Education Center, Carlisle Barracks, PA.

Records of the Army–Air Force Center for Low Intensity Conflict. U.S. Air Force Historical Research Agency, Maxwell Air Force Base, AL.

BOOKS AND REPORTS

Alam, Manzoor. *War on Terrorism or American Strategy for Global Dominance: Islamic Perspective on the Afghan-Iraq War.* New York: Vantage, 2009.

Allison, Graham. *Destined for War: Can America and China Escape Thucydides's Trap?* New York: Houghton Mifflin Harcourt, 2017. Kindle.

Bacevich, Andrew J. *America's War for the Greater Middle East: A Military History.* New York: Random House, 2016. Kindle.

Baker, James, and Lee H. Hamilton. *Iraq Study Group Report.* Washington, DC: U.S. Institute of Peace, 2006.

Banks, Arthur S., Thomas C. Miller, William R. Overstreet, and Judith F. Isacoff. *Political Handbook of the World 2009.* New York: CQ, 2009.

Baram, Amatzia. *Who Are the Insurgents? Sunni Arab Rebels in Iraq.* Special Report 134. Washington, DC: United States Institute of Peace, April 2005. https://www.usip.org /sites/default/files/sr134.pdf.

Barnett, Thomas P. M. *The Pentagon's New Map: War and Peace in the Twenty-First Century.* New York: Putnam, 2004.

Benjamin, Daniel, and Steve Simon. *The Next Attack: The Failure of the War on Terror and a Strategy for Getting It Right*. New York: Holt, 2006.

Boot, Max. *The Savage Wars of Peace: Small Wars and the Rise of American Power*. New York: Basic Books, 2002.

Bowden, Mark. *Black Hawk Down: A Story of Modern War*. New York: Grove, 1999.

——. *Guests of the Ayatollah: The Iran Hostage Crisis; The First Battle in America's War with Militant Islam*. New York: Grove/Atlantic, 2006.

Bremer, L. Paul, III. *My Year in Iraq: The Struggle to Build a Future of Hope*. New York: Simon and Schuster, 2006. Kindle.

Brigham, Robert Kendall. *Guerrilla Diplomacy: The NLF's Foreign Relations and the Viet Nam War*. Ithaca, NY: Cornell University Press, 1999.

Broadwell, Paula, and Vernon Loeb. *All In: The Education of General David Petraeus*. New York: Penguin, 2012.

Busch, Andrew. *Reagan's Victory: The Presidential Election of 1980 and the Rise of the Right*. Lawrence: University Press of Kansas, 2005.

Clancy, Tom, with Fred Franks Jr. *Into the Storm: A Study in Command*. New York: Berkley Books, 2004. Kindle.

Clancy, Tom, with Chuck Horner and Tony Koltz. *Every Man a Tiger: The Gulf War Air Campaign*. New York: Berkley Books, 2005. Kindle.

Clausewitz, Carl von. *On War*. Edited and translated by Michael Howard and Peter Paret. Princeton, NJ: Princeton University Press, 1976. Kindle.

Cordesman, Anthony H. *The Obama Strategy in Afghanistan: Finding a Way to Win*. Washington, DC: Center for Strategic and International Studies, July 7, 2016.

Craig, Campbell, and Fredrik Logevall, *America's Cold War: The Politics of Insecurity*. Cambridge, MA: Harvard University, 2009. Kindle.

Crane, Conrad C. *Cassandra in Oz: Counterinsurgency and Future War*. Annapolis, MD: Naval Institute Press, 2016. Kindle.

Cunningham, David E. *Barriers to Peace in Civil War*. New York: Cambridge University Press, 2011.

Downie, Richard Duncan. *Learning from Conflict: The U.S. Military in Vietnam, El Salvador, and the Drug War*. Santa Monica, CA: Praeger, 1998.

Dubik, James M. *The Army's "Twofer": The Dual Role of the Interim Force*. Washington, DC: Institute for Land Warfare, Association of the United States Army, October 2001.

Dunstan, Simon. *The Yom Kippur War 1973 (2): The Sinai*. Oxford: Osprey, 2003.

Dupuy, Alex. *The Prophet and Power: Jean-Bertrand Aristide, the International Community, and Haiti*. Lanham, MD: Rowman and Littlefield, 2007.

Echevarria, Antulio J., II. *Reconsidering the American Way of War: US Military Practice from the Revolution to Afghanistan*. Washington, DC: Georgetown University Press, 2014.

Elhadj, Elie. *The Islamic Shield: Arab Resistance to Democratic and Religious Reforms*. Boca Raton, FL: BrownWalker, 2007.

Engel, Jeffrey A. *When the World Seemed New: George H. W. Bush and the End of the Cold War*. New York: Houghton Mifflin Harcourt, 2017.

European Security Study. *Strengthening Conventional Deterrence in Europe: Proposals for the 1980s*. New York: Macmillan, 1983.

Farrell, Theo, Sten Rynning, and Terry Terriff. *Transforming Military Power since the Cold War: Britain, France, and the United States, 1991–2012*. New York: Cambridge University Press, 2013.

Ferguson, Charles H. *No End in Sight: Iraq's Descent into Chaos.* New York: PublicAffairs, 2008.

Filkins, Dexter. *The Forever War.* New York: Random House, 2008.

Fitzgerald, David. *Learning to Forget: US Army Counterinsurgency Doctrine and Practice from Vietnam to Iraq.* Stanford, CA: Stanford University Press, 2013. Kindle.

Ford, Ronnie E. *Tet 1968: Understanding Surprise.* New York: Frank Cass, 1995.

Friedman, Thomas L. *The Lexus and the Olive Tree: Understanding Globalization.* New York: Farrar, Straus and Giroux, 1999.

Fukuyama, Francis. *The End of History and the Last Man.* New York: Avon, 1992.

Gentile, Gian. *Wrong Turn: America's Deadly Embrace of Counterinsurgency.* New York: New Press, 2013. Kindle.

Gole, Henry G. *General William E. DePuy: Preparing the Army for Modern War.* Lexington: University Press of Kentucky, 2008. Kindle.

Gordon, Michael R., and Bernard E. Trainor. *Cobra II: The Inside Story of the Invasion and Occupation of Iraq.* New York: Pantheon, 2006. Kindle.

———. *The Endgame: The Inside Story of the Struggle for Iraq, from George W. Bush to Barack Obama.* New York: Pantheon, 2012.

Greenberg, Robert D. *Language and Identity in the Balkans: Serbo-Croatian and Its Disintegration.* New York: Oxford University Press, 2004.

Hagopian, Patrick. *The Vietnam War in American Memory: Veterans, Memorials, and the Politics of Healing.* Amherst: University of Massachusetts Press, 2011.

Haley, P. Edward. *Strategies of Dominance: The Misdirection of U.S. Foreign Policy.* Washington, DC: Woodrow Wilson Center Press, 2006.

Harsch, Michael F. *The Power of Dependence: NATO-UN Cooperation in Crisis Management.* New York: Oxford University Press, 2015.

Haskin, Jeanne M. *Bosnia and Beyond: The "Quiet" Revolution That Wouldn't Go Quietly.* New York: Algora, 2006.

Hastings, Max. *The Korean War.* New York: Simon and Schuster, 1987.

Heller, Charles E., and William A. Stofft. *America's First Battles, 1775–1965.* Lawrence: University of Kansas Press, 1986.

Herring, George C. *America's Longest War: The United States and Vietnam, 1950–1975.* Boston: McGraw-Hill, 2002.

Herzog, Chaim, and Shlomo Gazit. *The Arab-Israeli Wars: War and Peace in the Middle East from the 1948 War of Independence to the Present.* 2nd Vintage ed. New York: Vintage, 2006.

Hess, Gary R. *Presidential Decisions for War: Korea, Vietnam, the Persian Gulf, and Iraq.* 2nd ed. Baltimore: Johns Hopkins University Press, 2009.

Hofmann, George F., and Donn A. Starry, eds. *Camp Colt to Desert Storm: The History of U.S. Armored Forces.* Lexington: University Press of Kentucky, 1999.

Hosmer, Stephen. *The Conflict over Kosovo: Why Milosevic Decided to Settle When He Did.* MR-1351-AF. Santa Monica, CA: RAND Corporation, 2001.

Hudson, Walter M. *Army Diplomacy: American Military Occupation and Foreign Policy after World War II.* Lexington: University Press of Kentucky, 2015.

Hunt, Richard A. *Pacification: The American Struggle for Vietnam's Hearts and Minds.* New York: Routledge, 1995.

Isikoff, Michael, and David Corn. *Hubris: The Inside Story of Spin, Scandal, and the Selling of the Iraq War.* New York: Random House, 2006.

Kagan, Kimberly. *The Surge: A Military History.* New York: Encounter Books, 2009.

Kaplan, Fred. *The Insurgents: David Petraeus and the Plot to Change the American Way of War.* New York: Simon and Schuster, 2013.

Kennedy, David M. *The American People in World War II: Freedom from Fear, Part Two.* New York: Oxford University Press, 1999.

Kilcullen, David. *Counterinsurgency.* New York: Oxford University Press, 2010.

Kitfield, James. *Prodigal Soldiers: How the Generation of Officers Born of Vietnam Revolutionized the American Style of War.* Washington, DC: Brassey's, 1995. Kindle.

Komer, Robert W. *Bureaucracy Does Its Thing: Institutional Constraints on U.S.-GVN Performance in Vietnam.* R-967-ARPA. Santa Monica, CA: RAND Corporation, August 1972.

Krepinevich, Andrew F., Jr. *The Army and Vietnam.* Baltimore: Johns Hopkins University, 1986. Kindle.

Kuhn, Thomas S. *The Structure of Scientific Revolutions.* 2nd ed. Chicago: University of Chicago Press, 1970.

Langguth, A. J. *Our Vietnam: The War 1954–1975.* New York: Simon and Schuster, 2000.

Lansford, Tom. *A Bitter Harvest: U.S. Foreign Policy and Afghanistan.* Burlington, VT: Ashgate, 2003.

Larson, Eric, Gustav Lindstrom, Myron Hura, Ken Gardiner, Jim Keffer, and Bill Little. *Interoperability of U.S. and NATO Allied Air Forces: Supporting Data and Case Studies.* MR-1603-AF. Santa Monica, CA: RAND Corporation, 2003.

Lawrence, Christopher A. *America's Modern Wars: Understanding Iraq, Afghanistan, and Vietnam.* Philadelphia: Casemate, 2015. Kindle.

LeoGrande, William M. *Our Own Backyard: The United States in Central America, 1977–1992.* Chapel Hill: University of North Carolina Press, 1998.

Leonard, Thomas, Jurgen Buchenau, Kyle Longley, and Graeme Mount, eds. *Encyclopedia of U.S.-Latin American Relations.* Los Angeles: CQ, 2012.

Macgregor, Douglas A. *Breaking the Phalanx: A New Design for Land Power in the 21st Century.* Westport, CT: Praeger, 1997.

Mansoor, Peter R. *Surge: My Journey with General David Petraeus and the Remaking of the Iraq War.* New Haven, CT: Yale University Press, 2013. Kindle.

Mao Tse-Tung. *Selected Military Writings of Mao Tse-Tung.* Seattle: Praetorian, 2011.

Marquis, Susan. *Unconventional Warfare: Rebuilding U.S. Special Operation Forces.* Washington, DC: Brookings Institution Press, 1997.

Millett, Allan R., and Peter Maslowski. *For the Common Defense: A Military History of the United States from 1607 to 2012.* New York: Free Press, 2012. Kindle.

Morgenthau, Hans J. *The Restoration of American Politics.* 3 vols. Chicago: University of Chicago Press, 1962.

Moyar, Mark. *Oppose Any Foe: The Rise of America's Special Operations Forces.* New York: Basic Books, 2017.

Nagl, John. *Counterinsurgency Lessons from Malaya and Vietnam: Learning to Eat Soup with a Knife.* Westport, CT: Praeger, 2002.

Naland, John K. *Lessons from Embedded Provincial Reconstruction Teams in Iraq.* Washington, DC: U.S. Institute of Peace, October 2011.

National Commission on Terrorist Attacks upon the United States. *The 9/11 Commission Report: Final Report of the National Commission on Terrorist Attacks upon the United States.* New York: W. W. Norton, 2004.

Naylor, Sean. *Not a Good Day to Die: The Untold Story of Operation Anaconda.* New York: Berkley Books, 2005.

426

Nichiporuk, Brian, and Carl H. Builder. *Information Technologies and the Future of Land Warfare*. MR-560-A. Santa Monica, CA: RAND Corporation, 1995.

O'Hanlon, Michael E., and Ian Livingston. *Iraq Index: Tracking Variables of Reconstruction & Security in Post-Saddam Iraq*. Washington, DC: Brookings Institution, 2011.

Perritt, Henry H. *Kosovo Liberation Army: The Inside Story of an Insurgency*. Champaign: University of Illinois Press, 2008.

Perry, Walter L., and Marc Dean Millot. *Issues from the 1997 Army After Next Winter Wargame*. MR-988-A. Santa Monica, CA: RAND Corporation, 1998.

Porch, Douglas. *Counterinsurgency*. New York: Cambridge University Press, 2013. Kindle.

Powell, Colin, with Joseph E. Persico. *My American Journey*. New York: Random House, 1995.

Priest, Dana. *The Mission: Waging War and Keeping Peace with America's Military*. New York: W. W. Norton, 2004.

Proctor, Pat. *Containment and Credibility: The Ideology and Deceptions that Plunged America into the Vietnam War*. New York: Carrel Books, 2016. Kindle.

———. *Task Force Patriot and the End of Combat Operations in Iraq*. Lanham, MD: Government Institutes, 2012. Kindle.

Rashid, Ahmed. *Descent into Chaos: The U.S. and the Disaster in Pakistan, Afghanistan, and Central Asia*. New York: Penguin, 2008. Kindle.

Record, Jeffrey. *The Wrong War: Why We Lost in Vietnam*. Annapolis, MD: Naval Institute Press, 1998.

Ricks, Thomas E. *Fiasco: The American Military Adventure in Iraq*. New York: Penguin, 2006.

———. *The Gamble: General Petraeus and the American Military Adventure in Iraq*. New York: Penguin, 2010.

Roberts, Priscilla, ed. *Voices of World War II: Contemporary Accounts of Daily Life*. Santa Barbara, CA: Greenwood, 2012.

Robinson, Linda. *Tell Me How This Ends: General David Petraeus and the Search for a Way Out of Iraq*. New York: PublicAffairs, 2008.

Ruysdael, Salomon, comp. *Speeches of Deception: Selected Speeches of Saddam Hussein; A Story of Propaganda Which Began in Kuwait 10 Years Ago Today Is Not Over*. Lincoln, NE: Writers Club, 2003.

Schadlow, Nadia. *War and the Art of Governance: Consolidating Combat Success into Political Victory*. Washington, DC: Georgetown University Press, 2017.

Schein, Edgar H. *Organizational Culture and Leadership*. 4th ed. San Francisco: Jossey-Bass, 2010.

Schelling, Thomas C. *Arms and Influence*. New Haven, CT: Yale University Press, 1966.

Schomp, Virginia. *The Vietnam Era*. Tarrytown, NY: Marshall Cavendish, 2005.

Schwartz, Benjamin C. *American Counterinsurgency Doctrine and El Salvador: The Frustrations of Reform and the Illusions of Nation Building*. R-4042-USDP. Santa Monica, CA: RAND Corporation, 1991.

Schwarzkopf, H. Norman, with Peter Petre. *It Doesn't Take a Hero: The Autobiography*. New York: Bantam Books, 1992.

Shaw, John M. *The Cambodian Campaign: The 1970 Offensive and America's Vietnam War*. Lawrence: University Press of Kansas, 2005.

Shimko, Keith L. *The Iraq Wars and America's Military Revolution*. New York: Cambridge University Press, 2010. Kindle.

Sky, Emma. *The Unraveling: High Hopes and Missed Opportunities in Iraq*. New York: PublicAffairs, 2015. Kindle.

Smith, Rupert. *The Utility of Force: The Art of War in the Modern World*. New York: Alfred A. Knopf, 2007.

Sorley, Lewis. *Westmoreland: The General Who Lost Vietnam*. New York: Houghton Mifflin Harcourt, 2011.

Sturgill, Claude C. *Low-Intensity Conflict in American History*. Santa Barbara, CA: Greenwood, 1993.

Summers, Harry G. *New World Strategy: A Military Policy for America's Future*. New York: Simon and Schuster, 1995.

———. *On Strategy: A Critical Analysis of the Vietnam War*. New York: Random House, 1995. Kindle.

Thucydides. *The History of the Peloponnesian War*. Translated by Richard Crawley. Seattle: Amazon, 2013. Kindle.

Travis, Philip W. *Reagan's War on Terrorism in Nicaragua: The Outlaw State*. Lanham, MD: Lexington, 2017.

Viney, Mark A. *United States Cavalry Peacekeepers in Bosnia: An Inside Account of Operation Joint Endeavor, 1996*. New York: McFarland, 2012.

Walling, Michael G. *Enduring Freedom, Enduring Voices: US Operations in Afghanistan*. Oxford: Osprey, 2015.

Walravens, Hartmut, ed. *International Newspaper Librarianship for the 21st Century*. Munich: Saur, 2006.

West, Bing. *No True Glory: A Frontline Account of the Battle for Fallujah*. New York: Bantam Books, 2005.

Woodward, Bob. *State of Denial: Bush at War, Part III*. New York: Simon and Schuster, 2006.

———. *The War Within: A Secret White House History, 2006–2008*. New York: Simon and Schuster, 2008.

BOOK AND REPORT CHAPTERS

Al Marashi, Ibrahim. "Iraq." In *Guide to Islamist Movements*, vol. 1, edited by Barry Rubin, 263-82. Armonk, NY: Sharpe, 2010.

Crane, Conrad. "Peace Dividends and Benevolent Interventions." In *The West Point History of Warfare*, edited by Clifford J. Rogers and Ty Seidule. New York: Rowan Technology, 2001. Electronic multimedia. http://www.westpointhistoryofwarfare.com.

Dodge, Toby. "Grand Ambitions and Far-Reaching Failures: The United States in Iraq." In *America and Iraq: Policy-Making, Intervention and Regional Politics*, edited by David Ryan and Patrick Kiely, 92–102. New York: Routledge, 2009.

Gowan, Peter. "The War and Its Aftermath." In *Degraded Capability: The Media and the Kosovo Crisis*, edited by Philip Hammond and Edward S. Herman, 39–55. Sterling, VA: Pluto, 2000.

Kugler, Richard L. "Replacing the 2 MTW Standard: Can a Better Approach Be Found?" In *Revising the Two MTW Force Shaping Paradigm: A "Strategic Alternatives Report,"* edited by Steven Metz, 41–70. Carlisle Barracks, PA: Strategic Studies Institute, U.S. Army War College, April 2001.

Lovelace, Douglas C., Jr. "Foreword." In Conrad C. Crane, *Landpower and Crises: Army Roles and Missions in Smaller-Scale Contingencies during the 1990s*, iii-iv. Carlisle Barracks, PA: Strategic Studies Institute, U.S. Army War College, January 2001.

"The NATO Capability Gap." *Strategic Survey 1999*, vol. 100, 15–21. London: International Institute for Strategic Studies, 1999.

Norton, Augustus Richard, and Thomas G. Weiss. "Rethinking Peacekeeping." In *The United Nations and Peacekeeping: Results, Limitations and Prospects—The Lessons of 40 Years of Experience*, edited by Indar Jit Rikhye and Kjell Skjelsbaek, 22–31. New York: Palgrave Macmillan, 1991.

Rothstein, Hy. "America's Longest War." In *Afghan Endgames: Strategy and Policy Choices for America's Longest War*, edited by Hy Rothstein and John Arquilla, 59–82. Washington, DC: Georgetown University Press, 2012.

Steele, Robert David. "Threats, Strategy, and Force Structure: An Alternative Paradigm for National Security." In *Revising the Two MTW Force Shaping Paradigm: A "Strategic Alternatives Report,"* edited by Steven Metz, 139–64. Carlisle Barracks, PA: Strategic Studies Institute, U.S. Army War College, April 2001.

Stockton, Paul N. "When the Bear Leaves the Woods: Department of Defense Reorganization in the Post–Cold War Era." In *US Foreign Policy after the Cold War*, edited by Randall B. Ripley and James M. Lindsay, 106–31. Pittsburgh: University of Pittsburgh Press, 1997.

Swain, Richard M. "AirLand Battle." In *Camp Colt to Desert Storm: The History of U.S. Armored Forces*, edited by George F. Hofmann and Donn A. Starry, 360–402. Lexington: University Press of Kentucky, 1999.

Von Einsiedel, Sebastian, and David M. Malone. "Haiti." In *The UN Security Council: From the Cold War to the 21st Century*, edited by David M. Malone, 467–82. Boulder, CO: Lynne Rienner, 2004.

Wass de Czege, Huba, and Antulio J. Echevarria II. "A New Strategy and Military Logic for the 21st Century." In *Revising the Two MTW Force Shaping Paradigm: A "Strategic Alternatives Report,"* edited by Steven Metz, 71–80. Carlisle Barracks, PA: Strategic Studies Institute, U.S. Army War College, April 2001.

JOURNAL ARTICLES

Anderson, Joseph. "Military Operational Measures of Effectiveness for Peacekeeping Operations." *Military Review* 81, no. 5 (September–October 2001): 36–44.

Arnold, S. L. "Somalia: An Operation Other Than War." *Military Review* 73, no. 12 (December 1993): 26–35.

Bacevich, Andrew. "Learning from Aidid." *Commentary* 16, no. 6 (December 1993): 30–33.

Baker-Cristales, Beth. "Salvadoran Transformations." *Latin American Perspectives* 31, no. 5 (September 2004): 15–33.

Barnes, Rudolph C., Jr. "The Diplomat Warrior." *Military Review* 70, no. 5 (May 1990): 55–63.

Barto, Joseph C., III. "Armor Future: To Fight, Deter or Disappear." *Military Review* 70, no. 8 (August 1990): 78–80.

Benson, Kevin. "OIF Phase IV: A Planner's Reply to Brigadier Aylwin-Foster." *Military Review* 86, no. 2 (March–April 2006): 61–68.

Bunker, Robert J. "Epochal Change: War over Social and Political Organization." *Parameters* 27, no. 2 (Summer 1997): 15–25.

Butler, George L. "Adjusting to Post–Cold War Strategic Realities." *Parameters* 21, no. 1 (Spring 1991): 2–9.

Byman, Daniel L., and Matthew C. Waxman. "Kosovo and the Great Air Power Debate." *International Security* 24, no. 4 (Spring 2000): 5–38.

Caldera, Louis, and Eric K. Shinseki. "Army Vision: Soldiers on Point for the Nation . . . Persuasive in Peace, Invincible in War." *Military Review* 80, no. 5 (September–October 2000): 3–5.

Carr, Caleb. "The Consequences of Somalia." *World Policy Journal* 10, no. 3 (Fall 1993): 1–4.

Cherrie, Stanley F. "Task Force Eagle." *Military Review* 77, no. 4 (July–August 1997): 63–72.

Clarke, Walter S. "Testing the World's Resolve in Somalia." *Parameters* 23, no. 4 (Winter 1993–94): 42–58.

Cucolo, Tony. "Grunt Diplomacy: In the Beginning There Were Only Soldiers." *Parameters* 29, no. 1 (Spring 1999): 110–26.

Curtiss, Richard H. "In Somalia, the Goal Must Be 'Do No Harm.'" *Washington Report on Middle East Affairs* 12, no. 4 (December 31, 1993): 38.

Dubik, James. "IBCT at Fort Lewis." *Military Review* 90, no. 5 (September–October 2000): 17–23.

Fastabend, David A. "Checking the Doctrinal Map: Can We Get There from Here with FM 100-5?" *Parameters* 25, no. 2 (Summer 1995): 37–46.

Finch, Raymond C., III. "A Face of Future Battle: Chechen Fighter Shamil Basayev." *Military Review* 77, no. 3 (May–June 1997): 33–41.

Fishel, John T., and Richard D. Downie. "Taking Responsibility for Our Actions? Establishing Order and Stability in Panama." *Military Review* 72, no. 4 (April 1992): 66–72.

Freakley, Benjamin C., Kevin C. M. Benson, Frederick Rudesheim, and Brian J. Butcher. "Training for Peace Support Operations." *Military Review* 28, no. 4 (July–August 1998): 17–24.

Freeman, Waldo D., Robert B. Lambert, and Jason D. Mims. "Operation Restore Hope: A USCENTCOM Perspective." *Military Review* 73, no. 9 (September 1993): 61–72.

Goulding, Vincent J., Jr. "Back to the Future with Asymmetric Warfare." *Parameters* 30, no. 3 (Winter 2000–2001): 21–30.

Gaddis, John Lewis. "The Cold War, the Long Peace, and the Future." *Diplomatic History* 16, no. 2 (April 1, 1992): 234–46.

———. "The Long Peace: Elements of Stability in the Postwar International System." *International Security* 10, no. 4 (Spring 1986): 99–142.

Grau, Lester W., and Jacob W. Kipp. "Urban Combat: Confronting the Specter." *Military Review* 89, no. 3 (July–August 1999): 9–17.

Gray, Victor. "People Wars: Ruminations on Population and Security." *Parameters* 27, no. 2 (Summer 1997): 52–60.

Hahn, Robert F., II, and Bonnie Jezior. "Urban Warfare and the Urban Warfighter of 2025." *Parameters* 29, no. 2 (Summer 1999): 74–86.

Heal, Sid. "Crowds, Mobs and Nonlethal Weapons." *Military Review* 90, no. 2 (March–April 2000): 45–50.

High, Gil. "Talking with . . . the New Chief of Staff." *Soldiers*54, no. 8(August 1999): 2.

Hillen, John. "Peace(keeping) in Our Time: The UN as a Professional Military Manager." *Parameters* 26, no. 3 (Autumn 1996): 17–34.

Hodge, Carl Cavanagh. "Woodrow Wilson in Our Time: NATO's Goals in Kosovo." *Parameters* 31, no. 1 (Spring 2001): 125–35.

Holder, L. D. "Educating and Training for Theater Warfare." *Military Review* 70, no. 9 (September 1990): 85–99.

Holt, Jimmie F. "LIC in Central America: Training Implications for the US Army." *Military Review* 70, no. 3 (March 1990): 2–15.

House, Randolph W., Mark R. Pires, and Lester W. Grau. "PEACEKEEPER 95: 27th Guards Train with 'Big Red One.'" *Military Review* 76, no. 2 (March–April 1996): 5–11.

Hunt, John B. "OOTW: A Concept in Flux." *Military Review* 76, no. 5 (September–October 1996): 3–10.

Jablonsky, David. "Army Transformation: A Tale of Two Doctrines." *Parameters* 31, no. 3 (Autumn 2001): 43–62.

Jandora, John W. "Threat Parameters for Operations Other Than War." *Parameters* 25, no. 1 (Spring 1995): 55–67.

Johnston, Kyle. "U.S. Special Operations Forces and the Interagency in Phase Zero." *InterAgency Journal* 8, no. 1 (Winter 2017): 76–104.

Kretchik, Walter E. "Force Protection Disparities." *Military Review* 77, no. 4 (July–August 1997): 73–76.

Lind, William S. "An Operational Doctrine for Intervention." *Parameters* 25, no. 2 (Summer 1995): 128–33.

Lorenz, F. M. "Law and Anarchy in Somalia." *Parameters* 23, no. 4 (Winter 1993–94): 27–41.

Maloney, Sean M. "Insights into Canadian Peacekeeping Doctrine." *Military Review* 76, no. 2 (March–April 1996): 12–23.

Manwaring, Max G. "Peace and Stability Lessons from Bosnia." *Parameters* 28, no. 4 (Winter 1998): 28–38.

McDonough, James R. "Building the New FM 100-5: Process and Product." *Military Review* 71, no. 10 (October 1991): 2–12.

McMichael, Scott R. "The Soviet Army, Counterinsurgency, and the Afghan War." *Parameters* 19, no. 12 (December 1989): 21–35.

McNerney, Michael J. "Stabilization and Reconstruction in Afghanistan: Are PRTs a Model or a Muddle?" *Parameters* 35, no. 4 (Winter 2005–6): 32–46.

Mearsheimer, John J. "Nuclear Weapons and Deterrence in Europe." *International Security* 9, no. 3 (Winter 1984–85): 19–46.

Mehaffey, Michael. "Vanguard of the Objective Force." *Military Review* 80, no. 5 (September–October 2000): 6–16.

Metz, Steven. "A Flame Kept Burning: Counterinsurgency Support after the Cold War." *Parameters* 25, no. 3 (Autumn 1995): 31–41.

———. "Which Army After Next? The Strategic Implications of Alternative Futures." *Parameters* 27, no. 3 (Autumn 1997): 15–26.

Moreno, Rafael, and Juan Jose Vega. "Lessons from Somalia." *Peacekeeping and International Relations* 23, no. 3 (May 1994): 11–12.

Nagl, John A., and Elizabeth O. Young. "*Si Vis Pacem, Para Pacem*: Training for Humanitarian Emergencies." *Military Review* 90, no. 2 (March–April 2000): 31–37.

Olsen, Howard, and John Davis. *Training U.S. Army Officers for Peace Operations: Lessons from Bosnia*. Washington, DC: U.S. Institute of Peace, October 1, 1999. https://www .usip.org/sites/default/files/sr991029.pdf.

Packett, Virgil L., II, and Timothy M. Gilhool. "Diplomacy by Other Means: JTF Aquila Responds to Hurricane Mitch." *Military Review* 90, no. 2 (March–April 2000): 81–84.

Peters, Ralph. "Heavy Peace." *Parameters* 29, no. 1 (Spring 1999): 71–79.

———. "The Human Terrain of Urban Operations." *Parameters* 30, no. 1 (Spring 2000): 4–12.

Powell, Colin L. "U.S. Forces: Challenges Ahead." *Foreign Affairs* 71, no. 5 (Winter 1992–93): 38–40.

Record, Jeffrey. "Operation Allied Force: Yet Another Wake-Up Call for the Army?" *Parameters* 29, no. 4 (Winter 1999–2000): 15–23.

Reimer, Dennis J. "Challenge and Change: A Legacy for the Future." *Military Review* 77, no. 4 (July–August 1997): 108–16.

Rendina, Mark D. "An Officer Corps for the 1990s." *Military Review* 70, no. 10 (October 1990): 64–73.

Rice, Andrew. "Revolutions: Relevance to Contemporary Warfare." *Military Review* 28, no. 4 (July–August 1998): 80–82.

Schnaubelt, Christopher M. "Wither the RMA." *Parameters* 37, no. 3 (Autumn 2007): 95–107.

Schook, Steven P. "Paying the Price for Versatility." *Military Review* 77, no. 5 (September–October 1997): 19–25.

Shelton, H. Hugh, and Timothy D. Vane. "Winning the Information War in Haiti." *Military Review* 75, no. 6 (November–December 1995): 3–9.

Shin, David W. "Future War: Back to Basics." *Military Review* 89, no. 4 (September–October 1999): 63–79.

Shull, George D. "Correcting the Force Structure Mismatch." *Military Review* 90, no. 3 (May–June 2000): 31–39.

Sullivan, Gordon. "A Vision for the Future." *Military Review* 75, no. 3 (May–June 1995): 4–14.

Sullivan, Gordon, and James M. Dubik. "War in the Information Age." *Military Review* 74, no. 4 (April 1994): 46–62.

Sullivan, Gordon, and Andrew B. Twomey. "The Challenges of Peace." *Parameters* 24, no. 3 (Autumn 1994): 4–17.

Swannack, Charles H., Jr., and David R. Gray. "Peace Enforcement Operations." *Military Review* 77, no. 6 (November–December 1997): 3–10.

Thomas, Timothy L. "The Battle of Grozny: Deadly Classroom for Urban Combat." *Parameters* 29, no. 2 (Summer 1999): 87–102.

———. "Kosovo and the Current Myth of Information Superiority." *Parameters* 30, no. 1 (Spring 2000): 13–29.

———. "Russian Lessons Learned in Bosnia." *Military Review* 76, no. 5 (September–October 1996): 38–43.

Vinson, Mark E. "Structuring the Army for Full-Spectrum Readiness." *Parameters* 30, no. 2 (Summer 2000): 19–32.

Vuono, Carl E. "National Strategy and the Army of the 1990s." *Parameters* 21, no. 2 (Summer 1991): 2–12.

Yates, Lawrence A. "Joint Task Force Panama: Just Cause—Before and After." *Military Review* 71, no. 10 (October 1991): 58–71.

———. "Military Stability and Support Operations: Analogies, Patterns and Recurring Themes." *Military Review* 77, no. 4 (July–August 1997): 51–61.

MAGAZINE ARTICLES

Ackerman, Spencer. "Petraeus Pal McMaster Headed to Afghanistan." *Wired*, June 29, 2010. https://www.wired.com/2010/06/petraeus-buddy-mcmaster-headed-to-afghanistan/.

Baker, Aryn. "Pakistan Braces for a Backlash after Taliban Raid." *Time*, October 30, 2006. http://www.time.com/time/world/article/0,8599,1552281,00.html.

———. "TIME's Interview with General Stanley McChrystal." *Time*, July 8, 2009. http://www.time.com/time/world/article/0,8599,1909254,00.html.

Barry, John. "Anatomy of a Revolt." *Newsweek*, April 23, 2006. http://www.newsweek.com/anatomy-revolt-107813.

Brown, John Sloan. "Defense Transformation Redux." *Army*, November 2012.

Calamur, Krishnadev. "Oil Was Supposed to Rebuild Iraq." *Atlantic*, March 19, 2018. https://www.theatlantic.com/international/archive/2018/03/iraq-oil/555827/.

Dallaire, R. A., and B. Poulin. "UNAMIR: Mission to Rwanda." *Joint Force Quarterly*, Spring 1995.

Dickerson, John. "Playing the U.N. Card." *Time*, September 23, 2002. http://content.time
.com/time/magazine/article/0,9171,1003300,00.html.

Duffy, Michael. "The Surge at Year One." *Time*, January 31, 2008. http://www.time.com/time
/magazine/article/0,9171,1708843,00.html.

Elliott, Michael. "Bush Isn't as Lonely as He Looks." *Time*, September 9, 2002. http://www
.time.com/time/columnist/elliott/article/0,9565,349390,00.html.

Fallows, James. "The Right and Wrong Questions about the Iraq War." *Atlantic*,
May 19, 2015. https://www.theatlantic.com/politics/archive/2015/05/
the-right-and-wrong-questions-about-the-iraq-war/393497/.

———. "The Tragedy of the American Military." *Atlantic*, January–February
2015. https://www.theatlantic.com/magazine/archive/2015/01/
the-tragedy-of-the-american-military/383516/.

Finel, Bernard I. "An Alternative to COIN: It's Time to Adapt Our Security Strategy to
Leverage America's Conventional Strengths." *Armed Forces Journal*, February 1, 2010.
http://armedforcesjournal.com/an-alternative-to-coin/.

Hastings, Michael. "The Runaway General." *Rolling Stone*, June 22, 2010. http://www
.rollingstone.com/politics/news/the-runaway-general-20100622.

Holmes, James R. "Preserving History and the Long Peace." *Diplomat*, December 10, 2013.
https://thediplomat.com/2013/12/preserving-history-and-the-long-peace/.

Ignatius, David. "How ISIS Spread in the Middle East, and How to Stop It." *Atlantic*, October
29, 2015. http://www.theatlantic.com/international/archive/2015/10
/how-isis-started-syria-iraq/412042/.

Joint Staff. "Joint Vision 2020: America's Military—Preparing for Tomorrow." *Joint Force
Quarterly*, Summer 2000.

Kaplan, Fred. "Who Disbanded the Iraqi Army? And Why Was Nobody Held Accountable?"
Slate, September 7, 2007. http://www.slate.com/articles/news_and_politics/war
_stories/2007/09/who_disbanded_the_iraqi_army.html.

Kaplan, Robert D. "Five Days in Fallujah." *Atlantic*, July–August 2004. https://www
.theatlantic.com/magazine/archive/2004/07/five-days-in-fallujah/303450/.

Karon, Tony. "Can the Northern Alliance Control Kabul?" *Time*, November 12, 2001. http://
www.time.com/time/nation/article/0,8599,184221,00.html.

———. "What Powell Achieved." *Time*, February 5, 2003. http://content.time.com/time
/world/article/0,8599,419939,00.html.

Kennedy, Harold. "Army Approaches Decision on Interim Combat Vehicle." *National
Defense*, September 2000. https://www.nationaldefensemagazine.org
/articles/2000/9/1/2000september-army-approaches-decision-on-interim-combat-
vehicle.

Khalil, Ahmad Bilal. "The Tangled History of the Afghanistan-India-Pakistan Triangle."
Diplomat, December 16, 2016. https://thediplomat.com/2016/12
/the-tangled-history-of-the-afghanistan-india-pakistan-triangle/.

Khatchadourian, Raffi. "The Kill Company: Did a Colonel's Fiery Rhetoric Set the Conditions
for a Massacre?" *New Yorker*, July 6 and 13, 2009. https://www.newyorker.com
/magazine/2009/07/06/the-kill-company.

Klein, Joe. "Why Bush Is (Still) Winning the War at Home." *Time*, June 18, 2006. http://www
.time.com/time/columnist/klein/article/0,9565,1205323,00.html.

MacLeod, Scott. "Arabs to Cheney: 'Curb Sharon before Saddam.'" *Time*, March 18, 2002.
http://www.time.com/time/world/article/0,8599,218463,00.html.

———. "Time Exclusive: The Saudi Initiative Explained." *Time*, February 2, 2003. http://www.time.com/time/world/article/0,8599,419297,00.html.

Mahanta, Siddhartha. "Fighting Terrorism in the Age of Trump: The President-Elect Has Vowed to Kill the Families of ISIS Members and Bring Back Bush-Era Torture Tactics." *Atlantic*, November 12, 2016. https://www.theatlantic.com/international/archive/2016/11/trump-torture-soufan-fbi-al-qaeda-isis-islam/507380/.

McGirk, Tim, and Michael Ware. "Losing Control? The U.S. Concedes It Has Lost Momentum in Afghanistan, While Its Enemies Grow Bolder." *Time*, November 11, 2002. http://content.time.com/time/magazine/article/0,9171,388964,00.html.

"New Academy Will Train NCOs for Security Force Assistance Brigades." *NCO Journal*, February 28, 2017. https://www.armyupress.army.mil/Journals/NCO-Journal/Archives/2017/February-New-Academy-Will-Train-NCOs-for-Security-Force-Assistance-Brigades/.

Nordland, Rod. "A Monster on the Loose." *Newsweek*, April 21, 1996. http://www.newsweek.com/monster-loose-176518.

Norton, James. "The Loyal Jirga." *Christian Science Monitor*, June 10, 2002. http://www.csmonitor.com/2002/0610/p10s01-wosc.html.

Nowowiejski, Dean A. "Regaining the Edge in Combined Arms." *Army*, March 2013.

Quinn-Judge, Paul. "How the Northern Alliance Plans to Win the War." *Time*, October 19, 2001. http://www.time.com/time/world/article/0,8599,180452,00.html.

Rawe, Julie, "Iraq: The Sad Tale of Nick Berg." *Time*, May 24, 2004. http://content.time.com/time/magazine/article/0,9171,994232,00.html.

Reimer, Dennis J. "The Army After Next: Revolutionary Transformation." *Strategic Review*, Spring 1999. Reprinted in Reimer, *Soldiers Are Our Credentials: The Collected Works and Selected Papers of the Thirty-Third Chief of Staff, United States Army*, edited by James Jay Carafano. Washington, DC: Center for Military History, U.S. Department of the Army, 2000.

———. "Where We've Been—Where We're Headed: Maintaining a Solid Framework While Building for the Future." *Army*, October 1995.

Remnick, David. "Telling the Truth about ISIS and Raqqa." *New Yorker*, November 22, 2015. http://www.newyorker.com/news/news-desk/telling-the-truth-about-isis-and-raqqa.

Reuters. "Saddam's Promises and Threats." *Economist*, August 9, 2002. https://www.economist.com/node/1279615.

Ricks, Thomas E. "The Widening Gap between Military and Society." *Atlantic*, July 1997. https://www.theatlantic.com/magazine/archive/1997/07/the-widening-gap-between-military-and-society/306158/.

Shinseki, Eric K. "The Army—Intent of the Chief of Staff." *Joint Force Quarterly*, Summer 1999.

———. "The Army Vision: A Status Report." *Army*, annual "Green Book" issue, October 2001.

Smucker, Philip. "How bin Laden Got Away." *Christian Science Monitor*, March 4, 2002. http://www.csmonitor.com/2002/0304/p01s03-wosc.html.

———. "New Afghan Leader Faces a Rogues Gallery Government." *Christian Science Monitor*, June 21, 2002. http://www.csmonitor.com/2002/0621/p07s02-wosc.html.

Soergel, Andrew. "War on Terror Could Be Costliest Yet." *U.S. News and World Report*, September 9, 2016. https://www.usnews.com/news/articles/2016-09-09/war-on-terror-could-be-costliest-yet.

Soloway, Colin. "Tale of an American Talib." *Newsweek*, December 1, 2001.

434

Steele, Dennis. "JRTC Concept Is Set, but It Needs a Home." *Army*, March 1989.

———. "The National Training Center: Decisive-Action Training Rotations 'Old School Without Going Back in Time.'" *Army*, February 2013.

Thompson, Mark. "An Army Apart: The Widening Military-Civilian Gap." *Time*, November 10, 2011. http://time.com/4254696/military-family-business/.

———. "Here's Why the U.S. Military Is a Family Business." *Time*, March 10, 2016. https://time.com/4254696/military-family-business/.

———. "Obama Weighs the Cost of an Afghan Surge." *Time*, November 25, 2009. http://www.time.com/time/printout/0,8816,1942837,00.html.

Tumulty, Karen, and Mike Allen. "His Search for A New Groove." *Time*, December 11, 2005. http://content.time.com/time/magazine/article/0,9171,1139848,00.html.

Waller, Douglas. "Bush and Musharraf: Friends Again." *Time*, September 22, 2006. http://www.time.com/time/world/article/0,8599,1538476,00.html.

Zagorin, Adam, and Michael Duffy. "Inside the Interrogation of Detainee 063." *Time*, June 20, 2005. http://content.time.com/time/magazine/article/0,9171,1071284-1,00.html.

GOVERNMENT PUBLICATIONS

Allard, Kenneth. *Somalia Operations: Lessons Learned.* CCRP Publication Series. Washington, DC: National Defense University Press, 1995.

Applegate, Melissa. *Preparing for Asymmetry: As Seen through the Lens of Joint Vision 2020.* Carlisle Barracks, PA: Strategic Studies Institute, U.S. Army War College, September 2001.

Aspin, Les. *Report of the Bottom-Up Review.* Washington, DC: U.S. Department of Defense, 1993.

Bailey, Michael, Robert Maguire, and J. O'Neil G. Pouliot. "Haiti: Military-Police Partnership for Public Security." In *Policing the New World Disorder: Peace Operations and Public Security,* edited by Robert B. Oakley, Michael J. Dziedzic, and Eliot M. Goldberg, 251–52. Washington, DC: National Defense University Press, 1998.

Baumann, Robert F., George Walter Gawrych, and Walter Edward Kretchik. *Armed Peacekeepers in Bosnia.* Fort Leavenworth, KS: Combat Studies Institute Press, 2004.

Baumann, Robert F., and Lawrence A. Yates with Versalle F. Washington. *"My Clan against the World": US and Coalition Forces in Somalia 1992–1994.* Fort Leavenworth, KS: Combat Studies Institute Press, 2004.

Birtle, Andrew J. *U.S. Army Counterinsurgency and Contingency Operations Doctrine, 1942–1976.* Washington, DC: U.S. Army Center of Military History, 2007.

Blank, Stephen J., and Earl H. Tilford Jr. *Russia's Invasion of Chechnya: A Preliminary Assessment.* Carlisle Barracks, PA: Strategic Studies Institute, U.S. Army War College, January 13, 1995.

Brown, John Sloan. *Kevlar Legions: The Transformation of the U.S. Army, 1989–2005.* Washington, DC: U.S. Army Center of Military History, 2011. Kindle.

Bunker, Robert J. "Five-Dimensional (Cyber) Warfighting: Can the Army after Next Be Defeated through Complex Concepts and Technologies?" Paper presented at the U.S. Army War College's Annual Strategy Conference, Carlisle Barracks, PA, March 31–April 2, 1998.

Center for Army Lessons Learned. *Somalia: Operations Other Than War.* CALL Special Edition No. 93-1. Fort Leavenworth, KS: Center for Army Lessons Learned, 1993. https://www.globalsecurity.org/military/library/report/call/call_93-1_tblcon.htm.

———. *US Army Operations in Support of UNOSOM II: Operations Other Than War.* Fort Leavenworth, KS: Center for Army Lessons Learned, October 1994.

Chambers, John. *Countering Gray-Zone Hybrid Threats: An Analysis of Russia's "New Generation Warfare" and Implications for the US Army.* West Point, NY: Modern War Institute at West Point, October 18, 2016.

Chapman, Anne W. *The Army's Training Revolution, 1973–1990: An Overview.* TRADOC Historical Studies Series. Fort Monroe, VA: Office of the Command Historian, U.S. Army Training and Doctrine Command, 1991.

Chapman, Anne W., Carol J. Lilly, John L. Romjue, and Susan Canedy. *Prepare the Army for War: A Historical Overview of the Army Training and Doctrine Command 1973–1998.* Fort Monroe, VA: Military History Office, U.S. Army Training and Doctrine Command, 1998.

Christoff, Joseph A. *Provincial Reconstruction Teams in Afghanistan and Iraq.* GAO-09-86R. Washington, DC: U.S. Government Accountability Office, October 1, 2008.

Cole, Ronald H. *Operation Just Cause: The Planning and Execution of Joint Operations in Panama, February 1988–January 1990.* Washington, DC: Joint History Office, Office of the Chairman of the Joint Chiefs of Staff, 1995.

Combined Arms Center History Office. *1991 Annual Command History.* Fort Leavenworth, KS: Combined Arms Center History Office, U.S. Army Combined Arms Command, 1992.

———. *2001 Annual Command History.* Fort Leavenworth, KS: Combined Arms Center History Office, U.S. Army Combined Arms Center, 2002.

Commission on Roles and Missions of the Armed Forces. *Directions for Defense.* Washington, DC: U.S. Department of Defense, June 1995.

Congressional Budget Office. *Projected Costs of U.S. Nuclear Forces, 2015 to 2024.* Washington, DC: Congressional Budget Office, January 22, 2015.

Crane, Conrad C. *Alternative National Military Strategies for the United States.* Carlisle Barracks, PA: Strategic Studies Institute, U.S. Army War College, December 2000.

———. *Avoiding Vietnam: The U.S. Army's Response to Defeat in Southeast Asia.* Carlisle Barracks, PA: Strategic Studies Institute, U.S. Army War College, September 2002.

———. *Landpower and Crises: Army Roles and Missions in Smaller-Scale Contingencies during the 1990s.* Carlisle Barracks, PA: Strategic Studies Institute, U.S. Army War College, January 2001.

Crane, Conrad C., and W. Andrew Terrill. *Reconstructing Iraq: Challenges and Missions for Military Forces in a Post-Conflict Scenario.* Carlisle Barracks, PA: Strategic Studies Institute, U.S. Army War College, January 29, 2003.

Director of Central Intelligence. *El Salvador: Government and Insurgent Prospects.* Langley, VA: Central Intelligence Agency, February 1989.

Feickert, Andrew. *The Army's Future Combat System (FCS): Background and Issues for Congress.* Washington, DC: Congressional Research Service, 2006.

Fontenot, Gregory, E. J. Degen, and David Tohn. *On Point: U.S. Army in Operation Iraqi Freedom.* Washington, DC: Office of the Chief of Staff, U.S. Army, 2004.

General Officer Management Office. *General Dennis Joe Reimer.* Washington, DC: U.S. Department of the Army, n.d. [2000].

———. *General Eric K. Shinseki.* Washington, DC: General Officer Management Office, U.S. Department of the Army, August 28, 2008.

Gerras, Stephen, and Leonard Wong. *Changing Minds in the Army: Why It Is So Difficult and What to Do about It.* Carlisle Barracks, PA: Strategic Studies Institute and U.S. Army War College Press, 2013.

Grau, Lester W., ed. *The Bear Went over the Mountain: Soviet Combat Tactics in Afghanistan.* Washington, DC: National Defense University Press, 1996.

Hasskamp, Charles W. *Operations Other Than War: Who Says Warriors Don't Do Windows?* Maxwell Paper No. 13. Maxwell Air Force Base, AL: Air War College, March 1998.

Headquarters, U.S. Department of the Army. *Counterinsurgency.* FM 3-24. Washington, DC: U.S. Department of the Army, December 2006.

———. *Intelligence and Electronic Warfare Support to Low-Intensity Conflict Operations.* FM 34-7. Washington, DC: U.S. Department of the Army, May 18, 1993.

———. *Low Intensity Conflict.* FC 100-20. Washington, DC: U.S. Department of the Army, July 16, 1986.

———. *Operations.* FM 3-0. Washington, DC: U.S. Department of the Army, June 14, 2001.

———. *Operations.* FM 3-0. Washington, DC: U.S. Department of the Army, October 2017.

———. *Operations.* FM 100-5. Washington, DC: U.S. Department of the Army, July 1, 1976.

———. *Operations.* FM 100-5. Washington, DC: U.S. Department of the Army, 1982.

———. *Operations.* FM 100-5. Washington, DC: U.S. Department of the Army, May 1986.

———. *Operations.* FM 100-5. Washington, DC: U.S. Department of the Army, 14 June 1993.

———. *Operations in a Low-Intensity Conflict.* FM 7-98. Washington, DC: U.S. Department of the Army, October 19, 1992.

———. *Peace Operations.* FM 100-23. Washington, DC: U.S. Department of the Army, October 31, 1994.

———. *Training the Force.* FM 25-100. Washington, DC: U.S. Department of the Army, November 15, 1988.

Headquarters, U.S. Departments of the Army and the Air Force. *Military Operations in Low Intensity Conflict.* FM 100-20. Washington, DC: Headquarters, U.S. Departments of the Army and the Air Force, December 5, 1990.

Herbert, Paul H. *Deciding What Has to Be Done: General William E. DePuy and the 1976 Edition of FM 100-5, Operations.* Leavenworth Papers No. 16. Fort Leavenworth, KS: Combat Studies Institute, U.S. Army Command and General Staff College, 1988.

Hines, Scott M. *Joint Task Force—Bravo: The U.S. Military Presence in Honduras, U.S. Policy for an Evolving Region.* Washington, DC: Institute for National Strategic Studies, National Defense University, 1994.

Institute for National Strategic Studies. *Project 2025.* Washington, DC: Institute for National Strategic Studies, National Defense University, May 6, 1992.

Joint Chiefs of Staff. *Counterinsurgency.* JP 3-24. Washington, DC: Joint Staff, U.S. Department of Defense, April 25, 2018.

———. *Expanding Joint Vision 2010.* Washington, DC: Joint Staff, U.S. Department of Defense, May 1997.

———. *Joint Doctrine for Military Operations Other Than War.* JP 3-07. Washington, DC: Joint Staff, U.S. Department of Defense, June 16, 1995.

———. *Joint Vision 2010.* Washington, DC: Joint Staff, U.S. Department of Defense, n.d. [1996].

Joint Warfighting Center. *Joint Task Force Commander's Handbook for Peace Operations.* Fort Monroe, VA: Joint Warfighting Center, U.S. Department of Defense, June 16, 1997.

Manwaring, Max G. *Internal Wars: Rethinking Problem and Response.* Carlisle Barracks, PA: Strategic Studies Institute, U.S. Army War College, September 2001.

Mehaffey, Michael, ed. *Army Transformation Taking Shape . . . Interim Brigade Combat Team,* CALL newsletter no. 01-18. Fort Leavenworth, KS: Center for Army Lessons Learned, 2007.

Metz, Steven. *Disaster and Intervention in Sub-Saharan Africa: Learning from Rwanda.* Carlisle Barracks, PA: Strategic Studies Institute, U.S. Army War College, September 9, 1994.

———, ed. *Revising the Two MTW Force Shaping Paradigm: A "Strategic Alternatives Report."* Carlisle Barracks, PA: Strategic Studies Institute, U.S. Army War College, April 2001.

———. *Strategic Horizons: The Military Implications of Alternative Futures.* Carlisle Barracks, PA: Strategic Studies Institute, U.S. Army War College, March 7, 1997.

Metz, Steven, William T. Johnsen, Douglas V. Johnson II, James O. Kievit, and Douglas C. Lovelace Jr. *The Future of American Landpower: Strategic Challenges for the 21st Century Army.* Carlisle Barracks, PA: Strategic Studies Institute, U.S. Army War College, March 12, 1996.

Metz, Steven, and Douglas V. Johnson II. *Symmetry and U.S. Military Strategy: Definition, Background, and Strategic Concepts.* Carlisle Barracks, PA: Strategic Studies Institute, U.S. Army War College, January 2001.

Moyar, Mark. *Village Stability Operations and the Afghan Local Police.* MacDill Air Force Base, FL: Joint Special Operations University, October 2014.

Objective Force Task Force. *Objective Force Task Force . . . Beyond Relevancy and Readiness: After Action Report.* Washington, DC: Objective Force Task Force, U.S. Department of the Army, n.d. [2003].

Phillips, R. Cody. *Bosnia-Herzegovina: The U.S. Army's Role in Peace Enforcement Operations, 1995–2004.* Washington, DC: U.S. Army Center of Military History, 2014.

———. *Operation Joint Guardian: The U.S. Army in Kosovo.* Washington, DC: Center for Military History, U.S. Department of the Army, 2007.

Public Papers of the President of the United States: George Bush, 1992–1993. 2 vols. Washington, DC: Government Printing Office, 1993.

Public Papers of the Presidents of the United States: Lyndon B. Johnson, 1968–69. 2 vols. Washington, DC: Government Printing Office, 1970.

Ramsey, Robert D., III, "Advising Indigenous Forces: American Advisors in Korea, Vietnam, and El Salvador." Occasional Papers 18, Combat Studies Institute, U.S. Army Command and General Staff College, Fort Leavenworth, KS, n.d. [2006].

Reimer, Dennis. *Soldiers Are Our Credentials: The Collected Works and Selected Papers of the Thirty-Third Chief of Staff, United States Army.* Edited by James Jay Carafano. Washington, DC: Center for Military History, U.S. Department of the Army, 2000.

Record, Jeffrey. "Ready for What and Modernized against Whom? A Strategic Perspective on Readiness and Modernization." Paper presented to the Annual Strategy Conference, U.S. Army War College, Carlisle Barracks, PA, April 10, 1995.

Regional Reachback Center Iraq Support Team. *Post-SOFA Motivations of Iraqi Insurgents.* Oyster Point, VA: Human Terrain Systems Research Reachback Center, November 30, 2009.

Scales, Robert H., Jr. *Certain Victory: The US Army in the Gulf War.* Fort Leavenworth, KS: U.S. Army Command and General Staff College, 1994.

———. *Future Warfare: Anthology.* Carlisle Barracks, PA: Strategic Studies Institute, U.S. Army War College, April 1999.

Shinseki, Eric K. "Statement by General Eric K. Shinseki, Chief of Staff, United States Army, before the Committee on Armed Services, United States Senate, Second Session, 107th Congress, on the Crusader Self-Propelled Howitzer Program." Washington, DC: Senate Committee on Armed Services, May 16, 2002.

———. "Statement by General Eric K. Shinseki, Chief of Staff, United States Army, before the Committee on Armed Services, United States Senate, Second Session, 107th Congress, on the Fiscal Year 2003 Defense Authorization Request and Future Years Defense Program." Washington, DC: Senate Committee on Armed Services, March 7, 2002.

Shultz, Richard H., Jr., and Robert L. Pfaltzgraff Jr. *The Future of Air Power in the Aftermath of the Gulf War.* Maxwell Air Force Base, AL: Air University, July 1992.

Spiller, Roger J. *Sharp Corners: Urban Operations at Century's End.* Fort Leavenworth, KS: U.S. Army Command and General Staff College, June 2000.

Stewart, Richard W. *Operation Urgent Fury: The Invasion of Grenada, October 1983.* CMH Publication 70-114-1. Washington, DC: U.S. Army Center of Military History, 2008.

———. *The United States Army in Somalia, 1992–1994.* Washington, DC: U.S. Army Center of Military History, 2002.

Sullivan, Gordon R. *The Collected Works of the Thirty-Second Chief of Staff, United States Army: June 1991–June 1995.* Washington, DC: U.S. Department of the Army, 1996.

Sullivan, Gordon R., and Anthony M. Coroalles. *The Army in the Information Age.* Carlisle Barracks, PA, U.S. Army War College, March 31, 1995.

Sullivan, Gordon R., and James M. Dubik. *Envisioning Future War.* Fort Leavenworth, KS: Command and General Staff College, 1995.

———. "Land Warfare in the 21st Century." Paper presented at the U.S. Army War College Fourth Annual Strategy Conference, February 24–25, 1993, Carlisle Barracks, PA.

———. *War in the Information Age.* Carlisle Barracks, PA: Strategic Studies Institute, U.S. Army War College, June 6, 1994.

Tilford, Earl H., Jr. *The Revolution in Military Affairs: Prospects and Cautions.* Carlisle Barracks, PA: Strategic Studies Institute, U.S. Army War College, June 23, 1995.

———, ed. *World View: The 1995 Strategic Assessment from the Strategic Studies Institute.* Carlisle Barracks, PA: Strategic Studies Institute, U.S. Army War College, February 10, 1995.

Trybula, David C. *"Big Five" Lessons for Today and Tomorrow.* IDA Paper NS P-4889. Carlisle Barracks, PA: Institute for Defense Analyses, U.S. Army War College, May 2012.

Ullman, Harlan, and James Wade, with L. A. Edney, Fred Franks, Charles Horner, Jonathan Howe, and Keith Bradley. *Shock and Awe: Achieving Rapid Dominance.* Washington, DC: National Defense University Press, 1996.

U.S. Army Combined Arms Center. *U.S. Army Combined Arms Center 1987 Annual Historical Review.* Fort Leavenworth, KS: CAC History Office, U.S. Army Combined Arms Center, n.d. [1987].

U.S. Army Europe. *Military Operations: The U.S. Army in Bosnia and Herzegovina.* AE PAM 525-100. Heidelberg: U.S. Army Europe, October 7, 2003.

U.S. Army Training and Doctrine Command. *Multiservice Procedures for Humanitarian Assistance Operations.* FM 100-23-1. Fort Monroe, VA: U.S. Army Training and Doctrine Command, October 31, 1994.

———. *Knowledge and Speed: Battle Force and the U.S. Army of 2025.* Fort Monroe, VA: U.S. Army Training and Doctrine Command, December 1998.

———. *Force XXI Operations.* PAM 525-5. Fort Monroe, VA: U.S. Army Training and Doctrine Command, 1994.

U.S. Army War College. *Army Transformation Wargame 2001.* Carlisle Barracks, PA: U.S. Army War College, n.d. [May 2001].

U.S. Commission on National Security / 21st Century. *Seeking a National Strategy: A Concert for Preserving Security and Promoting Freedom, The Phase II Report on a U.S. National Security Strategy for the 21st Century.* Washington, DC: U.S. Commission on National Security/21st Century, April 15, 2000.

U.S. Department of the Army. *Army Vision 2010.* Washington, DC: U.S. Department of the Army, n.d. [November 12, 1996].

——. *Decisive Victory: America's Power Projection Army.* Washington, DC: U.S. Department of the Army, October 1994.

U.S. Department of Defense. *Measuring Stability and Security in Iraq.* Washington, DC: U.S. Department of Defense, March 2009.

——. *Report of the Quadrennial Defense Review.* Washington, DC: U.S. Department of Defense, May 1997. http://www.defenselink.mil/pubs/qdr/toc.html.

U.S. Department of State. *Ethnic Cleansing in Kosovo: An Accounting.* Washington, DC: U.S. Department of State, December 1999.

U.S. Government Accounting Office. *Combating Terrorism: Interagency Framework and Agency Programs to Address the Overseas Threat.* GAO 03-16. Washington, DC: U.S. Government Accounting Office, May 2003.

White, Thomas E., and Eric K. Shinseki. "A Statement on the Posture of the United States Army 2001 by the Honorable Thomas E. White and General Eric K. Shinseki Presented to The Committees and Subcommittees of the United States Senate and the House of Representatives First Session, 107th Congress." Washington, DC: Congressional Activities Division, Office of the Chief of Staff, U.S. Army, August 2001.

——. "A Statement on the Posture of the United States Army 2002." Washington, DC: Special Actions Branch, Office of the Chief of Staff, U.S. Army, February 2002.

White House. *A National Security Strategy of Engagement and Enlargement.* Washington, DC: U.S. Government Printing Office, July 1994.

Wright, Donald P., and Timothy R. Reese. *On Point II: Transition to the New Campaign.* Fort Leavenworth, KS: Combat Studies Institute, U.S. Army Command and General Staff College, June 2008.

Yarrison, James L. *The Modern Louisiana Maneuvers.* Washington, DC: U.S. Army Center of Military History, 1999.

Yates, Lawrence A. *The U.S. Military Intervention in Panama: Origins, Planning, and Crisis Management, June 1987–December 1989.* Washington, DC: U.S. Army Center of Military History, 2008.

UNPUBLISHED MANUSCRIPTS

Abbaszadeh, Nima, Mark Crow, Marianne El-Khoury, Jonathan Gandomi, David Kuwayama, Christopher MacPherson, Meghan Nutting, Nealin Parker, and Taya Weiss. "Provincial Reconstruction Teams: Lessons and Recommendations." Woodrow Wilson School of Public and International Affairs, Princeton University, January 2008.

Cleveland, Charles T. "Command and Control of the Joint Commission Observer Program: U.S. Army Special Forces in Bosnia." Master's monograph, U.S. Army War College, 2001.

Dubik, James M. "On the Foundations of National Military Strategy: Past and Present." Master's monograph, U.S. Army School of Advanced Military Studies, 1991.

Percival, Valerie, and Thomas Homer-Dixon. "Environmental Scarcity and Violent Conflict: The Case of Rwanda, Part 2." Occasional paper, Project on Environment, Population and Security, American Association for the Advancement of Science and the University of Toronto, Washington, DC, June 1995.

Petraeus, David H. "The American Military and the Lessons of Vietnam: A Study of Military Influence and the Use of Force in the Post-Vietnam Era." PhD diss., Princeton University, 1987.

Schultz, Tammy S. "Ten Years Each Week: The Warrior's Transformation to Win the Peace." PhD diss., Georgetown University, 2005.

Tinder, Alan J. "Low Intensity Conflict." Master's thesis, Air War College, 1990.

Townsend, Ian J. "Combined Action Platoons in the Vietnam War: A Unique Counterinsurgency Capability for The Contemporary Operating Environment." Master's thesis, U.S. Army School of Advanced Military Studies, 2013.

INDEX

Note: page numbers in italics refer to figures.

443

on imperatives of low-intensity conflict, 123–24

interim edition (1986), 51

low-intensity conflict doctrine in, 398

on political dominance in low-intensity conflict, 239

senior leadership's push for high-intensity conflict prioritization in, 123

Sullivan's efforts to elevate status of, 111

writing of, 51–52

Field Manual (FM) 100-20/Air Force Publication 3-20, *Military Operations in Low Intensity Conflict*, 52

addition of host nation responsibility to, 113

Field Manual (FM) 100-23, *Peace Operations* (1994)

on low-intensity conflict, military's role in, 169–70, 171

military's role in, 167

Field Manual (FM) 100-23-1, *Multiservice Procedures for Humanitarian Assistance Operations*

and Bosnia-Herzegovina intervention, 226

effect of Somalia defeat on, 159–61, *160*

on low-intensity conflict, military's role in, 167, 169–71

on MOEs in low-intensity conflict, 169

V Corps, and Iraq War, 72, 75, 83

5th Infantry Division, and Panama invasion, 55

1st Cavalry Division, 82

in Bosnia-Herzegovina, 249, 252

First Chechen War, urban warfare in, 272–74, 278

1st Infantry Division

in Bosnia-Herzegovina, 6, 223

in Kosovo, 6, 296, 299–300

and low-intensity conflict competence, development of, 6

I Marine Expeditionary Force, and Iraq War, 72

1st Marine Regiment, and Iraq War, 76

Fishel, John, 131, 132, 134

Flournoy, Michèle A., 342–43

FOBs. *See* forward operating bases

Fontenot, Gregory, 230

force protection

as objection to army's responsibility for

low-intensity conflict, 21–22

overemphasis on

in Bosnia-Herzegovina, 222, 231–32, 253–54

in Haiti, 176, 186–87

in Iraq, 77

in Kosovo, 299

subordination to political goal, need to incorporate into Army doctrine, 409–10

Forces Armées d'Haïti (FAdH)

Aristide's efforts to bring under control, 190

as center of anti-Ariistide movement, 179, 189

rebuilding and retraining of, 180, 189–90

skirmishes with U.S. forces, 174

and U.S. neutral position, 189

Forces Command (FORSCOM), and Louisiana Maneuvers, 118

Force XXI, 162–65

criticisms of technological focus of, 248

Decisive Victory white paper on, 166, 169

design and function of, 163–65

digitization as key technology in, 280

and diversity of forces, focus on, 166–67

Dubik and, 162–64, 165

"full-spectrum" forces in, 279

Interim Force as development of, 307–8

introduction of, 109

and lessons on low-intensity conflict, failure to learn, 163, 165, 166–67

limited development of, 306

logistical component of, 163–64

Louisiana Maneuvers and, 162

and low-intensity conflict, lack of focus on, 193

and Quadrennial Defense Review (1997), 248

Reimer's continuation of, 209, 211, 215, 240

Sullivan and, 162–64, 165, 240

testing of (AWEs), 165

TRADOC Pamphlet (PAM) 525-5 on, 164–65, 192–93

FORSCOM. *See* Forces Command

Fort Lewis, and IBCTs, 309

forward operating bases (FOBs) in Iraq

and force protection, prioritization of, 77

need for training in, 347

as outside Army's remit, in Army's view, 254–55, 349–50, 394–95

in FM 3-0, 361

and "mission creep" concept, 23–24, 350–51

in Panama occupation, 133–34

as responsibility of host government (Nixon Doctrine), 36, 45, 46, 58, 84, 110, 112–13, 130, 137–38, 153, 160, 215–16, 219, 229, 356, 361

in Somalia intervention, 148–56

UN responsibility for, advocates for, 152

urban warfare and, 272, 273, 274, 275, 277, 278, 397

and U.S. policy of neutrality, 171, 258

in Bosnia-Herzegovina, 237–38, 252–53

FM 7-98 on, 131

in Haiti, 189–91

necessity of purging from doctrine, 408–9

Panama occupation and, 134–35

supposed legitimacy derived from, 24–25, 134–35

See also Bosnia-Herzegovina intervention, political dimension of; Haiti intervention, political dimensions of; Somalia intervention, and political engagement

political dominance

as necessary for success in low-intensity conflict

FM 7-98 on, 129

FM 100-20 on, 123, 239

omission from 1993 FM 100-5, 137, 239

omission from JP 3-07, 215

political settlement to conflict

Army's inability to achieve

as persistent issue since 1990s, 393

reasons for, 393

as goal in low-intensity conflict

cognitive obstacles to recognition of, 21–22

metrics used as substitute for, 394–95

need for doctrinal focus on, 408

need for U.S. to choose a side in, 24, 85, 257–58, 351, 381, 393

Afghanistan War and, 91–92

and Army policy of neutrality, 24–25, 131, 134–35, 171, 189–91, 252–53, 258. 237–38

in Bosnia-Herzegovina intervention, 237–38, 257–58

cognitive obstacles to recognition of, 24–25, 85, 130–31

Haiti intervention and, 188–91

Iraq War and, 87–88

liberal Western notions as obstacle to, 124, 155

Panama occupation and, 134–35

need to incorporate into Army doctrine, 408–9

POMCUS. See pre-positioning of materiel configured to unit sets

populace. See civilians, Army's operations among

Post Conflict Strategic Requirements Workshop, 348

Powell, Colin

and defense budget cuts of 1990s, 114, 162

DePuy and, 36, 42

and Haiti intervention, 173

and Iraq War, 396

on low-intensity conflict

avoidance of, 64

as secondary to Army's mission, 168, 264

My American Journey, 29

in Reagan administration, 42

and Soviet Union collapse, 59

Powell Doctrine

criticisms of, 328

George W. Bush's support of, 352

Gulf War and, 396

Iraq War and, 396

as justification for avoiding low-intensity conflict, 64, 151, 159

Rice's support of, 351

power projection concept, development of, 118

precision weapons

as goal of revolution in military affairs, 44–45

and Gulf War, 65

"Preparing for the Wrong War?" (Vought), 46